Analysis and Design of Steel and Composite Structures

Analysis and Design of Steel and Composite Structures

Qing Quan Liang

CRC Press is an imprint of the
Taylor & Francis Group, an **informa** business

CRC Press
Taylor & Francis Group
6000 Broken Sound Parkway NW, Suite 300
Boca Raton, FL 33487-2742

Printed on acid-free paper
Version Date: 20140707

International Standard Book Number-13: 978-0-415-53220-4 (Paperback)

Library of Congress Cataloging-in-Publication Data

Liang, Qing Quan, 1965-
Analysis and design of steel and composite structures / Qing Quan Liang.
pages cm
Includes bibliographical references and index.
ISBN 978-0-415-53220-4 (paperback)
1. Building, Iron and steel. 2. Composite construction. I. Title.

TA684.L5176 2014
624.1'821--dc23 2014024460

Visit the Taylor & Francis Web site at
http://www.taylorandfrancis.com

and the CRC Press Web site at
http://www.crcpress.com

This book is dedicated to the memory of my parents, Bo Fen Liang (1928–1981) and Xing Zi He (1936–1987), and to my wife, Xiao Dan Cai, and my sons, Samuel Zhi De Liang, Matthew Zhi Cheng Liang and John Zhi Guo Liang.

Contents

Preface

Steel and composite steel–concrete structures are widely used in modern bridges, buildings, sport stadia, towers and offshore structures. The analysis and design of steel and composite structures require a sound understanding of the behaviour of structural members and systems. This book provides an integrated and comprehensive introduction to the analysis and design of steel and composite structures. It describes the fundamental behaviour of steel and composite members and structures and the latest design criteria and procedures given in Australian Standards AS/NZS 1170, AS 4100, AS 2327.1, Eurocode 4 and AISC-LRFD specifications. The latest research findings on composite members by the author's research teams are also incorporated in the book. Emphasis is placed on a sound understanding of the fundamental behaviour and design principles of steel and composite members and connections. Numerous step-by-step examples are provided to illustrate the detailed analysis and design of steel and composite members and connections.

This book is an ideal course textbook on steel and composite structures for undergraduate and postgraduate students of structural and civil engineering, and it is a comprehensive and indispensable resource for practising structural and civil engineers and academic researchers.

Chapter 1 introduces the limit state design philosophy, the design process and material properties of steels and concrete. The estimation of design actions on steel and composite structures in accordance with AS/NZS 1170 is described in Chapter 2. Chapter 3 presents the local and post-local buckling behaviour of thin steel plates under in-plane actions, including compression, shear and bending of steel plates in contact with concrete. The design of steel members under bending is treated in Chapter 4, which includes the design for bending moments and the shear and bearing of webs to AS 4100. Chapter 5 is devoted to steel members under axial load and bending. The analysis and design of steel members under axial compression, axial tension and combined axial load and bending to AS 4100 are covered. In Chapter 6, the design of bolted and welded steel connections, including bolted moment end plate connections and pinned column base plate connections, is presented. Chapter 7 introduces the plastic analysis and design of steel beams and frames.

The behaviour and design of composite slabs for strength and serviceability to Eurocode 4 and Australian practice are treated in Chapter 8. Chapter 9 presents the behaviour and design of simply supported composite beams for strength and serviceability to AS 2327.1. The design method for continuous composite beams is also covered. The behaviour and design of short and slender composite columns under axial load and bending in accordance with Eurocode 4 are given in Chapter 10. This chapter also presents the nonlinear inelastic analysis of thin-walled concrete-filled steel tubular short and slender beam-columns under axial load and biaxial bending. Chapter 11 introduces the behaviour and design of composite connections in accordance with AISC-LRFD specifications, including single-plate and tee shear connections, beam-to-composite column moment connections and semi-rigid composite connections.

Qing Quan Liang
Associate Professor
Victoria University
Melbourne, Victoria, Australia

Acknowledgements

The author thanks Professor Yeong-Bin Yang at National Taiwan University, Dr. Anne W. M. Ng at Victoria University in Melbourne, Benjamin Cheung, senior project engineer in Melbourne, and Associate Professor Yanglin Gong at Lakehead University for their invaluable and continued support. The author also thanks all his co-researchers for their contributions to the research work, particularly Associate Professor Muhammad N. S. Hadi at the University of Wollongong, Professor Brian Uy and Professor Mark A. Bradford at the University of New South Wales, Professor Yi-Min Xie at RMIT University, Emeritus Professor Grant P. Steven at the University of Sydney, Professor Jat-Yuen Richard Liew at the National University of Singapore, Emeritus Professor Howard D. Wright at the University of Strathclyde, Dr. Hamid R. Ronagh at the University of Queensland and Dr. Mostafa F. Hassanein and Dr. Omnia F. Kharoob at Tanta University. Thanks also go to Professor Jin-Guang Teng at The Hong Kong Polytechnic University, Professor Dennis Lam at the University of Bradford, Professor Ben Young at the University of Hong Kong, Professor Lin-Hai Han at Tsinghua University, Associate Professor Mario Attard and Professor Yong-Lin Pi and Dr. Sawekchai Tangaramvong at the University of New South Wales, Dr. Zora Vrcelj at Victoria University and Professor N. E. Shanmugam at the National University of Malaysia for their useful communications and support. Grateful acknowledgement is made to the author's former PhD student Dr. Vipulkumar I. Patel for his contributions to the research work on composite columns and to ME students Dr. Sukit Yindeesuk in the Department of Highways in Thailand and Hassan Nashid for their support. Finally, and most importantly, the author thanks his wife, Xiao Dan Cai, and sons, Samuel, Matthew and John, for their great encouragement, support and patience while he was writing this book.

Chapter 1

Introduction

1.1 STEEL AND COMPOSITE STRUCTURES

Steel and composite steel–concrete structures are widely used in modern bridges, buildings, sport stadia, towers and offshore structures. According to their intended functions, buildings can be classified into industrial, residential, commercial and institutional buildings. A steel structure is composed of steel members joined together by bolted or welded connections, which may be in the form of a pin-connected truss or a rigid frame. In comparison with reinforced concrete structures, steel structures have the advantages of lightweight, large-span, high ductility and rapid construction. The rapid steel construction attributes to the fact that steel members and connection components can be prefabricated in a shop. As a result, significant savings in construction time and costs can be achieved. Perhaps, steel portal frames as depicted in Figure 1.1 are the most commonly used steel structures in industrial buildings. They are constructed by columns, roof rafters and bracings, which are joined together by knee, ridge and column base connections. The design of steel portal frames is treated in this book.

The advantages of the rapid and economical steel construction of multistorey buildings can only be utilised by composite steel–concrete structures, which are efficient and cost-effective structural systems. Composite structures are usually constructed by composite columns or steel columns and steel beams supporting composite slabs or concrete slabs. It is noted that steel is the most effective in carrying tension and concrete is the most effective in resisting compression. Composite members make the best use of the effective material properties of both steel and concrete. A composite beam is formed by attaching a concrete slab to the top flange of a steel beam as shown in Figure 1.2. By the composite action achieved by welding shear connectors to the top flange of the steel beam, the steel beam and the concrete slab works together as one structural member to resist design actions. In a composite beam under bending, the concrete slab is subjected to compression, while the steel beam is in tension, which utilises the effective material properties of both steel and concrete. The common types of composite columns include concrete encased composite columns, rectangular concrete-filled steel tubular columns and circular concrete-filled steel tubular columns as presented in Figure 1.3. High-strength composite columns have increasingly been used in high-rise composite buildings due to their high structural performance such as high strength and high stiffness. The fundamental behaviour and the state-of-the-art analysis and design of composite slabs, composite beams, composite columns and composite connections are covered in this book.

The design of steel and composite structures is driven by the limited material resources, environmental impacts and technological competition which demand lightweight, low-cost and high-performance structures. These demands require that structural designers must have a sound understanding of the fundamental behaviour of steel and composite

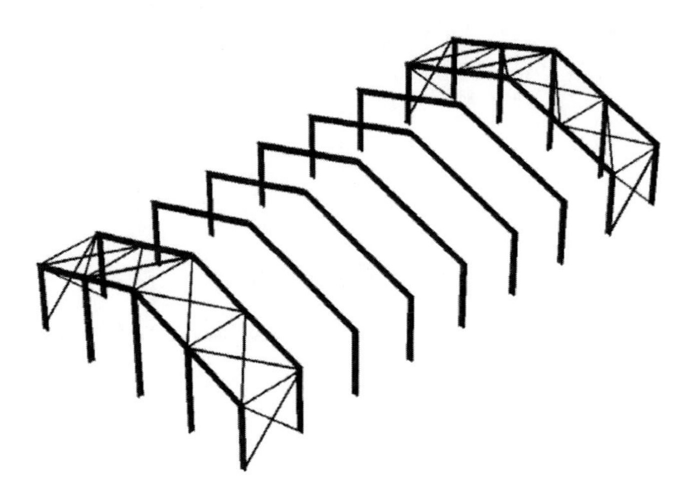

Figure 1.1 Steel portal frames.

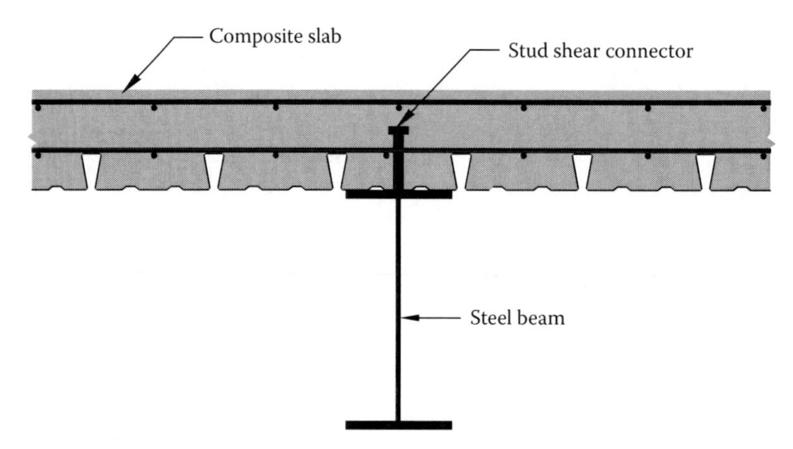

Figure 1.2 Cross section of composite beam.

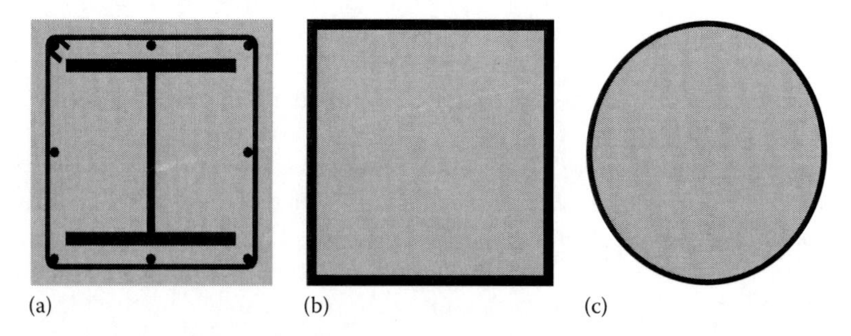

Figure 1.3 Cross sections of composite columns: (a) concrete encased composite column, (b) rectangular concrete-filled steel tubular column, and (c) circular concrete-filled steel tubular column.

structures and the latest design principles and technologies for the design of these structures. The forms of steel and composite structures have been evolving in the last few decades, and many innovative steel and composite structures have been designed and constructed around the world. Topology optimisation techniques can be used to find the optimal and innovative layouts of structures (Liang 2005). It is recognized that topology optimisation produces much more material savings and higher-performance optimal structures than shape and sizing optimisation.

This chapter introduces the limit state design philosophy, the structural design process and material properties of steels and concrete used in the construction of steel and composite structures.

1.2 LIMIT STATE DESIGN PHILOSOPHY

1.2.1 Basic concepts and design criteria

The limit state design philosophy has been adopted in the current codes of practice as the basic design method for the design of steel and composite structures as it is believed that this method is capable of yielding safer and more economical design solutions. The limit state is defined as the state beyond which the structure will not satisfy the design criteria. This limit state may be caused by the failure of one or more structural members, the instability of structural members or the whole structure, or excessive deformations of the structure. The limit state design is to design a structure or structural component that can perform the intended physical functions in its design lifetime. In the limit state design, the performance of a structure is evaluated by comparison of design action effects with a number of limiting conditions of usefulness. The limit states may include strength, stability, serviceability, fire, fatigue, earthquake and brittle fracture limit states.

Structural design criteria are expressed in terms of achieving multiple design objectives. There are usually multiple design objectives that must be considered by the structural designer when designing a structure. The main objectives are functionality, safety, economy and ease of construction. The safety is a structural design objective which is related to the strength and serviceability. Design codes and standards impose limitations on the serviceability and strength of a structure or structural members to ensure that the structure or structural members designed will perform normal functions. Functionality, which is the ability of a structure to perform its intended non-structural use, and economy are non-structural design objectives. However, they can be used to rank alternative designs that satisfy structural design criteria.

1.2.2 Strength limit state

The strength design criterion requires that the structure must be designed so that it will not fail in its design lifetime or the probability of its failure is very low. The strength limit state design is to design a structure including all of its members and connections to have design capacities in excess of their design action effects. This can be expressed in the mathematical form as follows:

$$E_a \le \phi R_n \tag{1.1}$$

where

E_a is the design action effect
ϕ is the capacity reduction factor
R_n is the nominal capacity or resistance of the structural member
ϕR_n is the design capacity or the design resistance of the structural member

The design action effect E_a represents an internal action such as axial force, shear force or bending moment, which is determined by structural analysis using factored combinations of design actions applied on the structure. In the strength limit state design, load factors are used to increase the nominal loads on structural members, while capacity reduction factors are employed to decrease the capacity of the structural member.

Table 1.1 Capacity reduction factor (ϕ) for strength limit states

Structural component	*Capacity reduction factor (ϕ)*	
Steel member	0.9	
Connection component (excluding bolt, pin or weld)	0.9	
Bolted or pin connection	0.8	
Ply in bearing	0.9	
Welded connection	SP category	GP category
Complete penetration butt weld	0.9	0.6
Longitudinal fillet weld in RHS (t < 3 mm)	0.7	—
Other welds	0.8	0.6

Source: AS 4100, *Australian Standard for Steel Structures*, Standards Australia, Sydney, New South Wales, Australia, 1998.

The use of load factors and capacity reduction factors in the strength limit state design is to ensure that the probability of the failure of a structure under the most adverse combinations of design actions is very small. These factors are used to account for the effects of errors and uncertainties encountered in the estimation of design actions on a steel or composite structure and of its behaviour. Errors made by the designer may be caused by simplified assumptions and lack of precision in the estimation of design actions, in structural analysis, in the manufacture and in the erection of the structure (Trahair and Bradford 1998). The design actions on a structure vary greatly. This may be caused by the estimation of the magnitude of the permanent actions (dead loads) owing to variations in the densities of materials. In addition, imposed actions (live loads) may change continually during the design life. Wind actions vary significantly and are usually determined by probabilistic methods. The uncertainties about the structure include material properties, residual stress levels, cross-sectional dimensions of steel sections and initial geometric imperfections of structural members. The aforementioned errors and uncertainties may lead to the underestimate of the design actions and the overestimate of the capacity of the structure. Load factors and capacity reduction factors are used to compensate these effects in the strength limit state design.

Probability methods are usually employed to determine load and capacity factors on the basis of statistical distributions of design actions and capacities of structural members. The load and capacity factors given in AS 4100 were derived by using the concept of safety index. The limit state design generally yields slightly safer designs with a safety index ranging from 3.0 to 3.5 in comparison with the traditional working stress design (Pham et al. 1986). The capacity reduction factor depends on the methods employed to determine the nominal capacities, nominal design actions and the values used for the load factors. Table 1.1 gives the capacity reduction factors for steel members and connections for the strength limit state design.

1.2.3 Stability limit state

The stability limit state is concentred with the loss of static equilibrium or of deformations of the structure or its members owing to sliding, uplifting or overturning. The stability limit state requires that the following condition be satisfied:

$$E_{a.dst} - E_{a.stb} \leq \phi R_n \tag{1.2}$$

where

$E_{a.dst}$ is the design action effect of destabilizing actions
$E_{a.stb}$ is the design action effect of stabilizing actions

1.2.4 Serviceability limit state

The serviceability limit state is the state beyond which a structure or a structural member will not satisfy the specified service design criteria. This means that beyond the limit state, the structure will not fit for the intended use under service load conditions. Serviceability limit states may include deformation, vibration and degradation limit states. The deformations of a structure are governed by the stiffness design requirements which are system performance criteria. For the stiffness limit state design, the deflections of the structure under most adverse service load conditions need to be limited so that the structure can perform the normal function without impairing its appearance, safety and public comfort. This can be expressed in the mathematical form as follows:

$$\delta_j \le \delta_j^* \tag{1.3}$$

where
- δ_j is the jth displacement or deflection of the structure under the most adverse service load combinations
- δ_j^* is the limit of the jth displacement or deflection

The deflections of a structure under service design actions are usually determined by performing a first-order linear elastic analysis or a second-order nonlinear elastic analysis. Only the most essential deflection limits are given in AS 4100 (1998). The structural designer needs to determine whether the structure designed satisfies the serviceability requirements.

1.3 STRUCTURAL DESIGN PROCESS

The overall purpose of the structural design is to develop the best feasible structural system that satisfies the design objectives in terms of the functionality, safety and economy. Structural design is a complex, iterative, trial-and-error and decision-making process. In the design process, a conceptual design is created by the designer based on his intuition, creativity and past experience. Structural analysis is then undertaken to evaluate the performance of the design. If the design does not satisfy the design objectives, a new design is then developed. This process is repeated until the design satisfies the multiple performance objectives. The main steps of the overall structural design process are illustrated in Figure 1.4.

The first step in the structural design process is to investigate the overall design problem. Firstly, the design engineers discuss the needs for the structure, its proposed function, requirements and constraints with the owner. The functionality is the ability of a structure to perform its intended non-structural use. It is one of the important design objectives that must be achieved for a structure and affects all stages of the structural design process. The site and geotechnical investigations are then followed. The structural designers also need to study similar structures and to consult authorities from whom permissions and approvals must be obtained. Multiple design objectives are then identified for the structure and selected by the owner who consults with the structural designers based on the consideration of his/her expectations, economic analysis and acceptable risk.

In the conceptual design stage, the structural designer develops the best feasible structural systems that appear to achieve the design objectives defined in the preceding stage. The selection of structural systems is generally iterative in nature based on the designer's creation, intuition and past experience. In order to obtain an optimal structure, a number of alternative structural systems must be invented and evaluated. The invention of structural

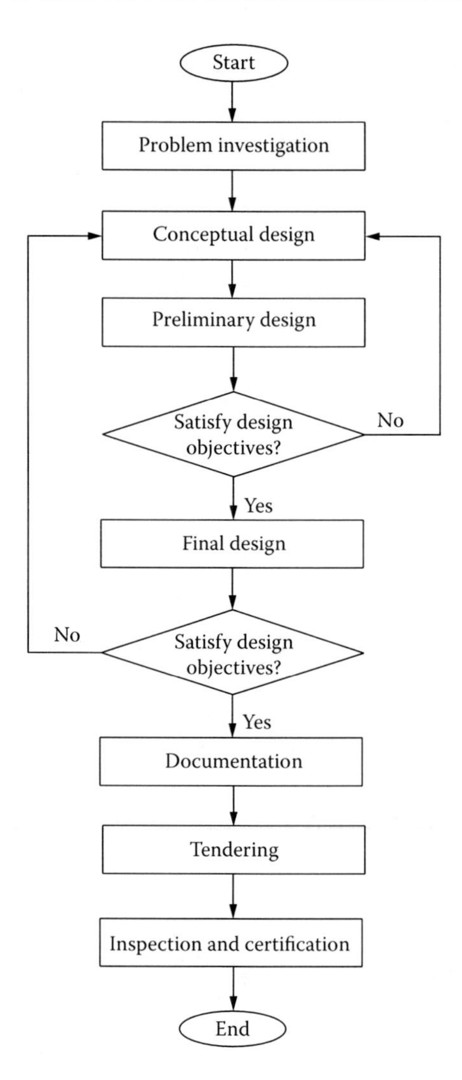

Figure 1.4 The structural design process.

systems is the most challenging task in structural design since it involves a large number of possibilities for the structural layouts. The traditional design process is highly time consuming and expensive. Since the development of structural systems is an optimal topology design problem, automated topology optimisation technique such as the performance-based optimisation (PBO) technique (Liang 2005) can be employed in the conceptual design stage to generate optimal structures. The optimal structural system is produced by topology optimisation techniques based on the design criteria and constraints but not on the past experience (Liang et al. 2000a). The designer also selects the materials of construction for the structure.

After the best feasible structure has been created, the preliminary design can be carried out. The design loads and load combinations applied to the structure are estimated in accordance with the loading codes. The structural analysis method or modern numerical technique such as the finite element method (Zienkiewicz and Taylor 1989, 1991) is then employed to analyse the structure to evaluate its structural performance. From the results of the structural analysis, structural members are preliminarily sized to satisfy the design criteria. The cost of the structure is also preliminarily estimated. If the structure does not satisfy the function, structural efficiency and cost design objectives, a new structural system must be developed and the design process is repeated, as depicted in Figure 1.4. It is obvious that shape and sizing optimisation techniques can be applied in the preliminary design stage to achieve cost-efficient designs.

Since the structure is approximately proportioned in the preliminary design stage, it must be checked against the design criteria and objectives in the final design stage. The loads applied to the structure are recalculated and the structure is reanalysed. The performance of the structure is then evaluated and checked with performance requirements. Any change in the member sizes may require a further reanalysis and resizing of the structure. The design and redesign process is repeated until no more modification can be made to the structure. The structure is evaluated for the design objectives such as function, serviceability, strength and cost. If these objectives are not satisfied, the structure may be modified or a new conceptual design may be generated. The design process is repeated as indicated in Figure 1.4. In the final design stage, the sizing of the structure is the main task. Therefore, sizing optimisation techniques can be employed to automate the design process. It is worth noting that topology optimisation techniques can also be used in the final design stage. Liang et al. (2000b, 2001, 2002) demonstrated that the automated PBO technique can be employed in the final design stage to generate optimal strut-and-tie models for the design and detailing of reinforced and prestressed concrete structures.

After the structure is finalised, the documentation such as the detailed drawings and specifications can be prepared and tenders for construction can be called for. At the final stage, the designers carry out inspection and certification during construction to ensure that all performance objectives defined are achieved in the structural design process.

1.4 MATERIAL PROPERTIES

1.4.1 Structural steel

Structural steel is usually hot rolled, welded from flat plates or cold formed from flat plates to form structural sections, such as I-sections, rectangular hollow sections (RHSs) and circular hollow sections (CHSs). Figure 1.5 depicts an idealised stress–strain curve for mild structural steel. It can be seen that the steel initially has a linear stress–strain relationship up to the elastic limit, which can be approximately defined by the yield stress f_y. The most important properties of mild structural steel are its Young's modulus of elasticity E_s ranging from 200 to 210 GPa and its yield stress ranging from 250 to 400 MPa. Beyond the

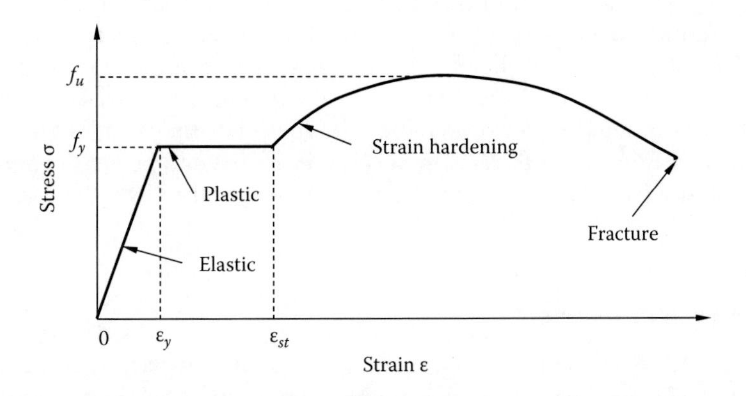

Figure 1.5 Idealised stress–strain curve for mild structural steel.

elastic limit, the steel undergoes large plastic flows without any increase in the stress until reaching the hardening strain ε_{st}, which is usually 10 or 11 times the yield strain ε_y. This plastic plateau indicates the ductility of the steel. After reaching the hardening strain ε_{st}, the stress increases above the yield stress with an increase in the strain until the ultimate tensile strength f_u is attained. This is followed by the necking of the cross section and decreasing in stress until the tensile fracture occurs. The steel normally follows the same stress–strain curve in tension and compression. In the elastic range, Poisson's ratio of steel is about 0.3. In AS 4100, Poisson's ratio is taken as 0.25 for Australian structural steels.

The yield stress is an important property of a structural steel, which depends on the chemical contents such as carbon and manganese, the heat treatment used and the amount of working induced during the rolling process. Cold working also increases the yield stress of the steel. The yield stress of a structural steel can be determined by standard tension tests. The minimum yield stress of the structural steel given in design codes for use in structural design is a characteristic value that is usually less than that determined from any standard tension test. This implies that the use of the yield stress given in design codes usually provides conservative designs.

1.4.2 Profiled steel

Profiled steel sheeting is used in composite slabs and beams as the permanent form work and part of reinforcement for the concrete. It is manufactured by cold rolling thin steel plate into shape with wide steel ribs. The yield stress may be increased by the cold-rolling process. The stress–strain curve for profiled steel is rounded without a well-defined yield stress as shown in Figure 1.5. A 0.2% proof stress of 550 MPa is usually used for profiled steel, while its elastic modulus is about 200 GPa.

The major types of profiled steel sheeting used in composite construction in Australia are LYSAGHT Bondek II, Comform and Condeck HP. Profiled steel sheeting might have an adverse influence on the behaviour of composite beams. AS 2327.1 imposes restrictions on the geometry of profiled steel sheeting so that composite slabs incorporating profiled steel sheeting can be treated as solid slabs when calculating the design capacity of shear connectors.

1.4.3 Reinforcing steel

The types of reinforcing steel commercially available in Australia are reinforcing bar, hard-drawn wire and welded wire fabric. Reinforcing bar can be classified into several grades, namely, 400Y high yield with a minimum guaranteed yield stress of 400 MPa, plain bar with a minimum guaranteed yield stress of 250 and 500 MPa steels with a characteristic yield stress of 500 MPa. The 500 MPa steels have three grades such as 500L, 500N and 500E grades, where the final letter denotes the level of ductility, with L indicating low ductility, N denoting normal ductility and E standing for special high-ductility steel for use in earthquake-resistant design. The stress–strain curve for reinforcing steel is assumed to be elastic–perfectly plastic in design. The elastic modulus of reinforcing steel is usually taken as 200 GPa.

1.4.4 Concrete

1.4.4.1 Short-term properties

The main properties of the hardened concrete are its compressive strength, elastic modulus in compression, tensile strength and durability. The characteristic compressive strength

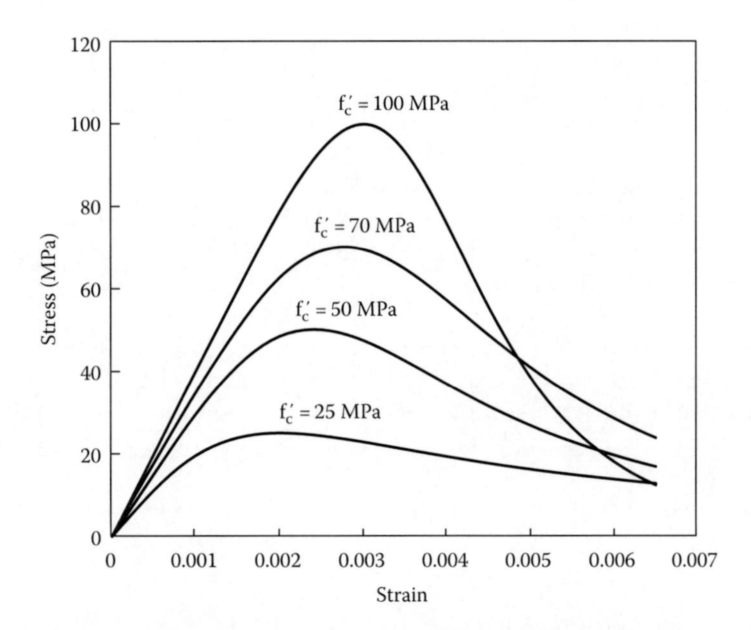

Figure 1.6 Stress–strain curves for normal- and high-strength concrete.

f'_c of concrete is commonly used for concrete due to the large variation of the concrete strength. It is determined as the strength attained at 28 days by 95% of the concrete as obtained by standard compression tests. The normal-strength concrete has a characteristic compressive strength f'_c up to 50 MPa; concrete with a compressive strength higher than 50 MPa is regarded as high-strength concrete (Warner et al. 1998), which can be made by using high-quality aggregates and superplasticizers, and the strength may exceed 100 MPa.

Figure 1.6 depicts the typical stress–strain curves for concrete in uniaxial compression with various compressive strengths. It appears from the figure that the stress–strain relationship is linear for stress up to $0.4f'_c$. However, at stress higher than $0.4f'_c$, the stress–strain relationship becomes nonlinear due to the effects of the formations and development of microcracks at the interfaces between the mortar and coarse aggregate. As shown in Figure 1.6, the shape of the stress–strain curve for concrete varies with the concrete compressive strengths and it is affected by the type of aggregate used and the strain rate applied in the compression tests. The stress–strain curve for high-strength concrete is steeper than for normal-strength concrete. The descending branch in the post-ultimate range decreases sharply with increasing the compressive strength of concrete. This indicates that high-strength concrete is very brittle.

Empirical equations have been proposed by various researchers based on experimental results to express the stress–strain curves for normal- and high-strength concrete. Mander et al. (1988) presented equations for modelling the stress–strain behaviour of unconfined concrete as follows:

$$\sigma_c = \frac{f'_c \lambda \left(\varepsilon_c / \varepsilon'_c\right)}{\lambda - 1 + \left(\varepsilon_c / \varepsilon'_c\right)^{\lambda}} \tag{1.4}$$

$$\lambda = \frac{E_c}{E_c - (f'_c / \varepsilon'_c)} \tag{1.5}$$

where

σ_c is the longitudinal compressive stress of concrete
ε_c is the longitudinal compressive strain of concrete
ε'_c is the strain at f'_c
E_c is Young's modulus of concrete

Young's modulus of concrete can be determined from the measured stress–strain curve as the secant modulus at a stress level equal to $0.45f'_c$. Young's modulus of concrete in tension is approximately the same as that of concrete in compression. In AS 3600 (2001), Young's modulus E_c of normal-strength concrete is calculated approximately by

$$E_c = 0.043\rho^{1.5}\sqrt{f_{cm}}\ \text{MPa} \tag{1.6}$$

where

ρ is the density of concrete in kg/m^3
f_{cm} is the mean compressive strength of concrete at any particular age

For normal- and high-strength concrete, the following equation suggested by ACI Committee 363 (1992) can be used to estimate Young's modulus:

$$E_c = 3320\sqrt{f'_c} + 6900\ \text{MPa} \tag{1.7}$$

It can be seen from Figure 1.6 that the strain ε'_c at the peak stress f'_c of concrete varies with the compressive strength of concrete. The value of strain ε'_c is between 0.002 and 0.003. For the compressive strength of concrete less than 28 MPa, the strain ε'_c is 0.002, while it can be taken as 0.003 for the compressive strength of concrete higher than 82 MPa. When the compressive strength of concrete is between 28 and 82 MPa, the strain ε'_c can be determined by linear interpolation. Poisson's ratio (ν) for concrete is in the range of 0.15–0.22 and can be taken as 0.2 in the analysis and design of practical structures.

The tensile strength of concrete appears to be much lower than its compressive strength and it may be ignored in some design calculations. However, it needs to be taken into account in the nonlinear inelastic analysis of composite beams and columns in order to capture the true behaviours. Tests such as direct tension tests, cylinder split tests or flexural tests can be conducted to determine the tensile strength of concrete. However, the tensile strength of concrete is often estimated from its compressive strength. In AS 3600 (2001), the characteristic flexural tensile strength at 28 days is given by

$$f'_{cf} = 0.6\sqrt{f'_c}\ \text{MPa} \tag{1.8}$$

In direct tension, the characteristic principal tensile strength of concrete at 28 days may be taken as

$$f'_{ct} = 0.4\sqrt{f'_c}\ \text{MPa} \tag{1.9}$$

An idealised stress–strain curve is usually assumed for concrete in tension in the nonlinear analysis (Liang 2009). The tension stress increases linearly with an increase in tensile strain up to concrete cracking. After concrete cracking, the tensile stress decreases linearly to zero as the concrete softens. The ultimate tensile strain is taken as 10 times the strain at cracking.

1.4.4.2 Time-dependent properties

The strain of a concrete member under a sustained load is not constant and rather, it gradually increases with time. This time-dependent behaviour of concrete is caused by creep and shrinkage. Creep strain is induced by the sustained stress and is both stress dependent and time dependent. Shrinkage strain is mainly caused by the loss of water in the drying process of the concrete and is stress independent and time dependent. Creep and shrinkage may induce axial and lateral deformations of composite sections, stress redistribution between the concrete and steel components and local buckling of steel sections in composite members. More details on the time-dependent properties of concrete can be found in books (Gilbert 1988; Gilbert and Mickleborough 2004).

Consider a concrete member under a constant sustained axial stress σ_o first applied at time τ_o. The total strain at any time greater than τ_o consists of the instantaneous strain $\varepsilon_{el}(\tau_o)$, creep strain $\varepsilon_{cr}(t, \tau_o)$ and shrinkage strain $\varepsilon_{sh}(t)$ as demonstrated in Figure 1.7 and can be expressed by

$$\varepsilon(t) = \varepsilon_{el}(\tau_o) + \varepsilon_{cr}(t, \tau_o) + \varepsilon_{sh}(t) \tag{1.10}$$

The instantaneous strain $\varepsilon_{el}(\tau_o)$ of the concrete at service loads is usually linear elastic and is given by

$$\varepsilon_{el}(\tau_o) = \frac{\sigma_o}{E_c} \tag{1.11}$$

The creep function or factor is usually used to evaluate the capacity of concrete to creep, which is defined as the ratio of the creep strain to the instantaneous strain as

$$\phi_c(t, \tau_o) = \frac{\varepsilon_{cr}(t, \tau_o)}{\varepsilon_{el}(\tau_o)} \tag{1.12}$$

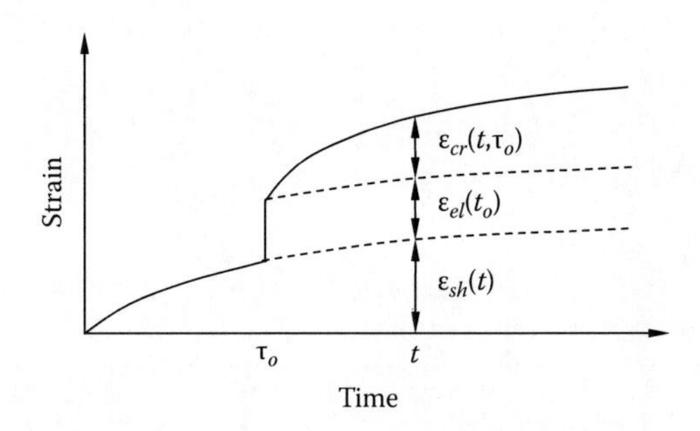

Figure 1.7 Time-dependent strain for concrete under constant stress.

From this equation, the creep strain can be written as

$$\varepsilon_{cr}(t,\tau_o)=\frac{\sigma_o}{E_c}\phi_c(t,\tau_o) \tag{1.13}$$

At the time infinity, the creep function approaches its final maximum value ϕ_c^*, which is usually in the range of 1.5–4.0. The strain at time t caused by a constant sustained stress σ_o consists of the elastic and creep components as follows (Gilbert and Mickleborough 2004):

$$\varepsilon_{el}(\tau_o)+\varepsilon_{cr}(t,\tau_o)=\frac{\sigma_o}{E_c}+\frac{\sigma_o}{E_c}\phi_c(t,\tau_o)=\frac{\sigma_o}{E_c}\left[1+\phi_c(t,\tau_o)\right]=\frac{\sigma_o}{E_{ce}(t,\tau_o)} \tag{1.14}$$

where $E_{ce}(t,\tau_o)$ is the effective modulus of concrete and is expressed by

$$E_{ce}(t,\tau_o)=\frac{E_c}{1+\phi_c(t,\tau_o)} \tag{1.15}$$

The compressive stress may be gradually applied to the concrete. This reduces significantly the creep strain of the concrete due to the aging of the concrete. For a stress increment $\Delta\sigma$, the stress-dependent strain is given by (Trost 1967; Bažant 1972)

$$\varepsilon_{el}(\tau_o)+\varepsilon_{cr}(t,\tau_o)=\frac{\Delta\sigma}{E_c}\left[1+\chi_a\phi_c(t,\tau_o)\right]=\frac{\Delta\sigma}{E_{ce}^*(t,\tau_o)} \tag{1.16}$$

where

χ_a is the aging coefficient (Trost 1967; Bažant 1972)

$E_{ce}^*(t,\tau_o)$ is the age-adjusted effective modulus for concrete, which is expressed by

$$E_{ce}^*(t,\tau_o)=\frac{E_c}{1+\chi_a\phi_c(t,\tau_o)} \tag{1.17}$$

The aging coefficient χ_a is in the range of 0.6–1.0 and is a function of the duration of loading and the age at the first loading.

The shrinkage strain decreases with time. At the time infinity, the shrinkage strain approaches its final value ε_{sh}^*. The shrinkage depends on all factors that influence the drying of concrete, including the relative humidity, the mix design and the size and shape of the concrete member. The basic shrinkage strain of concrete can be taken as 850×10^{-6} as suggested in AS 3600 (2001).

REFERENCES

ACI Committee 363 (1992) *State of the Art Report on High-Strength Concrete*, ACI Publication 363R-92, Detroit, MI: American Concrete Institute.

AS 3600 (2001) *Australian Standard for Concrete Structures*, Sydney, New South Wales, Australia: Standards Australia.

AS 4100 (1998) *Australian Standard for Steel Structures*, Sydney, New South Wales, Australia: Standards Australia.

Bažant, Z.P. (1972) Prediction of concrete creep effects using age-adjusted effective modulus method, *ACI Journal*, 69: 212–217.

Gilbert, R.I. (1988) *Time Effects in Concrete Structures*, Amsterdam, the Netherlands: Elsevier.

Gilbert, R.I. and Mickleborough, N.C. (2004) *Design of Prestressed Concrete*, London, U.K.: Spon Press.

Liang, Q.Q. (2005) *Performance-Based Optimization of Structures: Theory and Applications*, London, U.K.: Spon Press.

Liang, Q.Q. (2009) Performance-based analysis of concrete-filled steel tubular beam-columns, part I: Theory and algorithms, *Journal of Constructional Steel Research*, 65 (2): 363–372.

Liang, Q.Q., Uy, B. and Steven, G.P. (2002) Performance-based optimization for strut-tie modeling of structural concrete, *Journal of Structural Engineering*, ASCE, 128 (6): 815–823.

Liang, Q.Q., Xie, Y.M. and Steven, G.P. (2000a) Optimal topology design of bracing systems for multistory steel frames, *Journal of Structural Engineering*, ASCE, 126 (7): 823–829.

Liang, Q.Q., Xie, Y.M. and Steven, G.P. (2000b) Topology optimization of strut-and-tie models in reinforced concrete structures using an evolutionary procedure, *ACI Structural Journal*, 97 (2): 322–330.

Liang, Q.Q., Xie, Y.M. and Steven, G.P. (2001) Generating optimal strut-and-tie models in prestressed concrete beams by performance-based optimization, *ACI Structural Journal*, 98 (2): 226–232.

Mander, J.B., Priestley, M.J.N. and Park, R. (1988) Theoretical stress–strain model for confined concrete, *Journal of Structural Engineering*, ASCE, 114 (8): 1804–1826.

Pham, L., Bridge, R.Q. and Bradford, M.A. (1986) Calibration of the proposed limit stated design rules for steel beams and columns, *Civil Engineering Transactions*, Institution of Engineers Australia, 28 (3): 268–274.

Trahair, N.S. and Bradford, M.A. (1998) *The Behaviour and Design of Steel Structures to AS 4100*, 3rd edn. (Australian), London, U.K.: Taylor & Francis Group.

Trost, H. (1967) Auswirkungen des superpositionsprinzips auf kirech- und relaxations probleme bei beton und spannbeton, *Beton- und Stahlbetonbau*, 62: 230–238, 261–269.

Warner, R.F., Rangan, B.V., Hall, A.S. and Faulkes, K.A. (1998) *Concrete Structures*, Melbourne, Victoria, Australia: Addison Wesley Longman.

Zienkiewicz, O.C. and Taylor, R.L. (1989) *The Finite Element Method*, 4th edn., Vol. 1, *Basic Formulation and Linear Problems*, New York: McGraw-Hill.

Zienkiewicz, O.C. and Taylor, R.L. (1991) *The Finite Element Method*, 4th edn., Vol. 2, *Solid and Fluid Mechanics, Dynamics and Nonlinearity*, New York: McGraw-Hill.

Chapter 2

Design actions

2.1 INTRODUCTION

In order to design a steel or composite structure, the structural designer must estimate the design actions (loads) acting on the structure. Design actions on steel and composite structures may be divided into permanent actions, imposed actions, wind actions, snow actions, earthquake actions and other indirect actions caused by temperature, foundation settlement and concrete shrinkages. The structural designer must determine not only the types and magnitudes of design actions which will be applied to the structure but also the most severe combinations of these design actions for which the structure must be designed. The combinations of design actions are undertaken by multiplying the nominal design actions using load factors.

The accurate estimation of design actions on the structure is very important in structural design as it significantly affects the final design and objectives. Any error in the estimation of design actions may lead to wrong results of structural analysis on the structure and lead to the unrealistic sizing of its structural members or even collapse of the structure. AS/NZS 1170.0 (2002) provides specifications on the estimation of design actions based on statistical or probabilistic analyses owing to uncertainties about design actions on structures. The evaluation of permanent and imposed design actions is straightforward in accordance with AS/NZS 1170.1 (2002). However, the procedure given in the AS/NZS 1170.2 (2011) for determining the wind actions on buildings is quite complicated, particularly for irregular and sensitive structures. The detailed treatment of the calculation of wind actions is given in this chapter.

In this chapter, the estimation of design actions on steel and composite structures in accordance with AS/NZS 1170.0, AS/NZS 1170.1 and AS/NZS 1170.2 is presented. The discussion on permanent actions is given first. This is followed by the description of imposed actions for various structures. The basic procedure and underlining principals for determining wind actions are then provided. The combinations of actions for ultimate limit states and serviceability limit states are discussed. Finally, a worked example is provided to illustrate the procedure for calculating wind actions on an industrial building. This chapter should be read with AS/NZS 1170.0 (2002), AS/NZS 1170.1 (2002) and AS/NZS 1170.2 (2011).

2.2 PERMANENT ACTIONS

Permanent actions are actions acting continuously on a structure without significant changes in magnitude in its design life. Permanent actions are calculated as the self-weight of the structure including finishes, permanent construction materials, permanent equipments, fixed or movable partitions and stored materials. The self-weight of a structural member is calculated from its design or known dimensions and the unit weight, which is given in

Tables A1 and A2 of AS/NZS 1170.1 (2002). It should be noted that the unit weights of materials given in the code are average values for the specific materials.

The calculated self-weight of permanent partitions must be applied to the actual positions in the structure. If a structure is designed to allow for movable partitions, the calculated self-weight of the movable partitions can be applied to any probable positions where the partitions may be placed. The structure must be designed for the design actions. AS/NZS 1170.1 requires that a minimum uniformly distributed permanent load of 0.5 kPa shall be used to consider movable partitions. This is to ensure that the mass of the movable partitions is taken into account in designing the structure under an earthquake. In addition, the minimum load of 0.5 kPa is adequate to cover the self-weight of most partitions made of studs supporting glass, plywood and plasterboard.

2.3 IMPOSED ACTIONS

Imposed actions (or live loads) are loads on the structure which arise from the intended use of the structure, including gradually applied loads (static loads) and dynamic loads such as cyclic loads and impact loads. The live loads are characterised by their time-dependency and random distributions in space. The magnitudes and distributions of live loads vary significantly with the occupancy and function of the structure. Imposed actions on a structure vary from zero to the maximum values which occur rarely and are regarded as the maximum loads in the design life of the structure.

The imposed actions given in AS/NZS 1170.1 are characteristic loads, which represent the peak loads over a 50-year design life having a 5% probability of being exceeded. Imposed floor actions are given in Table 3.1 of AS/NZS 1170.1 (2002). The uniformly distributed loads (UDLs) and concentrated loads are listed in the table. The concentrated loads are used to represent the localised loads caused by heavy equipments or vehicles that may not be adequately covered by the UDLs. However, it should be noted that the distributed and concentrated live loads should be considered separately and the structure must be designed for the most adverse effect of design actions. The live loads given in the loading code consider the importance and design working life of the structure, which are assumed to be part of the occupancy description. This implies that once the occupancy of the structure has been determined, the imposed loads can be used to design the structure regardless of its importance and design working life.

AS/NZS 1170.1 allows for consideration of pattern loading for live loads. The purpose for this is to account for the most adverse effects of live loads on the structure. The consideration of pattern loading depends on the ratio of dead to live load and the type of structural member. For a structure subjected to wind, earthquake or fire loading, pattern imposed loading on continuous beams or slabs need not be considered.

In AS/NZS 1170.1, a reduction factor ψ_a is used to reduce the uniformly distributed live loads based on the results of load surveys. The reduction factor ψ_a is taken as 1.0 for areas used for occupancy types C3–C5 specified in Table 3.1 of AS/NZS 1170.1, storage areas subjected to imposed loads exceed 5 kPa, light and medium traffic areas and one-way slabs. For other areas, Clause 3.4.2 of AS/NZS 1170.1 (2002) provides the following formula for determining the reduction factor ψ_a:

$$\psi_a = 0.3 + \frac{3}{\sqrt{A_t}} \quad (0.5 \le \psi_a \le 1.0) \tag{2.1}$$

where A_t (m^2) is the sum of the tributary areas supported by the structural member under consideration. The reduction factor ψ_a must not be greater than 1.0 and not less than 0.5.

The roofs of industrial buildings are usually non-trafficable. For structural elements such as purlins and rafters and cladding providing direct support, the uniformly distributed live load is calculated by the following formula given in Clause 3.5.1 of AS/NZS 1170.1 (2002):

$$w_Q = \left(0.12 + \frac{1.8}{A_{pa}}\right) \geq 0.25\,\text{kPa} \tag{2.2}$$

where A_{pa} (m^2) is the plan projection of the surface area of the roof supported by the structural member. The aforementioned formula represents an imposed distributed load of 0.12 kPa plus a concentrated load of 1.8 kN which is distributed over the area A_{pa} supported by the structural member. The concentrated load of 1.8 kN is to account for the weight of a heavy worker standing on the roof. As shown in Table 3.2 of AS/NZS 1170.1, the structural elements of the roof must be designed to support a concentrated load of 1.4 kN at any point and the cladding must support a concentrated load of 1.1 kN.

2.4 WIND ACTIONS

Wind actions on structural members and structures or buildings are specified in AS/NZS 1170.2 (2011). The design of buildings, particularly industrial buildings, is influenced significantly by wind loads. Therefore, it is important to carefully estimate the wind loads in accordance with loading codes. Wind loads are both time dependent and space dependent. The estimation of wind loading is relatively complicated as it depends on the location and direction of the building being designed, site conditions related to terrain/height, shielding and topography, the shape of the building and the fundamental frequency of the structure (Holmes et al. 1990; Holmes 2001). The estimation of wind actions in accordance with AS/NZS 1170.2 is described in the subsequent sections.

2.4.1 Determination of wind actions

The main steps for determining wind actions on structural members or structures are given as follows:

1. Determine the site wind speeds.
2. Determine the design wind speed from the site wind speeds.
3. Calculate the design wind pressures and distributed forces.
4. Compute wind actions.

In Clause 2.2 of AS/NZS 1170.2 (2011), the site wind speeds are defined for the eight cardinal directions at the reference height above the ground and are calculated by

$$V_{sit,\beta} = V_R M_d (M_{z,cat} M_s M_t) \tag{2.3}$$

where

V_R is the regional 3 s gust wind speed (m/s) for annual probability of exceedance of $1/R$
M_d is the wind directional multipliers for the eight cardinal directions
$M_{z,cat}$ is the terrain/height multiplier
M_s is the shielding multiplier
M_t is the topographic multiplier

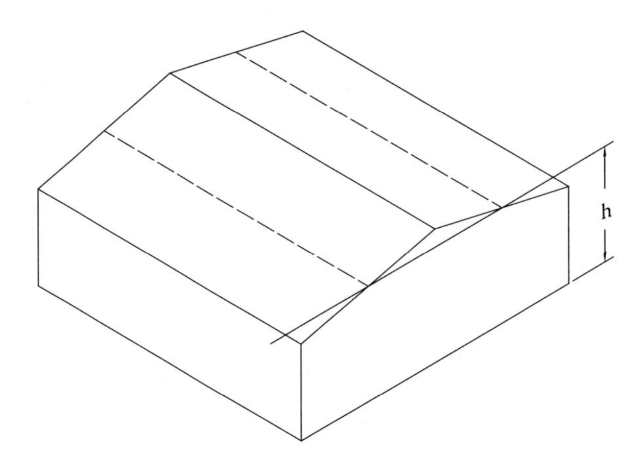

Figure 2.1 Average roof height of structure.

The wind speed is generally determined at the average roof height (h) of the building as shown in Figure 2.1. If the orientation of the building being designed is not known, the regional wind should be assumed to act from any cardinal directions and M_d can be conservatively taken as 1.0 for all directions.

The building orthogonal design wind speed ($V_{des,\theta}$) is determined as the maximum cardinal direction site wind speed ($V_{sit,\beta}$) within a sector of $\pm 45°$ to the orthogonal direction being considered. The design wind speed ($V_{des,\theta}$) may vary with the orthogonal direction. It is required that four orthogonal directions must be considered in the design of a building. The structure can be conservatively designed by using the site wind speed and multipliers for the worst direction. The minimum design wind speed ($V_{des,\theta}$) of 30 m/s is suggested in AS/NZS 1170.2 for the ultimate limit state design.

The design wind pressure acting normal to the surface of a structural member or building can be calculated in accordance with Clause 2.4.1 of AS/NZS 1170.2 (2011) as follows:

$$p = 0.5\rho_{air} V_{des,\theta}^2 C_{fig} C_{dyn} \tag{2.4}$$

where
- p is the design wind pressure (Pa)
- ρ_{air} is the density of air taken as 1.2 kg/m^2
- C_{fig} is the aerodynamic shape factor
- C_{dyn} is the dynamic response factor

The design wind frictional drag force per unit area (f) on structural members and structures can also be calculated using Equation 2.4.

Wind actions on a structure should be determined by considering the wind from no fewer than four orthogonal directions. The Clause 2.5.3.1 of AS/NZS 1170.2 specifies that the forces acting on structural members or surfaces are calculated by

$$F = \sum (p_z A_z) \tag{2.5}$$

where
- F denotes the force (N) derived from wind actions
- p_z stands for the design wind pressure (Pa) normal to the surface at height z
- A_z is the reference area (m^2) on which the wind pressure p_z acts at height z

For enclosed buildings, external pressures accounting for the effects of local pressure factors should be combined with internal pressures and the structure must be designed for the most severe combinations of wind actions.

2.4.2 Regional wind speeds

The regional wind speeds given in AS/NZS 1170.2 were determined from the analyses of long-term records of daily maximum wind speeds for each particular region in Australia. The regional wind speeds are a function of the standard site exposure, peak gust, annual probability of exceedance and wind direction. The standard site exposure represents 10 m height in terrain category 2, which is defined in Section 2.4.3.1. The annual probability of exceedance of the wind speed is the inverse of the return period or average recurrence interval, which is related to the importance level of the structure. In AS/NZS 1170.0, structures are classified into five importance levels according to their consequence of failure, which are given Table F1 of AS/NZS 1170.0 (2002). Once the design working life and importance level of the structure have been determined, the annual probability of exceedance of the wind speed can be obtained from Table F2 of AS/NZS 1170.0. The regional wind speeds (V_R) for all directions based on 3 s gust wind data are given in Table 3.1 of AS/NZS 1170.2 (2011). Wind regions in Australia are provided in AS/NZS 1170.2. The importance level of normal structures is 2. For normal structures, the annual probability of exceedance of the wind speed is 1/500. The regional wind speeds (V_R) for normal structures are provided in Table 2.1.

The wind directional multipliers (M_d) for regions *A* and *W* are provided in Table 3.2 of AS/NZS 1170.2 (2011). Theses multipliers can be used for strength and serviceability limit state designs and were derived based on the assumption that only the wind load within the two 45° directional sectors of the typical rectangular buildings contribute to the probability (Melbourne 1984). It can be seen from Table 3.2 of AS/NZS 1170.2 that the wind directional multiplier (M_d) varies from 0.8 to 1.0 for the eight cardinal directions. However, it should be noted that if the orientation of the building being designed in regions *A* and *W* is not known, the wind should be assumed to act in any direction so that M_d is taken as 1.0.

For buildings in regions *B–D*, Clause 3.3.2 of AS/NZS 1170.2 (2011) suggests that the wind directional multiplier (M_d) for all directions is taken as follows:

1. 0.95 for calculating the resultant forces and overturning moments on a complete building and wind loads on major structural framing members
2. 1.0 for all other design situations

The regions *C* and *D* are cyclonic regions where the directional multipliers are not used. This is because the maximum wind speed may occur in any direction. However, the wind directional multiplier of $M_d = 0.95$ can be applied to the wind speed when it is used to calculate the resultant forces and overtraining moments on a complete building and wind actions

Table 2.1 Regional wind speeds for R_{500} for normal structures

Wind region		*Regional wind speed (m/s)*
Non-cyclonic	A	45
	W	51
	B	57
Cyclonic	C	$66F_C$
	D	$80F_D$

on major structural members in regions B–D. This factor is used to account for the average probability of the design wind speed being exceeded for the building (Davenport 1977; Holmes 1981). For other design situations, such as non-major structural members including cladding and immediate supporting members, M_d is taken as 1.0.

Factors F_C and F_D applied to wind speeds in regions C and D as given in Table 2.1 are taken as $F_C = 1.05$ and $F_D = 1.1$ for $R \geq 50$ years and $F_C = F_D = 1.0$ for $R < 50$ years.

2.4.3 Site exposure multipliers

The exposure multipliers ($M_{z,cat}$, M_s, M_t) are used to account for the effects of the site conditions of the building on the site wind speeds ($V_{sit,}\beta$), which include terrain/height, shielding and topography. The terrain and surrounding buildings providing shielding may change in the design working life of the building due to new development in the area. Therefore, it is important to consider the known future changes to the terrain roughness in evaluating the terrain category and to the buildings that provide shielding in estimating the shielding multiplier.

2.4.3.1 Terrain/height multiplier ($M_{z,cat}$)

The terrain/height multiplier ($M_{z,cat}$) varies with the terrain roughness and height of the building. In Clause 4.2.1 of AS/NZS 1170.2 (2011), the terrain is divided into four categories as follows:

1. Terrain category 1 includes exposed open terrain with few or no obstructions and water surfaces at serviceability wind speeds.
2. Terrain category 2 covers open water surfaces, open terrain, grassland and airfields with few, well-scattered obstructions with heights generally from 1.5 to 10 m.
3. Terrain category 3 includes the terrain with numerous closely spaced obstructions with 3–5 m height, for example, in the areas of suburban housing and level wooded country.
4. Terrain category 4 covers large city centres and well-developed industrial areas with numerous large and closely spaced obstructions with heights from 10 to 30 m.

The terrain/height multipliers for gust wind speeds for fully developed terrains in all regions for serviceability limit state design and in regions $A1$–$A7$, W and B for ultimate limit state design are given in Table 4.1(A) of AS/NZS 1170.2. It appears from the table that the terrain/height multiplier ($M_{z,cat}$) is 1.0 for building height of 10 m in terrain category 2 as this condition is used as a reference for other categories and building heights. For the ultimate limit state design of buildings in regions C and D which are cyclonic regions, the terrain/height multipliers are provided in Table 4.1(B) of AS/NZS 1170.2. The terrain/height multipliers for buildings in terrain categories 1 and 2 having the same height are the same and this holds true for buildings in terrain categories 3 and 4. The design code allows for $M_{z,cat}$ to be taken as the weighted average value over the averaging distance upwind of the building when the terrain changes.

2.4.3.2 Shielding multiplier (M_s)

The shielding multiplier (M_s) is used to account for the effects of total and local wind actions on structures with a range of shielding configurations (Holmes and Best 1979; Hussain and Lee 1980). It depends on the shielding factors including the average spacing, roof height and breadth of shielding buildings normal to the wind direction, the average of roof height of the building being shielded and the number of upwind shielding buildings within a 45° sector of radius $20h$. It should be noted that only buildings located in a 45° sector of radius

20*h* symmetrically positioned about the direction and whose height is greater than or equal to the average roof height of the building being shielded can provide shielding as depicted in Figure 2.2. The shielding multiplier for buildings with various shielding parameters is provided in Table 4.3 of AS/NZS 1170.2. If the average upwind gradient is greater than 0.2 or no shielding in the wind direction, the shielding multiplier is taken as 1.0.

Clause 4.3.3 of AS/ZNS 1170.2 (2011) provides equations for calculating the shielding parameter given in Table 4.3 of AS/NZS 1170.2 as follows:

$$s = \frac{l_s}{\sqrt{h_s b_s}} \tag{2.6}$$

$$l_s = h\left(\frac{10}{n_s} + 5\right) \tag{2.7}$$

where

l_s, h_s and b_s are the average spacing, roof height and breadth of shielding buildings, respectively

h is the average roof height of the structure being shielded

n_s is the total number of upwind shielding buildings within a 45° sector of radius 20h

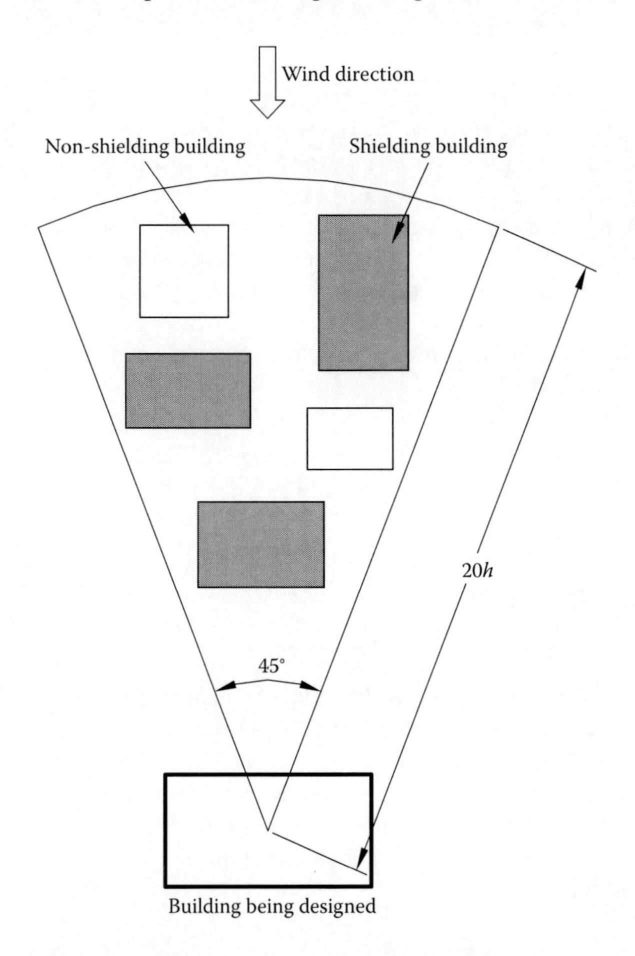

Figure 2.2 Shielding in complex urban areas.

2.4.3.3 *Topographic multiplier (M_t)*

The topographic multiplier (M_t) considers the effects of local topographic zones on the site wind speeds. In Clause 4.4.1 of AS/NZS 1170.2 (2011), for sites in Tasmania over 500 m above the sea level, M_t is taken as $M_t = M_h M_{lee}(1 + 0.00015E_{sl})$, where M_h is the hill-shape multiplier, the lee multiplier $M_{lee} = 1$ and E_{sl} is the site elevation above the mean sea level (m). For Australia sites, M_t is taken as the larger value of M_h and M_{lee}. The hill-shape multiplier is taken as 1.0 except that for the particular cardinal direction in the local topographic zones shown in Figures 4.2 through 4.4 of AS/NZS 1170.2. For the local topographic zones depicted in these figures, the hill-shape multiplier (M_h) is given in Clause 4.4.2 of AS/NZS 1170.2 (2011) as follows:

$$M_h = \begin{cases} 1.0 & \text{for } \dfrac{H}{2L_u} < 0.05 \\ 1 + \dfrac{H}{3.5(z + L_1)}\left(1 - \dfrac{|x|}{L_2}\right) & \text{for } 0.05 \le \dfrac{H}{2L_u} < 0.45 \\ 1 + 0.71\left(1 - \dfrac{|x|}{L_2}\right) & \text{for } \dfrac{H}{2L_u} > 0.45 \text{ (in separation zone)} \end{cases} \tag{2.8}$$

where

- H is the height of the hill, ridge or escarpment
- L_u is the horizontal distance upwind from the crest of the hill, ridge or escarpment to the level half the height below the crest
- z is the reference height on the structure from the average local ground level
- x is the horizontal distance from the structure to the crest of the hill, ridge or escarpment
- L_1 is the length scale (m) which is the larger of $0.36L_u$ and $0.4H$
- L_2 is the length scale (m) which is taken as $4L_1$ upwind for all types and downwind for hills and ridges or $10L_1$ downwind for escarpments

It should be noted that for $H/(2L_u) > 0.45$ and in zones other than the separation zone, M_h is taken as that for $0.05 \le H/(2L_u) < 0.45$ (Bowen 1983; Paterson and Holmes 1993).

The hill-shape multiplier (M_h) for the local topographic zones with x and z are zero is given in Table 4.4 of AS/NZS 1170.2. For Australia sites, the lee multiplier (M_{lee}) is taken as 1.0.

2.4.4 Aerodynamic shape factor

2.4.4.1 *Calculation of aerodynamic shape factor*

The aerodynamic shape factor (C_{fig}) considers the effects of the geometry of a structure on the surface local, resultant or average wind pressure. It is a function of the geometry and shape of the structure and the relative wind direction and speed (ISO 4354, 1997). The sign convention of the aerodynamic shape factor (C_{fig}) assumes that positive values indicate pressure acting towards the surface and negative values indicate pressure acting away from the surface.

For enclosed buildings, the aerodynamic shape factor is given in Clause 5.2 of AS/NZS 1170.2 (2011) as follows:

$$C_{fig} = C_{p,e} K_a K_{c,e} K_l K_p \quad \text{for external pressures} \tag{2.9}$$

$$C_{fig} = C_{p,i} K_{c,i} \quad \text{for internal pressures} \tag{2.10}$$

$$C_{fig} = C_f K_a K_c \quad \text{for frictional drag forces} \tag{2.11}$$

where

$C_{p,e}$ is the external pressure coefficient
K_a is the area reduction factor
K_c is the combination factor ($K_{c,e}$ for external pressures and $K_{c,i}$ for internal pressures)
K_l is the local pressure factor
K_p is the porous cladding reduction factor
$C_{p,i}$ is the internal pressure coefficient
C_f is the frictional drag force coefficient

2.4.4.2 Internal pressure coefficient

Internal pressure depends on the relative permeability of the external surfaces of a building. It may be positive or negative that depends on the position and size of the opening. The permeability of a surface is calculated as the sum of the areas of opening and leakage on that surface of the building. Open doors and windows, vents, ventilation systems and gaps in cladding are typical openings. The dominant opening means that it plays a dominant effect on the internal pressure in the building. If the sum of all openings in the surface is greater than the sum of openings in each of the other surfaces in the building, the surface is regarded as containing dominant openings. Internal pressure coefficients for enclosed rectangular buildings are given in Clause 5.3 of AS/NZS 1170.2. Table 5.1(A) of AS/NZS 1170.2 provides the internal pressure coefficients for buildings with open interior plan and having permeable walls without dominant openings. For buildings with open interior plan and having dominant openings on one surface, internal pressure coefficients are given in Table 5.1(B) of AS/NZS 1170.2.

Offices and houses with all windows closed usually have permeability between 0.01% and 0.2% of the wall area, which depends on the degree of draught proofing. The permeability of industrial and farm buildings can be up to 0.5% of the wall area. The walls of industrial buildings are usually considered to be permeable, while concrete, concrete masonry or other walls designed to prevent air passage may be treated as non-permeable.

2.4.4.3 External pressure coefficient

For enclosed rectangular buildings, external pressure coefficients are given in Tables 5.2(A) to 5.2(C) for walls and Tables 5.3(A) to 5.3(C) for roofs of AS/NZS 1170.2 (2011). It can be observed from these tables that in some cases, two values are given for the pressure coefficient. For these cases, roof surfaces may be subjected to either value owing to turbulence so that roofs should be designed for both values. Alternatively, external pressures are combined with internal pressures to obtain the most severe combinations of actions for the design of the structure. Discussions on external pressures on low-rise building and monoslope roofs are given by Holmes (1985) and Stathopoulos and Mohammadian (1985).

For crosswind roof slopes and for all roof pitches (α), the values given in Tables 5.3(A) and 5.3(B) should be used to determine the most severe action effects as follows:

1. Apply the more negative value of the two given in the table to both halves of the roof.
2. Apply the more positive value of the two given in the table to both halves of the roof.
3. Apply the more negative value to one half and more positive value to the other half of the roof.

2.4.4.4 Area reduction factor

The area reduction factor (K_a) for roofs and side walls depends on the tributary area (A_t), which is defined as the area contributing to the force under consideration. In Clause 5.4.2 of AS/NZS 1170.2, the area reduction factor K_a is taken as 1.0 for tributary area $A_t \leq 10$ m^2, 0.9 for $A_t = 25$ m^2 and 0.8 for $A_t \geq 100$ m^2. For intermediate areas, linear interpolation can be used to determine the area reduction factor. For all other cases, the area reduction factor is taken as 1.0. The values given in the code were determined by direct measurements of total roof loads in wind tunnel tests (Davenport et al. 1977; Kim and Mehta 1977).

2.4.4.5 Combination factor

The combination factor (K_c) accounts for the effects of non-coincidence of peak wind pressures on different surfaces of the building. For examples, wall pressures are well correlated with roof pressures. Table 5.5 of AS/NZS 1170.2 gives combination factors $K_{c,i}$ and $K_{c,e}$ for wind pressures on major structural elements of an enclosed building. However, it should be noted that the combination factors do not apply to cladding or purlins. When the area reduction factor (K_a) is less than 1.0, K_c for all surfaces must satisfy the following condition:

$$K_c \geq \frac{0.8}{K_a} \tag{2.12}$$

2.4.4.6 Local pressure factor

The wind pressures on small areas are evaluated using the local pressure factor (K_l). The peak wind pressures often occur on small areas near windward corners and roof edges of the building as depicted in Figure 2.3, where $a = \min(0.2b, 0.2d, h)$. The local pressure factor is applied to cladding, their fixings and the members that directly support the cladding and is given in Table 5.6 of AS/NZS 1170.2. For areas SA1, RA1, RA3 and WA1, $K_l = 1.5$. For areas SA2, RA2 and RA4, $K_l = 2$. For other cases which are not specified in this table or Figure 2.3, the local pressure factor is taken as 1.0.

2.4.4.7 Permeable cladding reduction factor

It has been found that negative surface pressures on permeable cladding are lower than those on a similar non-permeable cladding owing to the porous surface. This effect is taken into account in determining the aerodynamic shape factor by the permeable cladding reduction factor (K_p), which is given in Clause 5.4.5 of AS/NZS 1170.2. It should be noted that this factor is used for negative pressure only when external surfaces consisting of permeable cladding and the solidity ratio is less than 0.999 and greater than 0.99. The solidity ratio of the surface is defined as the ratio of solid area to the total area of the surface.

2.4.4.8 Frictional drag coefficient

The frictional drag forces on roofs and side walls of enclosed buildings with a ratio of d/h or d/b that is greater than 4 need to be considered when designing the roof and wall bracing systems. When determining frictional drag forces, the aerodynamic shape factor (C_{fig}) is taken as the frictional drag coefficient (C_f) in the direction of the wind, which is given in Clause 5.5 of AS/NZS 1170.2 (2011).

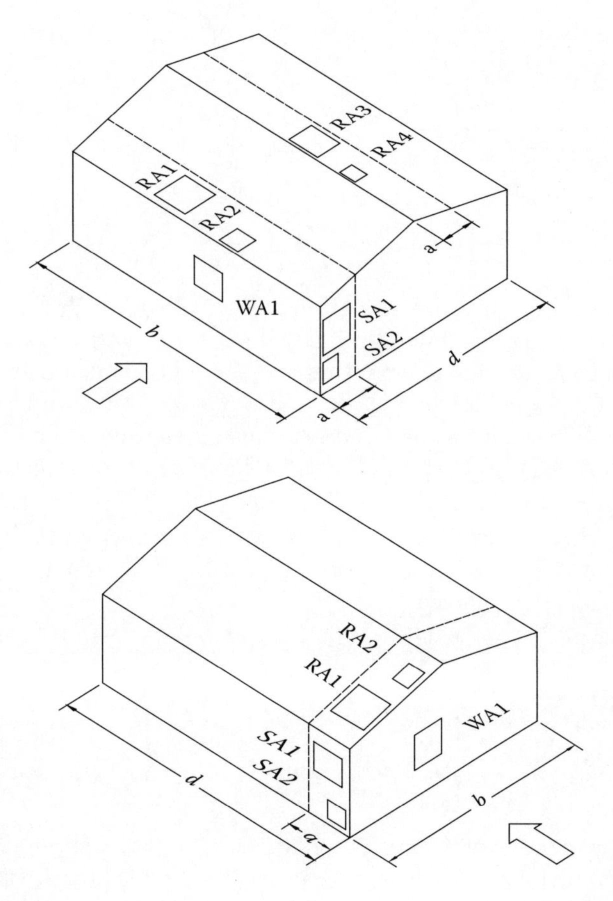

Figure 2.3 Local pressure areas.

2.4.5 Dynamic response factor

2.4.5.1 General

The dynamic response factor (C_{dyn}) accounts for the dynamic effects of wind on flexible, lightweight, slender or lightly damped structures. It considers the correlation effects of fluctuating along-wind forces on tall structures, effective pressures due to inertia forces, resonant vibrations and fluctuating pressures in the wake of the structure (ISO 4354, 1997). The dynamic response factor depends on the natural first mode fundamental frequencies of the structure. Most structures are not flexible, lightweight, slender or lightly damped so that they are not dynamically wind sensitive. The natural first mode fundamental frequencies of most structures are greater than 1.0 Hz and their dynamic response factor is taken as 1.0. For structures with natural first mode fundamental frequencies between 0.2 and 1.0 Hz, the dynamic response factor is determined for along-wind response and crosswind response discussed in the following sections.

2.4.5.2 Along-wind response

The along-wind response of most structures is due to the incident turbulence of the longitudinal component of the wind velocity (Davenport 1967; Vickery 1971). For tall buildings and towers, the dynamic response factor (C_{dyn}) is given in Clause 6.2.2 of AS/NZS 1170.2 (2011) as follows:

$$C_{dyn} = \frac{1 + 2I_h\sqrt{g_v^2 B_s + (H_s g_R^2 \beta_s S_t/\zeta)}}{1 + 2g_v I_h} \tag{2.13}$$

where

I_h is the turbulence intensity which is given in AS/NZS 1170.2 (2011)
g_v is the peak factor for the upwind velocity fluctuations and is taken as 3.7
H_s is the height factor for the resonant response and is calculated as $[1+(s/h)^2]$
s is the height of the level at which wind loads are determined for the structure
h is the average roof height of the structure above the ground
ζ is the ratio of structural damping to critical damping of the structure

The background factor B_s in Equation 2.13 is used to measure the slowly varying background component of the fluctuating response induced by low-frequency wind speed variations. This factor can be calculated by

$$B_s = \frac{1}{1 + \left(\sqrt{36(h-s)^2 + 64b_{sh}^2}/L_h\right)} \tag{2.14}$$

where

b_{sh} is the average breadth of the structure between height s and h
L_h is the integral turbulence length scale at height h and is taken as $85(h/10)^{0.25}$

The peak factor g_R for resonant response in 10 minutes period is expressed by

$$g_R = \sqrt{2\log_e(600 f_{nc})} \tag{2.15}$$

where f_{nc} is the first mode natural frequency of the structure in the crosswind direction in Hz.

The size reduction factor (β_s) is expressed as

$$\beta_s = \frac{1}{\left[1 + 3.5 f_{na} h(1 + g_v I_h)/V_{des,\theta}\right]\left[1 + 4 f_{na} b_{0h}(1 + g_v I_h)/V_{des,\theta}\right]} \tag{2.16}$$

where

f_{na} is the first mode natural frequency of the structure in the along-wind direction in Hz
b_{0h} is the average breadth of the structure between heights 0 and h

The spectrum of the turbulence of the structure is calculated by

$$S_t = \frac{\pi f_{nr}}{\left(1 + 70.8 f_{nr}^2\right)^{\frac{5}{6}}} \tag{2.17}$$

where f_{nr} is the reduced frequency, which is determined by

$$f_{nr} = \frac{f_{na}L_h(1+g_v I_h)}{V_{des,\theta}} \tag{2.18}$$

2.4.5.3 *Crosswind response*

Crosswind excitation of modern tall buildings and structures can be expressed by wake, incident turbulence and crosswind displacement mechanisms (Melbourne 1975). The equivalent crosswind static wind force per unit height (N/m) for tall enclosed buildings and towers of rectangular cross sections is a function of z. This wind force is given in Clause 6.3.2.1 of AS/NZS 1170.2 (2011) as follows:

$$w_{eq}(z) = 0.5\rho_{air}V_{des,\theta}^2 d(C_{fig}C_{dyn}) \tag{2.19}$$

in which the design wind speed $V_{des,}\theta$ is estimated at $z = h$ and d is the horizontal depth of the structure parallel to the wind direction. The product of the aerodynamic shape factor and the aerodynamic response factor is determined by

$$C_{fig}C_{dyn} = 1.5g_R\left(\frac{b}{d}\right)\frac{K_m}{(1+g_v I_h)^2}\left(\frac{z}{h}\right)^k\sqrt{\frac{\pi C_{fs}}{\zeta}} \tag{2.20}$$

where

- K_m is the mode shape correction factor for crosswind acceleration and is calculated as $(0.76 + 0.24k)$
- k is the mode shape power exponent for the fundamental mode

The power exponent is 1.5 for uniform cantilever, 0.5 for a slender framed structure, 1.0 for a building with central core and moment resisting curtain walls and 2.3 for a tower whose stiffness decreases with height or the value obtained from fitting $\phi_1(z) = (z/h)^k$ to the computed mode shape of the structure. The coefficient C_{fs} is the crosswind force spectrum coefficient for a linear mode shape.

The crosswind base overturning moment M_c (N m) can be determined by integrating the wind force $w_{eq}(z)$ from 0 to h. Clause 6.3.2.2 of AS/NZS 1170.2 (2011) provides a formula for calculating M_c as follows:

$$M_c = 0.5g_R b\left[\frac{0.5\rho_{air}V_{des,\theta}^2}{(1+g_v I_h)^2}\right]h^2\left(\frac{3}{k+2}\right)K_m\sqrt{\frac{\pi C_{fs}}{\zeta}} \tag{2.21}$$

The crosswind force spectrum coefficient (C_{fs}) generalized for a linear mode shape is a function of the aspect ratio of the cross section and height, turbulence intensity and reduced velocity (V_n). The reduced velocity (V_n) is provided in Clause 6.3.2.3 of AS/NZS 1170.2 (2011) as follows:

$$V_n = \frac{V_{des,\theta}}{f_{nc}b(1+g_v I_h)} \tag{2.22}$$

The crosswind force spectrum coefficient (C_{fs}) can be determined for the turbulence intensity evaluated at $2h/3$ in accordance with Clause 6.3.2.3 of AS/NZS 1170.2 (2011).

2.4.5.4 *Combination of long-wind and crosswind response*

In Clause 6.4 of AS/NZS 1170.2 (2011), the total combined peak scalar dynamic action effect such as an axial load in a column is calculated by

$$E_{a,t} = E_{a,m} + \sqrt{E_{a,p}^2 + E_{c,p}^2} \tag{2.23}$$

where

$E_{a,p}$ is the action effect caused by the peak along-wind response
$E_{c,p}$ is the action effect caused by the peak crosswind response
$E_{a,m}$ is the action effect caused by the mean along-wind response and is given by

$$E_{a,m} = \frac{E_{a,p}}{C_{dyn}(1 + 2g_v I_h)} \tag{2.24}$$

2.5 COMBINATIONS OF ACTIONS

2.5.1 Combinations of actions for strength limit state

Structures may be subjected to permanent action G (dead load), imposed action Q (live load), wind action W (wind load), earthquake E or a combination of them. The following basic combinations of actions for the strength limit state are suggested in Clause 4.4.2 of AS/NZS 1170.0 (2002):

1. $1.35G$
2. $1.2G + 1.5Q$
3. $1.2G + 1.5\psi_l Q$
4. $1.2G + \psi_c Q + W_u$
5. $0.9G + W_u$
6. $G + \psi_c Q + E_u$

In the aforementioned load combinations, ψ_l and ψ_c are the long-term and combination factors, respectively, and are given in Table 4.1 of AS/NZS 1170.0, E_u is the earthquake load and W_u is the ultimate wind load.

2.5.2 Combinations of actions for stability limit state

The stability limit state is an ultimate limit state which is concerned with the loss of the static equilibrium of structural members or the whole structure. The Clause 4.2.1 of AS/NZS 1170.0 (2002) specifies that if the permanent actions cause stabilizing effects, the combination is taken as $0.9G$. However, if the combinations of actions cause destabilizing effects, the code requires that combinations are taken as follows:

1. $1.35G$
2. $1.2G + 1.5Q$
3. $1.2G + 1.5\psi_l Q$
4. $1.2G + \psi_c Q + W_u$
5. $0.9G + W_u$
6. $G + \psi_c Q + E_u$

2.5.3 Combinations of actions for serviceability limit state

For the serviceability limit state, Clause 4.3 of AS/NZS 1170.0 (2002) states that appropriate combinations using the short-term and long-term factors should be used for the serviceability conditions considered. The following combinations of dead load, live load and service wind load (W_s) may be considered:

1. $G + \psi_s Q$
2. $G + W_s$
3. $G + \psi_s Q + W_s$
4. W_s

The short-term factor ψ_s is given in Table 4.1 of AS/NZS 1170.0.

Example 2.1: Calculation of wind actions on an industrial building

Figure 2.4 depicts a proposed steel portal framed industrial building of 28 m × 50 m. The height of the eave of the building is 5 m, while its ridge is 8.75 m. One of the walls contains a loading door of 4000 × 3600 mm, which is located in the second bay of the portal frames. There are no openings on other walls and roofs and the ridge is not vented as depicted in Figure 2.4. Internal steel frames are to be spaced at 5 m. The building is to be located on a flat exposed site in region *A*2, terrain category 2. There are no surrounding buildings and the orientation of the building has not been finalised. The design working life of the building is 50 years. It is required to determine the internal and external pressures on roofs and walls and the loading on the first internal frame.

1. Site wind speed

The building is a normal structure, which is designed for importance level 2. The building is located in region *A*2 which is a non-cyclonic wind region and its design working life is 50 years. The annual probability of exceedance of the wind event for this normal structure is 1/500.

The regional wind speed can be obtained from Table 2.1 as

$$V_R = 45\,\text{m/s}$$

As the orientation of the building has not been finalised, the wind directional multiplier for region *A*2 is $M_d = 1.0$.

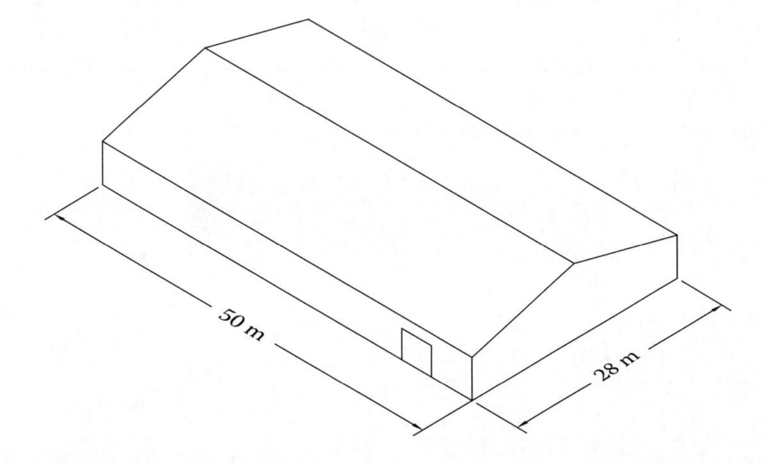

Figure 2.4 Steel-framed industrial building.

The average roof height of the building is

$h = 5 + 3.75/2 = 6.875$ m

The terrain/height multiplier can be obtained from Table 4.1(A) of AS/NZS 1170.2 using linear interpolation as follows:

$$M_{z,cat} = 0.91 + \frac{(6.875-5)(1-0.91)}{(10-5)} = 0.944$$

The building is located on a flat exposed site without upwind building so that there is no shielding. The shielding multiplier is taken as $M_s = 1.0$.

The topographic multiplier is $M_t = 1.0$.

The site wind speed can be calculated as

$$V_{sit,\beta} = V_R M_d \left(M_{z,cat} M_s M_t\right) = 45 \times 1.0 \times 0.944 \times 1.0 \times 1.0 = 42.48\,\text{m/s}$$

2. Design wind speed

The orientation of the building has not been finalised so that the design wind speed can be taken as the site wind speed:

$$V_{des,\theta} = 42.48\,\text{m/s}$$

3. Aerodynamic shape factor

3.1. External pressure coefficients under crosswind

Windward wall: $C_{p,e} = 0.7$ — Table 5.2(A) (AS/NZS 1170.2)

Leeward wall:

The roof pitch: $\alpha = 15° > 10°$, $d/b = 28/50 = 0.56$

Therefore, $C_{p,e} = -0.3$ — Table 5.2(B)

Roofs: $\alpha = 15° > 10°$, $h/d = 6.875/28 = 0.246$

For upwind slope: $C_{p,e} = -0.5$ — Table 5.3(B)

For downwind slope: $C_{p,e} = -0.5$ — Table 5.3(C)

The external pressure coefficients for sidewalls vary with the horizontal distance from the windward edge and can be obtained from Table 5.2(C) of AS/NZS 1170.2 as follows:

Horizontal distance from windward edge	*0–6.875 m*	*6.875–13.75 m*	*13.75–20.625 m*	*>20.625 m*
$C_{p,e}$	−0.65	−0.5	−0.3	−0.2

3.2. External pressure coefficients under longitudinal wind

Windward wall: $C_{p,e} = 0.7$ — Table 5.2(A)

Leeward wall:

The roof pitch: $\alpha = 15° > 10°$, $d/b = 50/28 = 1.786$

Therefore, $C_{p,e} = -0.3$ — Table 5.2(B)

The roof of the building is a gable roof. The external pressure coefficients for the gable roof vary with the horizontal distance from the windward edge and can be obtained from Table 5.3(A) of AS/NZS 1170.2 as follows:

$$\frac{h}{d} = \frac{6.875}{28} = 0.246$$

Horizontal distance from windward edge	*0–6.875 m*	*6.875–13.75 m*	*13.75–20.625 m*	*>20.625 m*
$C_{p,e}$	−0.9	−0.5	−0.3	−0.2

The external pressure coefficients for sidewalls vary with the horizontal distance from the windward edge and can be obtained from Table 5.2(C) of AS/NZS 1170.2 as follows:

Horizontal distance from windward edge	*0–6.875 m*	*6.875–13.75 m*	*13.75–20.625 m*	*>20.625 m*
$C_{p,e}$	−0.65	−0.5	−0.3	−0.2

3.3. Internal pressure coefficients under crosswind

Under crosswind, the windward or leeward wall contains a loading door which is a dominant opening. It needs to calculate the permeability ratio of the surfaces of the building in order to determine the internal pressure coefficients.

The area of the dominant opening is

$$A_{do} = 4 \times 3.6 = 14.4\,\mathrm{m}^2$$

Assume the building leakage is at 0.1% permeability. The total building leakage is 0.1% of the area of all other surfaces excluding the one containing the dominant opening, which is calculated as follows:

$$A_l = \left[2 \times \left(5 \times 28 + 28 \times \frac{3.75}{2}\right) + (2 \times 14.49 \times 50) + 50 \times 5\right] \times 0.1\% = 2.084\,\mathrm{m}^2$$

The permeability ratio is

$$\xi_p = \frac{A_o}{A_l} = \frac{14.4}{2.084} = 6.9 > 6.0$$

From Table 5.1(B), it can be seen that the internal pressure coefficient is equal to the external pressure coefficient: $C_{p,i} = C_{p,e}$.

The worst case for the internal pressure under crosswind is the windward wall door open so that the internal pressure coefficient is

$$C_{p,i} = +0.7$$

The worst case for the internal suction under crosswind is the leeward wall door open. For this case, the internal pressure coefficient is

$$C_{p,i} = -0.3$$

3.4. Internal pressure coefficients under longitudinal wind

The worst case for the internal pressure under longitudinal wind is the sidewall door closed. The internal pressure coefficient is

$$C_{p,i} = +0.0$$

The worst case for the suction under longitudinal wind is the sidewall door open. The internal pressure coefficient is

$$C_{p,i} = -0.65$$

3.5. Area reduction factor

The tributary area for rafter under crosswind is

$$A_t = 2 \times 14.49 \times 5 = 144.9\,\text{m}^2 > 100\,\text{m}^2$$

Therefore, for rafter, $K_a = 0.8$.

The tributary area for rafter and columns under longitudinal wind is

$$A_t = 2 \times 14.49 \times 5 + 2 \times 5 \times 5 = 194.9\,\text{m}^2 > 100\,\text{m}^2$$

Therefore, for rafter and columns, $K_a = 0.8$.

3.6. Local pressure factor

$h = 6.875$ m, $0.2b = 0.2 \times 50 = 10$ m, $0.2d = 0.2 \times 28 = 5.6$ m

Therefore, the dimension of the local pressure zone is

$$a = \min\left(0.2b, 0.2d, h\right) = 5.6\,\text{m}$$

The local pressure factor for local zone $a \times a = 5.6 \times 5.6$ m: $K_l = 1.5$.
The local pressure factor for local zone $(0.5a \times 0.5a) = 2.8 \times 2.8$ m: $K_l = 2.0$.

3.7. Combination factor

For the portal frame under external and internal wind loads, the combination factor is taken as $K_c = 1.0$ and satisfies the following condition:

$$K_c \geq \frac{0.8}{K_a} = \frac{0.8}{1.0} = 0.8, \text{OK}$$

3.8. Permeable cladding reduction factor

The cladding is not permeable, so that $K_p = 1.0$.

3.9. Aerodynamic shape factors

When calculating the wind pressures on surfaces rather than on the portal frame, the following factors are taken as 1.0:

$$K_a = K_{c,e} = K_{c,i} = K_l = 1.0$$

The aerodynamic shape factor for external pressures is calculated by

$$C_{fig} = C_{p,e} K_a K_c K_l K_p = C_{p,e} \times 1.0 \times 1.0 \times 1.0 \times 1.0 = C_{p,e}$$

The aerodynamic shape factor for internal pressures is calculated by

$$C_{fig} = C_{p,i} K_c = C_{p,e} \times 1.0 = C_{p,i}$$

4. Design wind pressures on surfaces

The industrial building is not sensitive to wind and its natural frequency is greater than 1.0 Hz so that its dynamic response factor can be taken as C_{dyn} = 1.0.

The design wind pressure is calculated as

$$p = (0.5\rho_{air})V_{des,\theta}^2 C_{fig} C_{dyn} = (0.5 \times 1.2) \times 42.48^2 \times C_{p,e} \times 1.0 = 1083C_{p,e} = 1.083C_{p,e} \text{ kPa}$$

The design wind pressures on surfaces for various pressure coefficients are calculated as follows:

$C_{p,e}$	*0.7*	*−0.9*	*−0.65*	*−0.5*	*−0.3*	*−0.2*
p (kPa)	0.758	−0.975	−0.704	−0.542	−0.325	−0.217

The external wind pressures on the surfaces of the building under cross and longitudinal winds are shown in Figures 2.5 and 2.6, respectively.

5. Loading on the first internal frame

a. Dead load (*G*)

Trimdek sheeting: 4.28 kg/m² = 0.0428 kPa.

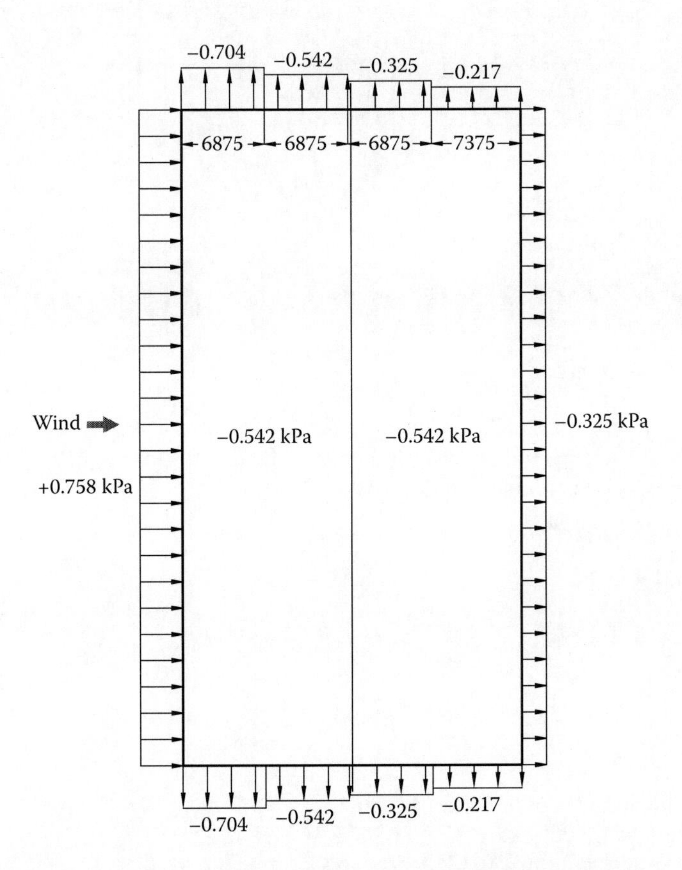

Figure 2.5 External wind pressures on surfaces of the industrial building under crosswind.

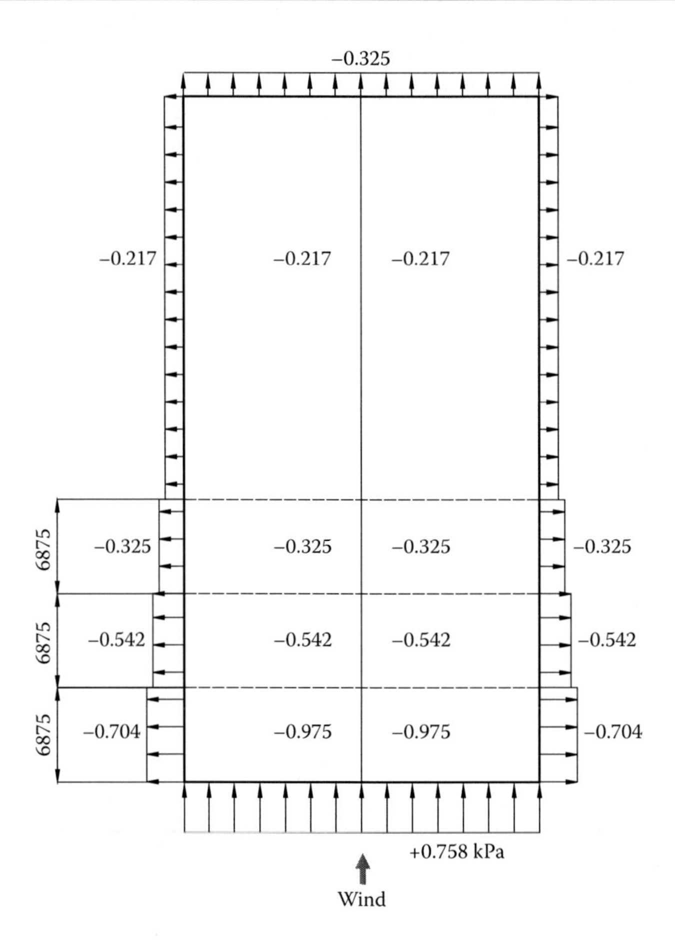

Figure 2.6 External wind pressures on surfaces of the industrial building under longitudinal wind.

Assuming that Z20019 LYSAGHT purlins (5.74 kg/m) at 1200 mm spacing are used, the self-weight of the purlin is

$$g_p = \frac{5.74 \times 9.8 \times 10^{-3}}{1.2} = 0.047\,\text{kPa}$$

Total weight of sheeting and purlin: $g = 0.0428 + 0.047 = 0.09 \approx 0.1$ kPa.

The sheeting and purlin load on rafter is

$G = 0.1 \times 5 = 0.5$ kN/m

b. Live load (Q)

$$w_L = 0.12 + \frac{1.8}{A} = 0.12 + \frac{1.8}{5 \times 28} = 0.133\,\text{kPa} < 0.25\,\text{kPa}$$

Live load on rafter: $Q = 0.25 \times 5 = 1.25$ kN/m.

c. Crosswind load

The area reduction factor for rafter: $K_a = 0.8$.

UDL on windward column = 0.758 × 5 = 3.79 kN/m.

UDL on leeward column = 0.325 × 5 = 1.63 kN/m.

UDL on rafter = 0.8 × 0.542 × 5 = 2.17 kN/m.

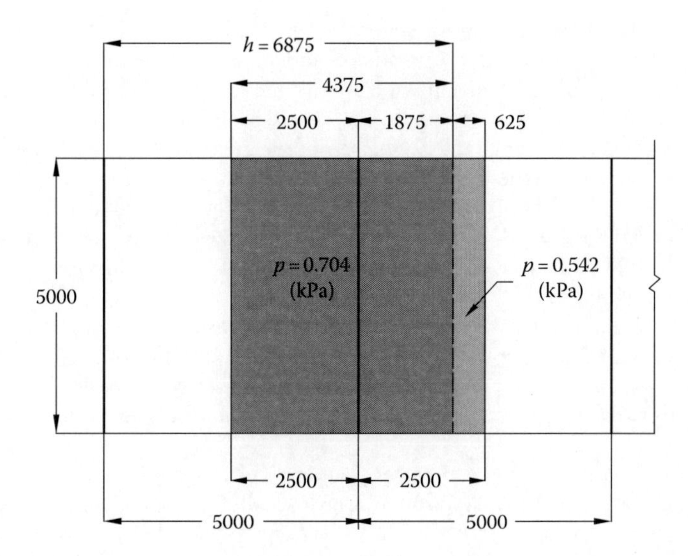

Figure 2.7 External wind pressures on the first internal frame column under longitudinal wind.

d. Longitudinal wind on first internal frame

The external wind pressures on the first internal frame column under longitudinal wind are shown in Figure 2.7. It is seen that the column supports wind pressures of 0.704 kPa on an area of 4375 × 5000 mm^2 and wind pressures of 0.542 kPa on an area of 625 × 5000 mm^2 because the wind pressures vary with the horizontal distance from the windward edge. The external wind pressures on the rafter also vary with the horizontal distance from the windward edge.

The area reduction factor for roof and walls is $K_a = 0.8$. The line loads on columns and rafters due to external wind pressures are calculated as follows:

UDL on columns = 0.8 × (0.704 × 4.375 + 0.542 × 0.625) = 2.74 kN/m.

UDL on rafter = 0.8 × (0.975 × 4.375 + 0.542 × 0.625) = 3.68 kN/m.

e. Internal pressure under crosswind

UDL on rafter and columns = 0.758 × 5 = 3.79 kN/m.

f. Internal pressure under longitudinal wind

UDL on rafter and columns = 0.0 × 5 = 0.0 kN/m.

g. Internal suction under crosswind

UDL on rafter and columns = 0.325 × 5 = 1.63 kN/m.

h. Internal suction under longitudinal wind

UDL on rafter and columns = 0.704 × 5 = 3.52 kN/m.

REFERENCES

AS/NZS 1170.0 (2002) *Australian/New Zealand Standard for Structural Design Actions, Part 0: General Principles*, Sydney, New South Wales, Australia: Standards Australia and Standards New Zealand.

AS/NZS 1170.1 (2002) *Australian/New Zealand Standard for Structural Design Actions, Part 1: Permanent, Imposed and Other Actions*, Sydney, New South Wales, Australia: Standards Australia and Standards New Zealand.

AS/NZS 1170.2 (2011) *Australian/New Zealand Standard for Structural Design Actions, Part 2: Wind Actions*, Sydney, New South Wales, Australia: Standards Australia and Standards New Zealand.
Bowen, A.J. (1983) The prediction of mean wind speeds above simple 2d hill shapes, *Journal of Wind Engineering and Industrial Aerodynamics*, 15 (1–3): 259–270.
Davenport, A.G. (1967) Gust loading factors, *Journal of the Structural Division*, ASCE, 93: 11–34.
Davenport, A.G. (1977) The prediction of risk under wind loading, Paper presented at the *Second International Conference on Structural Safety and Reliability*, Munich, Germany, pp. 169–174.
Davenport, A.G., Surry, D. and Stathopoulos, T. (1977) Wind loads on low-rise buildings, Final report of phases I and II, boundary layer wind tunnel report, BLWT SS8, University of Western Ontario, London, Ontario, Canada.
Holmes, J.D. (1981) Reduction factors for wind direction for use in codes and standards, Paper presented at the *Colloque, Design with the Wind*, Nantes, France, pp. VI.2.1–VI.2.15.
Holmes, J.D. (1985) Recent developments in the codification of wind loads on low-rise structures, Paper presented at the *Asia-Pacific Symposium on Wind Engineering*, Roorkee, Uttarakhand, India, pp. iii–xvi.
Holmes, J.D. (2001) *Wind Loading of Structures*, London, U.K.: Spon Press.
Holmes, J.D. and Best, R.J. (1979) A wind tunnel study of wind pressures on grouped tropical houses, Wind engineering report 5/79, James Cook University, Townsville, Queensland, Australia.
Holmes, J.D., Melbourne, W.H. and Walker, G.R. (1990) *A Commentary on the Australian Standard for Wind Loads: AS 1170 Part 2, 1989*, Melbourne, Victoria, Australia: Australian Wind Engineering Society.
Hussain, M. and Lee, B.E. (1980) A wind tunnel study of the mean pressure forces acting on large groups of low rise building, *Journal of Wind Engineering and Industrial Aerodynamics*, 6: 207–225.
ISO 4354 (1997) *Wind Actions on Structures*, International Organization for Standardization, Switzerland.
Kim, S.I. and Mehta, K.C. (1977) Wind loads on flat-roof area through full-scale experiment, Institute for Disaster Research Report, Texas Technology University, Lubbock, TX.
Melbourne, W.H. (1975) Cross-wind response of structures to wind action, Paper presented at the *Fourth International Conference on Wind Effects on Buildings and Structures*, Cambridge University Press, London, U.K.
Melbourne, W.H. (1984) Designing for directionality, Paper presented at the *First Workshop on Wind Engineering and Industrial Aerodynamics*, Highett, Victoria, Australia, pp. 1–11.
Paterson, D.A. and Holmes, J.D. (1993) Computation of wind flow over topography, *Journal of Wind Engineering and Industrial Aerodynamics*, 6: 207–225.
Stathopoulos, T. and Mohammadian, A.R. (1985) Code provisions for wind pressures on buildings with monoslope roofs, Paper presented at the *Asia-Pacific Symposium on Wind Engineering*, Roorkee, Uttarakhand, India, pp. 337–347.
Vickery, B.J. (1971) On the reliability of gust loading factors, *Civil Engineering Transactions*, Institute of Engineers Australia, 13: 1–9.

Chapter 3

Local buckling of thin steel plates

3.1 INTRODUCTION

Steel and composite members are usually made of thin-walled steel plate elements by hot rolling, welding or cold forming. Members composed of slender plate elements may fail prematurely owing to local buckling. Local buckling of thin steel plates remarkably reduces the ultimate strength and stiffness of steel and composite members. Therefore, it is important to understand the local buckling behaviour of thin steel plates under various loading and boundary conditions and to consider local buckling effects in the design of steel and composite members.

The elastic local buckling behaviour of a thin steel plate depends on its width-to-thickness ratio (slenderness ratio), material properties, geometric imperfections, loading and boundary conditions. A slender thin steel plate possesses significant post-local buckling reverse of strength. Because of this, slender steel plates will not fail by elastic local buckling. The post-local buckling strength of thin steel plates is influenced by their yield stress and residual stresses induced by the hot-rolling, welding or cold-forming process. In steel–concrete composite members such as concrete-filled steel tubular (CFST) columns and double skin composite panels, steel plates are restrained by concrete so that they can only buckle locally away from the concrete. The local buckling stress of thin steel plates in contact with concrete is much higher than that of the ones unrestrained by concrete.

This chapter describes the behaviour of rectangular thin steel plates that form steel or composite members. The plates considered are subjected to in-plane compression, shear, bending, bearing or combined states of stresses. The design of steel and composite cross sections composed of slender steel plates accounting for local buckling effects is discussed.

3.2 STEEL PLATES UNDER UNIFORM EDGE COMPRESSION

3.2.1 Elastic local buckling

3.2.1.1 Simply supported steel plates

Steel columns composed of slender plate elements under uniform compression, such as hollow steel box columns and I-section columns, may undergo local buckling. Figure 3.1 shows the buckled shape of a pin-ended hollow steel box short column under uniform compression. It can be seen from the figure that the two opposite sides of the box buckle locally outward while the other two sides buckle inward. It can be assumed that the plate elements are hinged along their common boundaries and can rotate freely about the four edges. The flanges and webs of the box column can be idealised as simply supported on their four edges.

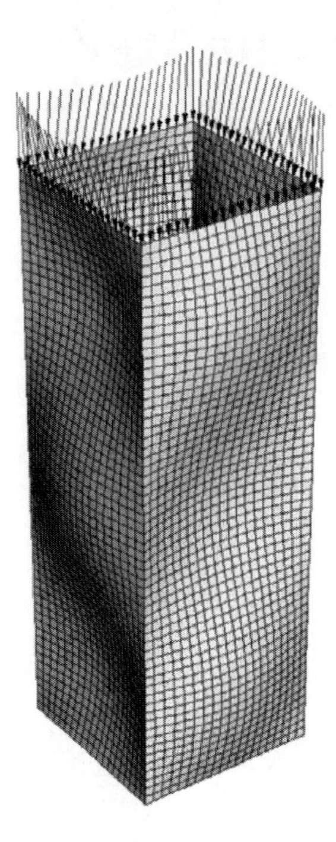

Figure 3.1 Buckled shape of a pin-ended hollow steel box short column under uniform compression.

Similarly, the four edges of the web in a pin-ended steel I-section column can be treated as simply supported.

A simply supported thin flat steel plate under uniform edge compression on two opposite edges is schematically depicted in Figure 3.2. The length of the plate is L, the width of the plate is b and its thickness is t. When the applied compressive load is equal to its elastic buckling load, the steel plate buckles locally by deflecting out of its plane. The elastic buckling load of the thin plate can be determined by the energy method (Bleich 1952; Timoshenko and Gere 1961; Bulson 1970) or the finite element method. Figure 3.3 shows the buckled shape of a simply supported long steel plate under uniform edge compression, which was modelled by finite elements. The local buckling displacements of the plate can be described by the following double series:

$$u = u_m \sin\left(\frac{n\pi x}{L}\right)\sin\left(\frac{m\pi y}{b}\right) \tag{3.1}$$

where

u_m is the undetermined deflection at the centre of the plate
m is the number of half waves across the width b
n is the number of half waves in the direction of the applied compressive load

The elastic buckling load can be calculated by the following equation (Bleich 1952; Timoshenko and Gere 1961; Bulson 1970):

$$P_{cr} = \frac{\pi^2 L^2 b D_r}{n^2}\left(\frac{n^2}{L^2} + \frac{m^2}{b^2}\right)^2 \tag{3.2}$$

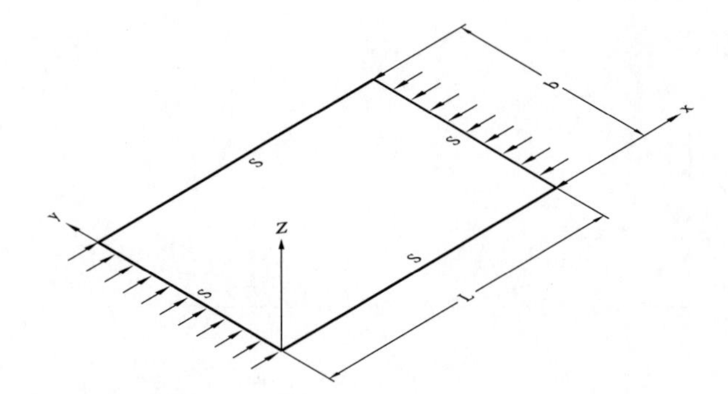

Figure 3.2 A simply supported steel plate under uniform edge compression.

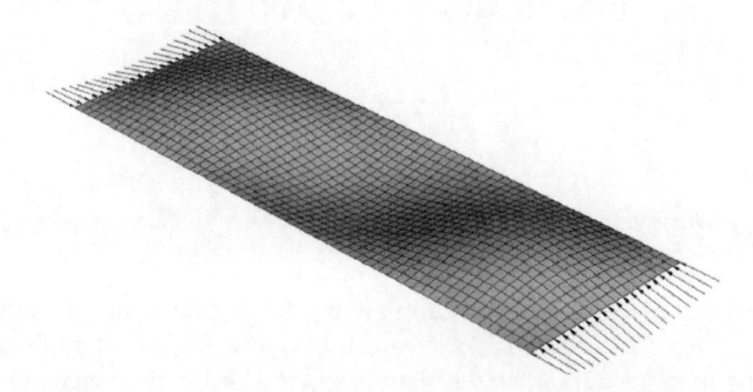

Figure 3.3 Buckled shape of a long simply supported steel plate under uniform edge compression.

where D_r is the plate flexural rigidity, which is written as

$$D_r = \frac{E_s t^3}{12(1-\nu^2)} \tag{3.3}$$

where
- E_s is Young's modulus of the steel material
- t is the thickness of the steel plate
- ν is Poisson's ratio

The lowest value of P_{cr} can be obtained by taking $m = 1$ in Equation 3.2. This implies that the buckled plate has only one half wave across its width b but several half waves in the direction of the applied loading. The elastic buckling stress of the plate is expressed by the following equation (Bleich 1952; Timoshenko and Gere 1961; Bulson 1970):

$$\sigma_{cr} = \frac{k_b \pi^2 E_s}{12(1-\nu^2)(b/t)^2} \tag{3.4}$$

where k_b is the elastic buckling coefficient, which is given by

$$k_b = \left(\frac{nb}{L} + \frac{L}{nb}\right)^2 \tag{3.5}$$

This equation indicates that the elastic buckling coefficient of a simply supported flat plate depends on its aspect ratio L/b and the number of half waves n along the plate in the direction

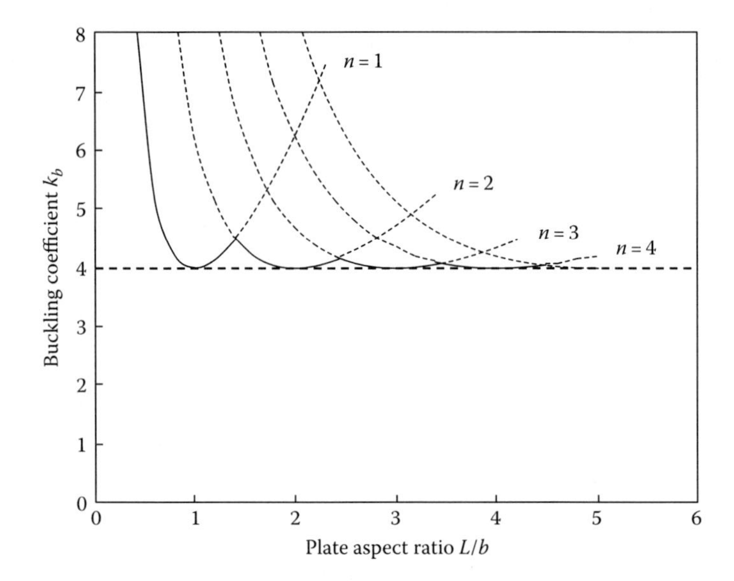

Figure 3.4 Buckling coefficients of simply supported steel plates under uniform edge compression.

of the applied compressive load. The buckling coefficients of simply supported steel plates under uniform edge compression are given in Figure 3.4. It can be seen from Figure 3.4 that the minimum buckling coefficient k_b is 4.0 regardless of the number of half waves. The minimum buckling coefficient occurs when the plate aspect ratio L/b is an even number such as 1, 2, 3, 4 and 5. The larger the number of half waves n, the flatter the buckling coefficient curve.

To prevent the elastic local buckling from occurring before steel yields, the limiting width-to-thickness ratio can be obtained from Equation 3.4 by setting the critical buckling stress to its yield stress. The calculated width-to-thickness ratio for simply supported plates under uniform compression is greater than the slenderness yield limit given in AS 4100 as the yield limit given in the code considers the effect of residual stresses.

It can be observed from Figure 3.3 that a simply supported long steel plate under uniform compression will buckle locally in several half waves in the direction of the loading with a length about the width b of the plate. As a result, the use of transverse stiffeners to reinforce the plate will have little effect on the local buckling stress unless the spacing of the transverse stiffeners is much less than the width of the plate (Trahair and Bradford 1998). An economical design can be achieved by welding one or more longitudinal stiffeners to the plate. The longitudinal stiffeners divide the plate into smaller panels, remarkably increasing the buckling stress of the plate according to Equation 3.4. In addition, the longitudinal stiffeners can withstand a portion of the compressive load.

To prevent the plate from deflecting at the stiffeners, intermediate longitudinal stiffeners must have adequate flexural rigidities. The required minimum second moment of area of an intermediate longitudinal stiffener placed at the centre line of a simply supported steel plate (Trahair and Bradford 1998) is given by

$$I_s = 4.5 b_1 t^3 \left[1 + \frac{2.3 A_s}{b_1 t} \left(1 + \frac{A_s}{2 b_1 t} \right) \right] \tag{3.6}$$

where

b_1 is taken as $b/2$

A_s is the cross-sectional area of the stiffener

Stiffeners are usually attached to one side of the plate rather than to both sides. It should be noted that a stiffener is usually made of steel strip, which may buckle locally when subjected to compression. Therefore, stiffeners must be proportioned to prevent from local buckling.

End stiffeners may be attached to the steel plate to increase the stiffness of the plate and to carry a portion of the compressive load. The required minimum second moment of area of an end longitudinal stiffener can be obtained by modifying Equation 3.6 as follows (Trahair and Bradford 1998):

$$I_s = 2.25bt^3\left[1+\frac{4.6A_s}{bt}\left(1+\frac{A_s}{2bt}\right)\right] \tag{3.7}$$

3.2.1.2 Steel plates free at one unloaded edge

The buckled patterns of a pin-ended steel I-section short column under uniform compression are presented in Figure 3.5. Local buckling is influenced by the relative stiffness of the connected elements in a steel section. The flange outstand of the steel I-section can be assumed to be simply supported by the web, while the opposite edge is free. As a result, the flange outstand is simply supported at two loaded edges and one unloaded edge and free at one unloaded edge as shown in Figure 3.6. The plate is subjected to uniform compressive edge stresses on two opposite edges. The buckled shape of a long steel plate free at one unloaded edge and modelled with finite elements is presented in Figure 3.7. The figure shows that the free unloaded edge causes the plate to buckle in one half wave in the direction of the compressive load. The elastic buckling stress for a steel plate free at one unloaded edge can be expressed by Equation 3.4. However, the elastic buckling coefficient k_b is given by (Bulson 1970)

$$k_b = 0.425+\left(\frac{b}{L}\right)^2 \tag{3.8}$$

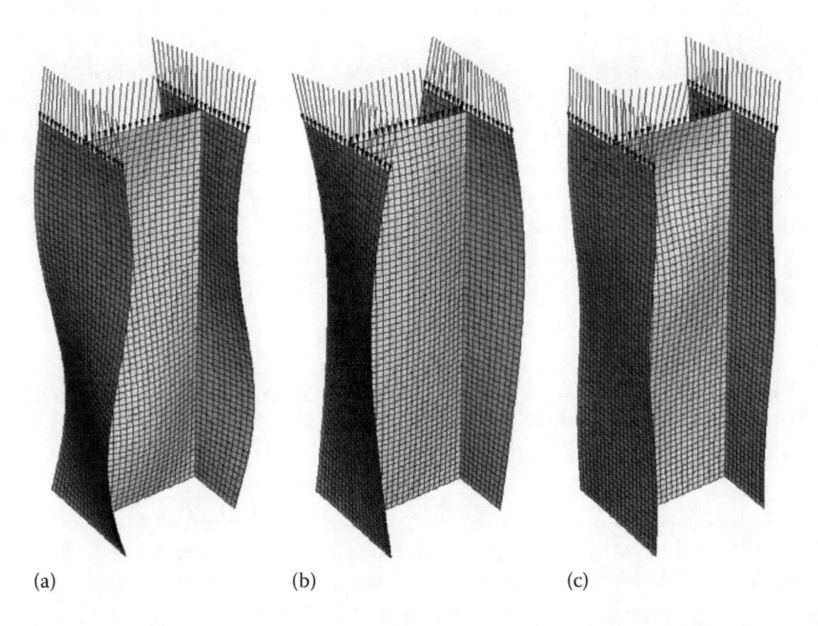

Figure 3.5 Buckled shapes of steel I-section short column under uniform compression: (a) mode 1, (b) mode 2 and (c) mode 3.

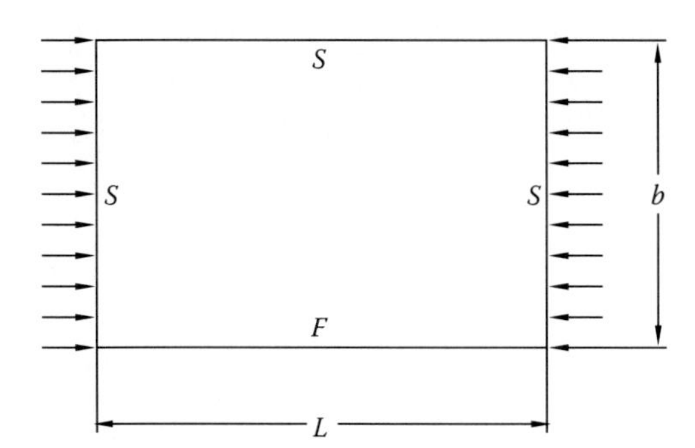

Figure 3.6 A steel plate with a free edge.

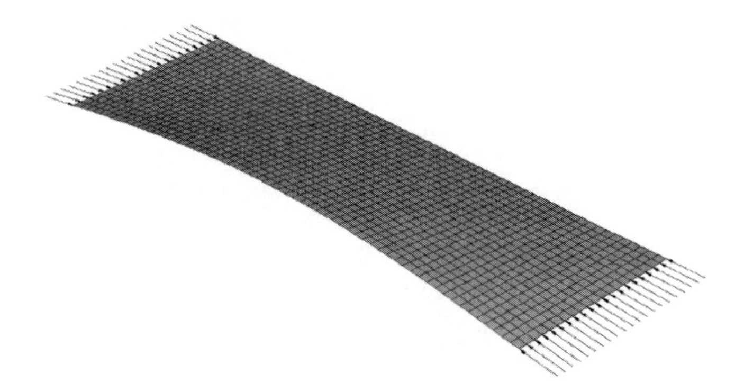

Figure 3.7 Buckled shape of a steel plate with a free unloaded edge.

Equation 3.8 indicates that the buckling coefficient depends on the plate aspect ratio L/b. The buckling coefficients of thin steel plates with one free unloaded edge and under uniform edge compression are demonstrated in Figure 3.8. It appears that when the plate aspect ratio is less than 2.0, the buckling coefficient decreases significantly with an increase in its L/b ratio. However, this decrease tends to be small when the L/b ratio is greater than 2.0. For long steel plates with large L/b ratios such as the flange outstands of I-section columns, the buckling coefficient approaches the minimum value of 0.425 as indicated in Equation 3.8. Therefore, the buckling coefficient k_b = 0.425 can be used in the design of flange outstands of I-sections in long steel columns under axial compression.

3.2.2 Post-local buckling

After initial local buckling, thin steel plates can still carry increased loads without failure. This behaviour of thin steel plates is called post-local buckling. The post-local buckling behaviour of a thin steel plate under edge compression is characterised by its transverse deflections and the in-plane stress redistribution within the buckled plate. The in-plane stress redistribution is associated with the in-plane boundary conditions of the plate (Trahair and Bradford 1998). The boundary lines of the loaded edges of the plate undergo a constant axial shortening, which is caused by both the transverse deflections and the axial strain. The axial shortening induced by the transverse deflections varies across the plate from a maximum at the centre to a minimum at the unloaded edges. This variation is compensated for by the axial shortening caused by the axial strain, varying from a minimum at the centre

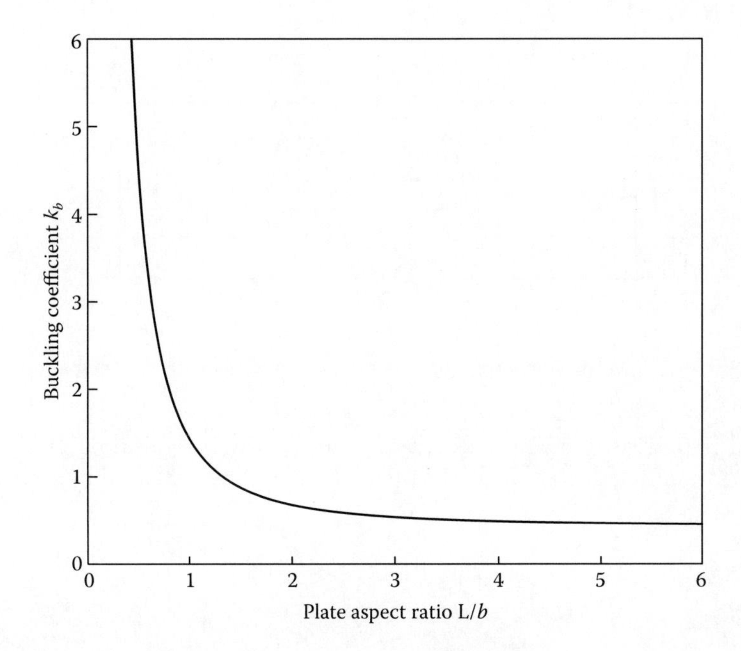

Figure 3.8 Buckling coefficients of steel plates with a free unloaded edge.

to a maximum at the unloaded edges. The in-plane stress distribution within the buckled plate in the loading direction must be the same as that of the axial strain. This implies that the central portion of the buckled plate carry relatively lower stresses, while the loaded edge strips withstand higher stresses. This was confirmed by the results of the finite element analysis carried out by Liang and Uy (1998).

The effective width concept is usually used to describe the post-local buckling behaviour of thin steel plates. Figure 3.9a depicts the in-plane ultimate stress distribution in a simply supported thin steel plate under uniform edge compression. This actual ultimate stress distribution is transformed into an idealised stress distribution within the buckled plate as illustrated in Figure 3.9b. The effective width concept assumes that the central portion of the buckled plate withstands zero stresses, while the effective width b_e carries the yield stress. The effective width of a thin steel plate can be evaluated by

$$\frac{b_e}{b} = \frac{\sigma_u}{f_y} \tag{3.9}$$

where

b_e is the effective width of the plate

σ_u is the average ultimate stress acting on the plate, which can be determined by experiments or nonlinear finite element analyses (Liang and Uy 2000; Liang et al. 2007)

The effective width of a simply supported thin steel plate under uniform edge compression was developed by von Karman et al. (1932) as

$$\frac{b_e}{b} = \sqrt{\frac{\sigma_{cr}}{f_y}} \tag{3.10}$$

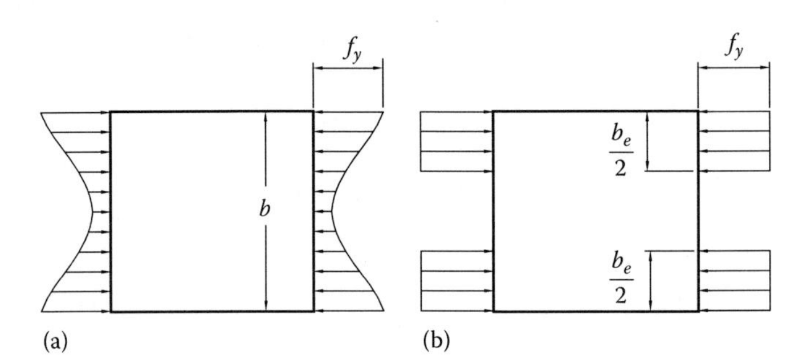

Figure 3.9 Effective width concept for simply support plates: (a) ultimate stress distribution and (b) effective width.

For hot-rolled and welded plates with initial curvatures and residual stresses, AS 4100 (1998) suggests that the effective width of the plates should be reduced by a reduction factor as follows:

$$\frac{b_e}{b} = \alpha\sqrt{\frac{\sigma_{cr}}{f_y}} \tag{3.11}$$

The reduction factor α accounts for the effect of initial curvatures and residual stresses on the ultimate strength of the plate. For hot-rolled plates, α is taken as 0.65 in AS 4100 (1998). Real steel plates have small initial curvatures which reduce the stiffness and strength of plates. It is noted that initial curvatures have little effect on the strength of thick plates but significantly reduce the strength of plates with intermediate slenderness ratios. Residual stresses presented in steel plates are usually caused by uneven cooling after rolling or welding. Tensile stresses are presented at the junctions of plate elements, while compressive stresses act at the remainder of the plate. Tensile stresses on a steel plate are balanced by compressive stresses acting on the same plate. Residual stresses cause premature buckling and yielding of the plate.

The effective widths of hot-rolled steel plates calculated by Equation 3.11 are presented in Figure 3.10, where the modified plate slenderness is defined as $\lambda_m = \sqrt{f_y/\sigma_{cr}}$. It appears from Figure 3.10 that when $\lambda_m \le 0.65$, the plate is fully effective in attaining its yield capacity. When $\lambda_m > 0.65$, the effective width of the plate decreases with increasing its slenderness.

For cold-formed members, the effective width of plate elements with initial curvatures can be expressed by the following equation (Winter 1947):

$$\frac{b_e}{b} = \sqrt{\frac{\sigma_{cr}}{f_y}}\left(1 - 0.22\sqrt{\frac{\sigma_{cr}}{f_y}}\right) \tag{3.12}$$

3.2.3 Design of slender sections accounting for local buckling

As discussed in the preceding sections, local and post-local buckling of steel plates reduces the ultimate strength of the cross sections of steel members under axial compression. The effect of local buckling is considered in the design of axially loaded steel members made of slender plate elements in AS 4100 by using the effective width concept (Bradford 1985, 1987; Bradford et al. 1987). The effective width of a plate element is calculated using its slenderness and yield limit given in Clause 6.2 of AS 4100 (1998). The plate element slenderness is defined as follows.

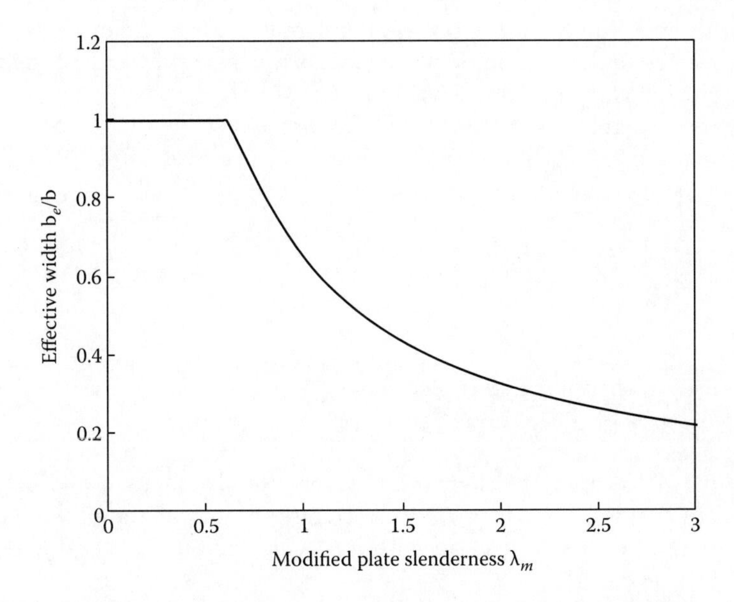

Figure 3.10 Effective widths of simply support plates under uniform edge compression.

The slenderness of a flat plate element is calculated as

$$\lambda_e = \frac{b}{t}\sqrt{\frac{f_y}{250}} \tag{3.13}$$

The slenderness of a circular hollow section is expressed by

$$\lambda_e = \left(\frac{d_o}{t}\right)\left(\frac{f_y}{250}\right) \tag{3.14}$$

where

d_o is the outside diameter of the circular section
t is the wall thickness of the section

Clause 6.2.4 of AS 4100 (1998) gives a simple method for determining the effective width of flat plate elements and circular hollow section. In this method, the effective width of a plate element is calculated by using the plate element slenderness and the element yield slenderness limits (λ_{ey}) (Bradford 1985, 1987; Bradford et al. 1987). The element yield slenderness limits depend on the plate type, support condition, stress distribution and residual stress level and are given in Table 5.2 of AS 4100.

The effective width for a flat plate element can be calculated as

$$b_e = b\left(\frac{\lambda_{ey}}{\lambda_e}\right) \le b \tag{3.15}$$

The effective outside diameter for a circular hollow section is determined by

$$d_e = \min\left[d_o\sqrt{\frac{\lambda_{ey}}{\lambda_e}},\ \ d_o\left(\frac{3\lambda_{ey}}{\lambda_e}\right)^2\right] \le d_o \tag{3.16}$$

The plate element under uniform compression is slender if $\lambda_e > \lambda_{ey}$. For a steel section made up of flat plate elements, the section slenderness λ_s is taken as the value of the plate element slenderness λ_e which has the greatest value of λ_e/λ_{ey}.

The form factor is used to account for local buckling effects on the ultimate axial strength of slender steel sections under axial compression (Rasmussen et al. 1989). Clause 6.2.3 of AS 4100 (1998) defines the form factor as

$$k_f = \frac{A_e}{A_g} \le 1.0 \tag{3.17}$$

where
 A_e is the effective area of the steel section
 A_g is the gross area of the section

The effective area A_e is calculated by summing the effective areas of individual elements. It should be noted that the form factor k_f is a strength reduction factor which must be less than or equal to 1.0. For a steel section without local buckling effects, the section is fully effective and $k_f = 1.0$.

The design section axial capacity of a steel member under axial compression can be determined in accordance with Clause 6.2.1 of AS 4100 (1998) as

$$\phi N_s = \phi k_f A_n f_y \tag{3.18}$$

where
 $\phi = 0.9$ is the capacity reduction factor
 A_n is the net area of the section which is usually taken as the gross area A_g of the section
 f_y is the minimum yield stress for the section

The design requirement for the section of a steel member under axial compression is

$$N^* \le \phi N_s \tag{3.19}$$

where N^* is design axial load acting on the section.

Example 3.1: Section capacity of a steel column under compression

Determine the design section axial capacity of the heavily welded steel I-section of a steel column under axial compression. The cross section of the column is shown in Figure 3.11. The yield stress of the steel section f_y is 320 MPa.

1. Plate element slenderness

The dimensions of the steel I-section are

$$b_f = 420\,\text{mm}, \quad t_f = 12\,\text{mm}, \quad d = 450\,\text{mm}, \quad t_w = 10\,\text{mm}$$

The slenderness of the flange outstands is

$$\lambda_{ef} = \frac{b}{t}\sqrt{\frac{f_y}{250}} = \frac{(b_f - t_w)/2}{t_f}\sqrt{\frac{f_y}{250}} = \frac{(420-10)/2}{12}\sqrt{\frac{320}{250}} = 19.33$$

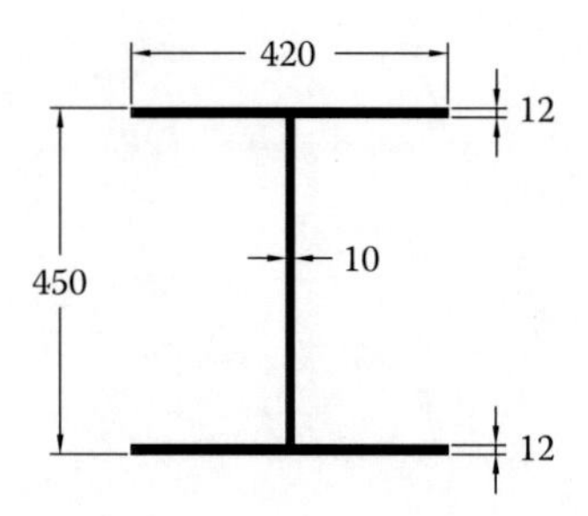

Figure 3.11 Section of compression member.

One of the longitudinal edges of the flange outstand is simply supported by the web and the opposite longitudinal edge is free. The top flange of the section is under uniform compression. From Table 5.2 of AS 4100, the yield slenderness limit can be obtained as $\lambda_{ey} = 14$.

$\lambda_{ef} = 19.33 > \lambda_{ey} = 14$, the flange is slender.

The slenderness of the web is

$$\lambda_{ew} = \frac{b}{t}\sqrt{\frac{f_y}{250}} = \frac{(d-2t_f)}{t_w}\sqrt{\frac{f_y}{250}} = \frac{(450-2\times 12)}{10}\sqrt{\frac{320}{250}} = 48.2$$

Both of the longitudinal edges of the web are simply supported by the flanges and are under uniform compression. From Table 5.2 of AS 4100, the yield slenderness limit can be obtained as $\lambda_{ey} = 35$.

$\lambda_{ew} = 48.2 > \lambda_{ey} = 35$, the web is slender.

2. Effective area of steel section

The effective width of the flange outstands is computed as

$$b_{ef} = b\left(\frac{\lambda_{ey}}{\lambda_e}\right) = \left(\frac{b_f - t_w}{2}\right)\left(\frac{\lambda_{ey}}{\lambda_{ef}}\right) = \left(\frac{420-10}{2}\right)\times\left(\frac{14}{19.33}\right) = 148.5\,\text{mm}$$

The effective width of the web is calculated as

$$b_{ew} = b\left(\frac{\lambda_{ey}}{\lambda_e}\right) = (d-2t_f)\left(\frac{\lambda_{ey}}{\lambda_{ew}}\right) = (450-2\times 12)\times\left(\frac{35}{48.2}\right) = 309.3\,\text{mm}$$

The effective area of the section can be calculated as

$$A_e = 2(2b_{ef} + t_w)t_f + b_{ew}t_w = 2\times(2\times 148.5 + 10)\times 12 + 309.3\times 10 = 10{,}461\,\text{mm}^2$$

The effective area of the steel I-section is illustrated in Figure 3.12.

The gross area of the steel section is

$$A_g = 2\times 420\times 12 + (450 - 2\times 12)\times 10 = 14{,}340\,\text{mm}^2$$

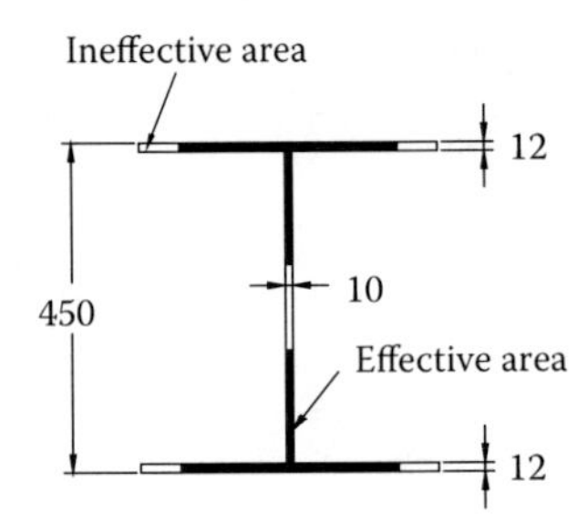

Figure 3.12 Effective area of steel I-section.

3. Design section axial capacity

The form factor can be calculated as

$$k_f = \frac{A_e}{A_g} = \frac{10{,}461}{14{,}340} = 0.73$$

The design section axial capacity is

$$\phi N_s = \phi k_f A_n f_y = 0.9 \times 0.73 \times 14{,}340 \times 320\,\text{N} = 3{,}014.8\,\text{kN}$$

3.3 STEEL PLATES UNDER IN-PLANE BENDING

3.3.1 Elastic local buckling

When a steel beam is under bending, the web of the beam is subjected to in-plane bending stresses and it may buckle locally. The beam web of length L, width d and thickness t is assumed to be simply supported on its four edges as schematically demonstrated in Figure 3.13. The plate is under in-plane linearly distributed bending stresses on two opposite edges. Local buckling occurs when the maximum bending stress acting on the plate reaches the elastic buckling stress of the plate. The typical buckled shape of a thin steel plate with an L/d ratio of 2 and subjected to bending stresses is presented in Figure 3.14. The figure shows that the portion of the plate under compressive stresses buckles out of the plane, while the portion under tensile stresses does not buckle. Solutions to the local buckling problem of thin steel plates in bending can be obtained by the energy method

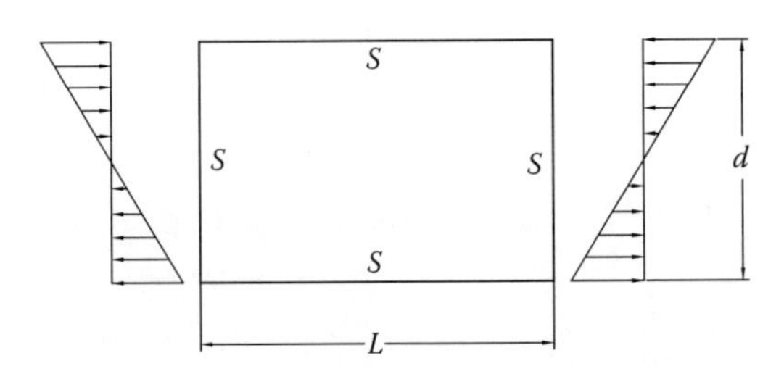

Figure 3.13 A simply supported steel plate in bending.

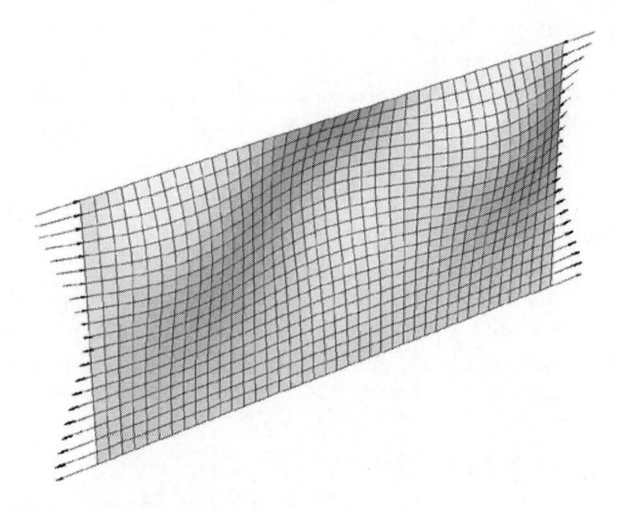

Figure 3.14 Buckled shape of a simply supported steel plate in bending.

(Bleich 1952; Timoshenko and Gere 1961; Bulson 1970) or the finite element method. The elastic local buckling stress can be determined by

$$\sigma_{of} = \frac{k_b \pi^2 E_s}{12(1-\nu^2)(d/t)^2} \tag{3.20}$$

in which the elastic buckling coefficient k_b is a function of the plate aspect ratio L/d and the number of buckles in plate. For long steel plates, the length of each buckle is about $2d/3$ and the minimum buckling coefficient is $k_b = 23.9$.

Like the simply supported steel plates, transverse stiffeners are not effective in preventing the local buckling of the plates subjected to in-plane bending stresses unless their spacing is less than $2d/3$. Longitudinal stiffeners attached to the plate under in-plane bending are effective in increasing the resistance to local buckling as they alter the buckled pattern of the plate. The longitudinal stiffener is most efficient when it is placed in the portion under compression at a distance $0.2d_2$ from the compression edge. The required minimum second moment of area for the longitudinal stiffener is specified in AS 4100.

3.3.2 Ultimate strength

The ultimate strength of a stocky steel plate under in-plane bending is determined by its plastic section modulus and yield stress. For a slender steel plate subjected to in-plane bending stresses, the elastic local buckling stress of the plate will be less than its yield stress. The post-local buckling behaviour of thin steel plates under in-plane bending stresses can also be described by the effective width concept (Bulson 1970; Usami 1982; Shanmugam et al. 1989; Liang et al. 2007). The effective width of the plate is located within the portion under compression, while the portion in tension is fully effective in carrying tensile stresses.

3.3.3 Design of beam sections accounting for local buckling

One of the flanges of a steel beam under bending such as a hollow steel box or a steel I-beam is subjected to compressive stresses, while the beam web is under in-plane bending stresses. In AS 4100, steel plate elements in a cross section are classified as compact, non-compact or slender based on their plate element slenderness ratio. The effective section modulus is used to account for local buckling effects on the section moment capacity of a steel beam under bending.

Compact elements under compression or in-plane bending do not undergo local buckling and can attain their full plastic capacities. A plate element is compact if its slenderness (λ_e) satisfies

$$\lambda_e \leq \lambda_{ep} \tag{3.21}$$

in which λ_{ep} is the plasticity slenderness limit given in Table 5.2 of AS 4100.

Non-compact elements under compression or in-plane bending can attain their first yield capacities but undergo local buckling before their full plastic capacities are reached. A plate element is non-compact if its slenderness (λ_e) satisfies

$$\lambda_{ep} < \lambda_e \leq \lambda_{ey} \tag{3.22}$$

Slender elements under compression or in-plane bending undergo elastic local buckling before yielding. A plate element is classified as slender if it satisfies

$$\lambda_e > \lambda_{ey} \tag{3.23}$$

The cross sections of steel beams are also classified as compact, non-compact or slender based on the classification of their elements in AS 4100. All elements must be compact in a compact steel section. There are no slender elements and at least one non-compact element in a non-compact steel section. There is at least one slender element in a slender steel section. The section slenderness (λ_s) of a steel section composed of flat plate elements is taken as the value of the plate element slenderness (λ_e) for the element of the section having the greatest value of λ_e/λ_{ey}.

In Clause 5.2.3 of AS 4100 (1998), the effective section modulus Z_e for a compact steel beam section is taken as

$$Z_e = Z_c = S \leq 1.5Z \tag{3.24}$$

where

Z_c is the effective section modulus of a compact section
S is the plastic section modulus defined in Section 7.2.2
Z is the elastic section modulus, which is defined in Section 4.3.4

However, for a non-compact steel beam section, Clause 5.2.4 of AS 4100 (1998) provides an equation based on linear interpolation for determining the effective section modulus as

$$Z_e = Z + (Z_c - Z)\left(\frac{\lambda_{sy} - \lambda_s}{\lambda_{sy} - \lambda_{sp}}\right) \tag{3.25}$$

where λ_s, λ_{sy} and λ_{sp} are the values of λ_e, λ_{ey} and λ_{ep} for the element of the section having the greatest value of λ_e/λ_{ey}.

Clause 5.2.5 of AS 4100 gives specifications for determining the effective section modulus for slender sections, which are described herein. For a beam with a slender flange under uniform compression, the effective section modulus can be calculated using the effective width or by the following equation:

$$Z_e = Z\left(\frac{\lambda_{sy}}{\lambda_s}\right) \tag{3.26}$$

For a beam consisting of a slender web, the effective section modulus can be determined by

$$Z_e = Z\left(\frac{\lambda_{sy}}{\lambda_s}\right)^2 \tag{3.27}$$

The effective section modulus for a slender circular hollow steel section is given by

$$Z_e = \min\left[Z\sqrt{\frac{\lambda_{sy}}{\lambda_s}},\quad Z\left(\frac{2\lambda_{sy}}{\lambda_s}\right)^2\right] \tag{3.28}$$

The nominal section moment capacity of a steel beam is calculated by

$$M_s = Z_e f_y \tag{3.29}$$

More details on the moment capacity of steel beams are provided in Section 4.4.

The design requirement for the section of a steel beam under bending is

$$M^* \le \phi M_s \tag{3.30}$$

in which $\phi = 0.9$ is the capacity reduction factor.

Example 3.2: Section moment capacity of a steel I-beam under bending

Determine the design section moment capacity of a hot-rolled 310UB32.0 steel I-beam bending about its principal x-axis as shown in Figure 3.15. The section properties are $f_y = 320\,\text{MPa}$, $Z_x = 424\times10^3\,\text{mm}^3$ and $S_x = 475\times10^3\,\text{mm}^3$.

1. Plate element slenderness

The dimensions of the steel I-section are

$$b_f = 149\,\text{mm}, \quad t_f = 8\,\text{mm}, \quad d = 298\,\text{mm}, \quad t_w = 5.5\,\text{mm}$$

The slenderness of the flange outstands is calculated as

$$\lambda_{ef} = \frac{b}{t}\sqrt{\frac{f_y}{250}} = \frac{(b_f - t_w)/2}{t_f}\sqrt{\frac{f_y}{250}} = \frac{(149-5.5)/2}{8}\sqrt{\frac{320}{250}} = 10.1$$

One of the longitudinal edges of the flange outstand is simply supported by the web and the opposite longitudinal edge is free. The top flange of the section is assumed to be in uniform compression. From Table 5.2 of AS 4100, the plasticity and yield slenderness limits can be obtained as $\lambda_{ep} = 9$ and $\lambda_{ey} = 16$.

$\lambda_{ep} = 9 < \lambda_{ef} = 10.1 < \lambda_{ey} = 16$, the flange is non-compact.

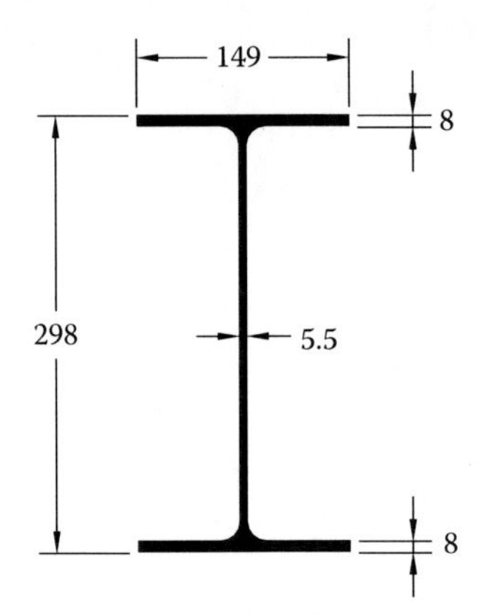

Figure 3.15 Hot-rolled steel I-section.

The slenderness of the web is computed as

$$\lambda_{ew} = \frac{b}{t}\sqrt{\frac{f_y}{250}} = \frac{(d-2t_{ef})}{t_w}\sqrt{\frac{f_y}{250}} = \frac{(298-2\times 8)}{5.5}\sqrt{\frac{320}{250}} = 58$$

Both of the longitudinal edges of the web are simply supported by the flanges and are under linear bending stresses. From Table 5.2 of AS 4100, the plasticity slenderness limits can be obtained as $\lambda_{ep} = 82$.

$\lambda_{ew} = 58 < \lambda_{ep} = 82$, the web is compact.

2. Effective section modulus

The section contains a non-compact flange so that the whole section is non-compact. For the non-compact section, the effective section modulus can be calculated by

$$Z_e = Z + (Z_c - Z)\left(\frac{\lambda_{sy}-\lambda_s}{\lambda_{sy}-\lambda_{sp}}\right) = 424\times 10^3 + (475-424)\times 10^3 \times \left(\frac{16-10.1}{16-9}\right) = 467\times 10^3\,\text{mm}^3$$

3. Design section moment capacity

The design section moment capacity is computed as

$$\phi M_s = \phi Z_e f_y = 0.9\times 467\times 10^3 \times 320\,\text{N mm} = 134.5\,\text{kN m}$$

3.4 STEEL PLATES IN SHEAR

3.4.1 Elastic local buckling

The web of a steel beam near the supports or zero bending moment may be subjected to pure shear stresses along its edges. Figure 3.16 depicts a simply supported steel plate with length L, depth d and thickness t and under shear stresses uniformly distributed along its four edges. Local buckling occurs when the shear stresses are equal to the elastic buckling stress of the plate. This local buckling problem of thin steel plates in shear can be solved by numerical methods such as the finite element method. The buckled shape of a simply supported steel plate under shear stresses on four edges is shown in Figure 3.17, where the plate aspect

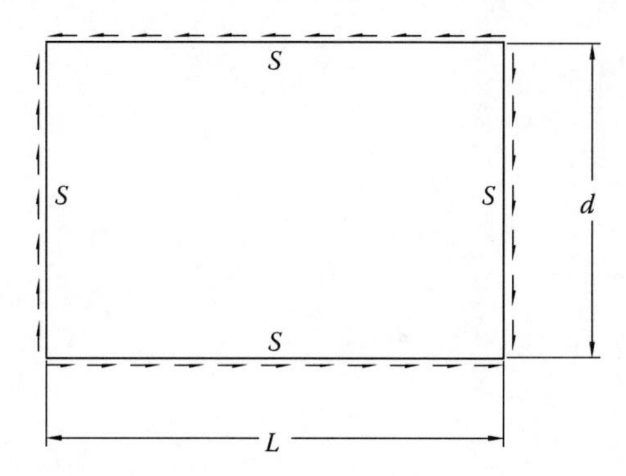

Figure 3.16 A simply supported steel plate in shear.

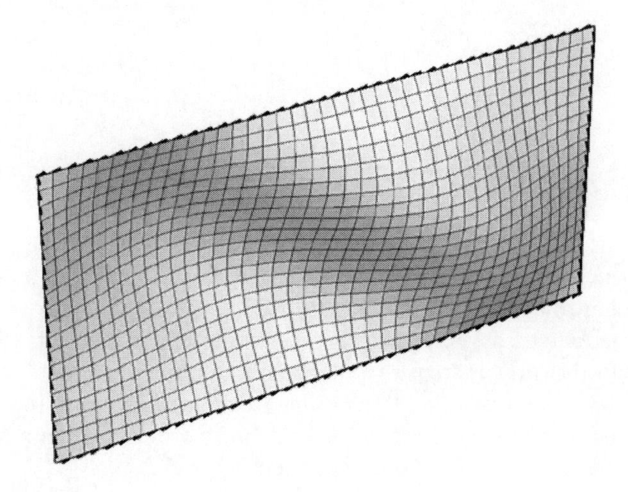

Figure 3.17 Buckled shape of a simply supported steel plate in shear.

ratio *L*/*d* is 2. Finite element analysis results show that increasing the plate aspect ratio *L*/*d* increases the number of buckles. The elastic local buckling stress can be expressed by

$$\sigma_{ov} = \frac{k_b \pi^2 E_s}{12(1-\nu^2)(d/t)^2} \tag{3.31}$$

where the buckling coefficient k_b is a function of the plate aspect ratio *L*/*d* (Timoshenko and Gere 1961) and can be determined by

$$k_b = \begin{cases} 5.35\left(\dfrac{d}{L}\right)^2 + 4 & \text{for } L \le d \\ 5.35 + 4\left(\dfrac{d}{L}\right)^2 & \text{for } L \ge d \end{cases} \tag{3.32}$$

Buckling coefficients calculated by Equation 3.32 are presented in Figure 3.18. The figure demonstrates that when $L \le d$, the buckling coefficient decreases significantly with increasing the *L*/*d* ratio. However, when $L \ge d$, increasing plate *L*/*d* ratio leads to only a

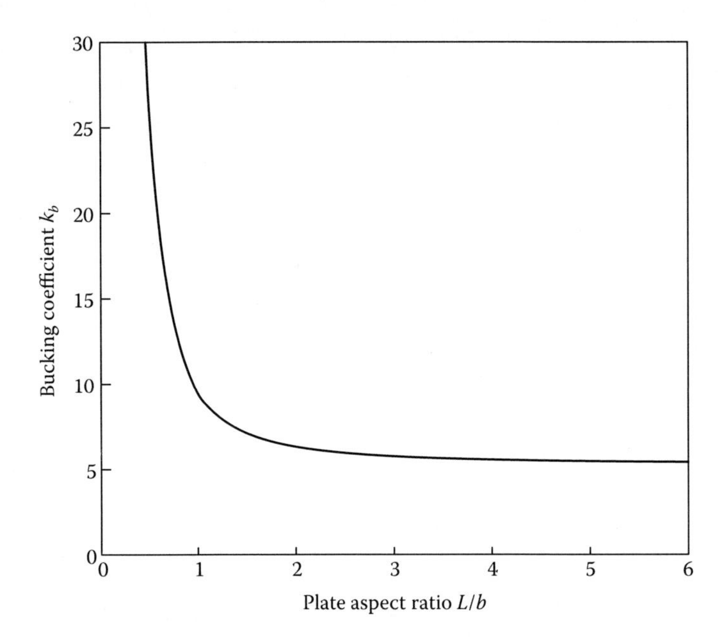

Figure 3.18 Buckling coefficients of simply supported steel plates in shear.

small decrease in the buckling coefficient. For a very long steel plate in shear, its buckling coefficient approaches to the minimum value of 5.35.

Attaching intermediate transverse stiffeners to the plate in pure shear to reduce the aspect ratio of L/d can significantly increase the buckling coefficient and buckling stress of the plate. The elastic buckling stress of a plate in shear can also be greatly increased by using the longitudinal stiffeners to reduce the d/t ratio. To achieve efficient designs, the aspect ratio of each panel divided by stiffeners should be between 0.5 and 2.

3.4.2 Ultimate strength

A stocky web in an I-section beam subjected to pure shear behaves elastically until first yield occurs at $\tau_y = f_y/\sqrt{3}$ and undergoes increasing plasticisation until it fully yields. The shear stress distribution in the web at first yield is nearly uniform and the shear shape factor is close to 1.0. Because stocky webs in steel beams in shear yield before buckling, they are usually unstiffened and their ultimate strengths are determined by the shear yield stress as follows:

$$V_w = d_w t_w \tau_y \tag{3.33}$$

where

d_w is the clear depth of the web
t_w is the thickness of the web

Slender webs with transverse stiffeners will buckle elastically before yielding occurs. The reserve of the post-local buckling strength of the slender webs is relatively high compared to stocky webs. The ultimate shear stress of a slender web can be estimated by its elastic local buckling stress with length equal to the stiffener spacing and the tension field contribution at yield (Basler 1961; Evans 1983).

3.5 STEEL PLATES IN BENDING AND SHEAR

3.5.1 Elastic local buckling

A simply supported thin flat steel plate of length L, depth d and thickness t under bending and shear is depicted in Figure 3.19. The elastic buckling stress of the thin plate can be determined from the following interaction equation (Bleich 1952; Timoshenko and Gere 1961; Bulson 1970):

$$\left(\frac{\sigma_f}{\sigma_{of}}\right)^2 + \left(\frac{\tau_v}{\tau_{ov}}\right)^2 = 1 \tag{3.34}$$

where

τ_{ov} is the elastic buckling stress of the plate in pure shear
σ_{of} is the elastic buckling stress of the plate in pure bending
τ_v and σ_f are the elastic buckling stresses of the plate under combined bending and shear

It can be found from the Hencky–von Mises yield criterion that the most severe loading condition for which elastic local buckling and yielding occur simultaneously is the pure shear.

3.5.2 Ultimate strength

In steel beams, stocky unstiffened webs yield before elastic local buckling occurs. The design capacities of stocky unstiffened webs can be estimated by the Hencky–von Mises yield criterion as

$$\left(\frac{V^*}{\phi V_u}\right)^2 + \left(\frac{M^*}{\phi M_u}\right)^2 = 1 \tag{3.35}$$

where

V^* and M^* are the design shear force and moment in the web
V_u is the nominal shear yield capacity of the web, which is calculated as

$$V_u = 0.6 f_y d_w t_w \tag{3.36}$$

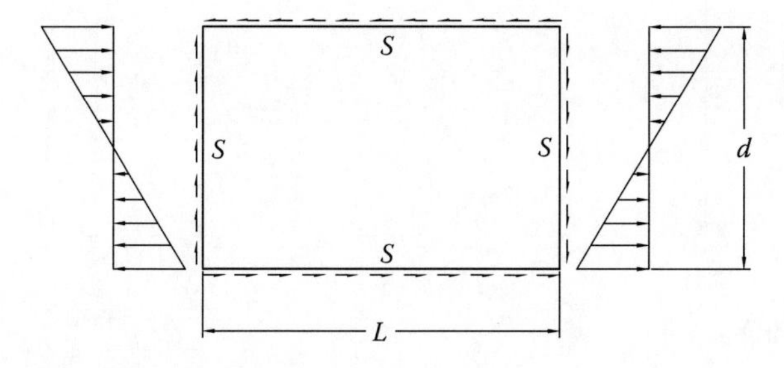

Figure 3.19 A simply supported steel plate in bending and shear.

In Equation 3.35, M_u is the nominal first yield moment capacity of the web, which is determined by

$$M_u = \frac{d_w^2 t_w f_y}{6} \tag{3.37}$$

For slender unstiffened webs under combined bending and shear, the reserve of post-local buckling is small so that their ultimate strength can be estimated approximately by their elastic buckling stresses satisfying Equation 3.34. The ultimate strength of a stiffened web in combined bending and shear is given in Clause 5.12.3 of AS 4100 and is discussed in Section 4.5.3.

3.6 STEEL PLATES IN BEARING

3.6.1 Elastic local buckling

Steel plate girders are often subjected to concentrated or locally distributed loads on their top flanges. The local load causes local bearing stresses in the web immediately beneath the load as depicted in Figure 3.20. These bearing stresses are resisted by vertical shear stresses at the transverse web stiffeners of a slender stiffened plate girder. Plate girders under transverse loads may be subjected to combined bending and shear or combined bending, shear and bearing at an interior support. For a panel of a stiffed web, the edges of the panel can be assumed to be simply supported. The elastic buckling stress of a panel under pure bearing can be calculated as

$$\sigma_{ob} = \frac{k_b \pi^2 E_s}{12(1-\nu^2)(d/t)^2} \tag{3.38}$$

in which the buckling coefficient k_b is a function of the panel aspect ratio s/d (Bulson 1970; Trahair and Bradford 1998).

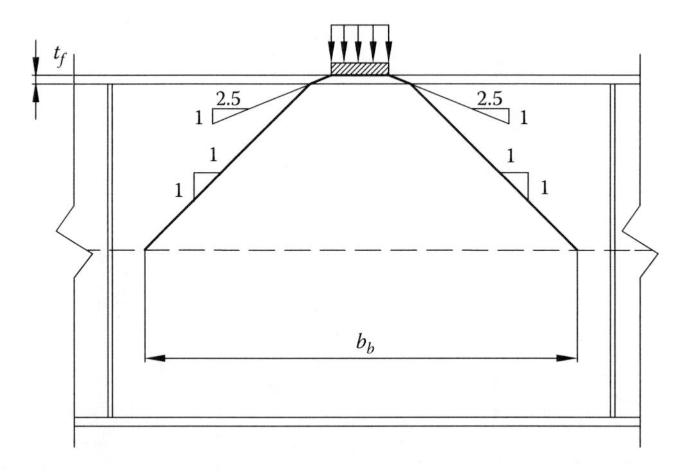

Figure 3.20 Beam web in bearing.

When a web panel is subjected to combined bending, shear and bearing, the elastic buckling stresses of the panel can be determined by the interaction equation (Rockey et al. 1972; Allen and Bulson 1980)

$$\left(\frac{\sigma_f}{\sigma_{of}}\right)^2 + \left(\frac{\sigma_v}{\tau_{ov}}\right)^2 + \frac{\sigma_b}{\sigma_{ob}} = 1 \tag{3.39}$$

where

σ_{of}, τ_{ov} and σ_{ob} are elastic buckling stresses of a plate under pure bending, shear or bearing only

σ_f, τ_v and σ_b are elastic buckling stresses of the plate under combined bearing, shear and bending

3.6.2 Ultimate strength

The point load or locally distributed load applied on the top of the flange is assumed to be dispersed uniformly through the flange at a slope of 1:2.5 and through the web at a slope of 1:1 as depicted in Figure 3.20. The general yielding of a thick web in bearing occurs when the web area defined by the dispersion of the applied load yields. The ultimate strength of a thick web in bearing depends on its yield stress. When the web is subjected to combined bearing, shear and bending, its ultimate strength can be determined from the Hencky–von Mises yield criterion. Thin stiffened web panels under bearing stresses have a considerable reserve of post-local buckling strength. This is attributed to its ability to redistribute in-plane stresses from the buckled region to the stiffeners.

3.7 STEEL PLATES IN CONCRETE-FILLED STEEL TUBULAR COLUMNS

3.7.1 Elastic local buckling

In a thin-walled CFST column as depicted in Figure 3.21, the steel tube walls are restrained to buckling locally outward by the concrete core. Figure 3.22 shows the buckled shape of the tested square CFST columns under axial loading or eccentric loading. The restraint of the concrete core considerably increases the local buckling stress of the steel tube and the ultimate strength of the CFST column (Ge and Usami 1992; Wright 1993; Uy and Bradford 1995; Bridge and O'Shear 1998; Liang and Uy 2000; Uy 2000; Liang et al. 2007). Steel plates in CFST beam–columns may be subjected to stress gradients caused by uniaxial bending or biaxial bending. This unilateral local buckling problem of steel plates can be solved by using the finite element method (Liang and Uy 2000; Liang et al. 2007). The four edges of the steel plate restrained by concrete are assumed to be clamped as illustrated in Figure 3.23. The buckled shape of steel plates restrained by concrete and under uniform edge compression predicted by the finite element method is given in Figure 3.24. The elastic local buckling stress of the clamped flat steel plate under compressive stress gradients as

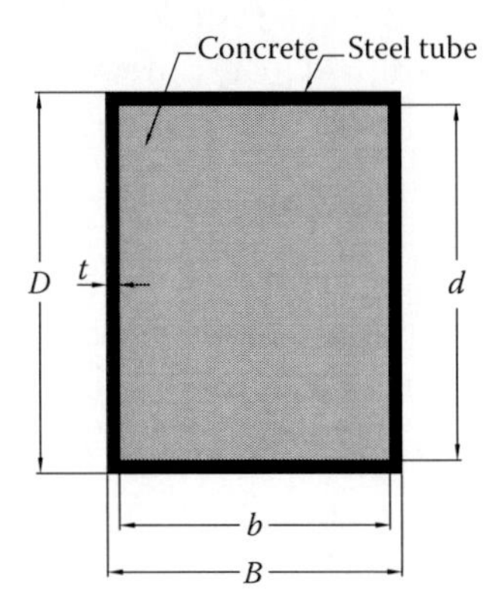

Figure 3.21 Cross section of rectangular CFST column.

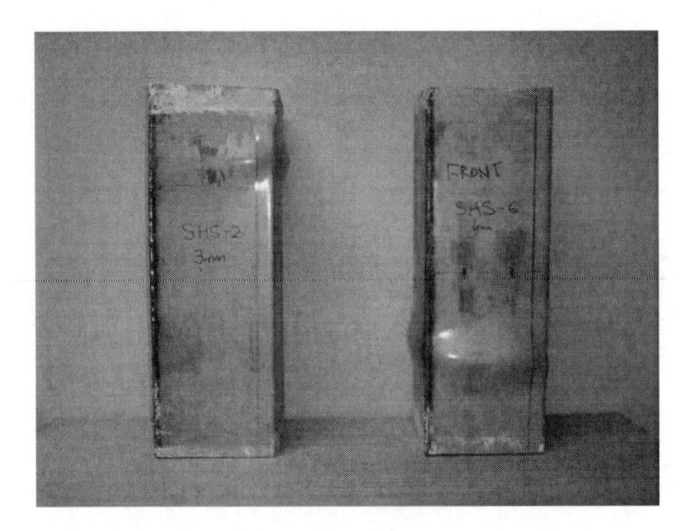

Figure 3.22 Local buckling of rectangular CFST short columns.

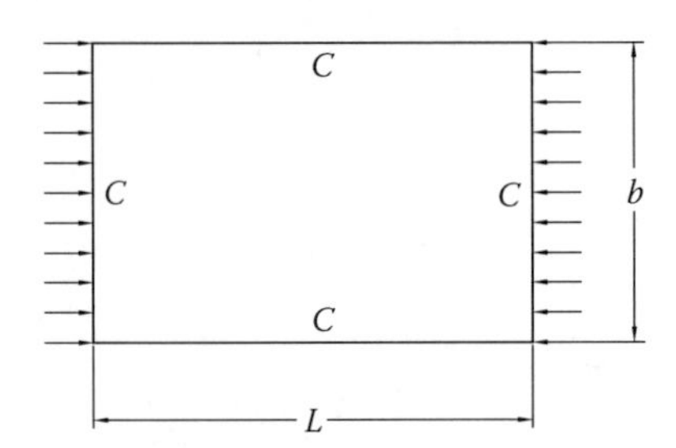

Figure 3.23 A clamped steel plate under uniform edge compression.

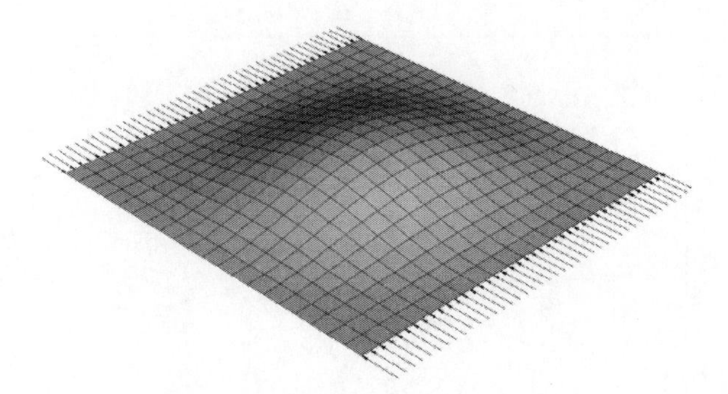

Figure 3.24 Buckled shape of a clamped square steel plate under uniform edge compression.

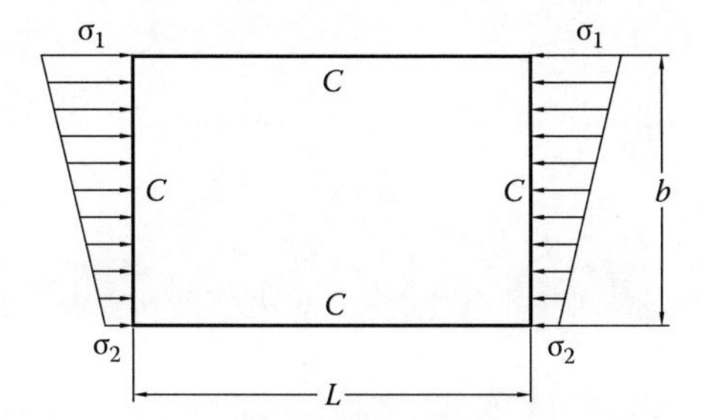

Figure 3.25 A clamped steel plate under compressive stress gradients.

depicted in Figure 3.25 can be determined by Equation 3.4 using the buckling coefficient given by Liang et al. (2007) as follows:

$$k_b = 18.89 - 14.38\alpha_s + 5.3\alpha_s^2 \tag{3.40}$$

where α_s is the stress gradient coefficient, which is defined as the ratio of the minimum edge stress (σ_2) to the maximum edge stress (σ_1) acting on the plate. Figure 3.26 shows the buckling coefficient as a function of the stress gradient coefficient. It appears that increasing the stress gradient coefficient decreases the buckling coefficient k_b. When the $\alpha_s = 1.0$, the plate is subjected to uniform compression and $k_b = 9.81$ (Liang and Uy 2000).

Real steel plates have initial imperfections including initial out-of-plane deflections and residual stresses, which are induced in the process of construction and hot rolling, cold forming or welding. These imperfections will reduce the stiffness and strength of steel plates. The maximum magnitude of initial geometric imperfections at a plate centre can be taken as $0.1t$. Figure 3.27 depicts the residual stress pattern in welded CFST columns. Tensile residual stresses that reach the steel yield stress are induced at the welded corners of the tubular cross section, while compressive residual stresses are present in the remainder of the tube walls. The tensile residual stresses are balanced by the compressive residual stresses in a tube wall. The compressive residual stress is usually about 25%–30% of the yield stress of the steel tube (Liang and Uy 2000).

The initial local buckling stress of a steel plate with prescribed geometric imperfections and residual stresses is a function of its plate width-to-thickness ratio, stress gradient

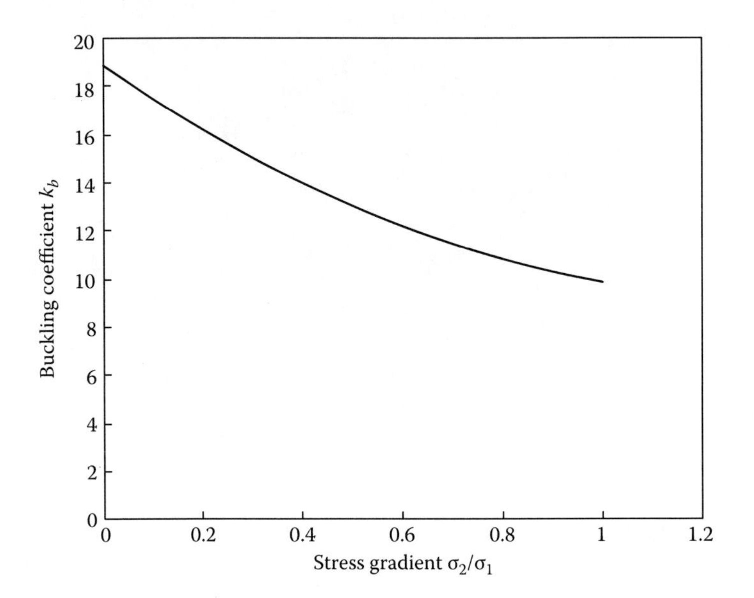

Figure 3.26 Buckling coefficients of clamped steel plates under compressive stress gradients.

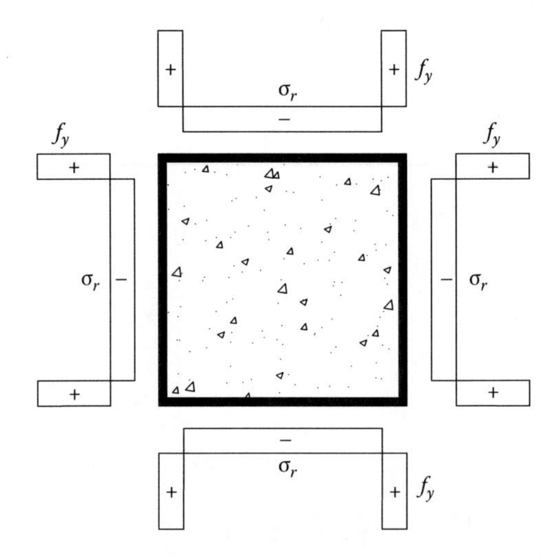

Figure 3.27 Residual stress pattern in welded CFST columns.

coefficient and yield stress. For thin steel plates with *b*/*t* ratios ranging from 30 to 100 and under linearly varying edge compression, their initial buckling stress can be determined by (Liang et al. 2007)

$$\frac{\sigma_{1c}}{f_y} = a_1 + a_2\left(\frac{b}{t}\right) + a_3\left(\frac{b}{t}\right)^2 + a_4\left(\frac{b}{t}\right)^3 \tag{3.41}$$

where

σ_{1c} is the initial local buckling stress of a plate with imperfections

a_1, a_2, a_3 and a_4 are constant coefficients which depend on the stress gradient coefficient α_s and are given in Table 3.1.

Table 3.1 Constant coefficients for determining the initial local buckling stresses of plate under stress gradients

α_s	a_1	a_2	a_3	a_4
0.0	0.6925	0.02394	-4.408×10^{-4}	1.718×10^{-6}
0.2	0.8293	0.01118	-2.427×10^{-4}	8.164×10^{-7}
0.4	0.6921	0.01223	-2.488×10^{-4}	8.676×10^{-7}
0.6	0.4028	0.02152	-3.742×10^{-4}	1.446×10^{-6}
0.8	0.5096	0.0112	-2.11×10^{-4}	7.092×10^{-7}
1.0	0.5507	0.005132	-9.869×10^{-5}	1.198×10^{-7}

Source: Adapted from Liang, Q.Q. et al., *J. Constr. Steel Res.*, 63(3), 396, 2007.

3.7.2 Post-local buckling

The post-local buckling strength of a steel plate with prescribed geometric imperfections and residual stresses depends on its b/t ratio, stress gradient coefficient (α_s) and yield stress (f_y) and can be calculated by (Liang et al. 2007)

$$\frac{\sigma_{1u}}{f_y} = c_1 + c_2\left(\frac{b}{t}\right) + c_3\left(\frac{b}{t}\right)^2 + c_4\left(\frac{b}{t}\right)^3 \tag{3.42}$$

where

σ_{1u} is the ultimate value of the maximum edge stress σ_1
c_1, c_2, c_3 and c_4 are constant coefficients which are given in Table 3.2

The ultimate strength of steel plates with stress gradient coefficients greater than zero can be approximately estimated by (Liang 2009)

$$\frac{\sigma_{1u}}{f_y} = (1+0.5\phi_s)\frac{\sigma_u}{f_y} \quad (0 \le \phi_s < 1.0) \tag{3.43}$$

where $\phi_s = 1-\alpha_s$ and σ_u is the ultimate stress of steel plates under uniform compression and can be calculated using Equation 3.42 with the stress gradient coefficient of $\alpha_s = 1.0$.

Table 3.2 Constant coefficients for determining the ultimate strengths of plate under stress gradients

α_s	c_1	c_2	c_3	c_4
0.0	1.257	−0.006184	1.608×10^{-4}	-1.407×10^{-6}
0.2	0.6855	0.02894	-4.89×10^{-4}	2.134×10^{-6}
0.4	0.6538	0.02888	-5.215×10^{-4}	2.424×10^{-6}
0.6	0.7468	0.01925	-3.689×10^{-4}	1.677×10^{-6}
0.8	0.6474	0.02088	-4.171×10^{-4}	2.058×10^{-6}
1.0	0.5554	0.02038	-3.944×10^{-4}	1.921×10^{-6}
−0.2	1.48	−0.01584	2.868×10^{-4}	-1.742×10^{-6}

Source: Adapted from Liang, Q.Q. et al., *J. Constr. Steel Res.*, 63(3), 396, 2007.

For the steel plates in CFST columns under uniform compression, their effective width can be expressed by the following equations given by Liang and Uy (2000):

$$\frac{b_e}{b} = 0.675\left(\frac{\sigma_{cr}}{f_y}\right)^{1/3} \quad \text{for } \sigma_{cr} \leq f_y \tag{3.44}$$

$$\frac{b_e}{b} = 0.915\left(\frac{\sigma_{cr}}{\sigma_{cr} + f_y}\right)^{1/3} \quad \text{for } \sigma_{cr} > f_y \tag{3.45}$$

where

b_e is the total effective width of the steel plate

σ_{cr} is the elastic critical buckling stress of the perfect steel plate under uniform edge compression

The effective widths of steel plates under stress gradients in CFST columns under biaxial bending are depicted in Figure 3.28. Effective width formulas of clamped steel plates under compressive stress gradients in CFST beam–columns with *b*/*t* ratios ranging from 30 to 100 are given by (Liang et al. 2007)

$$\frac{b_{e1}}{b} = 0.2777 + 0.01019\left(\frac{b}{t}\right) - 1.972\times10^{-4}\left(\frac{b}{t}\right)^2 + 9.605\times10^{-7}\left(\frac{b}{t}\right)^3 \quad \text{for } \alpha_s > 0.0 \tag{3.46}$$

$$\frac{b_{e1}}{b} = 0.4186 - 0.002047\left(\frac{b}{t}\right) + 5.355\times10^{-5}\left(\frac{b}{t}\right)^2 - 4.685\times10^{-7}\left(\frac{b}{t}\right)^3 \quad \text{for } \alpha_s = 0.0 \tag{3.47}$$

$$\frac{b_{e2}}{b} = (1 + \phi_s)\frac{b_{e1}}{b} \tag{3.48}$$

where b_{e1} and b_{e2} are the effective widths as shown in Figure 3.28. For the effective width $(b_{e1} + b_{e2}) > b$, the steel plate is fully effective in carrying loads and the ultimate strength of the steel plate can be determined using Equations 3.42 and 3.43.

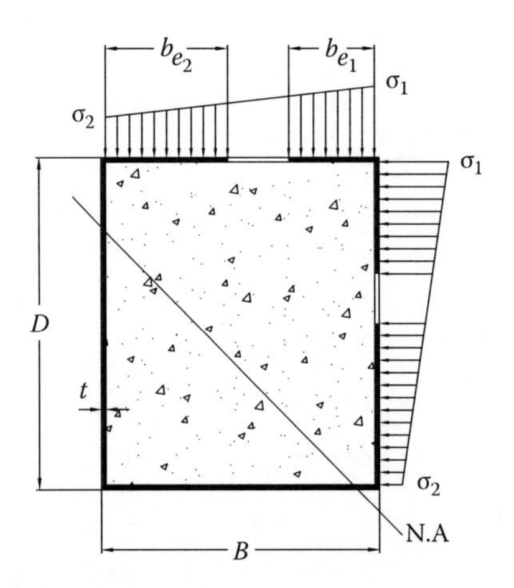

Figure 3.28 Effective widths of steel tube walls under stress gradients.

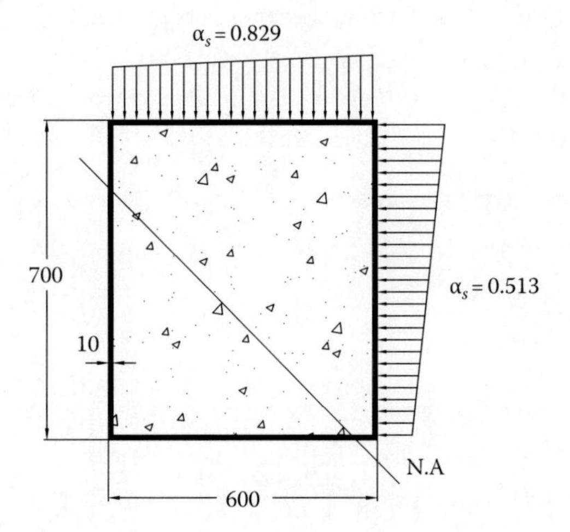

Figure 3.29 Cross section of CFST column under biaxial bending.

Example 3.3: Effective area of steel section of a CFST column

The cross section (600 × 700 mm) of a CFST column under biaxial bending is shown in Figure 3.29. One of the steel flanges is subjected to compressive stress gradient with a stress gradient coefficient of α_s = 0.829, while one of the webs is under compressive stresses with a stress gradient coefficient of α_s = 0.513. Calculate the effective cross-sectional area of the steel tube.

1. Effective width of the flange under compressive stress gradient

The clear width of the flange: $b = 600 - 2 \times 10 = 580$ mm.

The effective width b_{e1} of the flange under compressive stress gradient is calculated as

$$\frac{b_{e1}}{b} = 0.2777 + 0.01019\left(\frac{b}{t}\right) - 1.972 \times 10^{-4}\left(\frac{b}{t}\right)^2 + 9.605 \times 10^{-7}\left(\frac{b}{t}\right)^3$$

$$= 0.2777 + 0.01019\left(\frac{580}{10}\right) - 1.972 \times 10^{-4}\left(\frac{580}{10}\right)^2 + 9.605 \times 10^{-7}\left(\frac{580}{10}\right)^3 = 0.393$$

$$b_{e1} = 0.393 \times 580 = 227.9\,\text{mm}$$

The effective width of b_{e2} is computed as follows:

$$\phi_s = 1 - \alpha_s = 1 - 0.829 = 0.171$$

$$\frac{b_{e2}}{b} = (1 + \phi_s)\frac{b_{e1}}{b}$$

$$b_{e2} = (1 + \phi_s)b_{e1} = (1 + 0.171) \times 227.9 = 266.9\,\text{mm}$$

The total effective width of the flange is therefore

$$b_e = b_{e1} + b_{e2} = 227.9 + 266.9 = 494.8\,\text{mm} < b = 580\,\text{mm}$$

2. Effective width of the web under compressive stress gradient

The clear width of the web: $b = 700 - 2 \times 10 = 680$ mm.

The effective width b_{e1} of the web under compressive stress gradient is calculated as

$$\frac{b_{e1}}{b} = 0.2777 + 0.01019\left(\frac{b}{t}\right) - 1.972\times10^{-4}\left(\frac{b}{t}\right)^2 + 9.605\times10^{-7}\left(\frac{b}{t}\right)^3$$

$$= 0.2777 + 0.01019\left(\frac{680}{10}\right) - 1.972\times10^{-4}\left(\frac{680}{10}\right)^2 + 9.605\times10^{-7}\left(\frac{680}{10}\right)^3 = 0.361$$

$$b_{e1} = 0.361\times680 = 245.5\,\text{mm}$$

The effective width of b_{e2} is computed as follows:

$$\phi_s = 1 - \alpha_s = 1 - 0.513 = 0.487$$

$$b_{e2} = (1+\phi_s)b_{e1} = (1+0.487)\times245.5 = 365\,\text{mm}$$

The total effective width of the web is

$$b_e = b_{e1} + b_{e2} = 245.5 + 365 = 610.5\,\text{mm} < b = 680\,\text{mm}$$

3. Effective cross-sectional area of the steel tube

Assume that only the flange and web under compressive stress gradients will undergo local buckling. The ineffective cross-sectional area of the flange under compressive stress gradient is determined as

$$A_{nef} = (b - b_e)t = (580 - 494.8)\times10 = 852\,\text{mm}^2$$

The ineffective cross-sectional area of the web under compressive stress gradient is

$$A_{new} = (b - b_e)t = (680 - 610.5)\times10 = 695\,\text{mm}^2$$

The gross cross-sectional area of the steel tube is calculated as

$$A_g = 600\times700 - (600 - 2\times10)(700 - 2\times10) = 25{,}600\,\text{mm}^2$$

The effective cross-sectional area of the steel tube is

$$A_e = 25{,}600 - 852 - 695 = 24{,}053\,\text{mm}^2$$

The effective steel areas of the CFST column under biaxial bending are shown in Figure 3.30.

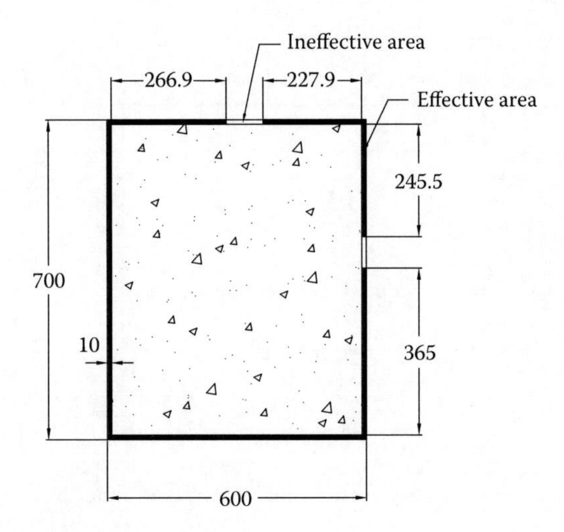

Figure 3.30 Effective steel area of CFST column under biaxial bending.

3.8 DOUBLE SKIN COMPOSITE PANELS

3.8.1 Local buckling of plates under biaxial compression

Double skin composite (DSC) panels are formed by filling concrete between two steel plates welded with stud or other type of shear connectors at a regular spacing as schematically depicted in Figure 3.31. The steel skins are used as permanent formwork and biaxial steel reinforcement for the concrete core, providing sound waterproofing in marine and freshwater environment. Stud shear connectors carry the longitudinal shear between the concrete core and the steel skins as well as separation at the interface. This composite system offers high strength, stiffness and ductility and is increasingly used in submerged tube tunnels, military shelters, nuclear installations, shear walls in buildings, liquid and gas containment structures and offshore structures. DSC panels exhibit two particular failure modes which include the local buckling of steel plate fields between stud shear connectors and the shear connection failure between the steel skins and the concrete core (Oduyemi and Wright 1989; Wright et al. 1991).

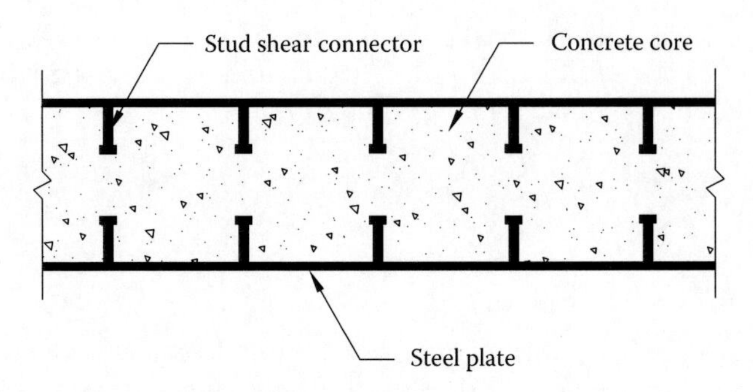

Figure 3.31 Cross section of double skin composite panel.

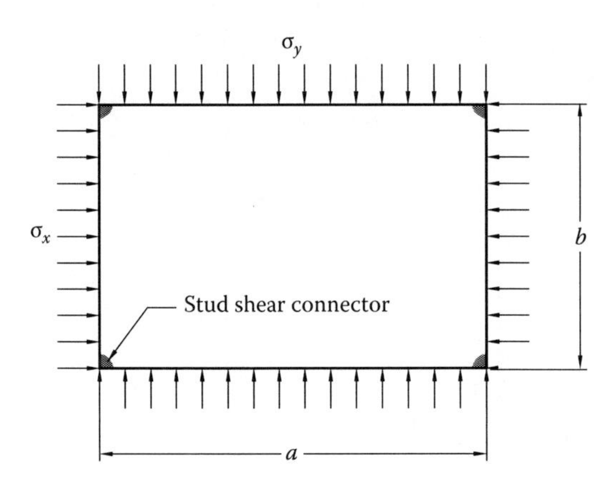

Figure 3.32 Single plate element restrained by stud shear connector under biaxial compression.

Figure 3.32 depicts a single plate field between stud shear connectors, which is restrained at the corners by stud shear connectors. It is assumed that the edges of the plate field between shear connectors are hinged and the rotations at the corners are restrained while the in-plane translations of stud shear connectors are defined by the shear–slip model (Liang et al. 2003). When the plate field is located at the edge of the panel, the edge of the plate field can be assumed to be clamped as the rotations are restrained. The elastic local buckling stress of a steel plate under biaxial compression depends on its aspect ratio (spacing of shear connectors in two directions), the plate thickness, compressive stresses in two directions ($\alpha_{cs} = \sigma_x/\sigma_y$) and boundary conditions including the restraint of shear connectors. The elastic bucking stress in x direction can be determined by

$$\sigma_{xcr} = \frac{k_{xo}\pi^2 E_s}{12(1-\nu^2)(b/t)^2} \tag{3.49}$$

where k_{xo} is the elastic buckling coefficient in the x direction. The elastic bucking stress σ_{ycr} in y direction can be obtained by substituting k_{yo} and a in Equation 3.49. Elastic buckling coefficients of plates with various boundary conditions and loading ratios of biaxial compressions were given by Liang et al. (2003).

Elastic buckling coefficients can be used to determine the limiting width-to-thickness ratios for steel plate fields under biaxial compression in DSC panels. The limiting width-to-thickness ratio of steel plate fields with $E_s = 200\,\text{GPa}$ and $\upsilon = 0.3$ can be obtained from the von Mises yield criterion as follows (Liang et al. 2003):

$$\frac{b}{t}\sqrt{\frac{f_y}{250}} = 26.89\left(k_{xo}^2 - \frac{k_{xo}k_{yo}}{\varphi^2} + \frac{k_{yo}^2}{\varphi^4}\right)^{1/4} \tag{3.50}$$

where $\varphi = a/b$ is the plate aspect ratio. For a square steel plate field under the same compressive stresses in two directions ($\alpha_{cs} = 1.0$), the local buckling coefficient is $k_{xo} = k_{yo} = 2.404$ (Liang et al. 2003). The limiting width-to-thickness ratio is 41.7. If the 16 mm thick steel plate of Grade 300 with a yield stress of 300 MPa is used, the maximum spacing of stud shear connectors in two directions is 609 mm.

3.8.2 Post-local buckling of plates under biaxial compression

Steel plate fields in DSC panels are restrained by stud shear connectors with a finite shear stiffness which considerably increases the resistance of plate fields against local buckling. Slender steel plate fields may buckle locally in a unilateral direction before shear connectors fail. In addition, shear connectors may fracture before stocky steel plate fields attain their full plastic capacities. Moreover, interaction modes between local buckling and shear connection failure may exist. The effect of stud shear connectors on the plate buckling can be taken into account in the nonlinear analysis by using the shear–slip model (Liang et al. 2003).

The post-local buckling behaviour of steel plate fields in a DSC panel can be described by biaxial strength interaction formulas derived from the von Mises yield ellipse as follows (Liang et al. 2003):

$$\left(\frac{\sigma_{xuo}}{f_y}\right)^{\zeta_c} + \eta_s\left(\frac{\sigma_{xuo}\sigma_{yuo}}{f_y^2}\right) + \left(\frac{\sigma_{yuo}}{f_y}\right)^2 = \gamma_n \quad (\gamma_n \le 1) \tag{3.51}$$

where

σ_{xuo} denotes the ultimate strength of a plate in x direction under biaxial compression
σ_{yuo} is the ultimate strength of a plate in y direction under biaxial compression
ζ_c is the shape factor of the interaction curve depending on the plate aspect ratio and slenderness
η_s is a function of the plate slenderness
γ_n is the uniaxial strength factor

The shape factor η_s can be used to define any shape of interaction curves from a straight line ($\eta_s = 2$) to the von Mises ellipse ($\eta_s = -1$). For square plates, the shape factor $\zeta_c = 2$ and the values of η_s and γ_n are given in Table 3.3.

3.8.3 Local buckling of plates under biaxial compression and shear

When DSC panels are used as slabs or shear walls, steel plate fields between stud shear connectors may be subjected to biaxial compression and in-plane shear. Figure 3.33 schematically depicts a plate field under combined biaxial compression and shear. This local buckling

Table 3.3 Parameters of strength interaction formulas for square plates in biaxial compression

b/t	ζ_c	η_s	γ_n
100	2.0	1.4	0.14
80	2.0	1.47	0.211
60	2.0	1.45	0.353
40	2.0	0.8	0.65
20	2.0	0.0	0.846

Source: Adapted from Liang, Q.Q. et al., *Proc. Inst. Civil Eng., Struct. Build.*, U.K., 156(2), 111, 2003.

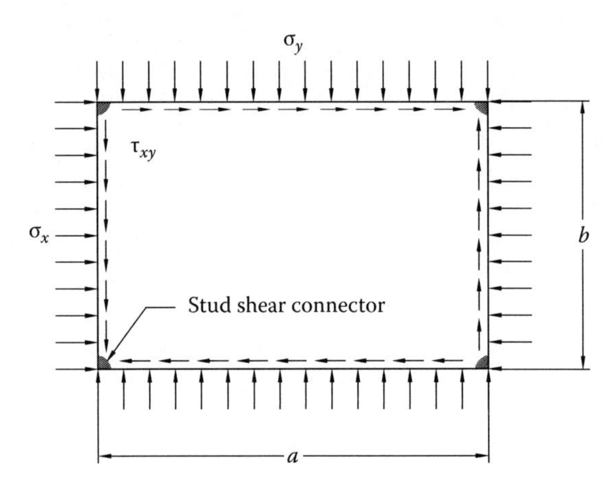

Figure 3.33 Single plate element restrained by stud shear connector under biaxial compression and shear.

problem of plate fields can be solved by using the finite element method (Liang et al. 2004). The elastic buckling coefficients can be calculated by the following equations:

$$\sigma_{xcr} = \frac{k_x \pi^2 E_s}{12(1-\nu^2)(b/t)^2} \tag{3.52}$$

$$\sigma_{ycr} = \frac{k_y \pi^2 E_s}{12(1-\nu^2)(a/t)^2} \tag{3.53}$$

$$\tau_{xycr} = \frac{k_{xy} \pi^2 E_s}{12(1-\nu^2)(b/t)^2} \tag{3.54}$$

where

σ_{xcr} stands for the elastic buckling stress in the x direction
σ_{ycr} represents the elastic buckling stress in the y direction
τ_{xycr} denotes the elastic shear buckling stress
k_x stands for the elastic buckling coefficient in the x direction
k_y denotes the elastic buckling coefficient in the y direction
k_{xy} is the elastic shear buckling coefficient

The buckling coefficient of plates under combined states of stresses accounts for the effects of plate aspect ratio, boundary condition including restraints by shear connectors and interaction between biaxial compression and shear on the critical buckling stress.

The interaction formula for determining the elastic buckling coefficients for square plates under biaxial compression and shear is expressed by (Liang et al. 2004)

$$\left(\frac{k_x}{k_{xo}}\right)^{\zeta_b} + \left(\frac{k_{xy}}{k_{xyo}}\right)^2 = 1 \tag{3.55}$$

where

k_{xo} denotes the buckling coefficient in the x direction in the absence of shear stresses
k_{xyo} stands for the shear buckling coefficient in the absence of biaxial compression
ζ_b is the buckling shape factor defining the shape of a buckling interaction curve

Table 3.4 Parameters of buckling interaction formulas for plates in biaxial compression and shear

	k_{xo}						
Boundary condition	$\alpha_{cs} = 1.5$	$\alpha_{cs} = 1.0$	$\alpha_{cs} = 0.5$	$\alpha_{cs} = 0.25$	$\alpha_{cs} = 0$	k_{xyo}	ζ_b
C-C-S-S+SC	3.362	4.216	5.514	6.56	7.797	18.596	2
C-S-S-S+SC	2.589	3.168	4.06	4.705	5.552	14.249	1.7
S-S-S-S+SC	1.923	2.404	3.204	3.84	4.782	10.838	1.1

Source: Adapted from Liang, Q.Q. et al., *J. Struct. Eng.*, ASCE, 130(3), 443, 2004.

The values of buckling coefficients k_{xo} and k_{xyo} for steel plates with different boundary conditions are given in Table 3.4 for design.

Buckling coefficients presented can be used to determine the limiting width-to-thickness ratios for steel plates under biaxial compression and shear in DSC panels. This ensures that the elastic local buckling of steel plates between stud shear connectors will not occur before steel yielding. The relationship between critical buckling stress components at yield can be expressed by the von Mises yield criterion as

$$\sigma_{xcr}^2 - \sigma_{xcr}\sigma_{ycr} + \sigma_{ycr}^2 + 3\tau_{xycr}^2 = f_y^2 \tag{3.56}$$

If the material properties $E = 200$ GPa and $\nu = 0.3$ and the plate aspect ratio $\varphi = a/b$ are used, the limiting width-to-thickness ratio can be derived by substituting Equations 3.52 through 3.54 into Equation 3.56 as (Liang et al. 2004)

$$\frac{b}{t}\sqrt{\frac{f_y}{250}} = 26.89\left(k_x^2 - \frac{k_x k_y}{\varphi^2} + \frac{k_y^2}{\varphi^4} + 3k_{xy}^2\right)^{1/4} \tag{3.57}$$

Stresses acting at the edges of a plate field in a DSC panel can be determined by undertaking a global stress analysis on the DSC panel. It is assumed that a square plate field ($\varphi = 1$) with the S-S-S-S+SC boundary condition is under biaxial compressive stresses ($\alpha_{cs} = 1$) and shear stress $\tau_{xy} = 0.5\sigma_x$. This gives $k_x = k_y$ and $k_{xy} = 0.5k_x$ according to Equations 3.52 through 3.54. From Table 3.4, parameters for buckling interactions can be obtained as $k_{xo} = 2.404$, $k_{xyo} = 10.838$ and $\zeta_b = 1.1$. By substituting these parameters into Equation 3.55, buckling coefficients are obtained as $k_x = 2.38$ and $k_{xy} = 1.19$. By using Equation 3.57, the limiting

Table 3.5 Parameters of strength interaction formulas for plates in biaxial compression and shear

b/t	ζ_s	σ_{xuo}/f_y	τ_{xyuo}/τ_0
100	0.8	0.205	0.875
80	1.1	0.248	0.984
60	1.3	0.321	1.0
40	1.6	0.481	1.0
20	2.0	0.658	0.927

Source: Adapted from Liang, Q.Q. et al., *J. Struct. Eng.*, ASCE, 130(3), 443, 2004.

width-to-thickness ratio for this plate field with a yield stress of 300 MPa is 48. If the compression steel skin with a thickness of 16 mm is used, the maximum stud spacing in two directions in this DSC panel is 700 mm.

3.8.4 Post-local buckling of plates under biaxial compression and shear

The shape of strength interaction curves strongly depends on the plate slenderness. The post-local buckling strength of plate fields in DSC panels can be described by the following strength interaction formulas (Liang et al. 2004):

$$\left(\frac{\sigma_{xu}}{\sigma_{xuo}}\right)^{\zeta_s} + \left(\frac{\tau_{xyu}}{\tau_{xyuo}}\right)^2 = 1 \tag{3.58}$$

where

σ_{xu} denotes the ultimate strength of a plate in x direction under biaxial compression and shear

σ_{xuo} is the ultimate strength of a plate in x direction under biaxial compression only

τ_{xyu} represents the ultimate shear strength of a plate

τ_{xyuo} denotes the ultimate strength of a plate under pure shear only

ζ_s is the strength shape factor of the ultimate strength interaction curve

Table 3.5 gives the ultimate strength of square steel plates under either biaxial compression or shear alone and the strength shape factors for plates with various slenderness ratios.

REFERENCES

Allen, H.G. and Bulson, P.S. (1980) *Background to Buckling*, London, U.K.: McGraw-Hill.

AS 4100 (1998) *Australian Standard for Steel Structures*, Sydney, New South Wales, Australia: Standards Australia.

Basler, K. (1961) Strength of plate girders in shear, *Journal of the Structural Division*, ASCE, 87 (ST7): 151–180.

Bleich, F. (1952) *Buckling Strength of Metal Structures*, New York: McGraw-Hill.

Bradford, M.A. (1985) Local and post-local buckling of fabricated box members, *Civil Engineering Transactions*, Institution of Engineers, Australia, CE27 (4): 391–396.

Bradford, M.A. (1987) Inelastic local buckling of fabricated I-beams, *Journal of Constructional Steel Research*, 7 (5): 317–334.

Bradford, M.A., Bridge, R.Q., Hancock, G.J., Rotter, J.M., and Trahair, N.S. (1987) Australian limit state design rules for the stability of steel structures, Paper presented at *the International Conference on Steel and Aluminium Structures*, Cardiff, UK, pp.11–23.

Bridge, R.Q. and O'Shear, M.D. (1998) Behaviour of thin-walled steel box sections with or without internal restraint, *Journal of Constructional Steel Research*, 47: 73–91.

Bulson, P.S. (1970) *The Stability of Flat Plates*, London, U.K.: Chatto and Windus.

Evans, H.R. (1983) Longitudinally and transversely reinforced plate girders, Chapter 1 in *Plated Structures: Stability and Strength*, R. Narayanan (ed.), Applied Science Publishers, London, U.K., pp. 1–37.

Ge, H.B. and Usami, T. (1992) Strength of concrete-filled thin-walled steel box columns: Experiment, *Journal of Structural Engineering*, ASCE, 118 (11): 3036–3054.

Liang, Q. Q. (2009) Performance-based analysis of concrete-filled steel tubular beam-columns, Part I: Theory and algorithms, *Journal of Constructional Steel Research*, 65(2): 363–372.

Liang, Q.Q. and Uy, B. (1998) Parametric study on the structural behaviour of steel plates in concrete-filled fabricated thin-walled box columns, *Advances in Structural Engineering*, 2 (1): 57–71.

Liang, Q.Q. and Uy, B. (2000) Theoretical study on the post-local buckling of steel plates in concrete-filled box columns, *Computers and Structures*, 75 (5): 479–490.

Liang, Q.Q., Uy, B. and Liew, J.Y.R. (2007) Local buckling of steel plates in concrete-filled thin-walled steel tubular beam-columns, *Journal of Constructional Steel Research*, 63 (3): 396–405.

Liang, Q.Q., Uy, B., Wright, H.D. and Bradford, M.A. (2003) Local and post-local buckling of double skin composite panels, *Proceedings of the Institution of Civil Engineers, Structures and Buildings*, U.K., 156 (2): 111–119.

Liang, Q.Q., Uy, B., Wright, H.D. and Bradford, M.A. (2004) Local buckling of steel plates in double skin composite panels under biaxial compression and shear, *Journal of Structural Engineering*, ASCE, 130 (3): 443–451.

Oduyemi, T.O.S. and Wright, H.D. (1989) An experimental investigation into the behaviour of double skin sandwich beams, *Journal of Constructional Steel Research*, 14: 197–220.

Rasmussen, K.J.R., Hancock, G.J. and Davids, A.J. (1989) Limit state design of columns fabricated from slender plates, *Civil Engineering Transactions*, Institution of Engineers, Australia, 27 (3): 268–274.

Rockey, K.C., El-Gaaly, M.A. and Bagchi, D.K. (1972) Failure of thin-walled members under patch loading, *Journal of the Structural Division*, ASCE, 98 (ST12): 2739–2752.

Shanmugam, N.E., Liew, J.Y.R. and Lee, S.L. (1989) Thin-walled steel box columns under biaxial loading, *Journal of Structural Engineering*, ASCE, 115 (11): 2706–2726.

Timoshenko, S.P. and Gere, J.M. (1961) *Theory of Elastic Stability*, 2nd edn., New York: McGraw-Hill.

Trahair, N.S. and Bradford, M.A. (1998) *The Behaviour and Design of Steel Structures to AS 4100*, 3rd edn. (Australian), London, U.K.: Taylor & Francis Group.

Usami, T. (1982) Effective width of locally buckled plates in compression and bending, *Journal of Structural Engineering*, ASCE, 119 (5): 1358–1373.

Uy, B. (2000) Strength of concrete-filled steel box columns incorporating local buckling, *Journal of Structural Engineering*, ASCE, 126 (3): 341–352.

Uy, B. and Bradford, M.A. (1995) Local buckling of thin steel plates in composite construction: Experimental and theoretical study, *Proceedings of the Institution of Civil Engineers, Structures and Buildings*, U.K., 110: 426–440.

von Karman, T., Sechler, E.E. and Donnel, L.H. (1932) Strength of thin plates in compression, *Transactions of ASME*, 54: 53–57.

Winter, G. (1947) *Strength of Thin Steel Compression Flanges*, Cornell University Eng. Exp. Stn., Reprint No. 32.

Wright, H.D. (1993) Buckling of plates in contact with a rigid medium, *The Structural Engineer*, 71 (12): 209–215.

Wright, H.D., Oduyemi, T.O.S. and Evans, H.R. (1991) The experimental behaviour of double skin composite elements, *Journal of Constructional Steel Research*, 19 (2): 97–110.

Chapter 4

Steel members under bending

4.1 INTRODUCTION

Steel members under bending are flexural members (beams) which are used to transfer transverse loads to the supports. The transverse loads acting on a beam may induce the actions of bending, shear and bearing in the beam. Therefore, steel beams need to be designed for bending, shear and bearing. Steel beams are often made of thin-walled elements by hot rolling, welding and cold forming. Typical sections for steel beams are given in Figure 4.1. The behaviour of a steel beam depends on its section slenderness, material properties and member slenderness. Lateral and torsional restraints along the steel beam significantly increase its member moment capacity. As a result, the use of lateral and torsional restraints leads to significant economies. Steel plate girders are often made of slender webs which may undergo shear and bearing buckling. Transverse web and load-bearing stiffeners are attached to the webs of steel plate girders to increase their buckling capacities. The design of a steel beam for strength includes the verification of its section and member moment capacities, web shear and bearing capacities and the design of web stiffeners and restraints.

This chapter presents the behaviour and design of steel members under bending to AS 4100 (1998). The fundamental behaviour of steel beams under bending is discussed first. The basic principles for determining the elastic section properties of thin-walled members are described. Methods for calculating the section moment and member moment capacities of steel beams are presented. The design of steel beam webs with or without stiffeners for shear and bearing is also given.

4.2 BEHAVIOUR OF STEEL MEMBERS UNDER BENDING

The behaviour of a steel member under bending is influenced by its material properties, section slenderness, member slenderness and lateral and torsional restraints. For flexural members composed of slender steel elements, local buckling of the compression flange or bending web may occur before steel yields. As discussed in Chapter 3, local plate buckling remarkably reduces the ultimate section moment capacity of steel members in bending. Under a high shear force, the web of a steel beam may fail by shear buckling or yielding. This results in a further reduction in the moment capacity of the steel beam. Under concentrated loads or reactions at the supports, the web of a steel flexural member is subject to bearing stresses, which may cause the web bearing buckling or yielding. The aforementioned local failures prevent steel members subjected to transverse loads from attaining their full plastic moment capacities. Steel beams made of compact steel sections restrained laterally and torsionally would not fail until well after yielding. These beams of compact sections can attain their full plastic moment capacities beyond the yield moments.

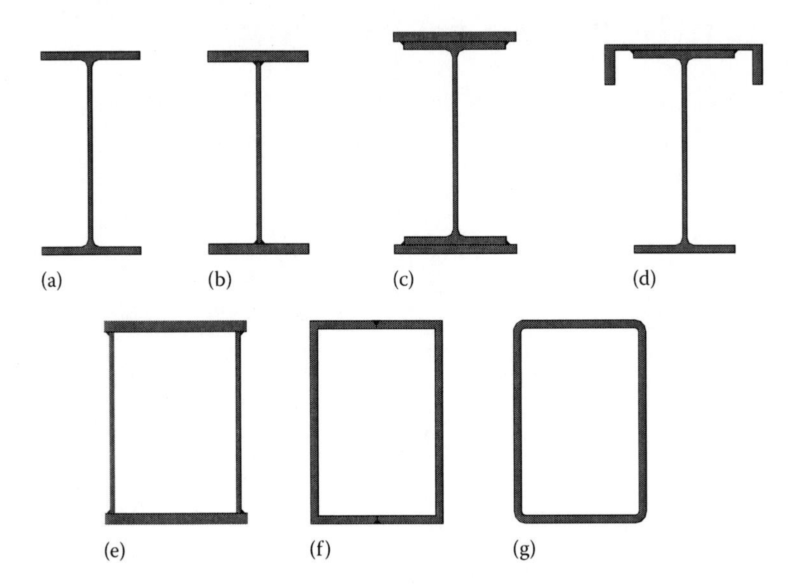

Figure 4.1 Typical steel sections for beams: (a) hot-rolled section, (b) welded section, (c) built-up section, (d) hot-rolled section with flange plates, (e) welded box section, (f) welded box section from channels and (g) cold-formed hollow section.

If a steel beam under in-plane loading does not have sufficient lateral stiffness or lateral and torsional supports, it may buckle out of its plane of the loading by deflecting laterally and twisting as illustrated in Figure 4.2. This behaviour is called flexural–torsional buckling, which significantly reduces the in-plane load-carrying capacity of the beam (Trahair 1993a). When the applied moment reaches the elastic buckling moment of the beam, the elastic flexural–torsional buckling occurs. Long and unrestrained steel I-beams have such low resistances to bending and torsion that their capacities are governed by the elastic flexural–torsional buckling. A perfectly straight beam with an intermediate slenderness may yield before the elastic flexural–torsional buckling occurs. Stocky steel beams are not affected by

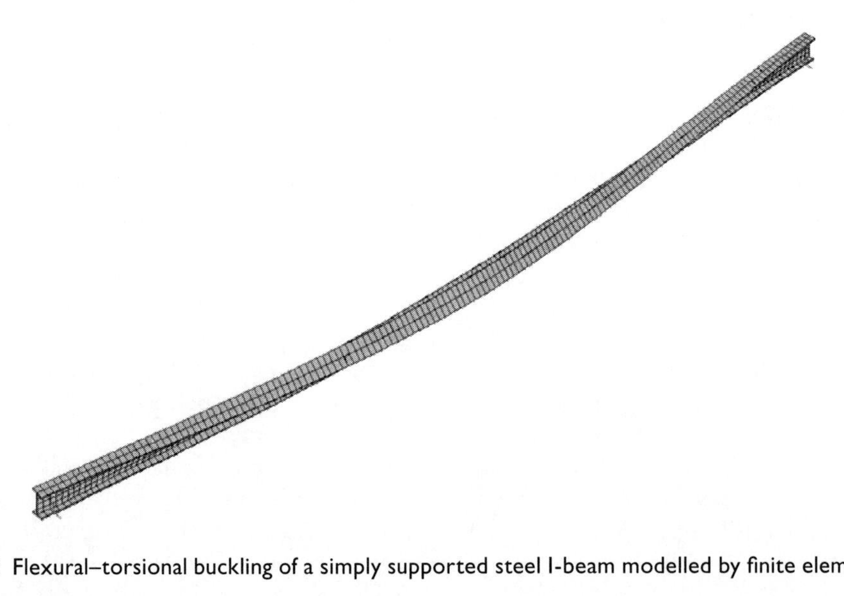

Figure 4.2 Flexural–torsional buckling of a simply supported steel I-beam modelled by finite elements.

lateral buckling, and their inelastic buckling moments are higher than the in-plane plastic collapse moments. Lateral and torsional restraints are often used in steel beams in practice to prevent the flexural–torsional buckling.

4.3 PROPERTIES OF THIN-WALLED SECTIONS

4.3.1 Centroids

The centroid of a compound thin-walled section is defined as the geometric centre of the cross section. If the section is composed of uniform or homogeneous material, the centroid of the section coincides with its centre of mass or its centre of gravity. For a thin-walled steel section composed of n elements, the coordinates of the centroid position (x_c, y_c) about the reference axes can be determined by

$$x_c = \frac{\sum_{j=1}^{n} A_j x_j}{\sum_{j=1}^{n} A_j} \tag{4.1}$$

$$y_c = \frac{\sum_{j=1}^{n} A_j y_j}{\sum_{j=1}^{n} A_j} \tag{4.2}$$

where

A_j is the area of element j

x_j and y_j are the centroidal coordinates of element j measured from the reference axes

4.3.2 Second moment of area

The second moment of area of a compound thin-walled steel section about its centroidal axes can be calculated using the parallel axis theorem as follows:

$$I_x = \sum_{j=1}^{n} [I_{ox\cdot j} + A_j (y_j - y_c)^2] \tag{4.3}$$

$$I_y = \sum_{j=1}^{n} [I_{oy\cdot j} + A_j (x_j - x_c)^2] \tag{4.4}$$

where

$I_{ox\cdot j}$ is the second moment of area of the jth element about its centroidal axis ox

$I_{oy\cdot j}$ is the second moment of area of the jth element about its centroidal axis oy

4.3.3 Torsional and warping constants

The torsional loading acting on a steel beam is resisted by two shear stress components. When a steel beam is subjected to uniform torsion, the rate of change in the angle of twist rotation and the longitudinal warping deflections is constant along the beam (Kollbrunner and Basler 1969; Trahair and Bradford 1998). A single set of shear stresses distributed around the cross section resists the torque acting at the cross section. The stiffness of the beam associated with these shear stresses is referred to the torsional rigidity GJ of the beam, where G is the shear modulus and J is the torsional constant. When a steel beam is subjected to non-uniform torsion, the longitudinal warping deflections vary along the beam. An additional set of shear stresses may act together with those induced by uniform torsion to resist the torque acting at the cross section. The stiffness of the beam associated with these additional shear stresses is referred to the warping rigidity EI_w of the beam, where I_w is the warping constant. Torsional and warping constants are needed in the determination of the elastic buckling moments of steel beams.

The torsional constant (J) of a section is the polar moment of inertia of the cross-sectional area. For circular hollow sections, the torsional constant (J) is calculated by

$$J = \frac{\pi}{32}\left(d_o^4 - d_i^4\right) \tag{4.5}$$

where d_o and d_i are the outer and inner diameters of the circular section, respectively.

For thin-walled open sections, the torsional constant can be approximately computed as the sum of the torsional constant of individual rectangular element by neglecting the contribution of the fillet region where elements are joined:

$$J \approx \sum \frac{bt^3}{3} \tag{4.6}$$

where

b is the length

t is the thickness of each rectangular element that forms the cross section

For I-beams with equal flanges, the warping constant is given by

$$I_w = \frac{I_y d_{fc}^2}{4} \tag{4.7}$$

where d_{fc} is the distance between the centroids of the two flanges.

For monosymmetric I-sections as depicted in Figure 4.3, the warping constant is calculated by (Kitipornchai and Trahair 1980; Trahair and Bradford 1998)

$$I_w = \frac{q_m b_1^3 t_1 d_{fc}^2}{12} \tag{4.8}$$

where q_m is given by

$$q_m = \frac{1}{1 + (b_1/b_2)^3 (t_1/t_2)} \tag{4.9}$$

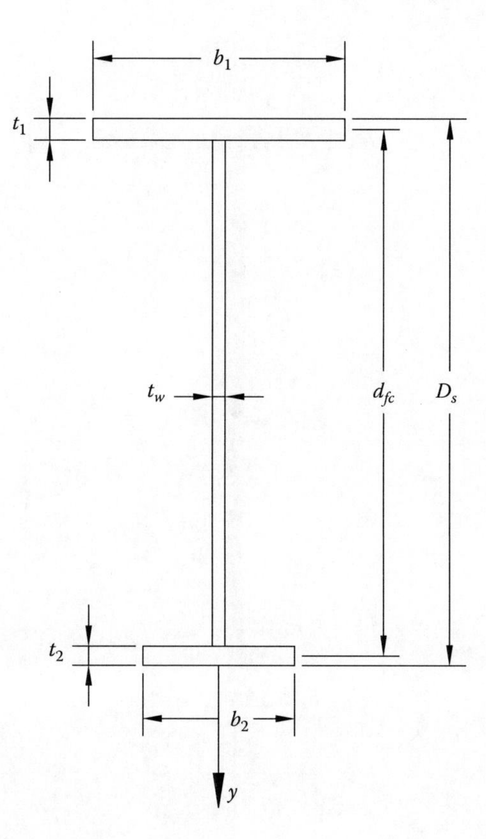

Figure 4.3 Dimensions of monosymmeric I-section.

4.3.4 Elastic section modulus

The elastic section modulus can be determined from the second moment of area as follows:

$$Z_x = \frac{I_x}{y_{\max}} \tag{4.10}$$

$$Z_y = \frac{I_y}{x_{\max}} \tag{4.11}$$

where

Z_x and Z_y are the elastic section moduli about its centroidal x- and y-axes, respectively
$x_{\max}$ and $y_{\max}$ are the maximum distances from the centroidal x- and y-axes of the section to its extreme fibres, respectively

The elastic section modulus is used in the calculation of elastic stresses in steel members under bending. It is noted that the effective section modulus (Z_e) is used in the calculation of the section moment capacities of steel beams. As discussed in Chapter 3, the effective section modulus of a non-compact or slender steel section is determined by accounting for local buckling effects.

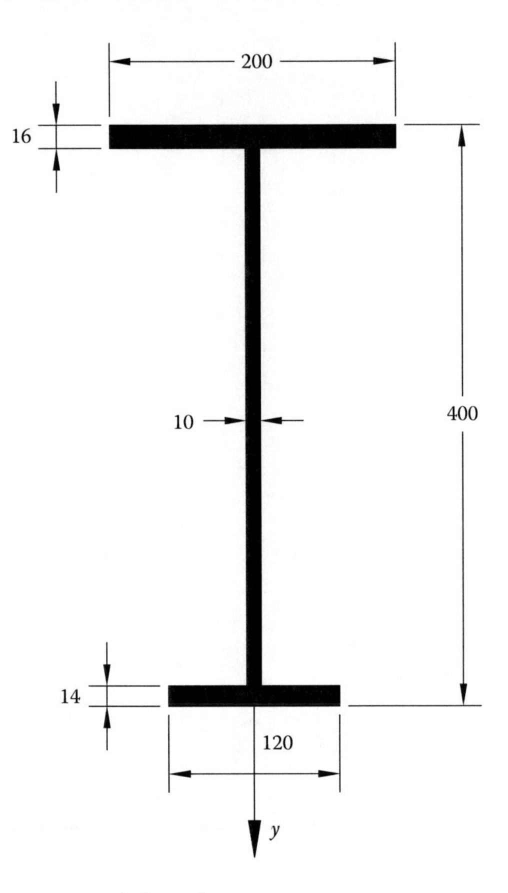

Figure 4.4 Dimensions of monosymmetric I-section.

Example 4.1: Calculation of elastic properties of a monosymmetric I-section

The monosymmetric steel I-section is depicted in Figure 4.4. The section is symmetric about its minor principal y-axis. Calculate the elastic properties of the monosymmetric section.

1. Centroid of the section

The clear depth of the section is $d_1 = 400-16-14 = 370\,\text{mm}$.

The centroid position of the section measured from the top fibre is calculated as

$$y_c = \frac{\sum_{j=1}^{n} A_j y_j}{\sum_{j=1}^{n} A_j}$$

$$= \frac{200\times16\times(16/2)+370\times10\times(370/2+16)+120\times14\times(400-14/2)}{200\times16+370\times10+120\times14}$$

$$= 166.6\,\text{mm}$$

2. Second moment of area

The second moment of area about the major principal x-axis is

$$I_x = \sum_{j=1}^{n} \left[I_{ox \cdot j} + A_j (y_j - y_c)^2 \right]$$

$$= \left[\frac{200 \times 16^3}{12} + 200 \times 16 \times \left(166.6 - \frac{16}{2} \right)^2 \right]$$

$$+ \left[\frac{10 \times 370^3}{12} + 10 \times 370 \times \left(\frac{370}{2} + 16 - 166.6 \right)^2 \right]$$

$$+ \left[\frac{120 \times 14^3}{12} + 120 \times 14 \times \left(400 - \frac{14}{2} - 166.6 \right)^2 \right] = 213.29 \times 10^6 \text{ mm}^4$$

The second moment of area about the minor principal y-axis is

$$I_y = \sum_{j=1}^{n} \left[I_{oy \cdot j} + A_j (x_j - x_c)^2 \right]$$

$$= \left[\frac{16 \times 200^3}{12} \right] + \left[\frac{370 \times 10^3}{12} \right] + \left[\frac{14 \times 120^3}{12} \right] = 12.71 \times 10^6 \text{ mm}^4$$

3. Torsion and warping constants

The torsion constant can approximately be calculated as

$$J \approx \sum \frac{bt^3}{3} = \frac{200 \times 16^3}{3} + \frac{370 \times 10^3}{3} + \frac{120 \times 14^3}{3} = 506.16 \times 10^3 \text{ mm}^4$$

The warping constant can be calculated as follows:

$$q_m = \frac{1}{1 + (b_1/b_2)^3 (t_1/t_2)} = \frac{1}{1 + (200/120)^3 \times (16/14)} = 0.159$$

$$I_w = \frac{q_m b_1^3 t_1 d_{cf}^2}{12} = \frac{0.159 \times 200^3 \times 16 \times (400 - 16/2 - 14/2)^2}{12} = 251.39 \times 10^9 \text{ mm}^6$$

4. Section modulus

The section modulus about its principal x-axis is

$$Z_x = \frac{I_x}{y_{\max}} = \frac{213.29 \times 10^6}{400 - 166.6} = 913.8 \times 10^3 \text{ mm}^3$$

The section modulus about its minor principal y-axis is

$$Z_y = \frac{I_y}{x_{\max}} = \frac{12.71 \times 10^6}{200/2} = 127.1 \times 10^3 \text{ mm}^3$$

4.4 SECTION MOMENT CAPACITY

The section moment capacity of a steel section can be derived from the stress distribution shown in Figure 4.5. For the rectangular section, the second moment of area about its section major principal x-axis is $I_x = BD^3/12$. The effective section modulus of this section which is assumed to be fully effective is determined as

$$Z_{ex} = \frac{I_x}{y_{\max}} = \frac{BD^3/12}{D/2} = \frac{BD^2}{6} \tag{4.12}$$

The extreme fibre of the section depicted in Figure 4.5 is assumed to reach the yield stress (f_y) of the steel. The compression and tension forces in the section are $C = T = \frac{1}{2}B(D/2)f_y = \frac{1}{4}BDf_y$. The nominal moment capacity of the section for bending about the section major principal x-axis can be obtained by taking moments about its centroid as

$$M_{sx} = \frac{1}{4}BDf_y \times \left(\frac{2}{3}D\right) = \left(\frac{BD^2}{6}\right)f_y \tag{4.13}$$

The earlier equation can be rewritten as

$$M_{sx} = Z_{ex}f_y \tag{4.14}$$

where f_y is taken as the minimum yield stress for the steel section.

When a steel beam is subjected to bending about its section major principal x-axis, all sections of the beam must satisfy the following design requirement:

$$M_x^* \le \phi M_{sx} \tag{4.15}$$

where

M_x^* is the factored design bending moment about the x-axis
$\phi = 0.9$ is the capacity reduction factor

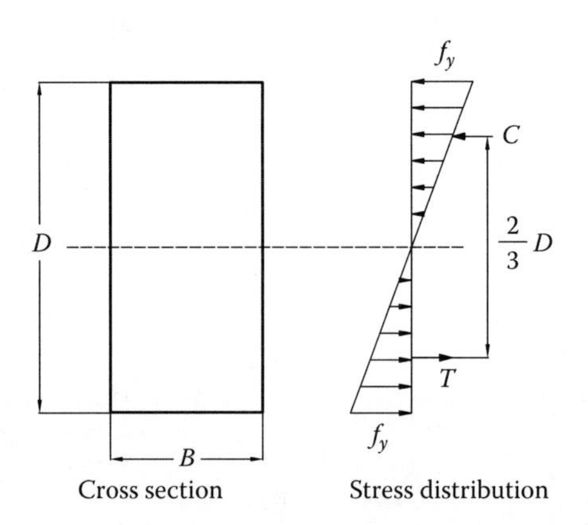

Figure 4.5 Stress distributions in rectangular steel section under bending.

Similarly, for a steel beam bending about its section minor principal y-axis, all sections of the beam must satisfy

$$M_y^* \leq \phi M_{sy} \tag{4.16}$$

where

M_y^* is the factored design bending moment about the section minor principal y-axis
M_{sy} is the nominal section moment capacity for bending about the section minor principal y-axis and is determined as

$$M_{sy} = Z_{ey} f_y \tag{4.17}$$

in which Z_{ey} is the effective section modulus for bending about the section minor principal y-axis.

4.5 MEMBER MOMENT CAPACITY

4.5.1 Restraints

The member moment capacity of a steel beam under bending depends on the lateral and torsional restraints at its ends and along the beam. The restraint such as an element, support or connection is used to prevent a beam from lateral deflection and/or lateral rotation about the minor axis and/or twist about the centre line of the beam. Various restraint conditions for cross sections are defined in Clause 5.4 of AS 4100 (1998) and briefly described herein. All supports are assumed to fully or partially restrain the cross sections against deflections and twist out of the plane of loading.

If the lateral deflection of the critical flange is effectively prevented and the twist rotation of the section is either effectively prevented or partially prevented, the cross section is considered to be *fully restrained* (F). If the lateral deflection of some points in the cross section rather than the critical flange and the twist rotation of the section is effectively suppressed, the cross section is also fully restrained. The critical flange is the flange that would deform further if the restraint is removed. This is the compression flange for a simply supported beam and the top flange for a cantilever under gravity loads. Some of the fully restrained cross sections are illustrated in Figure 4.6.

A *partially restrained* (P) cross section is the section where the lateral deflection of some points in the cross section rather than the critical flange is effectively suppressed while the twist rotation of the section is partially prevented. Figure 4.7 schematically depicts partially restrained cross sections.

If the lateral deflection of the critical flange is effectively prevented by the restraint which ineffectively suppresses the twist rotation of the section, the cross section is considered to be *laterally restrained* (L), as shown in Figure 4.8.

If the rotation of the critical flange about the section's minor axis in a fully or partially restrained cross section is prevented, the cross section is treated as *rotationally restrained* as demonstrated in Figure 4.9.

To be effective in restraining a segment in a steel beam, the restraining elements at the ends of the segment must be able to transfer a transverse force acting at the critical flange as specified in Clause 5.4.3 of AS 4100. The nominal transverse design force (N_R^*) transferred by the restraint against lateral deflection or twist rotation is

$$N_R^* = 0.025 N_f^* \tag{4.18}$$

where N_f^* is the maximum force in the critical flanges of the adjacent segments.

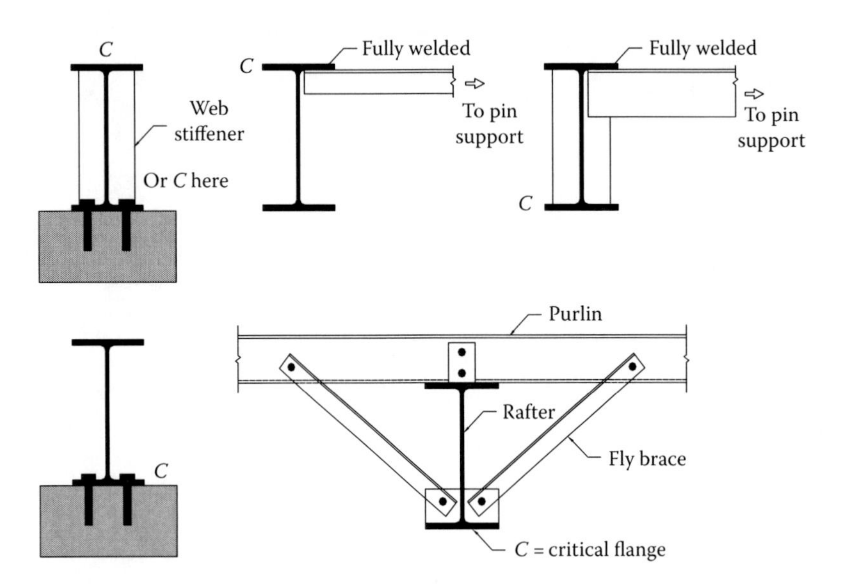

Figure 4.6 Fully restrained cross sections.

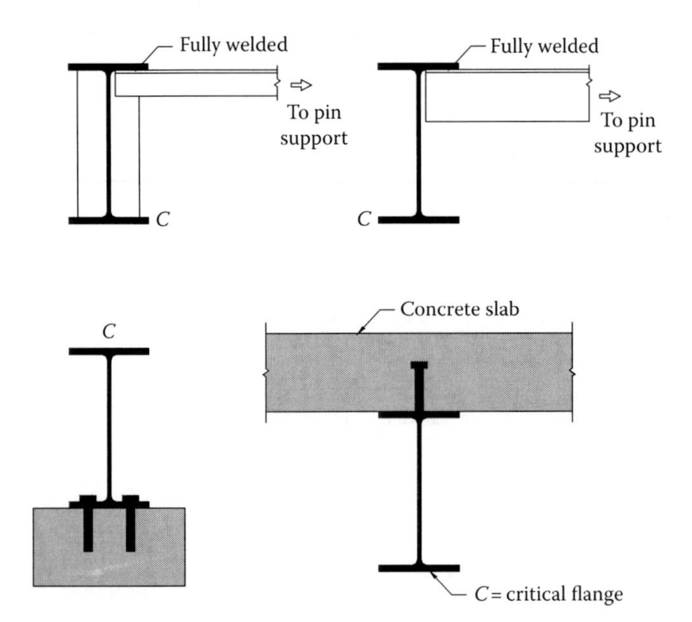

Figure 4.7 Partially restrained cross sections.

When parallel members are restrained by a continuous restraining element, each restraining element should be designed to carry a transverse force equal to the sum of $0.025N_f^*$ from the connected member and 0.0125 times the sum of flange forces in the connected members beyond.

4.5.2 Members with full lateral restraint

The flexural–torsional buckling of a steel beam with full lateral restraint is effectively prevented by the restraint. This implies that the nominal member moment capacity (M_b) of a

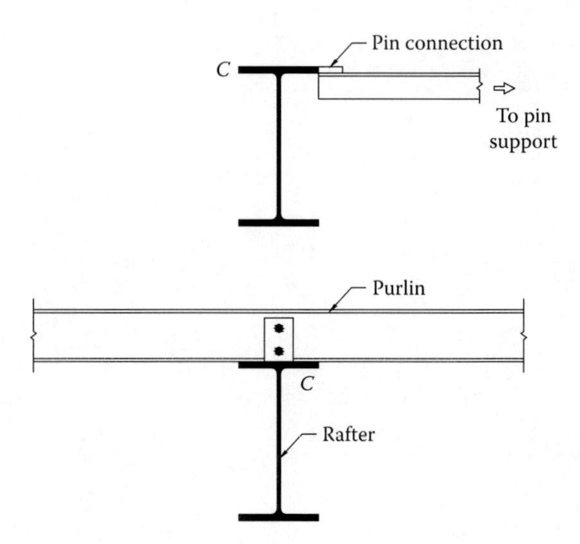

Figure 4.8 Laterally restrained cross sections.

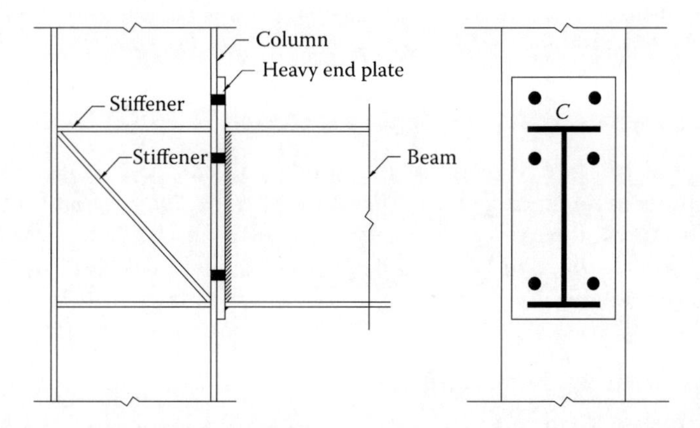

Figure 4.9 Rotationally restrained cross sections.

steel member with full lateral restraint can be taken as the nominal section moment capacity (M_s) of the critical section. The critical section in a segment or member is defined as the cross section having the largest ratio of M^*/M_s.

As specified in Clause 5.3.2 of AS 4100 (1998), a segment fully or partially restrained at both ends is considered to have full lateral restraint if it satisfies one of the following restraint conditions:

a. The segment has continuous restraints at the critical flange.
b. The segment has intermediate lateral restraints at the critical flange and the length of each sub-segment satisfies the slenderness requirements given in (c).
c. The segment satisfies the slenderness (l/r_y) requirements given in Table 4.1, where r_y is the radius of gyration about the section minor principal y-axis.

The moment ratio β_m given in Table 4.1 is taken as –1.0 or –0.8 for segments subjected to transverse loads or $\beta_m = \pm M_2^*/M_1^*$ for segments without transverse loads, where M_1^* and M_2^* ($M_1^* \geq M_2^*$) are design bending moments at the segment ends. The moment ratio β_m is taken as positive for bending in reverse curvature and negative for bending in single curvature.

Table 4.1 Slenderness requirements for full lateral restraint for segments fully or partially restrained at both ends

Segment section	*Slenderness limits*
I-section with equal flanges	$\frac{l}{r_y} \leq (80 + 50\beta_m)\sqrt{\frac{250}{f_y}}$
Equal channel	$\frac{l}{r_y} \leq (60 + 40\beta_m)\sqrt{\frac{250}{f_y}}$
I-section with unequal flanges	$\frac{l}{r_y} \leq (80 + 50\beta_m)\left(\sqrt{\frac{2I_{cy}Ad_{fc}}{2.5I_yZ_{ex}}}\right)\sqrt{\frac{250}{f_y}}$
RHS or square hollow section (SHS)	$\frac{l}{r_y} \leq (1800 + 1500\beta_m)\left(\frac{b_f}{d_w}\right)\left(\frac{250}{f_y}\right)$
Angle	$\frac{l}{r_y} \leq (210 + 175\beta_m)\sqrt{\frac{b_2}{b_1}}\left(\frac{250}{f_y}\right)$

Source: AS 4100, *Australian Standard for Steel Structures*, Standards Australia, Sydney, New South Wales, Australia, 1998.

4.5.3 Members without full lateral restraint

Steel beams without full lateral restraint may undergo flexural–torsional buckling, which reduces their member moment capacities. Therefore, steel beams without full lateral restraint must be designed against flexural–torsional buckling (Trahair 1993a,b; Trahair et al. 1993; Trahair and Bradford 1998). The effect of flexural–torsional buckling is taken into account by using a slenderness reduction factor α_s.

4.5.3.1 Open sections with equal flanges

In Clause 5.6.1.1 of AS 4100, the nominal member moment capacity (M_b) for open section segments with equal flanges and full or partial restraints at both ends is computed by

$$M_b = \alpha_m \alpha_s M_s \leq M_s \tag{4.19}$$

where

α_m is the moment modification factor which accounts for the effect of non-uniform moment distribution along the segment
α_s is the slenderness reduction factor which considers the effect of the segment slenderness on the member moment capacity
M_s is the nominal section moment capacity

It is noted that the member moment capacity should not be greater than the section moment capacity.

The moment modification factor (α_m), which is usually greater than 1.0, may increase the member moment capacity. Economical designs can be achieved by using α_m for members with high moment gradients along the segments. This factor can be obtained from Table 5.6.1 of AS 4100 or calculated from the design bending moment distribution determined by structural analysis within the segment as follows:

$$\alpha_m = \frac{1.7M_m^*}{\sqrt{\left(M_2^*\right)^2 + \left(M_3^*\right)^2 + \left(M_4^*\right)^2}} \leq 2.5 \tag{4.20}$$

where

M_m^* is the maximum design bending moment within the segment considered
M_2^* and M_4^* are design bending moments at the quarter points of the segment
M_3^* is the design bending moment at the midpoint of the segment

The member slenderness reduction factor (α_s), which is usually less than 1.0, may reduce the member moment capacity (M_b) below the section moment capacity (M_s). This factor is a function of the section moment capacity and the elastic buckling moment (M_{oa}) which reflects the slenderness of the member and is determined by

$$\alpha_s = 0.6\left[\sqrt{\left(\frac{M_s}{M_{oa}}\right)^2 + 3} - \left(\frac{M_s}{M_{oa}}\right)\right] \leq 1.0 \tag{4.21}$$

where M_{oa} can be either taken as the reference buckling moment M_o or determined from an elastic buckling analysis. Figure 4.10 shows the relationship between α_s and the moment ratio of M_s/M_{oa}. It appears that the slenderness reduction factor decreases with increasing the moment ratio of M_s/M_{oa}.

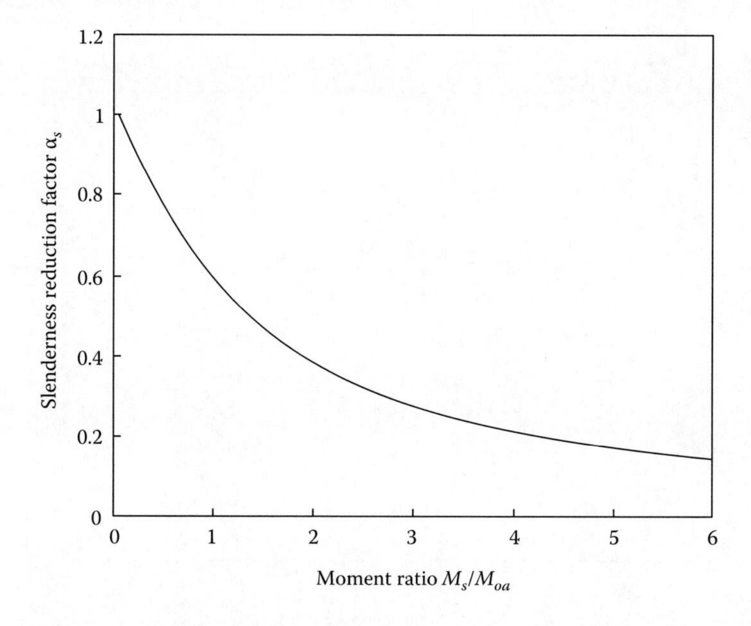

Figure 4.10 Slenderness reduction factor.

The reference buckling moment (M_o), which is the theoretical elastic lateral–torsional buckling strength of the beam under uniform bending moment (Timoshenko and Gere 1961), is given in Clause 5.6.1.1 of AS 4100 (1998) as follows:

$$M_o = \sqrt{\frac{\pi^2 E_s I_y}{L_e^2}\left(GJ + \frac{\pi^2 E_s I_w}{L_e^2}\right)} \tag{4.22}$$

where
 E_s is Young's modulus
 G is the shear modulus of steel (80,000 MPa)
 J is the torsional constant
 I_w is the warping constant
 L_e is the effective length of the segment

Figure 4.11 presents the reference elastic buckling moments with various slenderness ratios of L_e/r_y. It can be seen that increasing the member slenderness ratio significantly reduces the elastic buckling moment (M_o). In other words, the elastic buckling moment can be increased by decreasing L_e and increasing I_y and I_w.

The effective length (L_e) of a segment depends on its twist restraint, load height position and lateral rotational restraint (Bradford and Trahair 1983). Clause 5.6.3 of AS 4100 (1998) suggests that the effective length of a segment should be determined by

$$L_e = k_t k_l k_r l \tag{4.23}$$

where
 l is the actual length of the segment
 k_t is the twist restraint factor that accounts for the effect of partial torsional restraint

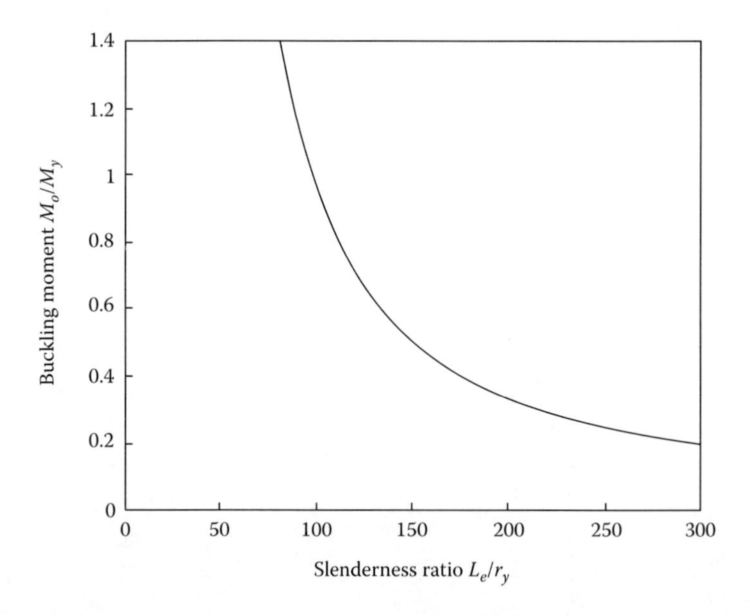

Figure 4.11 Elastic buckling moments of simply supported I-beams.

For segment ends with restraint conditions of FP, PL or PU, the twist restraint factor is determined by (Bradford and Trahair 1983)

$$k_t = 1 + \frac{(d_1/l)(t_f/2t_w)^3}{n_w} \tag{4.24}$$

For segment ends with restraint conditions of PP, k_t is calculated by (Bradford and Trahair 1983)

$$k_t = 1 + \frac{2\left(\frac{d_1}{l}\right)\left(\frac{t_f}{2t_w}\right)^3}{n_w} \tag{4.25}$$

where n_w is the number of web in the segment section. For other restraint conditions not mentioned earlier, k_t is taken as 1.0.

The load height factor k_l is used to consider the destabilizing effect of gravity loads at the top flange in comparison with the loading at the shear centre. The load height factor k_l is taken as 1.4 for gravity loads within the segment and on the top flange of segment and 2.0 for gravity loads on the top flange of cantilever. For other restraint conditions at segment ends and loading at segment ends and for shear centre loads, k_l is taken as 1.0.

The lateral rotation restraint factor k_r is taken as 0.85 for segment ends with restraint conditions of FF, FP or PP and with lateral rotation restraint at one end and 0.7 for segments with lateral rotation restraints at both ends (Trahair and Bradford 1998). For other cases, k_r is taken as 1.0.

4.5.3.2 I-sections with unequal flanges

As specified in Clause 5.6.1.2 of AS 4100 (1998), the nominal member moment capacities of steel I-sections with unequal flanges symmetrical about the minor axis can also be calculated using Equation 4.19 and the reference buckling moment (M_o) determined either by an elastic buckling analysis or by the following equation:

$$M_o = \sqrt{\frac{\pi^2 E_s I_y}{L_e^2}}\left[\sqrt{GJ + \frac{\pi^2 E_s I_w}{L_e^2} + \frac{\beta_x^2}{4}\frac{\pi^2 E_s I_y}{L_e^2}} + \frac{\beta_x}{2}\sqrt{\frac{\pi^2 E_s I_y}{L_e^2}}\right] \tag{4.26}$$

where β_x is the monosymmetric section constant, which can be determined by (Kitipornchai and Trahair 1980)

$$\beta_x = 0.8 d_{fc}\left(\frac{2I_{cy}}{I_y} - 1\right) \tag{4.27}$$

The nominal member moment capacity (M_b) of an angle section member or a rectangular hollow section (RHS) member can be determined using Equation 4.19 with the warping constant of $I_w = 0$.

4.5.4 Design requirements for members under bending

For a steel member subjected to a bending moment M_x^* about its section major principal x-axis which is determined by the elastic method of structural analysis, Clause 5.1 of AS 4100 (1998) requires that both the section and member moment capacities shall be checked as follows:

$$M_x^* \leq \phi M_{sx} \tag{4.28}$$

$$M_x^* \leq \phi M_{bx} \tag{4.29}$$

in which M_{bx} is the member moment capacity bending about the major principal x-axis.

For a steel member subjected to a bending moment M_y^* about its section minor principal y-axis which is determined by elastic method of structural analysis, the member will not undergo lateral–torsional buckling so that only its in-plane section moment capacity needs to be checked as follows:

$$M_y^* \leq \phi M_{sy} \tag{4.30}$$

Example 4.2: Design of steel beam without intermediate lateral restraints

A simply supported steel I-beam is depicted in Figure 4.12. The beam is subject to a uniformly distributed dead load of 4.4 kN/m and a live load of 5.3 kN/m and a concentrated dead load of 32 kN and a concentrated live load of 37 kN. All loads are applied to the top flange of the beam. The beam is partially restrained at the ends where the lateral deflections are effectively prevented and twist rotations are partially suppressed. There are no intermediate lateral restraints between the supports. Check the adequacy of the beam with a 610UB113 section of Grade 300 steel.

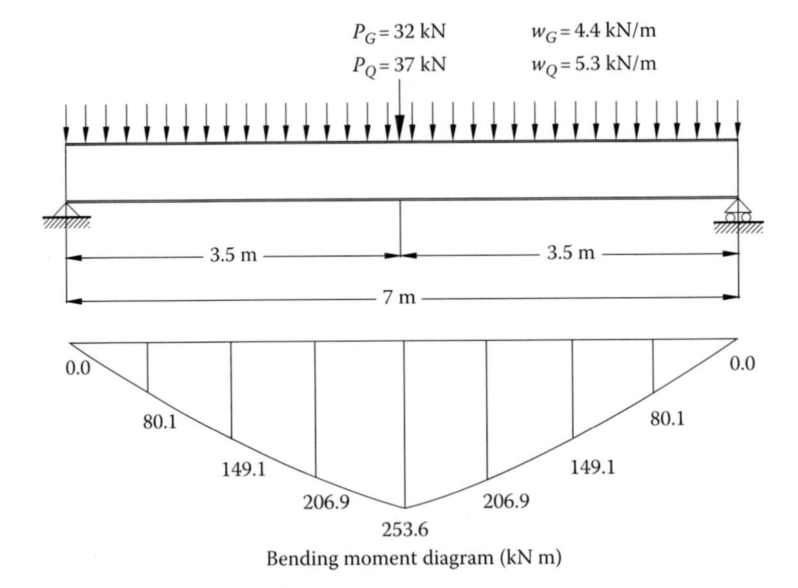

Figure 4.12 Steel beam without intermediate lateral restraints.

1. Design actions

The uniformly distributed dead load is 4.4 kN/m.
The self-weight of the steel beam is $113\times9.81\times10^{-3}$ = 1.11 kN/m.
The uniformly distributed live load is 5.3 kN/m.
The uniformly distributed design load $w^* = 1.2G + 1.5Q = 1.2\times(4.4 + 1.11) + 1.5\times5.3 = 14.56$ kN/m.
The concentrated design load $P^* = 1.2G + 1.5Q = 1.2\times32 + 1.5\times37 = 93.9$ kN.
The design bending moment diagram of the beam is shown in Figure 4.12.

2. Section moment capacity

The section properties of 610UB113 of Grade 300 steel are

$$d_1 = 572\,\text{mm}, \quad t_f = 17.3\,\text{mm}, \quad t_w = 11.2\,\text{mm}$$

$$I_y = 34.3\times10^6\,\text{mm}^4, \quad I_w = 2980\times10^9\,\text{mm}^6, \quad J = 1140\times10^3\,\text{mm}^4$$

$$Z_{ex} = 3290\times10^3\,\text{mm}^3, \quad G = 80\times10^3\,\text{MPa}, \quad E_s = 200{,}000\,\text{MPa}, \quad f_y = 280\,\text{MPa}$$

The nominal section moment capacity can be calculated as

$$M_{sx} = Z_{ex}f_y = 3290\times10^3\times280\,\text{N mm} = 921.2\,\text{kN m}$$

The design section moment capacity is

$$\phi M_{sx} = 0.9\times921.2 = 892\,\text{kN m} > M_x^* = 253.6\,\text{kN m}$$

3. Moment modification factor

As shown in Figure 4.13, the design bending moments are

$$M_m^* = 253.6\,\text{kN m}, \quad M_2^* = 149.1\,\text{kN m}, \quad M_3^* = 253.6\,\text{kN m}, \quad M_4^* = 149.1\,\text{kN m}$$

The moment modification factor is calculated as

$$\alpha_m = \frac{1.7M_m^*}{\sqrt{\left(M_2^*\right)^2 + \left(M_3^*\right)^2 + \left(M_4^*\right)^2}} = \frac{1.7\times253.6}{\sqrt{(149.1)^2 + (253.6)^2 + (149.1)^2}} = 1.307 \le 2.5$$

4. Slenderness reduction factor

The beam is partially restrained at both supports (PP) so that the twist restraint factor can be calculated as

$$k_t = 1 + \frac{2(d_1/l)(t_f/2t_w)^3}{n_w} = 1 + \frac{2\times(572/7000)(17.3/2\times11.2)^3}{1} = 1.075$$

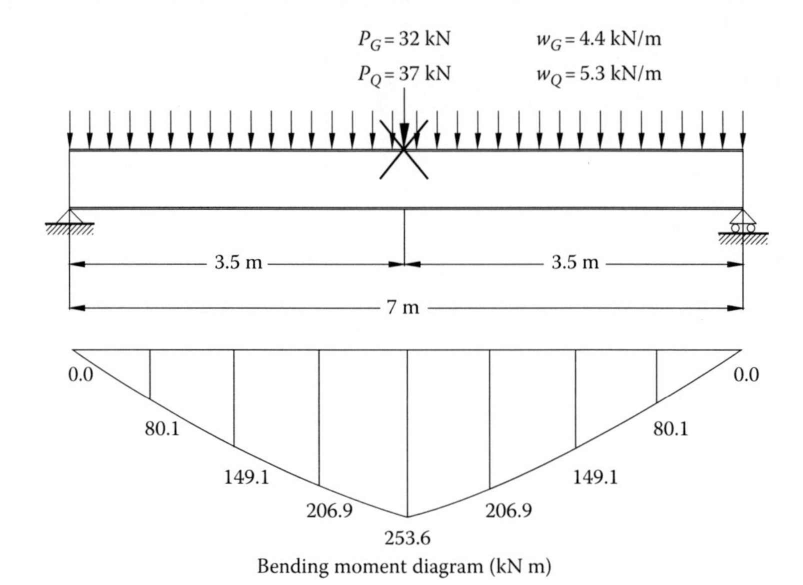

Figure 4.13 Steel beam with an intermediate lateral restraint.

The loads are applied to the top flange within the beam with PP restraints at the supports. The load height factor is taken as $k_l = 1.4$.

Since none of the ends of the beam are restrained rotationally, the lateral rotation restraint factor is $k_r = 1.0$.

The effective length of the beam is determined as

$$L_e = k_t k_l k_r l = 1.075 \times 1.4 \times 1.0 \times 7{,}000 = 10{,}535\,\text{mm}$$

The reference buckling moment is calculated as follows:

$$M_o = \sqrt{\frac{\pi^2 E_s I_y}{L_e^2}\left(GJ + \frac{\pi^2 E_s I_w}{L_e^2}\right)}$$

$$= \sqrt{\frac{\pi^2 \times 200 \times 10^3 \times 34.3 \times 10^6}{10{,}535^2}\left(80 \times 10^3 \times 1{,}140 \times 10^3 + \frac{\pi^2 \times 200 \times 10^3 \times 2{,}980 \times 10^9}{10{,}535^2}\right)}\ \text{N mm}$$

$$= 296.6\,\text{kN m}$$

The slenderness reduction factor is determined by

$$\alpha_s = 0.6\left[\sqrt{\left(\frac{M_s}{M_{oa}}\right)^2 + 3} - \left(\frac{M_s}{M_{oa}}\right)\right] = 0.6\left[\sqrt{\left(\frac{921.2}{296.6}\right)^2 + 3} - \left(\frac{921.2}{296.6}\right)\right] = 0.27 < 1.0$$

5. Member moment capacity

The nominal member moment capacity of the beam is

$$M_{bx} = \alpha_m \alpha_s M_{sx} = 1.307 \times 0.27 \times 921.2 = 325\,\text{kN m}$$

The design member moment capacity of the beam is

$$\phi M_{bx} = 0.9 \times 325 \ = 292.5\,\text{kN m} > M_x^* = 253.6\,\text{kN m, OK}$$

Example 4.3: Design of steel beam with an intermediate lateral restraint

Redesign the steel beam presented in Example 4.2 by incorporating one lateral restraint at the mid-span of the beam as depicted in Figure 4.13.

1. Design actions

The design actions have been calculated in Example 4.2 and the bending moment diagram is shown in Figure 4.13.

2. Section moment capacity

Try section 460UB67.1 of Grade 300 steel. The properties of the 460UB67.1 are

$$d_1 = 428\,\text{mm}, \quad t_f = 12.7\,\text{mm}, \quad t_w = 8.5\,\text{mm}$$

$$I_y = 14.5 \times 10^6\,\text{mm}^4, \quad I_w = 708 \times 10^9\,\text{mm}^6, \quad J = 378 \times 10^3\,\text{mm}^4$$

$$Z_{ex} = 1{,}480 \times 10^3\,\text{mm}^3, \quad G = 80 \times 10^3\,\text{MPa}, \quad E_s = 200{,}000\,\text{MPa}, \quad f_y = 300\,\text{MPa}$$

The nominal section moment capacity can be calculated as

$$M_{sx} = Z_{ex} f_y = 1480 \times 10^3 \times 300\,\text{N mm} = 444\,\text{kN m}$$

The design section moment capacity is

$$\phi M_{sx} = 0.9 \times 444 \ = 399.6\,\text{kN m} > M_x^* = 253.6\,\text{kN m}$$

3. Moment modification factor

As shown in Figure 4.13, the design bending moments acting on the segment between the mid-span and the support are

$$M_m^* = 253.6\,\text{kN m}, \quad M_2^* = 80.1\,\text{kN m}, \quad M_3^* = 149.1\,\text{kN m}, \quad M_4^* = 206.9\,\text{kN m}$$

The moment modification factor is calculated as

$$\alpha_m = \frac{1.7 M_m^*}{\sqrt{\left(M_2^*\right)^2 + \left(M_3^*\right)^2 + \left(M_4^*\right)^2}} = \frac{1.7 \times 253.6}{\sqrt{(80.1)^2 + (149.1)^2 + (206.9)^2}} = 1.613 \le 2.5$$

4. Slenderness reduction factor

The segment between the support and mid-span is partially restrained at the support (P) and laterally restrained at the mid-span (L) so that the twist restraint factor is calculated as

$$k_t = 1 + \frac{(d_1/l)\,(t_f/2t_w)^3}{n_w} = 1 + \frac{(428/3500)(12.7/2 \times 8.5)^3}{1} = 1.051$$

The loads are applied to the top flange within the segment with PL restraints at the ends. The load height factor is taken as $k_l = 1.4$.

Since none of the ends of the segment are restrained rotationally, the lateral rotation restraint factor is $k_r = 1.0$.

The effective length of the segment is determined as

$$L_e = k_t k_l k_r l = 1.051 \times 1.4 \times 1.0 \times 3500 = 5150\,\text{mm}$$

The reference buckling moment is calculated as

$$M_o = \sqrt{\frac{\pi^2 E_s I_y}{L_e^2}\left(GJ + \frac{\pi^2 E_s I_w}{L_e^2}\right)}$$

$$= \sqrt{\frac{\pi^2 \times 200 \times 10^3 \times 14.5 \times 10^6}{5150^2}\left(80 \times 10^3 \times 378 \times 10^3 + \frac{\pi^2 \times 200 \times 10^3 \times 708 \times 10^9}{5150^2}\right)}\,\text{N mm}$$

$$= 299\,\text{kN m}$$

The slenderness reduction factor is computed as follows:

$$\alpha_s = 0.6\left[\sqrt{\left(\frac{M_s}{M_{oa}}\right)^2 + 3} - \left(\frac{M_s}{M_{oa}}\right)\right] = 0.6\left[\sqrt{\left(\frac{444}{299}\right)^2 + 3} - \left(\frac{444}{299}\right)\right] = 0.478 < 1.0$$

5. Member moment capacity

The nominal member moment capacity of the segment is

$$M_{bx} = \alpha_m \alpha_s M_{sx} = 1.613 \times 0.478 \times 444 = 342\,\text{kN m}$$

The design member moment capacity of the segment or the beam is

$$\phi M_{bx} = 0.9 \times 342 = 308\,\text{kN m} > M_x^* = 253.6\,\text{kN m, OK}$$

4.6 SHEAR CAPACITY OF WEBS

4.6.1 Yield capacity of webs in shear

The web of a steel beam under bending is subjected to shear. The capacity of a steel web in shear depends on its depth-to-thickness ratio and the spacing of transverse web stiffeners (Bradford 1987; Trahair and Bradford 1998). Clause 5.10 of AS 4100 (1998) provides requirements on the minimum thickness of beam webs including any transverse or longitudinal stiffeners, which are given in Table 4.2. For webs with the stiffener spacing to depth ratio s/d_p greater than 3.0, the webs should be considered to be unstiffened.

The design requirement for a steel beam web under a design shear force (V^*) is

$$V^* \leq \phi V_v \tag{4.31}$$

where V_v is the nominal shear capacity of the web.

The shear distribution in webs of most I-section members is approximately uniform. For a web with a approximately uniform shear stress distribution, Clause 5.11.2 of AS 4100

Table 4.2 Minimum web thickness

Arrangement of webs	*Required thickness t_w*
Unstiffened web bounded by two flanges	$t_w \geq \frac{d_1}{180}\sqrt{\frac{f_y}{250}}$
Unstiffened web bounded by one free edge	$t_w \geq \frac{d_1}{90}\sqrt{\frac{f_y}{250}}$
Transversely stiffened webs	
$\frac{s}{d_1} \leq 0.74$	$t_w \geq \frac{d_1}{270}\sqrt{\frac{f_y}{250}}$
$0.74 < \frac{s}{d_1} \leq 1.0$	$t_w \geq \frac{s}{200}\sqrt{\frac{f_y}{250}}$
$1.0 \leq \frac{s}{d_1} \leq 3.0$	$t_w \geq \frac{d_1}{200}\sqrt{\frac{f_y}{250}}$
Webs with one longitudinal and transverse stiffeners	
$\frac{s}{d_1} < 0.74$	$t_w \geq \frac{d_1}{340}\sqrt{\frac{f_y}{250}}$
$0.74 \leq \frac{s}{d_1} \leq 1.0$	$t_w \geq \frac{s}{250}\sqrt{\frac{f_y}{250}}$
$1.0 \leq \frac{s}{d_1} \leq 2.4$	$t_w \geq \frac{d_1}{250}\sqrt{\frac{f_y}{250}}$
Webs with two longitudinal stiffeners and $\frac{s}{d_1} < 1.5$	$t_w \geq \frac{d_1}{400}\sqrt{\frac{f_y}{250}}$
Webs containing plastic hinges	$t_w \geq \frac{d_1}{82}\sqrt{\frac{f_y}{250}}$

Source: AS 4100, *Australian Standard for Steel Structures*, Standards Australia, Sydney, New South Wales, Australia, 1998.

allows V_v be taken as the nominal shear capacity of the web (V_u) with a uniform shear stress distribution, which is given by

$$V_u = \begin{cases} V_w & \text{for } \frac{d_p}{t_w}\sqrt{\frac{f_y}{250}} \leq 82 \\ V_b & \text{for } \frac{d_p}{t_w}\sqrt{\frac{f_y}{250}} > 82 \end{cases} \tag{4.32}$$

where

d_p is the clear depth of the web panel
t_w is the thickness of the web
V_b is the shear buckling capacity of the web
V_w is the nominal shear yield capacity, which is determined by

$$V_w = 0.6 f_y A_w \tag{4.33}$$

where A_w is the cross-sectional area of the web, which is taken as $A_w = D_s t_w$ for hot-rolled steel I-sections where D_s is the total depth of the hot-rolled I-section and $A_w = d_1 t_w$ for welded or built-up steel I-sections.

It is noted that for a stocky web with $(d_p/t_w)\sqrt{f_y/250} \le 82$, the web yields before elastic local buckling so that V_u is taken as the shear yield capacity V_w. In contrast, for a slender web with $(d_p/t_w)\sqrt{f_y/250} > 82$, the web buckles elastically before yielding so that V_u is taken as the shear buckling capacity V_b.

The shear stress distribution in the web of a steel beam with unequal flanges, varying web thickness or holes not used for fasteners is non-uniform. In Clause 5.11.3 of AS 4100, the shear capacity of the web with non-uniform shear stress distribution is determined from the shear capacity of the web with uniform shear stress distribution (V_u) by considering the effect of the shear stress ratio in the web as

$$V_v = \frac{2V_u}{0.9 + \left(f_{vm}^* / f_{va}^*\right)} \le V_u \tag{4.34}$$

in which f_{vm}^* and f_{va}^* are the maximum and average design shear stresses in the web, respectively, and are determined by an elastic analysis.

4.6.2 Shear buckling capacity of webs

As specified in Clause 5.11.5.1 of AS 4100, the nominal shear buckling capacity (V_b) of a slender unstiffened web is based on its elastic local buckling stress and is calculated by

$$V_b = \left[\frac{82}{(d_p/t_w)\sqrt{f_y/250}}\right]^2 V_w \le V_w \tag{4.35}$$

If the design shear buckling capacity (ϕV_b) of a slender unstiffened web is less than the design shear force (V^*), intermediate transverse stiffeners may be welded to the web to increase the shear buckling capacity of the web. The nominal shear buckling capacity of a slender stiffened web with a spacing-to-depth ratio of $s/d_p \le 3.0$ is given in Clause 5.11.5.2 of AS 4100 (1998) as follows:

$$V_b = \alpha_v \alpha_d \alpha_f V_w \le V_w \tag{4.36}$$

where α_v is the stiffening factor which accounts for the effects of the increased elastic buckling resistance due to transverse stiffeners and is given by

$$\alpha_v = \left[\frac{82}{(d_p/t_w)\sqrt{f_y/250}}\right]^2 \left[\frac{0.75}{(s/d_p)^2} + 1.0\right] \le 1.0 \quad \text{when } 1.0 \le s/d_p \le 3.0 \tag{4.37}$$

$$\alpha_v = \left[\frac{82}{(d_p/t_w)\sqrt{f_y/250}}\right]^2 \left[\frac{1}{(s/d_p)^2} + 0.75\right] \le 1.0 \quad \text{when } s/d_p \le 1.0 \tag{4.38}$$

where s is the stiffener spacing. The effect of transverse stiffeners on the shear buckling capacity is incorporated in factor α_v, and it depends on the stiffener spacing to depth ratio of the web panel.

In Equation 4.36, α_d is the tension field contribution factor which considers the contribution of the tension field to the shear buckling capacity and is expressed by

$$\alpha_d = 1 + \frac{1-\alpha_v}{1.15\alpha_v\sqrt{1+(s/d_p)^2}} \tag{4.39}$$

Factor α_f in Equation 4.36 is the flange restraint factor reflecting the increase in the shear buckling capacity of the web due to the restraining effects provided by the flanges and is given by

$$\alpha_f = 1.6 - \frac{0.6}{\sqrt{\left(1+\left(40b_{fo}t_f^2/d_1^2t_w\right)\right)}} \tag{4.40}$$

where b_{fo} and t_f are the width and thickness of the flange outstand, respectively. For webs without longitudinal stiffeners, b_{fo} is taken as the least of $(12t_f)/\sqrt{f_y/250}$, the distance from the mid-plane of the web to the nearer edge of the flange or half the clear distance between the webs.

4.6.3 Webs in combined shear and bending

When the beam is subjected to both high design moment M^* and shear force V^*, it must be designed for combined bending and shear. Clause 5.12 of AS 4100 permits two methods for the design of beam webs under combined bending and shear: the proportioning and interaction methods.

In the proportioning method, the bending moment is assumed to carry only by the flanges and the web resists the whole shear force. The design bending moment and shear force must satisfy

$$M^* \le \phi A_{fm} d_{fc} f_y \tag{4.41}$$

$$V^* \le \phi V_v \tag{4.42}$$

where

- A_{fm} is the lesser of the flange effective areas for the compression flange and the lesser of the gross area of the flange and $0.85A_{fn}f_u/f_y$ for the tension flange, in which A_{fn} is the net area of the flange
- V_v is the nominal shear capacity of the web

The proportioning method is used to design beams with slender webs. The flanges of these beams should be at least non-compact to achieve better designs.

In the interaction method, the bending moment is assumed to be carried by the whole cross section. This method is used to design beams with less slender webs and applies to both stiffened and unstiffened webs. The bending and shear interaction diagram is schematically depicted in Figure 4.14, which is expressed by equations given in Clause 5.12.3 of AS 4100 (1998) as follows:

$$V^* \le \phi V_v \qquad \text{for } M^* \le 0.75\phi M_s \tag{4.43}$$

$$V^* \le \phi V_v\left(2.2 - \frac{1.6M^*}{\phi M_s}\right) \qquad \text{for } 0.75\phi M_s \le M^* \le \phi M_s \tag{4.44}$$

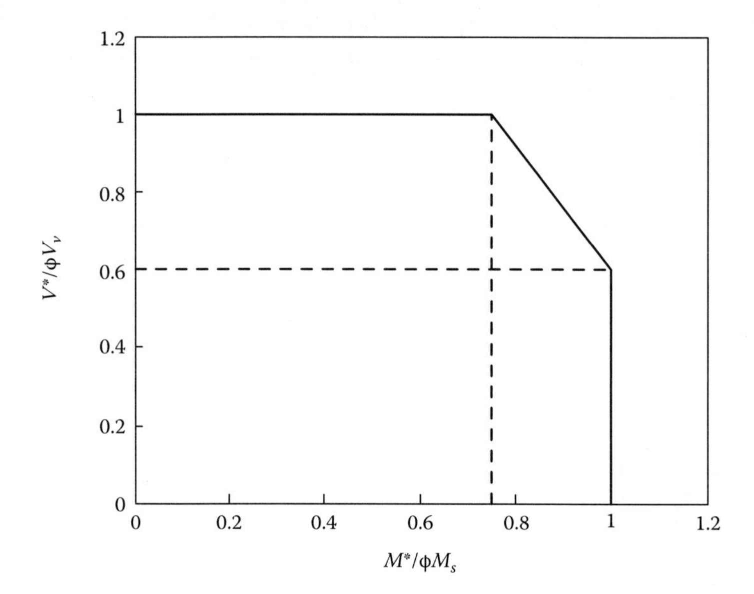

Figure 4.14 Strength interaction diagram for sections in bending and shear.

4.6.4 Transverse web stiffeners

Intermediate transverse web stiffeners can be used to prevent local buckling of the web in shear. These web stiffeners must have not only adequate stiffness to ensure that the elastic buckling stress of a panel can be attained but also adequate strength to carry the tension field stiffener force. Transverse web stiffeners are usually not connected to the tension flange and can be attached to either one side or both sides of the web as illustrated in Figure 4.15. The spacing of the transverse web stiffeners should be less than $3d_1$ in order to effectively resist the shear force. For the strength design, Clause 5.15.3 of AS 4100 (1998) gives the following minimum area of an intermediate web stiffener:

$$A_s \geq 0.5\gamma_w(1-\alpha_v)\left(\frac{V^*}{\phi V_u}\right)\left[\frac{s}{d_p}-\frac{(s/d_p)^2}{\sqrt{1+(s/d_p)^2}}\right]A_w \tag{4.45}$$

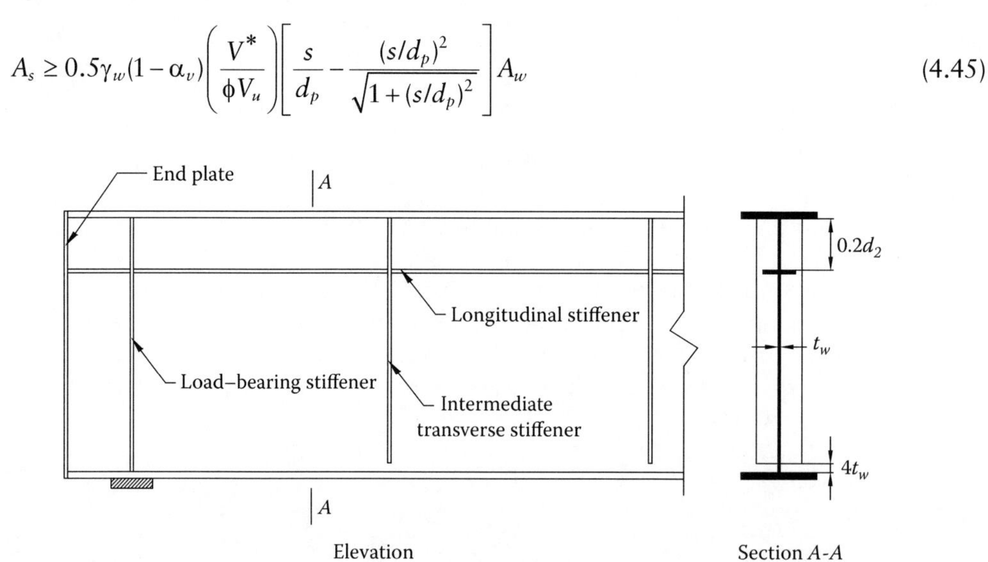

Figure 4.15 Web stiffeners of a steel plate girder.

where γ_w is 1.0 for a pair of stiffeners, 1.8 for a single angle stiffener and 2.4 for a single plate stiffener. This minimum area is to ensure the stiffener has the yield capacity that is sufficient to transmit the force caused by the tension field.

The design shear force (V^*) acting on the slender stiffened web of a steel plate girder as illustrated in Figure 4.15 is resisted by the web shear buckling and the stiffener–web buckling. As a result, an intermediate web stiffener must satisfy the buckling strength requirement given in Clause 5.15.4 of AS 4100 (1998) as follows:

$$V^* \leq \phi(R_{sb} + V_b) \tag{4.46}$$

where

R_{sb} is the buckling capacity of the stiffener–web compression member as a whole
V_b is the shear buckling capacity of the stiffened web given in Equation 4.36

The buckling capacity of the web and the intermediate web stiffener as a whole (R_{sb}) is determined as the axial load capacity of the stiffener–web compression member in accordance with Clause 6.3.3 of AS 4100. The effective cross-sectional area of the stiffener–web strut is taken as the area of the stiffener plus the web area having an effective width on each side of the centreline of the stiffener considered as schematically depicted in Figure 4.16. The effective width of the web (b_{ew}) as part of the stiffener–web compression member is taken as

$$b_{ew} = \min\left(\frac{17.5t_w}{\sqrt{f_y/250}}, \frac{s}{2}\right) \tag{4.47}$$

where

t_w is the thickness of the web
s is the web panel width or spacing of the stiffeners

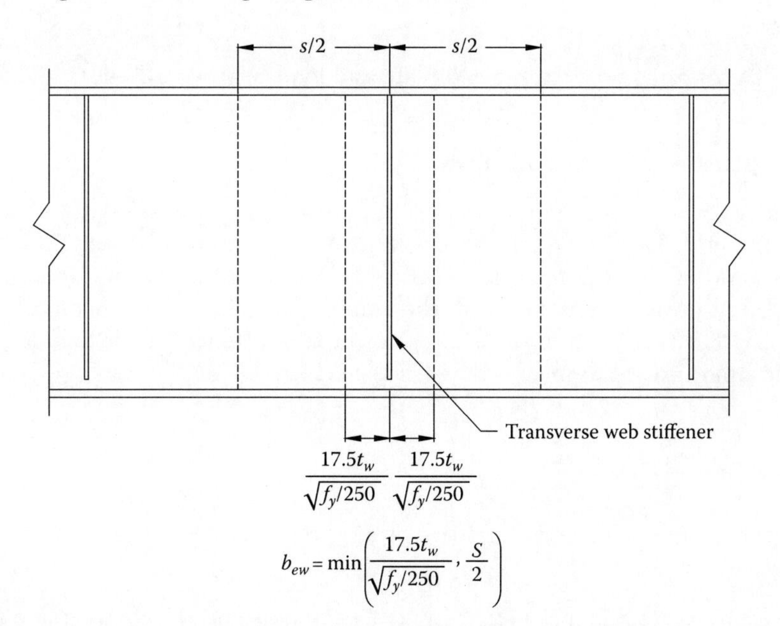

Figure 4.16 Effective width of the web as part of stiffener–web compression member.

The effective length of the stiffener–web strut is taken as d_1. The second moment of area of the stiffener–web section is calculated about the axis parallel to the web. The slenderness reduction factor α_c is determined by taking $\alpha_b = 0.5$ and $k_f = 1.0$ in accordance with Clause 6.3.3 of AS 4100.

The design buckling capacity of the web–stiffener compression member is expressed by

$$\phi R_{sb} = \phi \alpha_c A_{ws} f_y \tag{4.48}$$

where A_{ws} is the cross-sectional area of the stiffener–web compression member.

For an intermediate web stiffener that is not subjected to external loads or moments, the minimum second moment of area (I_s) about the centreline of the web is given in Clause 5.15.5 of AS 4100 as

$$I_s \geq \begin{cases} 0.75 d_1 t_w^3 & \text{for } s/d_1 \leq \sqrt{2} \\ 1.5 d_1^3 t_w^3 / s^2 & \text{for } s/d_1 > \sqrt{2} \end{cases} \tag{4.49}$$

At the end of a plate girder, the end stiffener must resist the horizontal component of the tension field in the end panel. To avoid this, the length of the end panel can be reduced so that the contribution of the tension field to the ultimate stress is not required. This can be achieved by designing the end panel with $\alpha_d = 1.0$ in Equation 4.36. Alternatively, an end post consisting of a load-bearing stiffener and a parallel end plate can be used to transfer the tension field action on the end of a plate girder as illustrated in Figure 4.15. Clause 5.15.9 of AS 4100 requires that the area of the end plate must satisfy the following condition:

$$A_{ep} \geq \frac{d_1 (V^*/\phi - \alpha_v V_w)}{8 s_{ep} f_y} \tag{4.50}$$

where s_{ep} is the distance between the end plate and load-bearing stiffener.

4.6.5 Longitudinal web stiffeners

Longitudinal web stiffeners are continuous and attached to the transverse web stiffeners to increase the effectiveness of the web in resisting shear and bending. When the depth-to-thickness ratio of the web $(d_1/t_w)\sqrt{f_y/250}$ is greater than 200, a first longitudinal stiffener is needed at a distance $0.2d_2$ from the compression flange as schematically depicted in Figure 4.15. The second moment of area (I_s) of this stiffener about the face of the web (Trahair and Bradford 1998) must satisfy Clause 5.16.2 of AS 4100 as

$$I_s \geq 4 d_2 t_w^3 \left[1 + \frac{4A_s}{d_2 t_w} \left(1 + \frac{A_s}{d_2 t_w} \right) \right] \tag{4.51}$$

where

d_2 is twice the clear distance between the neutral axis and the compression flange
A_s is the stiffener area (Bradford 1987, 1989)

When the depth-to-thickness ratio of the web $(d_1/t_w)\sqrt{f_y/250}$ is greater than 250, a second longitudinal stiffener is required at the neutral axis of the section, and its second moment of area about the face of the web must satisfy

$$I_s \geq d_2 t_w^3 \tag{4.52}$$

Example 4.4: Design of a stiffened plate girder web for shear

The cross section of a plate girder of Grade 300 steel is shown in Figure 4.17. Intermediate transverse web stiffeners 100 × 14 mm of Grade 300 steel are spaced at 1500 mm. The width of the end panel is 1200 mm. There is no end post. The flanges of the plate girder are restrained by other structural members against rotation. A pair of load-bearing stiffeners 100 × 14 mm is used above the support of the plate girder, which is supported by stiff bearing 200 mm long. The design reaction is 1200 kN. The yield stresses of the web and stiffeners are f_y = 310 MPa and f_{ys} = 300 MPa, respectively:

a. Determine the design shear capacity of the web.
b. Determine the design shear capacity of the end panel.
c. Check the adequacy of the intermediate transverse web stiffeners.

a. Shear capacity of the stiffened web

1. Slenderness of the web

The dimensions of the steel I-section are

$$b_f = 350\,\text{mm}, \quad t_f = 20\,\text{mm}, \quad d = 1200\,\text{mm}, \quad t_w = 10\,\text{mm}$$

The slenderness of the web is

$$\lambda_{ew} = \frac{d_p}{t_w}\sqrt{\frac{f_y}{250}} = \frac{(d-2t_f)}{t_w}\sqrt{\frac{f_y}{250}} = \frac{(1200-2\times 20)}{10}\sqrt{\frac{310}{250}} = 129.2 > 82$$

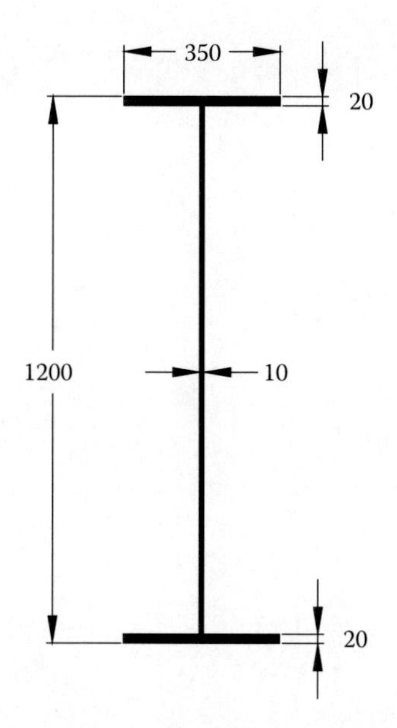

Figure 4.17 Section of plate girder.

The web of the plate girder subjected to uniform shear stresses will undergo shear buckling. Thus, $V_u = V_b$.

Since $\lambda_{ew} = 129.2 < 200$, the longitudinal web stiffener is not required.

2. Partial factors

The spacing-to-depth ratio is

$s/d_p = 1500/(1200 - 2\times 20) = 1.29 < 3.0$; the web is treated as a stiffened one.

The stiffening factor α_v is calculated as

$$\alpha_v = \left[\frac{82}{(d_p/t_w)\sqrt{f_y/250}}\right]^2 \left[\frac{0.75}{(s/d)^2} + 1.0\right] = \left(\frac{82}{129.2}\right)^2 \left(\frac{0.75}{1.29^2} + 1.0\right) = 0.584 < 1.0$$

The tension field contribution factor α_d is determined as

$$\alpha_d = 1 + \frac{1-\alpha_v}{1.15\alpha_v\sqrt{1+(s/d_p)^2}} = 1 + \frac{1-0.584}{1.15\times 0.584\times\sqrt{1+1.29^2}} = 1.38$$

The width of the flange outstand is $b_{fo} = (b_f - t_w)/2 = (350-10)/2 = 170$.

The flange restraint factor α_f can be computed as

$$\alpha_f = 1.6 - \frac{0.6}{\sqrt{\left(1+\left(40b_{fo}t_f^2/d_1^2t_w\right)\right)}} = 1.6 - \frac{0.6}{\sqrt{\left(1+(40\times 170\times 20^2/1160^2\times 10)\right)}} = 1.053$$

3. Shear capacity of the web

The shear yield capacity of the web is

$$V_w = 0.6f_yA_w = 0.6\times 310\times(1200 - 2\times 20)\times 10\,\text{N} = 2157.6\,\text{kN}$$

The nominal shear buckling capacity of the stiffened web is computed as

$$V_b = \alpha_v\alpha_d\alpha_f V_w = 0.584\times 1.38\times 1.053\times 2157.6\,\text{kN} = 1831\text{ kN} < V_w = 2157.6\,\text{kN}$$

The design shear capacity of the web is determined as

$$\phi V_u = 0.9\times 1831 = 1648\,\text{kN}$$

b. Shear capacity of the end panel

1. Slenderness of the end panel

The slenderness of the end panel is the same as that of the web, $\lambda_{ew} = 129.2 > 82$. The end panel will undergo shear buckling before yielding; take $V_u = V_b$.

2. Partial factors

The width of the end panel s is 1200 mm. The spacing-to-depth ratio of the end panel is

$$\frac{s}{d_p} = \frac{1200}{1200 - 2\times 20} = 1.034 < 3.0$$

The stiffening factor α_v for the end panel is calculated as

$$\alpha_v = \left[\frac{82}{(d_p/t_w)\sqrt{f_{sy}/250}}\right]^2 \left[\frac{0.75}{(s/d)^2}+1.0\right] = \left(\frac{82}{129.2}\right)^2 \left(\frac{0.75}{1.034^2}+1.0\right) = 0.685 < 1.0$$

The width of the end panel has been reduced from 1500 mm stiffener spacing to 1200 mm and the end post is not required by taking α_d = 1.0. The flange restraint factor has been calculated as α_f = 1.053.

3. Shear capacity of the web

The nominal shear buckling capacity of the end panel is calculated as

$$V_b = \alpha_v \alpha_d \alpha_f V_w = 0.685 \times 1.0 \times 1.053 \times 2157.6\,\text{kN} = 1556.3\,\text{kN} < V_w = 2157.6\,\text{kN}$$

The design shear capacity of the end panel can be determined by

$$\phi V_u = 0.9 \times 1556.3 = 1401\,\text{kN}$$

c. Intermediate transverse stiffeners

1. Minimum area of transverse web stiffener

The dimensions and properties of the transverse stiffeners are

$$b_s = 100\,\text{mm}, \quad t_s = 14\,\text{mm}, \quad f_{ys} = 300\,\text{MPa}, \quad \gamma_w = 1.0\ \text{(for a pair of stiffeners)}$$

$$A_s = 2 \times 100 \times 14 = 2800\,\text{mm}^2, \quad s/d_p = 1.29 < 3.0$$

Taking $V^* = \phi V_u$, the minimum area of the intermediate web stiffener is calculated as

$$A_s \geq 0.5\gamma_w(1-\alpha_v)\left(\frac{V^*}{\phi V_u}\right)\left[\frac{s}{d_p} - \frac{(s/d_p)^2}{\sqrt{1+(s/d_p)^2}}\right] A_w$$

$$\geq 0.5 \times 1.0 \times (1-0.584) \times 1.0 \times \left[1.29 - \frac{1.29^2}{\sqrt{1+1.29^2}}\right] \times 1160 \times 10$$

$$\geq 652\,\text{mm}^2 < A_s = 2800\,\text{mm}^2,\ \text{OK}$$

2. Section properties of the stiffener–web compression member

The effective length of the stiffened web on each side of the stiffener is

$$b_{ew} = \frac{17.5 t_w}{\sqrt{f_y/250}} = \frac{17.5 \times 10}{\sqrt{310/250}} = 157.2\,\text{mm} < \frac{s}{2} = \frac{1500}{2} = 750\,\text{mm}$$

Take b_{ew} = 157.2 mm.

The section properties of the web–stiffener compression member are

$$A_{ws} = 2 \times 157.2 \times 10 + 2 \times 100 \times 14 = 5943\,\text{mm}^2$$

$$I_{ws}=2\left(\frac{b_{ew}t_w^3}{12}\right)+\frac{t_s(2b_s+t_w)^3}{12}=2\left(\frac{154.7\times10^3}{12}\right)+\frac{14\times(2\times100+10)^3}{12}=10.83\times10^6\text{ mm}^6$$

$$r_{ws}=\sqrt{\frac{I_{ws}}{A_{ws}}}=\sqrt{\frac{10.83\times10^6}{5943}}=42.7$$

3. Slenderness reduction factor

The effective length of the stiffener–web compression member is

$$L_e=d_1=1200-2\times20=1160\text{ mm}$$

The modified slenderness of the web–stiffener can be calculated as

$$\lambda_n=\frac{L_e}{r_{ws}}\sqrt{k_f}\sqrt{\frac{f_y}{250}}=\frac{1160}{42.7}\sqrt{1.0}\sqrt{\frac{300}{250}}=29.8$$

From Table 5.3 of Chapter 5 with α_b = 0.5 and k_f = 1.0, the slenderness reduction factor can be obtained as α_c = 0.918.

4. Buckling capacity of the stiffener–web compression member

The nominal buckling capacity of the web–stiffener strut is calculated as

$$R_{sb}=\alpha_c A_{ws} f_y=0.918\times5943\times300\text{ N}=1637\text{ kN}$$

The design buckling capacity of the stiffener–web compression member can be determined as

$$\phi(R_{sb}+V_b)=0.9\times(1637+1831)=3121.2\text{ kN}>\phi V_u=1648\text{ kN, OK}$$

For $s/d_1=1.29<\sqrt{2}$, the required minimum second moment of area of the transverse web stiffener is

$$I_s\geq0.75d_1t_w^3=0.75\times1160\times10^3=0.87\times10^6\text{ mm}^6<I_{ws}=10.83\times10^6\text{ mm}^6\text{, OK}$$

4.7 BEARING CAPACITY OF WEBS

4.7.1 Yield capacity of webs in bearing

Concentrated loads or locally distributed loads on the top flange of a steel beam and reactions on the supports induce bearing stresses in the web as schematically depicted in Figure 4.18. These bearing stresses may cause yielding or buckling of the web. Therefore, the web in bearing must be designed for yielding and buckling limit states. The web of a steel beam in bearing must satisfy the following strength requirement:

$$R^*\leq\phi R_b \tag{4.53}$$

where

R^* is the design bearing force on the web

$\phi=0.9$ is the capacity reduction factor

R_b is the nominal bearing capacity of the web, which is taken as the lesser of its nominal bearing yield capacity (R_{by}) and bearing buckling capacity (R_{bb})

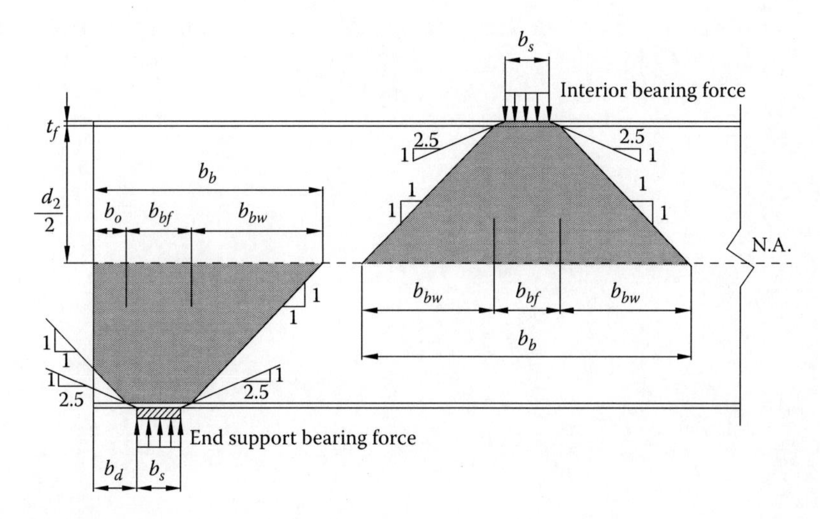

Figure 4.18 Bearing force dispersions in the flanges and web of a steel I-beam.

Stocky webs can attain their bearing yield capacities as the elastic buckling of the webs will not occur. Clause 5.13.3 of AS 4100 (1998) suggests that the nominal bearing yield capacity of a web should be calculated by

$$R_{by} = 1.25 b_{bf} t_w f_y \tag{4.54}$$

in which b_{bf} is the bearing width of the flange of the I-section beam as depicted in Figure 4.18 and is determined as the lesser of the following calculated values:

$$b_{bf} = b_s + 5t_f \tag{4.55}$$

$$b_{bf} = b_s + 2.5t_f + b_d \tag{4.56}$$

where b_d is the remaining distance to the end of the beam as shown in Figure 4.18.

In Clause 5.13.3 of AS 4100, the nominal bearing yield capacity of both webs in square and RHSs is determined by

$$R_{by} = 2 b_b t \alpha_p f_y \tag{4.57}$$

where

b_b is the bearing width and is taken as $b_b = b_s + 5r_e + d_5$, where r_e is the outside radius of the section and d_5 is the flat width of the web

t is the thickness of the hollow section

α_p is a reduction factor which is different for interior bearing and end bearing (Zhao et al. 1996) and is given as follows:

For interior bearing with $b_d \geq 1.5d_5$, α_p is given by

$$\alpha_p = \frac{1}{2k_s}\left[1 + \left(1 - \alpha_{pm}^2\right)\left(1 + \frac{k_s}{k_v} - \left(1 - \alpha_{pm}^2\right)\frac{1}{4k_v^2}\right)\right] \tag{4.58}$$

in which d_5 is the flat width of web and is taken as $d_5 = d - 2r_e$, and the coefficient α_{pm} and ratios k_s and k_v are given by

$$\alpha_{pm} = \frac{1}{k_s} + \frac{1}{2k_v} \tag{4.59}$$

$$k_s = \frac{2r_e}{t} - 1 \tag{4.60}$$

$$k_v = \frac{d_5}{t} \tag{4.61}$$

For end bearing with $b_d < 1.5d_5$, α_p is given by

$$\alpha_p = \sqrt{2 + k_s^2} - k_s \tag{4.62}$$

The bearing width of end bearing is calculated as

$$b_b = b_s + 2.5r_e + \frac{d_5}{2} \tag{4.63}$$

4.7.2 Bearing buckling capacity of webs

The buckling capacity of an unstiffened web under bearing stresses is determined as the axial load capacity of an equivalent compression member with an area taken as $A_w = b_b t_w$ and a slenderness ratio (L_e/r) taken as $(2.5d_1)/t_w$. The total bearing width of the web (b_b) is obtained by dispersions at a slope of 1:1 from b_{bf} to the neutral axis as illustrated in Figure 4.18. For end bearing, the total bearing width is given by

$$b_b = b_o + b_{bf} + b_{bw} \tag{4.64}$$

where $b_o = b_d - 2.5t_f$, $b_{bf} = b_s + 5t_f$ and $b_{bw} = d_2/2$. For interior bearing, b_o in Equation 4.64 is replaced by b_{bw} as shown in Figure 4.18.

For square and RHSs, however, the slenderness ratio (L_e/r) is taken as $(3.5d_5)/t_w$ for interior bearing with $b_d \geq 1.5d_5$ and equals to $(3.8d_5)/t_w$ for end bearing with $b_d < 1.5d_5$.

The slenderness reduction factor α_c is determined by taking $\alpha_b = 0.5$ and $k_f = 1.0$ in accordance with Clause 6.3.3 of AS 4100, and the bearing buckling capacity of the web can be calculated by

$$\phi R_{bb} = \phi \alpha_c A_w f_y \tag{4.65}$$

4.7.3 Webs in combined bearing and bending

The ultimate strengths of square and RHS beams under combined bearing and bending are influenced by the interaction between bearing and bending. The presence of bending moment reduces the bearing strength, while the presence of bearing force reduces the bending strength of the sections. Interaction equations are given in Clause 5.13.5 of AS 4100 (1998) for determining the capacities of square and RHSs under combined bearing and bending (Zhao et al. 1996):

$$1.2\left(\frac{R^*}{\phi R_b}\right)+\left(\frac{M^*}{\phi M_s}\right)\leq 1.5 \quad \text{for } \frac{b_s}{b}\geq 1.0 \text{ and } \frac{d_1}{t_w}\leq 30 \tag{4.66}$$

$$0.8\left(\frac{R^*}{\phi R_b}\right)+\left(\frac{M^*}{\phi M_s}\right)\leq 1.0 \quad \text{for } \frac{b_s}{b}< 1.0 \text{ and } \frac{d_1}{t_w}> 30 \tag{4.67}$$

where

$\phi = 0.9$ and b is the total width of the section
M_s is the nominal section moment capacity

4.7.4 Load-bearing stiffeners

When the web has insufficient capacity to withstand the imposed concentrated loads, it may be strengthened by welding bearing stiffeners to the web adjacent to the loads as depicted in Figure 4.15. The design rules for load-bearing stiffeners are provided in Clause 5.14 of AS 4100, which specifies that the outstands of the stiffener from the face of the web must satisfy the following condition:

$$b_{es}\leq \frac{15t_s}{\sqrt{f_{ys}/250}} \tag{4.68}$$

where

t_s is the thickness of the stiffener
f_{ys} is the yield stress of the stiffener

The load-bearing stiffener and part of the web in the vicinity of the stiffener considered are treated as a compression member. The load-bearing stiffener–web compression member must be checked for its yield and buckling capacities against the design bearing force or design reaction (R^*) acting on the bearing stiffener as follows:

$$R^*\leq \phi R_{sy} \tag{4.69}$$

$$R^*\leq \phi R_{sb} \tag{4.70}$$

where $\phi = 0.9$ and R_{sy} is the yield capacity of the stiffener–web compression member, which is given in Clause 5.14.1 of AS 4100 (1998) as

$$R_{sy} = R_{by} + A_s f_{ys} \tag{4.71}$$

where

R_{by} is the bearing yield capacity of the web given in Equation 4.54
A_s is the cross-sectional area of the stiffener
f_{ys} is the yield stress of the stiffener

The design buckling capacity of the web and load-bearing stiffener as a whole (ϕR_{sb}) is given by Equation 4.48.

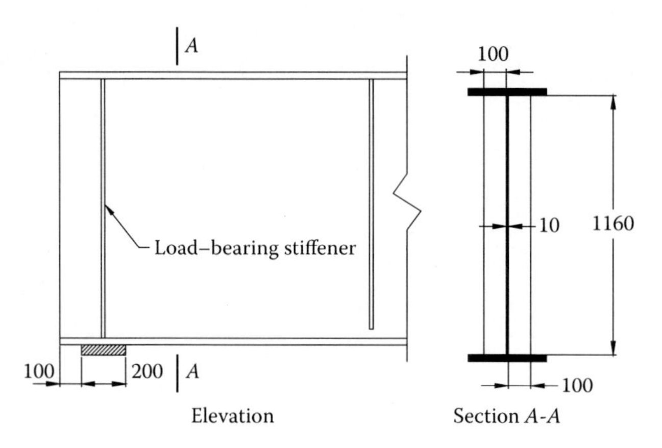

Figure 4.19 Steel plate girder.

Example 4.5: Design of a stiffened plate girder web for bearing

The steel plate girder presented in Example 4.4 is to be designed for bearing at the supports as shown in Figure 4.19. There is no end post. A pair of load-bearing stiffeners 100 × 14 mm is to be used above the support of the plate girder, which is supported by stiff bearing 200 mm long. The distance from the support to the end of the beam is 100 mm as illustrated in Figure 4.19. The design reaction is 1200 kN. The yield stresses of the web and stiffeners are f_y = 310 MPa and f_{ys} = 300 MPa, respectively:

a. Check whether the load-bearing stiffeners are required at the supports.
b. Check the adequacy of the load-bearing stiffeners at the supports.

a. Bearing capacity of the web without load-bearing stiffeners

1. Bearing yield capacity of the web

The bearing width of the flange is calculated as

$$b_{bf} = b_s + 5t_f = 200 + 5 \times 20 = 300\,\text{mm}$$

$$b_{bf} = b_s + 2.5t_f + b_d = 200 + 2.5 \times 20 + 100 = 350\,\text{mm}$$

Take b_f = 300 mm.

The nominal bearing yield capacity is determined by

$$R_{by} = 1.25 b_{bf} t_w f_y = 1.25 \times 300 \times 10 \times 310\,\text{N} = 1162.5\,\text{kN}$$

The design bearing yield capacity is

$$\phi R_{by} = 0.9 \times 1162.5\,\text{kN} = 1046.3\,\text{kN} < R^* = 1200\,\text{kN, NOT OK!}$$

Load-bearing stiffeners are required in the web at the supports to transfer the bearing force.

2. Bearing buckling capacity

The total bearing width at the support can be determined as

$$b_b = b_o + b_{bf} + b_{bw} = b_d - 2.5t_f + b_{bf} + (d - 2t_f)/2$$

$$= 100 - 2.5 \times 20 + 300 + (1200 - 2 \times 20)/2 = 930\,\text{mm}$$

The cross-sectional area of the web is $A_w = b_b t_w = 930 \times 20 = 18{,}600\,\text{mm}^2$.

The effective section of the web is treated as column section. The modified slenderness ratio is

$$\lambda_n = \frac{L_e}{r}\sqrt{k_f}\sqrt{\frac{f_y}{250}} = \frac{2.5d_1}{t_w}\sqrt{k_f}\sqrt{\frac{f_y}{250}} = \frac{2.5 \times 1160}{20}\sqrt{1.0}\sqrt{\frac{310}{250}} = 161.5$$

Taking $\alpha_b = 0.5$, the slenderness reduction factor can be obtained as $\alpha_c = 0.242$.

The bearing buckling capacity of the web compression member is

$$\phi R_{bb} = \phi \alpha_c A_w f_y = 0.9 \times 0.242 \times 18{,}600 \times 310 = 1{,}255.8\,\text{kN} > R^* = 1{,}200\,\text{kN, OK}$$

b. Load-bearing stiffener

1. Bearing yield capacity of the stiffener–web compression member

The bearing length at the junction of the web and flange is

$$b_{bf} = b_s + 5t_f = 200 + 5 \times 20 = 300\,\text{mm}$$

The nominal yield capacity of the stiffener–web compression member can be computed as

$$R_{sy} = R_{by} + A_s f_{ys} = 1.25 b_{bf} t_w f_y + A_s f_{ys}$$

$$= 1.25 \times 300 \times 10 \times 310 + 2800 \times 300\,\text{N} = 2002.5\,\text{kN}$$

The design yield capacity of the stiffener–web member is

$$\phi R_{sy} = 0.9 \times 2002.5 = 1802\,\text{kN} > R^* = 1200\,\text{kN, OK}$$

2. Buckling capacity of the stiffener–web compression member

The nominal buckling capacity of the stiffener–web compression member has been determined in Example 4.4 as $R_{sb} = 1637\,\text{kN}$. The design buckling capacity of the member is

$$\phi R_{sb} = 0.9 \times 1637 = 1473.3\,\text{kN} > R^* = 1200\,\text{kN, OK}$$

4.8 DESIGN FOR SERVICEABILITY

The design of steel beams for serviceability needs to check for deflections, bolt slips or vibrations. In service conditions, it is required to check for the deflections of the steel beams under service loads defined in Section 2.5.3. Under service loads, steel beams are usually assumed to behave elastically. Therefore, the first-order linear elastic analysis can be performed to determine the deflections of steel beams under service loads. For this purpose, modern interactive computer software such as Strand7, Multiframe and Space Gass can be used.

Deflection limits on steel beams are given in AS 4100 as follows:

1. For all beams, the total deflection is limited to *L*/250 for spans and *L*/125 for cantilever.
2. For beams supporting masonry partitions, the incremental deflection, which occurs after the attachment of partitions, is limited to *L*/500 for spans and *L*/250 for cantilever where provision is provided to reduce the effect of movement; otherwise, the incremental deflection is limited to *L*/1000 for spans and *L*/500 for cantilever.

REFERENCES

AS 4100 (1998) *Australian Standard for Steel Structures*, Sydney, New South Wales, Australia: Standards Australia.

Bradford, M.A. (1987) Inelastic local buckling of fabricated I-beams, *Journal of Constructional Steel Research*, 7: 317–334.

Bradford, M.A. (1989) Buckling of longitudinally stiffened plates in bending and compression, *Canadian Journal of Civil Engineering*, 16 (5): 607–614.

Bradford, M.A. and Trahair, N.S. (1983) Lateral stability of beam on seats, *Journal of Structural Engineering*, ASCE, 109 (9): 2212–2215.

Kitipornchai, S. and Trahair, N.S. (1980) Buckling properties of monosymmetric I-beams, *Journal of the Structural Division*, ASCE, 106 (ST5): 941–957.

Kollbrunner, C.F. and Basler, K. (1969) *Torsion in Structures*, 2nd edn., Berlin, Germany: Springer-Verlag.

Timoshenko, S.P. and Gere, J.M. (1961) *Theory of Elastic Stability*, 2nd edn., New York: McGraw-Hill.

Trahair, N.S. (1993a) *Flexural-Torsional Buckling of Structures*, London, U.K.: Spon Press.

Trahair, N.S. (1993b) Design of unbraced cantilevers, *Steel Construction*, Australian Institute of Steel Construction, 27 (3): 2–10.

Trahair, N.S. and Bradford, M.A. (1998) *The Behaviour and Design of Steel Structures to AS 4100*, 3rd edn. (Australian), London, U.K.: Taylor & Francis Group.

Trahair, N.S., Hogan, T.J. and Syam, A. (1993) Design of unbraced beams, *Steel Construction*, Australian Institute of Steel Construction, 27 (1): 2–26.

Zhao, X.L., Hancock, G.J. and Sully, R. (1996) Design of tubular members and connections using amendment number 3 to AS 4100, *Steel Construction*, Australian Institute of Steel Construction, 30 (4): 2–15.

Chapter 5

Steel members under axial load and bending

5.1 INTRODUCTION

Members in steel trusses under point loads at joints are subjected to either axial compression or axial tension. In contrast, members in steel frames may be subjected to the combined axial load and bending, which may be caused by lateral loads, eccentric loading or frame actions. The axial load and bending may include the combined actions of axial load and uniaxial bending and of axial load and biaxial bending. Members under compressive axial load and bending are regarded as beam–columns, which combine the functions of beams and columns.

This chapter deals with the behaviour, analysis and design of steel members under axial load and bending in accordance with AS 4100. The behaviour and design of steel members in axial compression are described first. This is followed by the discussions of the design of members in axial tension. The behaviour and design of steel members under combined actions of axial load and uniaxial bending are then presented, including methods for calculating the section moment capacity reduced by axial forces, in-plane member capacity and out-of-plane member capacity. In Section 5.6, the analysis and design of steel members under combined actions of axial load and biaxial bending are given in detail.

5.2 MEMBERS UNDER AXIAL COMPRESSION

5.2.1 Behaviour of members in axial compression

The behaviour of a steel member in axial compression depends on its material properties, section slenderness and member slenderness, initial geometric imperfections and residual stresses. The design of a very stocky member is governed by its section capacity, which depends on the yield stress, slenderness and residual stresses of the cross section. For a compression member mode of slender steel elements, local buckling may occur before steel yields. Local buckling may significantly reduce the ultimate axial section capacity of steel members and must be taken into account in design. Residual stresses induced by hot rolling or welding may cause a significant reduction in the axial section capacity due to premature yielding.

The ultimate strength of an axially loaded steel member decreases with an increase in its length. This is caused by the applied axial load which induces bending actions and lateral deflections in the member with initial geometric imperfections. The lateral deflections and bending actions of the member increase with increasing the member slenderness, which leads to a decrease in the strength of the member. An elastic steel member without initial imperfections will not deflect until the applied axial compressive force reaches its elastic buckling load, which is called the Euler buckling load. The elastic buckling load gives an indication of the slenderness of an axially loaded member, while the squash load reflects on its resistance to yielding and local buckling.

Practical compression members usually have initial imperfections which include geometric imperfections, residual stresses and initial loading eccentricity. These initial imperfections reduce the strengths of intermediate and slender compression members below the elastic buckling loads of the members. The load–deflection behaviour of practical compression members is nonlinear inelastic. The strengths of steel members in axial compression are found to decrease with an increase in the initial imperfections. Geometric imperfections are always present in steel members because it is difficult to manufacture a steel column with an initial geometric imperfection less than $L/1000$ at its mid-length. The effect of initial geometric imperfection of $L/1000$ at the mid-length of steel columns has been taken into account in the design codes.

5.2.2 Section capacity in axial compression

Axially loaded steel members composed of slender plate elements may buckle locally before the ultimate axial load is attained. The effect of local buckling on the section capacity of compression members is taken into account by the section form factor (k_f), which was discussed by Rasmussen et al. (1989). In Clause 6.2 of AS 4100 (1998), the nominal section capacity of a steel member subject to axial compression is expressed by

$$N_s = k_f A_n f_y \tag{5.1}$$

in which A_n is the net area of the cross section taking as $A_n = A_g - \sum d_h t$, where A_g is the gross cross-sectional area, d_h is the diameter of a hole and t is the thickness of the member at the hole. For sections with unfilled holes or penetrations that reduce the section area by less than $100\{1-[f_y/(0.85f_u)]\}\%$, A_n is taken as the gross area (A_g). The form factor k_f is expressed by

$$k_f = \frac{A_e}{A_g} \tag{5.2}$$

where A_e is the effective cross-sectional area of the section as given in Chapter 3.

5.2.3 Elastic buckling of compression members

The elastic buckling load of a perfectly straight pin-ended member under axial compression as depicted in Figure 5.1 can be determined by ascertaining the deflected equilibrium position, which is defined by the displacement function as follows:

$$u = u_m \sin\left(\frac{\pi z}{L}\right) \tag{5.3}$$

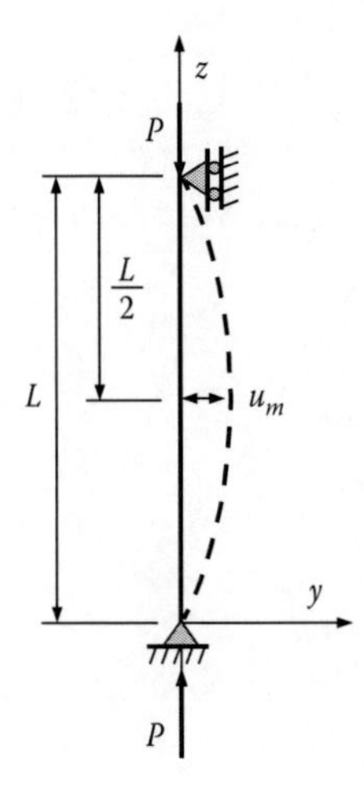

Figure 5.1 Pin-ended compression member.

in which u_m is the deflection at the mid-length of the member. The elastic buckling load or the Euler buckling load (Timoshenko and Gere 1961; Bulson 1970) can be obtained as

$$P_{cr} = \frac{\pi^2 E_s I}{L^2} \tag{5.4}$$

where
- I is the second moment of area of the column cross section about the principal axis
- L is the member length

The elastic buckling load can be expressed by the column slenderness ratio (L/r) as

$$P_{cr} = \frac{\pi^2 E_s A}{(L/r)^2} \tag{5.5}$$

where
- A is the cross-sectional area
- $r = \sqrt{I/A}$ is the radius of gyration

The elastic buckling stress can be determined as

$$\sigma_{cr} = \frac{P_{cr}}{A} = \frac{\pi^2 E_s}{(L/r)^2} \tag{5.6}$$

It should be noted that Equation 5.5 is valid only for perfectly straight pin-ended members without residual stresses and loaded at the centre of gravity. It can be seen from Figure 5.2 that Equation 5.5 overestimates the capacity of compression columns with

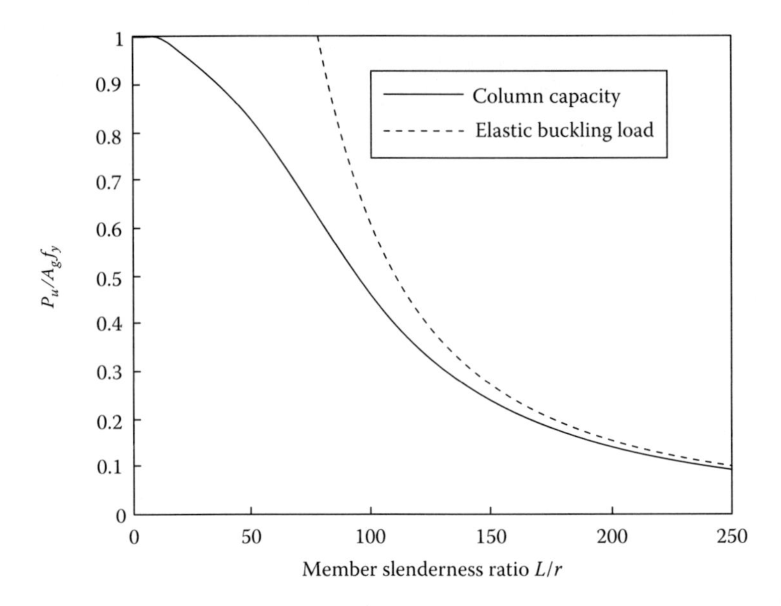

Figure 5.2 Capacity of compression members.

a slenderness ratio less than 200 so that it cannot be used to calculate the capacity of intermediate length columns.

In general, the elastic buckling load of compression members with end restraints can be expressed by

$$P_{cr} = \frac{\pi^2 E_s I}{(k_e L)^2} \tag{5.7}$$

where

k_e is the member effective length factor

$L_e = k_e L$ is the effective length of a compression member, which is the unsupported distance between the zero moment points

The member effective length factor (k_e) depends on the translational and rotational restraints at the ends of the member. For members with idealised end restraints, the values of k_e are given in Figure 5.3 as provided in AS 4100. For braced compression members in a steel frame with rigid connections, the effective length factor (k_e) can be determined from the following equation (Duan and Chen 1988; Trahair and Bradford 1998):

$$\frac{\gamma_1\gamma_2}{4}\left(\frac{\pi}{k_e}\right)^2 + \left(\frac{\gamma_1+\gamma_2}{2}\right)\left[1-\frac{\pi}{k_e}\cot\left(\frac{\pi}{k_e}\right)\right] + \frac{\tan(\pi/2k_e)}{(\pi/2k_e)} - 1 = 0 \tag{5.8}$$

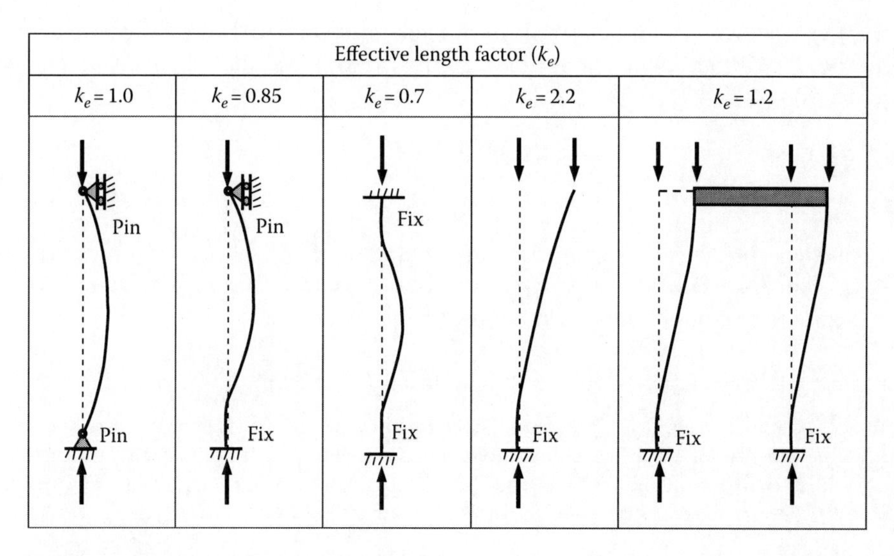

Figure 5.3 Effective length factors for idealised columns. (Adapted from AS 4100, Australian standard for steel structures, Standards Australia, Sydney, New South Wales, Australia, 1998.)

where γ_1 and γ_2 denote the stiffness ratios of a compression member at end 1 and end 2, respectively. The stiffness ratio of a compression member in a rectangular frame with negligible axial forces in beams is given in Clause 4.6.3.4 of AS 4100 as

$$\gamma_i = \frac{\sum (I/L)_c}{\sum \beta_e (I/L)_b} \tag{5.9}$$

where

$\sum (I/L)_c$ is the sum of the stiffness in the plane of bending of all compression members rigidly connected at the end of the member considered, including the member itself

$\sum (I/L)_b$ is the sum of the stiffness in the plane of bending of all beams rigidly connected at the end of the member considered

The stiffness of any beams pin-connected to the member is not considered. The γ_i value of a compression member that is not rigidly connected to a footing should be taken as greater than or equal to 10. For a compression member that is rigidly connected to a footing, the γ_i value should be taken as greater than or equal to 0.6.

The modifying factor (β_e) is used to account for the conditions at the far ends of the beam, which is given in Clause 4.6.3.4 of AS 4100 (1998) as follows:

1. If the far end of the beam is pinned, β_e = 1.5 when the beam restrains a braced member and β_e = 0.5 when the beam restrains a sway member.
2. If the far end of the beam is rigidly connected to a column, β_e = 1.0 when the beam restrains a braced member or a sway member.
3. If the far end of the beam is fixed, β_e = 2.0 when the beam restrains a braced member and β_e = 0.67 when the beam restrains a sway member.

For a sway compression member in an unbraced rigid-jointed frame, the effective length factor can be determined from the following equation (Duan and Chen 1989; Trahair and Bradford 1998):

$$\frac{\gamma_1\gamma_2(\pi/k_e)^2-36}{6(\gamma_1+\gamma_2)}-\frac{\pi}{k_e}\cot\left(\frac{\pi}{k_e}\right)=0 \tag{5.10}$$

The effective length factors for braced members and sway members in frames are given in Figure 4.6.3.3 in AS 4100. The effective length is also used in the calculation of the member capacity of practical compression members with imperfections.

Example 5.1: Calculation of effective length factors for columns in frame

Figure 5.4 shows a rigid-jointed plane steel frame whose base is fixed to the foundation. The out-of-plane behaviour of the frame is prevented. All beams are subjected to negligible axial forces. The sections and their properties used in the analysis are given in Table 5.1. Determine the effective length factors for all columns.

1. Column 1-4

The base of column 1-4 is fixed; the stiffness ratio of the column at end 1 can be taken as $\gamma_1 = 0.6$ according to Clause 4.6.4.4(a) of AS 4100.

At column end 4:

The far end of beam 4-5 is rigidly connected to a column, and the beam 4-5 restrains a sway column 1-4; thus, $\beta_e = 1.0$.

The stiffness ratio of column 1-4 at end 4 can be calculated as

$$\gamma_4=\frac{\sum(I/L)_c}{\sum(\beta_e I/L)_b}=\frac{61.3/4.0+61.3/3.6}{1.0\times142/6.5}=1.481$$

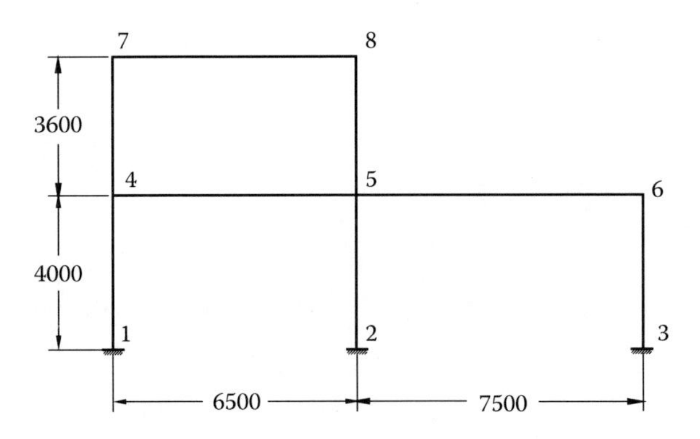

Figure 5.4 Rigid-jointed plane steel frame.

Table 5.1 Section properties

Member	*Section*	I_x*(mm⁴)*
1-4, 4-7	200UC59.5	61.3×10^6
2-5, 5-8	250UC72.9	114×10^6
3-6	200UC46.2	45.9×10^6
Beams	360UB50.7	142×10^6

The effective length factor can be obtained by solving Equation 5.10 or from Figure 4.6.3.3 in AS 4100 as $k_e = 1.321$.

2. Column 5-8

At column end 7:

The far end of beam 7-8 is rigidly connected to a column, and the beam 7-8 restrains a sway column 4-7; thus, $\beta_e = 1.0$.

The stiffness ratio of column 4-7 at the end 7 can be calculated as

$$\gamma_7 = \frac{\sum (I/L)_c}{\sum (\beta_e I/L)_b} = \frac{61.3/3.6}{1.0 \times 142/6.5} = 0.779$$

The effective length factor for column 4-7 with $\gamma_4 = 1.481$ and $\gamma_7 = 0.779$ can be obtained by solving Equation 5.10 or from Figure 4.6.3.3 in AS 4100 as $k_e = 1.35$.

3. Column 2-5

The base of column 2-5 is fixed; its stiffness ratio at end 2 is $\lambda_1 = 0.6$ according to Clause 4.6.4.4(a) of AS 4100.

At column end 5:

The far ends of beam 4-5 and beam 5-6 are rigidly connected to columns, and these two beams restrain a sway column 2-5. This gives $\beta_e = 1.0$.

The stiffness ratio of column 2-5 at end 5 is computed as

$$\gamma_5 = \frac{\sum (I/L)_c}{\sum (\beta_e I/L)_b} = \frac{114/4.0 + 114/3.6}{1.0 \times 142/6.5 + 1.0 \times 142/7.5} = 1.475$$

The effective length factor can be obtained by solving Equation 5.10 or from Figure 4.6.3.3 in AS 4100 as $k_e = 1.321$.

4. Column 5-8

At column end 8:

The far end of beam 7-8 is rigidly connected to a column, and beam 7-8 restrains a sway column 5-8, which gives $\beta_e = 1.0$.

The stiffness ratio of column 5-8 at end 8 is

$$\gamma_8 = \frac{\sum (I/L)_c}{\sum (\beta_e I/L)_b} = \frac{114/3.6}{1.0 \times 142/6.5} = 1.45$$

The effective length factor for column 5-8 with $\gamma_5 = 1.475$ and $\gamma_8 = 1.45$ can be obtained by solving Equation 5.10 or from Figure 4.6.3.3 in AS 4100 as $k_e = 1.448$.

5. Column 3-6

The base of the column 3-6 is fixed so that its stiffness ratio at end 3 is taken as $\lambda_1 = 0.6$.

At column end 6:

The far end of beam 5-6 is rigidly connected to a column, and beam 5-6 restrains a sway column 3-6 so that $\beta_e = 1.0$.

The stiffness ratio of column 3-6 at end 6 can be calculated as

$$\gamma_6 = \frac{\sum (I/L)_c}{\sum (\beta_e I/L)_b} = \frac{45.9/4.0}{1.0 \times 142/7.5} = 0.606$$

The effective length factor can be obtained by solving Equation 5.10 or from Figure 4.6.3.3 in AS 4100 as $k_e = 1.197$.

5.2.4 Member capacity in axial compression

A long compression steel member with residual stresses and geometric imperfections has a lower axial strength than its section capacity. The effects of member slenderness, residual stress pattern and geometric imperfections on the member capacity of compression members are accounted for by the member slenderness reduction factor (Hancock 1982; Rotter 1982; Hancock et al. 1987; Galambos 1988; Trahair and Bradford 1998). In Clause 6.3.3 of AS 4100 (1998), the nominal member capacity of a compression member with constant cross section is calculated by

$$N_c = \alpha_c N_s \leq N_s \tag{5.11}$$

where α_c is the member slenderness reduction factor. A set of equations for calculating the member slenderness reduction factor (α_c) given by Rotter (1982) is provided in Clause 6.3.3 of AS 4100 (1998) and is described as follows:

The modified member slenderness is expressed by

$$\lambda_n = \frac{L_e}{r} \sqrt{k_f} \sqrt{\frac{f_y}{250}} \tag{5.12}$$

The slenderness modifier is computed as

$$\alpha_a = \frac{2100(\lambda_n - 13.5)}{\lambda_n^2 - 15.3\lambda_n + 2050} \tag{5.13}$$

The combined slenderness is written as

$$\lambda = \lambda_n + \alpha_a \alpha_b \tag{5.14}$$

where α_b is the member section constant that accounts for the effects of residual stress pattern on the capacity of a column and is given in Table 5.2. The section constant is influenced by the section type, manufacturing and fabricating methods that induce residual stresses, thickness of main elements and section form factor (Davids and Hancock 1985; Key et al. 1988; Rasmussen and Hancock 1989).

The imperfection parameter (η) is calculated as

$$\eta = 0.00326(\lambda - 13.5) \geq 0 \tag{5.15}$$

Table 5.2 Member section constant (α_b)

Section	*Section constant* α_b $k_f = 1.0$	$k_f < 1.0$
Hot-formed RHS and circular hollow section (CHS)	−1.0	−0.5
Cold-formed RHS and CHS (stress relieved)		
Cold-formed RHS and CHS (non-stress relieved)	−0.5	−0.5
Hot-rolled UB and UC sections ($t_f \leq 40$)	0	0
Welded box sections		
Welded H- and I-sections	0	–
Tees flame-cut from UB and UC, angles	0.5	–
Hot-rolled channels		
Welded H- and I-sections ($t_f \leq 40$)	0.5	0.5
Hot-rolled UB and UC sections ($t_f > 40$)	1.0	–
Welded H- and I-sections ($t_f > 40$)		
Other sections not listed in this table	0.5	1.0

Source: Adapted from AS 4100, Australian standard for steel structures, Standards Australia, Sydney, New South Wales, Australia, 1998.

The factor ξ is a function of the combined slenderness and imperfection parameter, which is determined as

$$\xi = \frac{(\lambda/90)^2 + 1 + \eta}{2(\lambda/90)^2} \tag{5.16}$$

The member slenderness reduction factor (α_c) is therefore calculated by

$$\alpha_c = \xi\left[1 - \sqrt{1 - \left(\frac{90}{\xi\lambda}\right)^2}\right] \leq 1.0 \tag{5.17}$$

The member slenderness reduction factor (α_c) can be calculated by either the formulas given earlier or linear interpolation from Table 5.3, in which all α_c values were calculated using the earlier equations. It can be seen from Table 5.3 that there are five values of the section constant (α_b), representing five residual stress levels and imperfections. The value of $\alpha_b = -1.0$ represents the lowest imperfection and residual stress. Figure 5.5 demonstrates the effects of the modified member slenderness (λ_n) and section constant (α_b) on the member slenderness reduction factor (α_c). The member slenderness reduction factor is shown to decrease with increasing either the modified slenderness ratio or the residual stress level.

Steel members of varying cross sections are sometimes used in portal frames as tapered columns and rafters. Clause 6.3.4 of AS 4100 states that the nominal section capacity of a compression member with varying cross sections can be taken as the minimum section capacity of all cross sections along the length of the member. The member capacity is calculated using the following modified member slenderness:

$$\lambda_n = 90\sqrt{\frac{N_s}{N_{om}}} \tag{5.18}$$

Table 5.3 Member slenderness reduction factor (α_c)

Modified member slenderness (λ_n)	*Member slenderness reduction factor* (α_c)				
	$\alpha_b = -1.0$	$\alpha_b = -0.5$	$\alpha_b = 0$	$\alpha_b = 0.5$	$\alpha_b = 1.0$
≤10	1.000	1.000	1.000	1.000	1.000
15	1.000	0.998	0.995	0.992	0.990
20	1.000	0.989	0.978	0.967	0.956
25	0.997	0.979	0.961	0.942	0.923
30	0.991	0.968	0.943	0.917	0.888
35	0.983	0.955	0.925	0.891	0.853
40	0.973	0.940	0.905	0.865	0.818
50	0.944	0.905	0.861	0.808	0.747
60	0.907	0.862	0.809	0.746	0.676
65	0.886	0.837	0.779	0.714	0.642
70	0.861	0.809	0.748	0.680	0.609
75	0.835	0.779	0.715	0.646	0.576
80	0.805	0.746	0.681	0.612	0.545
90	0.737	0.675	0.610	0.547	0.487
95	0.700	0.638	0.575	0.515	0.461
100	0.661	0.600	0.541	0.485	0.435
110	0.584	0.528	0.477	0.431	0.389
115	0.546	0.495	0.448	0.406	0.368
120	0.510	0.463	0.421	0.383	0.348
125	0.476	0.434	0.395	0.361	0.330
130	0.445	0.406	0.372	0.341	0.313
140	0.389	0.357	0.330	0.304	0.282
150	0.341	0.316	0.293	0.273	0.255
155	0.320	0.298	0.277	0.259	0.242
160	0.301	0.281	0.263	0.246	0.231
170	0.267	0.251	0.236	0.222	0.210
175	0.252	0.238	0.224	0.212	0.200
180	0.239	0.225	0.213	0.202	0.192
185	0.226	0.214	0.203	0.193	0.183
190	0.214	0.203	0.193	0.184	0.175
200	0.194	0.185	0.176	0.168	0.161
205	0.184	0.176	0.168	0.161	0.154
210	0.176	0.168	0.161	0.154	0.148
215	0.167	0.161	0.154	0.148	0.142
220	0.160	0.154	0.148	0.142	0.137
225	0.153	0.147	0.142	0.137	0.132
230	0.146	0.141	0.136	0.131	0.127
240	0.134	0.130	0.126	0.122	0.118
245	0.129	0.125	0.121	0.117	0.114
250	0.124	0.120	0.116	0.113	0.110

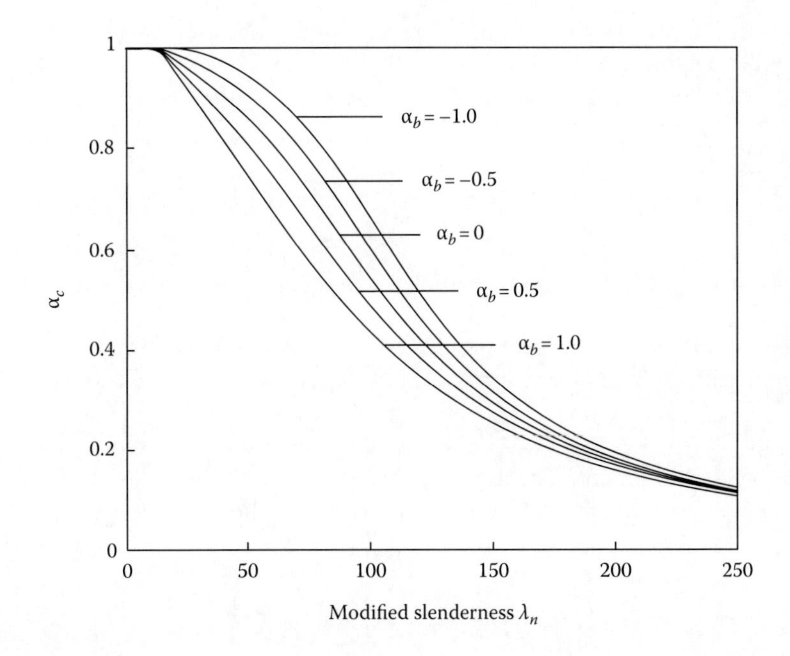

Figure 5.5 Member slenderness reduction factor α_c.

in which N_{om} is the elastic buckling load of the compression member predicted by the elastic buckling analysis (Galambos 1988).

For members under axial compression, the following requirements must be satisfied:

$$N^* \leq \phi N_s \tag{5.19}$$

$$N^* \leq \phi N_c \tag{5.20}$$

where

N^* is the design axial compression force

$\phi = 0.9$ is the capacity factor

5.2.5 Laced and battened compression members

In practice, two or more parallel steel components may be tied together by lacing or battening to form a single compression member to carry heavy axial compression loads, including laced, battened and back-to-back compression members. Design rules for the design of these compression members are given in Clauses 6.4 and 6.5 of AS 4100. The main components and their connections must be designed to resist a design transverse shear force which is applied at any point along the length of the member. The design transverse shear force (V^*) (McGuire 1968) is determined as

$$V^* = \frac{\pi\left(\frac{N_s}{N_c} - 1\right)N^*}{\lambda_n} \tag{5.21}$$

where

N_s and N_c are the nominal section and member capacity of the compression member, respectively
N^* is the total design axial force applied to the compression member
λ_n is the modified member slenderness given in Clauses 6.4 and 6.5 of AS 4100

The batten and its connections must be designed to resist a design longitudinal shear force and a design bending moment. The design longitudinal shear force (V_l^*) and design bending moment (M^*) are given in Clause 6.4.3.7 of AS 4100 as follows:

$$V_l^* = \frac{V^* s_b}{n_b d_b} \tag{5.22}$$

$$M^* = \frac{V^* s_b}{2n_b} \tag{5.23}$$

where

V^* is the design transverse shear forcc
s_b is the longitudinal centre-to-centre distance between the battens
n_b is the number of parallel planes of battens
d_b is the lateral distance between the centroids of the welds of fasteners

Example 5.2: Checking the capacity of a compression steel column

A 360UB44.7 Grade 300 steel column of 8 m length is fixed at its base and pinned at its top. The column is braced against buckling about the y-axis by struts that are pin-connected to its mid-height. The struts prevent lateral deflections in the minor principal plane. The column is subjected to compressive forces including a nominal dead load of 200 kN and a nominal live load of 250 kN. The section properties of 360UB44.7 Grade 300 steel shown in Figure 5.6 are A_g = 5720 mm^2, r_x = 146 mm, r_y = 37.6 mm and f_y = 320 MPa. Check the capacity of the compression column.

1. Design axial load

The design axial load is

$$N^* = 1.2G + 1.5Q = 1.2 \times 200 + 1.5 \times 250 = 615 \text{ kN}$$

2. Check the slenderness of elements

The slenderness of the flange is

$$\lambda_{ef} = \frac{b}{t}\sqrt{\frac{f_y}{250}} = \frac{(171-6.9)/2}{9.7}\sqrt{\frac{320}{250}} = 9.57 < \lambda_{ey} = 16 \quad \text{Table 5.2 of AS4100}$$

Hence, the flange is not slender.

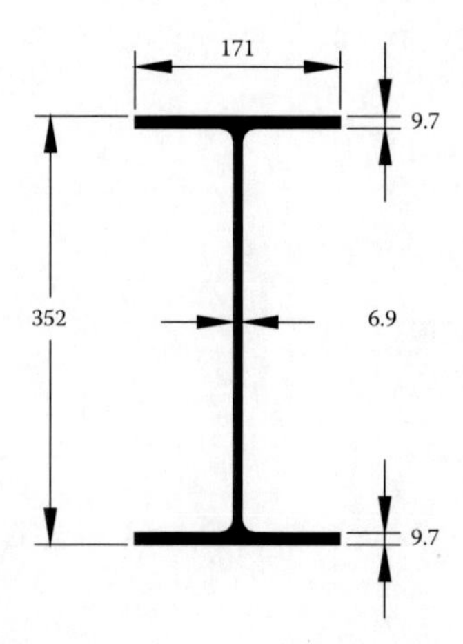

Figure 5.6 Cross section of steel column.

The slenderness of the web is

$$\lambda_{ew} = \frac{b}{t}\sqrt{\frac{f_y}{250}} = \frac{(352-2\times 9.7)}{6.9}\sqrt{\frac{320}{250}} = 54.54 > \lambda_{ey} = 45 \quad \text{Table 5.2 of AS4100}$$

Hence, the web is slender.

3. Section capacity

The effective width of the web can be calculated as

$$b_{ew} = b\left(\frac{\lambda_{ey}}{\lambda_e}\right) = (352-2\times 9.7)\left(\frac{45}{54.54}\right) = 274.4 \text{ mm}$$

The effective area of the section is

$$A_e = 2\times(171\times 9.7) + 274.4\times 6.9 = 5210.76 \text{ mm}^2$$

The gross area of the section can be calculated as

$$A_g = 2\times(171\times 9.7) + (352-2\times 9.7)\times 6.9 = 5612.34 \text{ mm}^2$$

The form factor is determined as

$$k_f = \frac{A_e}{A_g} = \frac{5210.76}{5612.34} = 0.928$$

The section design capacity is calculated as

$$\phi N_s = \phi k_f A_n f_y = 0.9\times 0.928\times 5720\times 320 = 1529 \text{ kN} > N^* = 615 \text{ kN, OK}$$

4. Member capacity

The column is fixed at its base and pinned at its top so that the effective length factor for buckling about the x-axis is $k_e = 0.85$. The effective length is

$$L_{ex} = k_e L = 0.85\times 8000 = 6800 \text{ mm}$$

The ends of the upper segment of the column are pinned. The effective length of the upper segment buckling about y-axis is

$$L_{ey} = k_e L = 1.0 \times 4000 = 4000 \text{ mm}$$

The modified member slenderness is calculated as follows:

$$\lambda_{nx} = \frac{L_{ex}}{r_x}\sqrt{k_f}\sqrt{\frac{f_y}{250}} = \frac{6800}{146}\sqrt{0.928}\sqrt{\frac{320}{250}} = 50.76$$

$$\lambda_{ny} = \frac{L_{ey}}{r_y}\sqrt{k_f}\sqrt{\frac{f_y}{250}} = \frac{4000}{37.6}\sqrt{0.928}\sqrt{\frac{320}{250}} = 115.94 > \lambda_{nx}$$

Hence, the column will buckle about the y-axis, $\lambda_n = 115.94$.
For hot-rolled I-sections $k_f < 0$, the section constant is obtained from Table 5.2 as $\alpha_b = 0$.
The member slenderness reduction factor can be obtained from Table 5.3 as

$$\alpha_c = 0.448 - \frac{(0.448 - 0.421)(15.94 - 115)}{(120 - 115)} = 0.443$$

The design capacity of the column is

$$\phi N_c = \phi\alpha_c N_s = 0.443 \times 1529 = 677 \text{ kN} > N^* = 615 \text{ kN, OK}$$

Example 5.3: Checking the capacity of an RHS compression column

The pin-ended rectangular hollow section (RHS) column 200 × 100 × 4.0 RHS of Grade C350 steel as depicted in Figure 5.7 is 4 m length. The column is subjected to axial compression forces including a nominal dead load of 100 kN and a nominal live load of 120 kN. The section properties of the RHS column are A_g = 2280 mm², r_x = 72.1 mm, r_y = 42.3 mm and f_y = 350 MPa. Check the capacity of the column.

1. Design axial load

The design axial load is

$$N^* = 1.2G + 1.5Q = 1.2 \times 100 + 1.5 \times 120 = 300 \text{ kN}$$

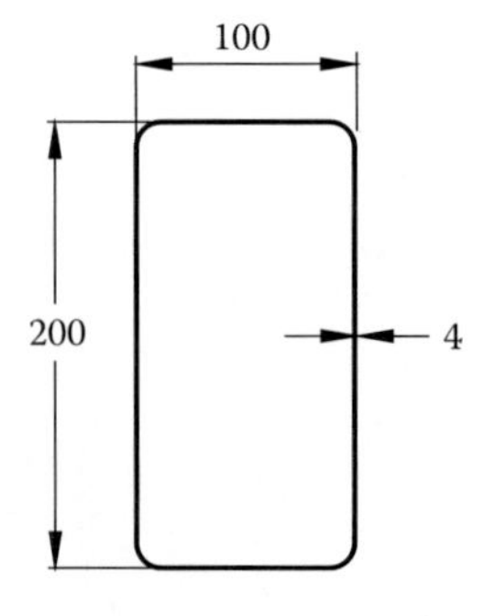

Figure 5.7 Cross section of RHS steel column.

2. Check the slenderness of elements

The slenderness of the flange is determined as

$$\lambda_{ef} = \frac{b}{t}\sqrt{\frac{f_y}{250}} = \frac{(100-2\times 4)}{4}\sqrt{\frac{350}{250}} = 27.2 < \lambda_{ey} = 40 \quad \text{Table 5.2 of AS4100}$$

Hence, the flange is not slender.

The slenderness of the web is calculated as

$$\lambda_{ew} = \frac{b}{t}\sqrt{\frac{f_y}{250}} = \frac{(200-2\times 4)}{4}\sqrt{\frac{350}{250}} = 56.79 > \lambda_{ey} = 40$$

Hence, the web is slender.

3. Section capacity

The effective width of the web can be calculated as

$$b_{ew} = b\left(\frac{\lambda_{ey}}{\lambda_e}\right) = (200-2\times 4)\left(\frac{40}{56.79}\right) = 135.2 \text{ mm}$$

The effective area of the section is determined as

$$A_e = 2\times(100-2\times 4)\times 4 + 2\times 135.2\times 4 = 1817.6 \text{ mm}^2$$

The gross area of the section is calculated as

$$A_g = 2\times(100-2\times 4)\times 4 + (200-2\times 4)\times 4 = 2272 \text{ mm}^2$$

The form factor is $k_f = A_e/A_g = 1817.6/2272 = 0.8$.

The section design capacity is determined as

$$\phi N_s = \phi k_f A_n f_y = 0.9\times 0.8\times 2280\times 350 = 574.56 \text{ kN} > N^* = 300 \text{ kN, OK}$$

4. Member capacity

The effective length of column buckling about the minor principal y-axis is

$$L_{ey} = k_e L = 1.0\times 4000 = 4000 \text{ mm}$$

The modified member slenderness is computed as

$$\lambda_{ny} = \frac{L_{ey}}{r_y}\sqrt{k_f}\sqrt{\frac{f_y}{250}} = \frac{4000}{42.3}\sqrt{0.8}\sqrt{\frac{350}{250}} = 100$$

For cold-formed RHS with $k_f < 1.0$, the section constant can be obtained from Table 5.2 as $\alpha_b = -0.5$. The member slenderness reduction factor can be obtained from Table 5.3 as $\alpha_c = 0.6$.

The design capacity of the column is calculated as

$$\phi N_c = \phi \alpha_c N_s = 0.6\times 574.56 = 334.7 \text{ kN} > N^* = 300 \text{ kN, OK}$$

5.3 MEMBERS IN AXIAL TENSION

5.3.1 Behaviour of members in axial tension

The load–extension behaviour of steel tension members without holes, geometric imperfections and residual stresses is found to follow the material stress–strain relationship. These idealised tension members display ductile behaviour and can attain the gross yield capacity ($A_g f_y$) of the sections. Residual stresses in tension members cause local early yielding and strain hardening. Holes in tension members lead to early yielding around the holes. Consequently, the load–deflection behaviour of tension members with holes is nonlinear. When the holes are large, the member under axial tension may fail by fracturing at the holes, and its strength is governed by the facture capacity of its cross section (Dhalla and Winter 1974a,b; Bennetts et al. 1986). The strength and behaviour of steel members in axial tension may be governed by either the gross yield capacity or the fracture capacity of their cross sections. Therefore, the design of tension members must consider these two failure criteria.

5.3.2 Capacity of members in axial tension

A steel member subject to a design axial tension force (N^*) must satisfy the following strength design requirement:

$$N^* \leq \phi N_t \tag{5.24}$$

where

$\phi = 0.9$ is the capacity reduction factor
N_t is the nominal section capacity in axial tension

Clause 7.2 of AS 4100 (1998) specifies that N_t is taken as the lesser of the gross yield capacity (N_{ty}) and fracture capacity (N_{ta}):

$$N_{ty} = A_g f_y \tag{5.25}$$

$$N_{ta} = 0.85 k_{ct} A_n f_u \tag{5.26}$$

where

k_{ct} is the correction factor considering the effect of non-uniform force distributions induced by the end connections of the tension member
A_n is the net cross-sectional area
f_u is the design tensile strength

The factor of 0.85 in the earlier equation is used to account for the sudden failure by local brittle facture behaviour at the net section. The nominal member capacity of a steel member under axial tension is taken as its nominal section capacity.

Guidelines for determining the correction factor are given in Clause 7.3 of AS 4100. The correction factor (k_{ct}) is taken as 1.0 for a tension member whose end connections are designed to provide uniform force distribution in the member. To achieve this condition, the end connections must be symmetrically placed about the centroidal axis of the member, and each part of the connection must be capable of resisting the design force in

Table 5.4 Correction factor (k_{ct})

Configuration	k_{ct}
Single angle connection or twin angles on same side	
Unequal angle connected by short leg	0.75
Otherwise	0.85
Single channel (web connected to the plate)	0.85
Single tee (flange connected to the plate)	0.9
Back-to-back symmetric connection	1.0

Source: Adapted from AS 4100, Australian standard for steel structures, Standards Australia, Sydney, New South Wales, Australia, 1998.

its connected part. If the end connections of a tension member induce non-uniform force distributions in the member, the member should be designed for combined actions of axial tension and bending. However, for eccentrically connected angles, channels and tees, the correction factor (k_{ct}) given in Table 5.4 can be used in Equation 5.26 to determine its capacity (Bennetts et al. 1986). The correction factor (k_{ct}) is taken as 0.85 for symmetrical I-sections and channel sections connected by both flanges only if the length between the first and last lows of bolts is greater than the depth of the member section.

Tension members with staggering holes may fail by fracture along a zigzag path ABCDEF as depicted in Figure 5.8 rather than along the path perpendicular to the applied axial tensile force. The failure line may be along the path *DG* or the diagonal path *DE* as shown in Figure 5.8. The critical failure path is the one that has the minimum net area. The difference between the path *DG* and the path *DE* is represented by a length correction (Cochrane 1922) as

$$l_c = \frac{s_p^2}{4s_g} \tag{5.27}$$

where

s_p stands for the staggered pitch, which is the centre-to-centre distance parallel to the direction of the tensile force in the member

s_g represents the gauge that is the centre-to-centre distance of holes measured at right angle to the direction of the tensile force in the member

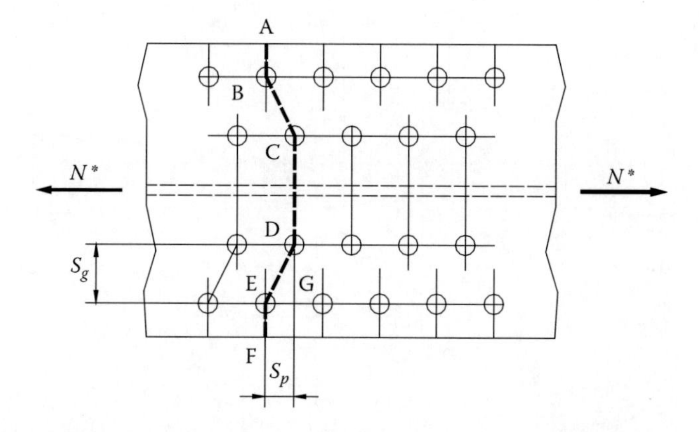

Figure 5.8 Failure paths on net section.

The area correction is calculated by multiplying the length correction and the thickness.

The net cross-sectional area along the zigzag path can be calculated by

$$A_n = A_g - \sum d_h t + \sum \frac{s_p^2 t}{4 s_g} \tag{5.28}$$

where

d_h is the diameter of a hole
t is the thickness of the holed material

It is noted that the net area (A_n) must be less than or equal to the gross area (A_g). This gives $s_p \le \sqrt{4 s_g d_h}$. This means that if the stagger spacing $s_p > \sqrt{4 s_g d_h}$, the hole does not reduce the area of the member (Trahair and Bradford 1998). If holes are not staggered or $s_p > \sqrt{4 s_g d_h}$, s_p is taken as zero in Equation 5.28.

Example 5.4: Capacity of a bolted steel member in axial tension

Both flanges of a steel I-section member under axial tension are bolted as depicted in Figure 5.9. The depth of the I-section is 339 mm. The diameter of each bolt hole is 24 mm. The gross area of the member section is 29,300 mm² and the thickness of the flange is 32 mm. The yield stress of the section is 280 MPa, while its tensile strength is 430 MPa. Determine the capacity of the tension member.

1. Net area of the section

The minimum stagger is calculated as

$$s_{pm} = \sqrt{4 s_g d_h} = \sqrt{4 \times 70 \times 24} = 82 \text{ mm} > s_p = 35 \text{ mm}$$

Thus, the failure path at each flange is staggered as indicated by the path *ABCDEF* shown in Figure 5.9. This failure path at each flange includes four holes and two staggers. The net area of the section can be calculated as

$$A_n = A_g - \sum d_h t + \sum \frac{s_p^2 t}{4 s_g} = 29{,}300 - 2 \times \left[4 \times (24 \times 32)\right] + 2 \times \left[2 \times \left(\frac{35^2 \times 32}{4 \times 70}\right)\right] = 23{,}716 \text{ mm}^2$$

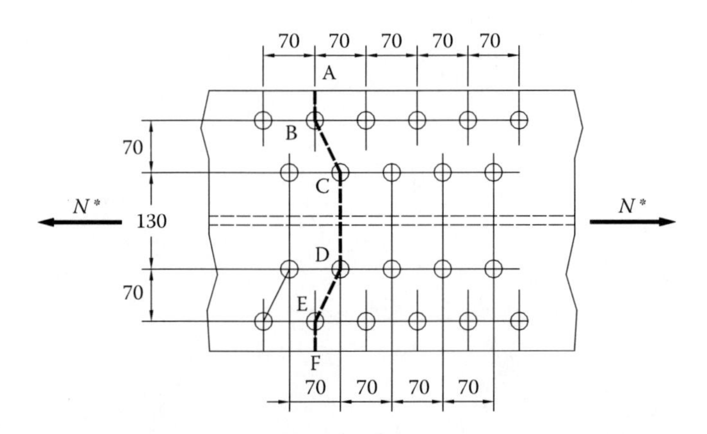

Figure 5.9 Bolted flange of a welded I-section member.

2. Capacity of tension member

The gross yield capacity can be calculated as

$$N_{ty} = A_g f_y = 29{,}300 \times 280 \times 10^{-3} \text{ kN} = 8{,}204 \text{ kN}$$

This I-section tension member is connected by both flanges only. The length between the first and last lows of bolts is $l_{cb} = 5 \times 70 = 350$ mm $> D = 339$ mm. Thus, the correction factor of this I-section member satisfying the requirement of Clause 7.3.2 of AS 4100 is $k_{ct} = 0.85$.

The fracture capacity can be calculated as

$$N_{ta} = 0.85 k_{ct} A_n f_u = 0.85 \times 0.85 \times 23{,}716 \times 430 \text{ N} = 7{,}368 \text{ kN} < 8{,}204 \text{ kN}$$

Thus, $N_t = 7368$ kN.
The design capacity can be determined as

$$\phi N_t = 0.9 \times 7368 \text{ kN} = 6331 \text{ kN}$$

5.4 MEMBERS UNDER AXIAL LOAD AND UNIAXIAL BENDING

5.4.1 Behaviour of members under combined actions

Steel members subject to combined actions of axial load and bending are called beam–columns. The bending moments are induced by the loading eccentricity, the lateral loads applied to the columns and the overall frame actions. The behaviour of steel members under combed actions is characterised by the in-plane, out-of-plane and biaxial bending. When a steel beam–column is constrained to bend about its major principal axis or when it is bent about its minor principal axis, its deformations occur in the plane of bending. This is the in-plane behaviour, which is characterised by the bending of beams and by the buckling of compression members in the plane. When a steel beam–column that is not restrained laterally is bent about its principal axis, it may undergo flexural–torsional buckling.

The ultimate strength of a steel beam–column under combined axial compression force and bending moment is influenced by the interaction between the axial force and bending moment. The axial compression force reduces the moment capacity of the beam–column, while the bending moment reduces the member axial load capacity. The interaction between the axial compression force and deformations leads to second-order effects which amplify the bending moments.

For a steel member subject to combined actions of axial tension and bending, the axial tensile force reduces the section moment capacity of the member but increases its out-of-plane member capacity when bending about its major principal x-axis.

5.4.2 Section moment capacity reduced by axial force

The design rules for section moment capacity reduced by axial force are given in Clause 8.3 of AS 4100. Further information can be found in publications by Woolcock and Kitipornchai (1986), Bradford et al. (1987), Bridge and Trahair (1987) and Trahair and Bradford (1998). When a steel member is subjected to an axial force (N^*) and a design bending moment (M_x^*) about its section major principal x-axis, the nominal section moment capacity (M_{rx}) reduced by the axial force is given in Clause 8.3.2 of AS 4100 (1998) as follows:

$$M_{rx} = M_{sx}\left(1 - \frac{N^*}{\phi N_s}\right) \tag{5.29}$$

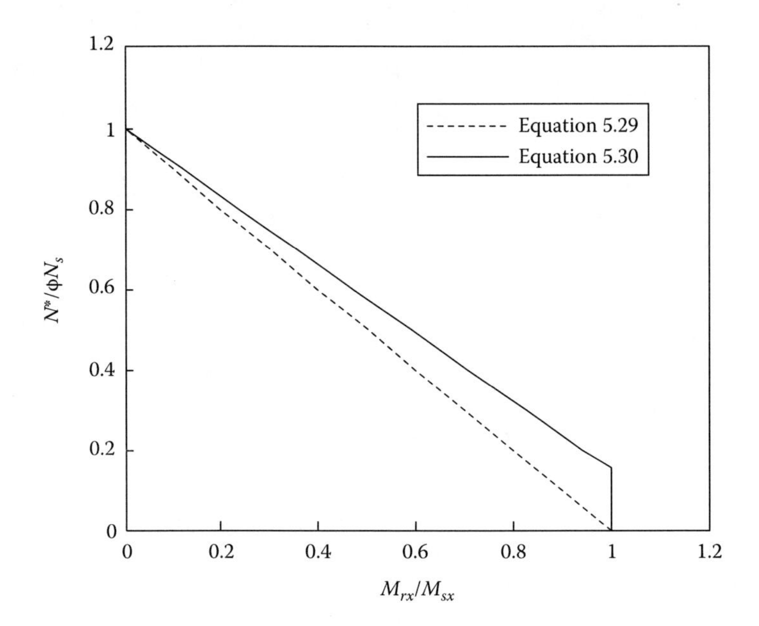

Figure 5.10 Strength interaction curves for compact doubly symmetric I-sections under axial force and uniaxial bending about principal y-axis.

where N_s stands for the nominal section axial load capacity for axial compression or axial tension. Equation 5.29 is based on the simple straight-line interaction curve as depicted in Figure 5.10 and is conservative for compact doubly symmetric I-sections and rectangular and square hollow sections.

For compression members with $k_f = 1.0$ and tension members that are of compact doubly symmetric I-sections and rectangular and square hollow sections, Clause 8.3.2 of AS 4100 (1998) provides a more accurate expression for calculating M_{rx} as follows:

$$M_{rx} = 1.18 M_{sx}\left(1 - \frac{N^*}{\phi N_s}\right) \le M_{sx} \quad (5.30)$$

The strength interaction curve representing Equation 5.30 for compact doubly symmetric I-sections and rectangular and square hollow sections is shown in Figure 5.10. It can be seen from the figure that the earlier strength interaction formula gives higher section capacities than Equation 5.29.

If compression members with $k_f < 1.0$ are of compact doubly symmetric I-sections and rectangular and square hollow sections, Clause 8.3.2 of AS 4100 (1998) provides the following more accurate formula for determining the reduced section moment capacity:

$$M_{rx} = M_{sx}\left(1 - \frac{N^*}{\phi N_s}\right)\left[1 + 0.18\left(\frac{82 - \lambda_w}{82 - \lambda_{wy}}\right)\right] \le M_{sx} \quad (5.31)$$

in which λ_w and λ_{wy} are the slenderness and slenderness yield limit of the web, respectively.

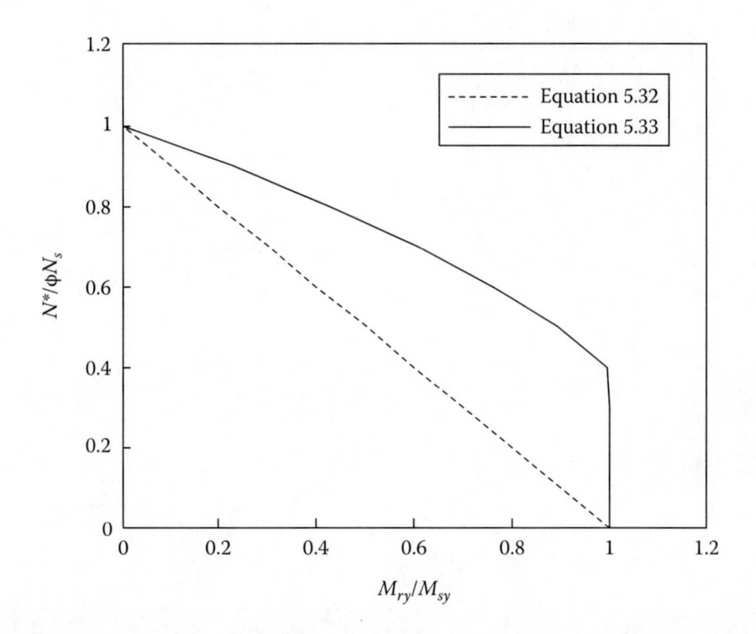

Figure 5.11 Strength interaction curves for compact doubly symmetric I-sections under axial force and uniaxial bending about principal y-axis.

As given in Clause 8.3.3 of AS 4100 (1998), for a member subject to axial force (N^*) and design bending moment (M_y^*) about its section minor principal y-axis, the nominal section moment capacity (M_{ry}) reduced by the axial tension or compression force is expressed by

$$M_{ry} = M_{sy}\left(1 - \frac{N^*}{\phi N_s}\right) \tag{5.32}$$

where M_{sy} denotes the nominal section moment capacity for bending about the minor principal y-axis. Equation 5.32 represents a straight-line interaction curve as shown in Figure 5.11 and is conservative for compact doubly symmetric I-sections and rectangular and square hollow sections.

For compact doubly symmetric I-sections, M_{ry} can be more accurately calculated by the following formula given in AS 4100 (1998)

$$M_{ry} = 1.19M_{sy}\left[1 - \left(\frac{N^*}{\phi N_s}\right)^2\right] \leq M_{sy} \tag{5.33}$$

The strength interaction curve which represents Equation 5.33 is also shown in Figure 5.11. It is shown that the straight-line interaction curve based on Equation 5.32 is very conservative, and significant economy can be achieved by using Equation 5.33 for compact doubly symmetric I-sections.

For compact rectangular and square hollow sections, the following expression given in Clause 8.3.3 of AS 4100 (1998) provides more accurate prediction of M_{ry} as illustrated in Figure 5.12:

$$M_{ry} = 1.18M_{sy}\left(1 - \frac{N^*}{\phi N_s}\right) \leq M_{sy} \tag{5.34}$$

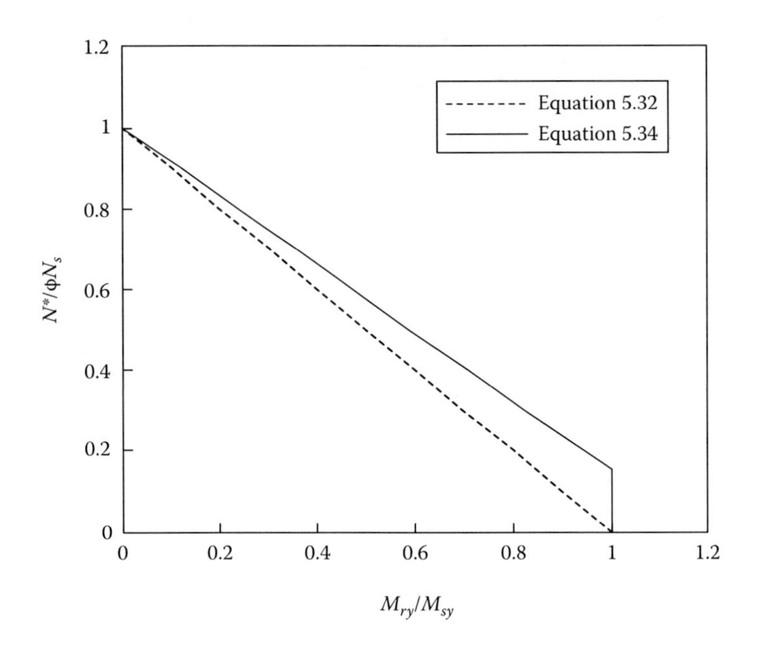

Figure 5.12 Strength interaction curves for compact rectangular and square sections under axial force and uniaxial bending about principal y-axis.

For strength design, all sections of a steel member under axial force and uniaxial bending must satisfy the following conditions:

$$M_x^* \leq \phi M_{rx} \tag{5.35}$$

$$M_y^* \leq \phi M_{ry} \tag{5.36}$$

5.4.3 In-plane member capacity

The design of members under axial load and uniaxial bending for in-plane bending and buckling is given in Clause 8.4.2 of AS 4100 for members analysed by the elastic method. The nominal in-plane member moment capacity (M_i) of a compression member is given in Clause 8.4.2.2 of AS 4100 (1998) as

$$M_i = M_s\left(1 - \frac{N^*}{\phi N_c}\right) \tag{5.37}$$

where

N^* is the design axial compressive force

N_c is the nominal member capacity in axial compression for buckling about the same principal axis determined using the effective length factor of $k_e = 1.0$ for both braced and sway members, unless a lower value of k_e can be determined for braced members

The reason for taking k_e as 1.0 for a sway member is that the effects of end restraints on the member buckling have been considered in the second-order elastic analysis. However, this

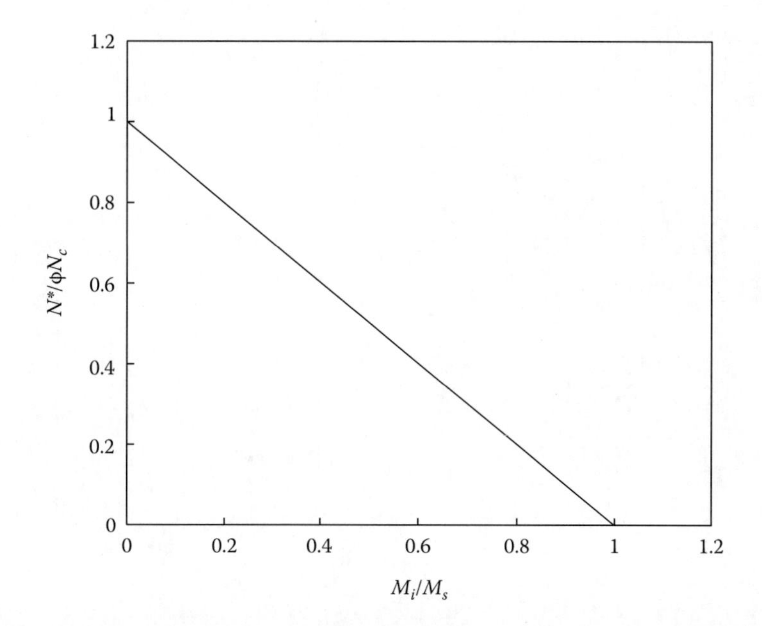

Figure 5.13 Strength interaction curves for compression members under in-plane bending.

may result in unsafe designs for some sway compression members under small bending moments. Therefore, if the effective length factor determined in accordance with Clause 4.6.3 is used in the calculation of the compression member capacity, the design axial compression force alone must be less than the section and member capacity in axial compression. Figure 5.13 demonstrates the strength interaction curve for in-plane bending and axial compression.

For compact doubly symmetric I-sections and rectangular and square hollow sections with $k_f = 1.0$, Clause 8.4.2 of AS 4100 provides the following formula for calculating the in-plane member moment capacity (M_i) (Trahair 1986):

$$M_i = M_s\left\{\left[1-\left(\frac{1+\beta_m}{2}\right)^3\right]\left(1-\frac{N^*}{\phi N_c}\right)+1.18\left(\frac{1+\beta_m}{2}\right)^3\sqrt{\left(1-\frac{N^*}{\phi N_c}\right)}\right\} \leq M_{rx} \text{ or } M_{ry} \quad (5.38)$$

where $\beta_m = 1.0$ for uniform bending.

The in-plane member moment capacity of members under axial tension and bending is not reduced by axial tension so that their design is governed by the section capacities.

The design of a steel member for in-plane bending and axial force must satisfy

$$M_x^* \leq \phi M_{ix} \quad \text{or} \quad M_y^* \leq \phi M_{iy} \quad (5.39)$$

where M_{ix} and M_{iy} are the nominal in-plane member moment capacities for bending about the principal axes, respectively.

5.4.4 Out-of-plane member capacity

A steel member subject to axial load and uniaxial bending about its major principal x-axis may fail by buckling out of the plane of bending. The design of these members for

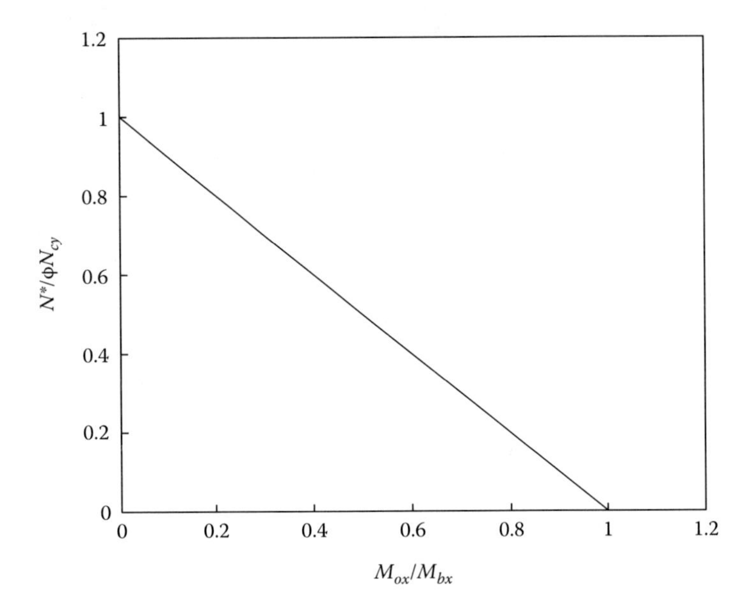

Figure 5.14 Strength interaction curves for compression members for out-of-plane buckling.

out-of-plane buckling is given in Clause 8.4.4 of AS 4100. As provided in Clause 8.4.4.1 of AS 4100 (1998), the nominal out-of-plane member moment capacity (M_{ox}) of a compression member is calculated by

$$M_{ox} = M_{bx}\left(1 - \frac{N^*}{\phi N_{cy}}\right) \tag{5.40}$$

where

M_{bx} is the nominal member moment capacity of the member without full lateral restraint
N_{cy} is the nominal member capacity in axial compression for buckling about the minor principal y-axis

Equation 5.40 is plotted as a straight-line strength interaction curve in Figure 5.14.

For members with compact double symmetric I-sections fully or partially restrained at both ends and with $k_f = 1.0$, a more accurate expression (Cuk et al. 1986) is given in Clause 8.4.4.1 of AS 4100 as

$$M_{ox} = \alpha_{bc} M_{bxo}\sqrt{\left(1 - \frac{N^*}{\phi N_{cy}}\right)\left(1 - \frac{N^*}{\phi N_{oz}}\right)} \le M_{rx} \tag{5.41}$$

where the factor α_{bc} accounts for the effects of the moment ratio (β_m) and the axial force (N^*) and can be determined by

$$\alpha_{bc} = \frac{1}{\dfrac{1-\beta_m}{2} + \left(\dfrac{1+\beta_m}{2}\right)^3\left(0.4 - 0.23\dfrac{N^*}{\phi N_{cy}}\right)} \tag{5.42}$$

In Equation 5.41, M_{bxo} is the nominal member moment capacity without full lateral restrain and under uniform design moment distribution, and N_{oz} is the elastic torsional bucking capacity of the member, which is given by

$$N_{oz} = \frac{GJ + \left(\frac{\pi^2 E_s I_w}{l_z^2}\right)}{(I_x + I_y)/A} \tag{5.43}$$

in which l_z is the distance between partial or full torsional restraints.

For a steel member under, an axial tensile force and a design bending moment, the nominal out-of-plane member moment capacity (M_{ox}) is given in Clause 8.4.4.2 of AS 4100 (1998) as

$$M_{ox} = M_{bx}\left(1 + \frac{N^*}{\phi N_t}\right) \le M_{rx} \tag{5.44}$$

where N_t is the nominal section capacity in axial tension. It can be seen from the earlier equation that the tensile force increases the out-of-plane member moment capacity.

The design of a steel member under an axial compressive force and a design bending moment about its major principal x-axis must check for its in-plane and out-of-plane member moment capacities as follows:

$$M_x^* \le \phi M_i \tag{5.45}$$

$$M_x^* \le \phi M_{ox} \tag{5.46}$$

However, for tension members, only the out-of-plane member moment capacity needs to be checked.

5.5 DESIGN OF PORTAL FRAME RAFTERS AND COLUMNS

Rafters and columns in portal frames are subjected to combined axial force and uniaxial bending. The design of rafters and columns in portal frames may be governed by the strength criteria or by the deflection criteria. For design for the strength criteria, the economical designs of rafters and columns can be achieved by designing the member capacity as close as possible to the section capacity. This can be done by providing adequate fly braces to laterally restrain the inside compression flanges of the rafters and columns. For large span portal frames, deflection usually governs the design. For this case, haunches can be added to the rafters to reduce the deflections. A typical portal frame is depicted in Figure 5.15.

5.5.1 Rafters

Purlins, which are bolted to the top flange of a rafter in a steel portal frame, provide lateral but not rotational restraint to the top flange because the bolted connection between purlins and the flange allows for rotation. Under dead and live loads, most of the top flange of a rafter is subjected to compression. As a result, the effective length can be taken as the distance between the purlins when calculating the member moment capacity (M_{bx}) of the rafter. Under upward wind loads, however, most of the bottom flange of the rafter is in

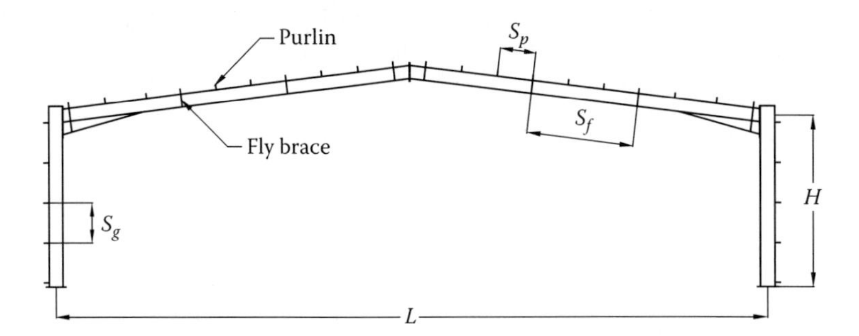

Figure 5.15 Portal frame.

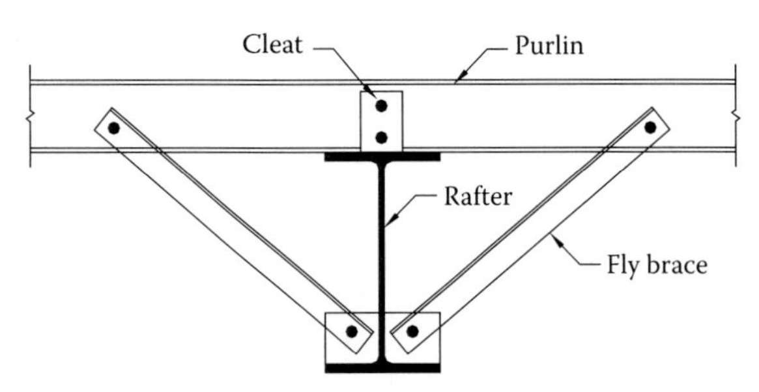

Figure 5.16 Double fly braces.

compression. To increase the rafter member capacity, fly braces in the form of small angle section members that connect the bottom flange to the purlins are usually used to brace the rafter as depicted in Figure 5.16. As fly braces provide full restraint to the bottom flange of the rafter, the effective length of the rafter is taken as the distance between the fly braces in the calculation of its member moment capacity (M_{bx}). It is recommended that fly braces should be provided within the first quarter of the total rafter span, at the inside corner of the knee joint and near the ridge (Woolcock et al. 2003). When calculating the in-plane member capacity of a rafter under combined actions, the nominal member capacity in axial compression (N_{cx}) is required. For this purpose, the effective length (L_{ex}) is taken as the actual rafter length measured from the centreline of the column to the ridge. For columns under combined actions, however, the nominal member capacity (N_{cy}) in axial compression for buckling about the y-axis can be computed using the distance between purlins as the effective length (L_{ey}). This is because purlins and roof sheeting act as a rigid diaphragm between roof bracing nodes, which force the rafter to buckle between the purlins.

5.5.2 Portal frame columns

At the bottom of a portal frame column, the base plate and bolts offer full lateral and torsional restraint and nearly some minor axis and warping restraint. At the top of the column, the wall bracing and the fly brace at the inside corner of the haunch provide full lateral restraint. It is noted that the rafter does not offer minor axis and warping restraint to the column. When the inside flange of the portal column without fly braces is in compression, the effective length of the column can be taken as distance from the base plate to the underside of the haunch for determining its member moment capacity (M_{bx}). For columns with fly braces, the effective length can be taken as the distance between fly braces in the calculation of M_{bx}. When the outside flange of a portal column is in compression, its member moment capacity (M_{bx}) should be calculated using the spacing of girts as the effective length.

The nominal member capacity in axial compression (N_{cx}) is used to calculate the in-plane member capacity of a portal frame column under combined actions. For this purpose, the effective length (L_{ex}) is taken as the actual column length. When determining the out-of-plane member capacity of the column under combined actions, the effective length (L_{ey}) is taken as the distance between girts in the calculation of the nominal member capacity (N_{cy}) in axial compression for buckling about the y-axis. The reason for this is that girts and wall sheeting act as a rigid diaphragm between wall bracing nodes which should be effective in ensuring the column buckle between girts. However, when designing heavily loaded columns, the effective length should be taken as the distance between fly braces.

Example 5.5: Design of a steel portal frame column under axial force and uniaxial bending

A steel portal frame column of 460UB74.6 of Grade 300 steel is schematically depicted in Figure 5.17. The height of the portal frame column measured from the floor to the centreline of rafter is 6000 mm, while the height of the underside of the haunch is 5364 mm. The column is pinned at its base and braced by girts with a spacing of 1400 mm. The second-order elastic analysis of the portal frame under various load combinations calculated in Example 2.1 has been performed. Check the capacities of the column under the following design actions obtained from the second-order elastic analysis:

a. $M_x^* = 330$ kN m, $N_c^* = 87$ kN (compression)
b. $M_x^* = 420$ kN m, $N_t^* = 115$ kN (tension)

1. Section properties

The dimensions and properties of 460UB74.6 are

$d = 457$ mm, $b_f = 190$ mm, $t_f = 14.5$ mm, $t_w = 9.1$ mm, $A_g = 9520$ mm^2

$I_x = 335 \times 10^6$ mm^4, $r_x = 188$ mm, $Z_{ex} = 1660 \times 10^3$ mm^3, $I_y = 16.6 \times 10^6$ mm^4

$r_y = 41.8$ mm, $Z_{ey} = 262 \times 10^3$ mm^3, $I_w = 815 \times 10^9$ mm^6, $J = 530 \times 10^3$ mm^3

$f_y = 300$ MPa, $f_u = 440$ MPa, $k_f = 0.948$, $E_s = 200 \times 10^3$ MPa, $G = 80 \times 10^3$ MPa

Compactness about the x-axis = compact

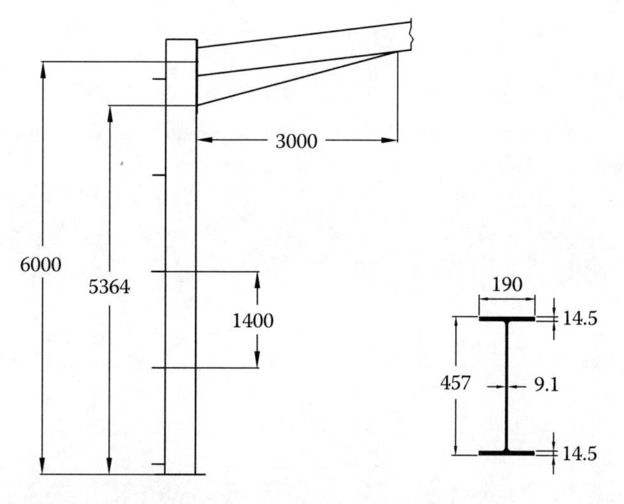

Figure 5.17 Steel portal frame column.

2. Axial section capacities

The section capacity in axial compression is calculated as

$$N_{sc} = k_f A_n f_y = 0.948 \times 9520 \times 300 \text{ N} = 2707.5 \text{ kN}$$

$$\phi N_{sc} = 0.9 \times 2707.5 = 2436.7 \text{ kN} > N_c^* = 87 \text{ kN, OK}$$

The gross yield capacity of the section is

$$N_{ty} = A_g f_y = 9520 \times 300 \text{ N} = 2856 \text{ kN}$$

The fracture capacity of the section is

$$N_{ta} = 0.85 k_t A_n f_u = 0.85 \times 1.0 \times 8520 \times 440 \text{ N} = 3186.5 \text{ kN} > 2856 \text{ kN}$$

$$\therefore N_t = 2856 \text{ kN}$$

The design capacity in axial tension is

$$\phi N_t = 0.9 \times 2856 = 2570.4 \text{ kN} > N_t^* = 115 \text{ kN, OK}$$

3. Section moment capacities

3.1. Section moment capacity without axial force

The section moment capacity is computed as

$$M_{sx} = Z_{ex} f_y = 1660 \times 10^3 \times 300 \times 10^{-6} = 498 \text{ kN m}$$

$$\phi M_{sx} = 0.9 \times 498 = 448.2 \text{ kN m}$$

3.2. Reduced section moment capacity due to axial compression

460UB74.6 is a compact doubly symmetric I-section with $k_f = 0.948 < 1.0$ so that its section moment capacity reduced by axial compression can be calculated using Equation 5.31.

The slenderness of the web is

$$\lambda_w = \frac{b}{t}\sqrt{\frac{f_y}{250}} = \frac{457 - 2 \times 14.5}{9.1}\sqrt{\frac{300}{250}} = 51.5$$

The slenderness limit for the web supported by two flanges under uniform compression can be obtained from Table 5.2 of AS 4100 as $\lambda_{wy} = 45$.

The section moment capacity reduced by axial compression is calculated as

$$M_{rx} = M_{sx}\left(1 - \frac{N^*}{\phi N_s}\right)\left[1 + 0.18\left(\frac{82 - \lambda_w}{82 - \lambda_{wy}}\right)\right] \le M_{sx}$$

$$= 498 \times \left(1 - \frac{87}{2436.7}\right)\left[1 + 0.18 \times \left(\frac{82 - 51.5}{82 - 45}\right)\right] = 551.5 \text{kN m} > M_{sx} = 498 \text{ kN m}$$

$$\therefore M_{rx} = 498 \text{ kN m}$$

$$\phi M_{rx} = 0.9 \times 498 = 448.2 \text{ kN m} > M_x^* = 330 \text{ kN m, OK}$$

3.3. Reduced section moment capacity due to axial tension

For compact doubly symmetric I-section, the section moment capacity reduced by axial tension can be calculated as

$$M_{rx} = 1.18M_{sx}\left(1 - \frac{N^*}{\phi N_s}\right) \le M_{sx}$$

$$= 1.18 \times 498 \times \left(1 - \frac{115}{2570.4}\right) = 561.3 \text{ kN m} > M_{sx} = 498 \text{ kN m}$$

$$\therefore M_{rx} = 498 \text{ kN m}$$

$$\phi M_{rx} = 0.9 \times 498 = 448.2 \text{ Nm} > M_x^* = 420 \text{ kN m, OK}$$

4. Axial member capacities

4.1. Axial member capacity N_{cx}

The effective length factor k_e is taken as 1.0 as required for combined action (Clause 8.4.2.2). The effective length is

$$L_e = k_e L = 1.0 \times 6000 = 6000 \text{ mm}$$

The modified member slenderness can be calculated as

$$\lambda_{nx} = \frac{L_{ex}}{r_x}\sqrt{k_f}\sqrt{\frac{f_y}{250}} = \frac{6000}{188}\sqrt{0.948}\sqrt{\frac{300}{250}} = 34$$

For hot-rolled UB (universal beam) section with $k_f < 1.0$, $\alpha_b = 0$, the slenderness reduction factor can be obtained from Table 5.3 using linear interpolation as follows:

$$\alpha_c = 0.943 - \frac{(0.943 - 0.925)(34 - 30)}{(35 - 30)} = 0.928$$

The design axial member capacity is

$$\phi N_{cx} = \phi \alpha_c N_s = 0.9 \times 0.928 \times 2707.5 = 2261.3 \text{ kN}$$

4.2. Axial member capacity N_{cy}

The portal frame column is braced by girts with a spacing of 1400 for buckling about the minor principal y-axis so that the effective length is taken as the girt spacing L_{ey} = 1400 mm.

The modified member slenderness is

$$\lambda_{ny} = \frac{L_{ey}}{r_y}\sqrt{k_f}\sqrt{\frac{f_y}{250}} = \frac{1400}{41.8}\sqrt{0.948}\sqrt{\frac{300}{250}} = 35.7$$

For hot-rolled UB section $k_f < 1.0$, $\alpha_b = 0$, the slenderness reduction factor can be obtained from Table 5.3 using linear interpolation as

$$\alpha_c = 0.925 - \frac{(0.925 - 0.905)(35.7 - 35)}{(40 - 35)} = 0.922$$

The design axial member capacity is determined as

$$\phi N_{cy} = \phi \alpha_c N_s = 0.9 \times 0.922 \times 2707.5 = 2246.7 \text{ kN}$$

5. In-plane member moment capacity

The in-plane member moment capacity of the column of k_f <1.0 under compression and bending can be determined as

$$M_i = M_s \left(1 - \frac{N^*}{\phi N_c}\right) = 498 \times \left(1 - \frac{87}{2436.7}\right) = 478.8 \text{ kN m}$$

$$\phi M_i = 0.9 \times 478.8 = 431 \text{ kN m} > M_x^* = 330 \text{ kN m, OK}$$

6. Out-of-plane member capacity

6.1. Member moment capacity without full lateral restraints

The portal frame column is treated as the one without fly braces. The length of the column for flexural–torsional buckling is taken as $L = 5364$ mm.

As the column is fully restrained against twist at both ends, the twist restraint factor is $k_t = 1.0$.

The column is subjected to only moments; the load height factor is $k_l = 1.0$.

As the column is restrained by base plate against lateral rotation, the lateral rotational restraint factor is $k_r = 0.85$.

The effective length factor is therefore

$$L_e = k_t k_l k_r L = 1.0 \times 1.0 \times 0.85 \times 5364 = 4559.4 \text{ mm}$$

The elastic buckling moment is calculated as

$$M_o = \sqrt{\frac{\pi^2 E_s I_y}{L_e^2}\left(GJ + \frac{\pi^2 E_s I_w}{L_e^2}\right)}$$

$$= \sqrt{\frac{\pi^2 \times 200 \times 10^3 \times 16.6 \times 10^6}{4559.4^2}\left(80 \times 10^3 \times 530 \times 10^3 + \frac{\pi^2 \times 200 \times 10^3 \times 815 \times 10^9}{4559.4^2}\right)}$$

$$= 434.5 \text{ kN m}$$

The member slenderness reduction factor is computed as

$$\alpha_s = 0.6\left[\sqrt{\left(\frac{M_s}{M_{oa}}\right)^2 + 3} - \frac{M_s}{M_{oa}}\right] = 0.6\left[\sqrt{\left(\frac{498}{434.5}\right)^2 + 3} - \frac{498}{434.5}\right] = 0.558$$

The bending moment distribution along the column is linear with zero moment at the bottom and maximum moment at the top. The moment modification factor is $\alpha_m = 1.75$.

The member moment capacity is

$$M_{bx} = \alpha_m \alpha_s M_{sx} \leq M_{sx}$$

$$= 1.75 \times 0.558 \times 498 = 486.3 \text{ kN m} < M_{sx} = 498 \text{ kN m}$$

$$\therefore M_{bx} = 486.3 \text{ kN m}$$

6.2. Out-of-plane member capacity in axial compression and bending

The out-of-plane member moment capacity of the column under axial compression and bending can be calculated as

$$M_{ox} = M_{bx}\left(1 - \frac{N^*}{\phi N_{cy}}\right) \leq M_{rx}$$

$$= 486.3 \times \left(1 - \frac{87}{2246.3}\right) = 467.3 \text{ kN m} < M_{rx} = 498 \text{ kN m}$$

$$\therefore M_{ox} = 467.5 \text{ kN m}$$

$$\phi M_{ox} = 0.9 \times 467.5 = 421 \text{ kN m} > M_x^* = 330 \text{ kN m, OK}$$

6.3. Out-of-plane member capacity in axial tension and bending

For axial tension and bending, the out-of-plane member capacity is determined as

$$M_{ox} = M_{bx}\left(1 + \frac{N^*}{\phi N_t}\right) \leq M_{rx}$$

$$= 486.3 \times \left(1 + \frac{115}{2570.4}\right) = 508 \text{ kN m} > M_{rx} = 498 \text{ kN m}$$

$$\therefore M_{ox} = 498 \text{ kN m}$$

$$\phi M_{ox} = 0.9 \times 498 = 448.2 \text{ kN m} > M_x^* = 420 \text{ kN m, OK}$$

Therefore, the capacity of the portal frame column is adequate.

5.6 MEMBERS UNDER AXIAL LOAD AND BIAXIAL BENDING

5.6.1 Section capacity under biaxial bending

Clause 8.3.4 of AS 4100 provides a simple linear expression for conservatively estimating the axial load and bending moment interaction section capacities of steel members under biaxial bending. The design axial tensile or compressive force (N^*) and design bending moments (M_x^*) and (M_y^*) about the major and minor principal axes must satisfy the following condition:

$$\frac{N^*}{\phi N_s} + \frac{M_x^*}{\phi M_{sx}} + \frac{M_y^*}{\phi M_{sy}} \le 1 \tag{5.47}$$

For compact doubly symmetric I-sections and rectangular and square hollow sections under biaxial bending, the design bending moments must satisfy the following power law expression given in Clause 8.3.4 of AS 4100:

$$\left(\frac{M_x^*}{\phi M_{rx}}\right)^{\gamma_b} + \left(\frac{M_y^*}{\phi M_{ry}}\right)^{\gamma_b} \le 1 \tag{5.48}$$

where γ_b is given as

$$\gamma_b = 1.4 + \frac{N^*}{\phi N_s} \le 2.0 \tag{5.49}$$

When there is no axial force ($N^* = 0$), the section is subjected to biaxial bending moments and $\gamma_b = 1.4$. When the axial force (N^*) is greater than $0.6N_s$, the exponent γ_b is taken as 2.0. Figure 5.18 illustrates the strength interaction curves for sections under axial load

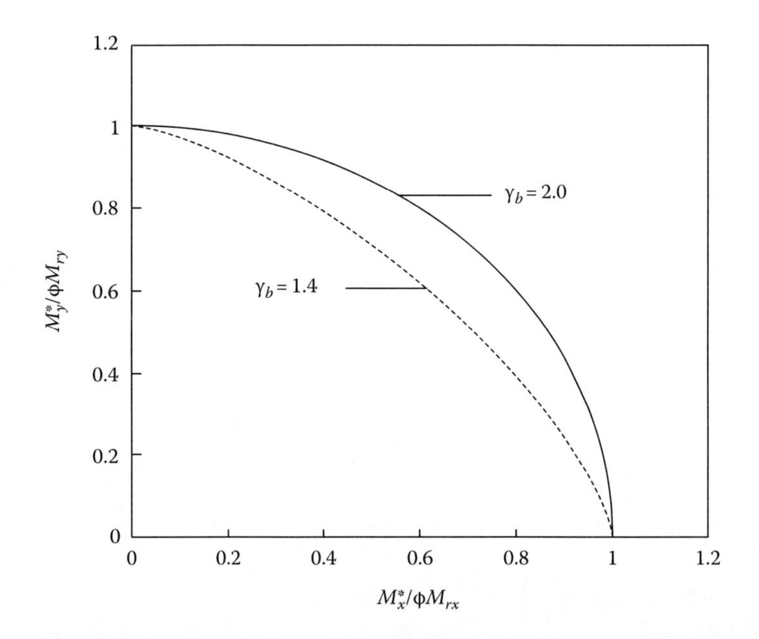

Figure 5.18 Strength interaction curves for compact symmetric I-sections under biaxial bending.

and biaxial bending. It is seen that any significant design bending moment M_y^* remarkably reduces the design bending moment capacities ϕM_{rx} of the section.

5.6.2 Member capacity under biaxial bending

The member capacity of a steel beam–column under axial compression and biaxial bending moments depends on its in-plane and out-of-plane member moment capacities. Clause 8.4.5 of AS 4100 specifies that a steel beam–column under axial compression and biaxial bending must satisfy the following strength interaction formula:

$$\left(\frac{M_x^*}{\phi M_{cx}}\right)^{1.4} + \left(\frac{M_y^*}{\phi M_{iy}}\right)^{1.4} \leq 1 \tag{5.50}$$

where

M_{cx} is taken as the lesser of the in-plane member moment capacity (M_{ix}) and the nominal out-of-plane member moment capacity (M_{ox}) for bending about the major principal x-axis

M_{iy} stands for the nominal in-plane member moment capacity about the minor principal y-axis

Figure 5.19 shows the strength interaction curve for steel member under axial compression and biaxial bending.

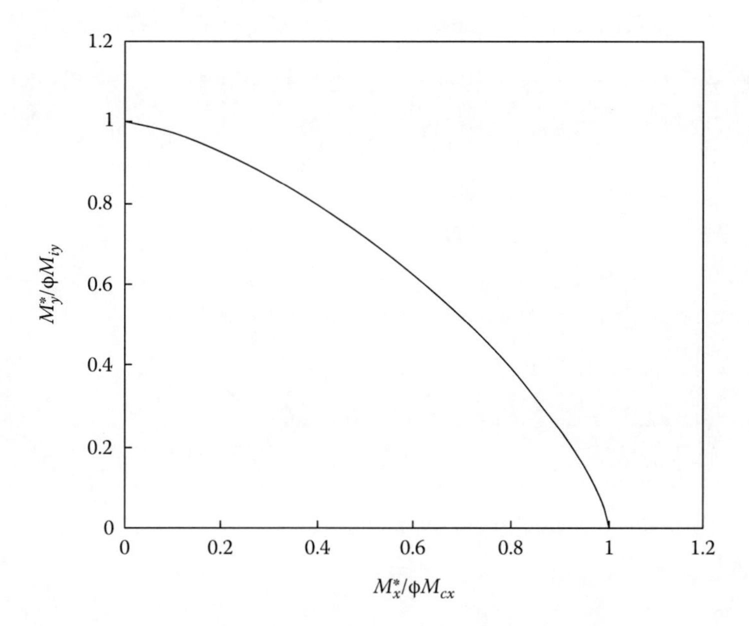

Figure 5.19 Strength interaction curves for compression members under biaxial bending.

Similarly, as noted in Clause 8.4.5.2 of AS 4100, a steel tension member subject to biaxial bending must satisfy the following interaction expression:

$$\left(\frac{M_x^*}{\phi M_{tx}}\right)^{1.4} + \left(\frac{M_y^*}{\phi M_{ry}}\right)^{1.4} \leq 1 \qquad (5.51)$$

where

M_{tx} is taken as the lesser of the nominal section moment capacity (M_{rx}) reduced by axial tension and the nominal out-of-plane member moment capacity (M_{ox}) for bending about the major principal x-axis

M_{ry} denotes the nominal section moment capacity reduced by axial tension about the minor principal y-axis

Example 5.6: Design of a steel beam–column under axial compression and biaxial bending

An 8 m length steel beam–column of 200UC59.5 Grade 300 is fixed at its base and pinned at its top. The column is braced about the y–y-axis by struts that are pin-connected to its mid-height. The struts prevent lateral deflections in the minor principal plane. The column is subjected to an axial compressive force of $N^* = 450$ kN and biaxial bending moments of $M_x^* = 85$ kN m and $M_y^* = 15$ kN m, which have been determined from the second-order elastic analysis. Check the capacity of the beam–column.

1. Section properties

The dimensions and properties of 200UC59.5 are

$$d = 210 \text{ mm}, \quad b_f = 205 \text{ mm}, \quad t_w = 9.3 \text{ mm}, \quad A_g = 7620 \text{ mm}^2, \quad I_x = 61.3\times10^6 \text{ mm}^4$$

$$r_x = 89.7 \text{ mm}, \quad Z_{ex} = 656\times10^3 \text{ mm}^3, \quad I_y = 20.4\times10^6 \text{ mm}^4, \quad r_y = 51.7 \text{ mm}$$

$$Z_{ey} = 299\times10^3 \text{ mm}^3, \quad I_w = 195\times10^9 \text{ mm}^6, \quad J = 477\times10^3 \text{ mm}^3, \quad f_y = 300 \text{ MPa}$$

$$E_s = 200\times10^3 \text{ MPa}, \quad G = 80\times10^3 \text{ MPa}, \quad k_f = 1.0$$

Compactness about the x-axis = compact
Compactness about the y-axis = compact

2. Axial section capacity

The section capacity in axial compression is calculated as follows:

$$N_{sc} = k_f A_n f_y = 1.0\times7620\times300 \text{ N} = 2286 \text{ kN}$$

$$\phi N_{sc} = 0.9\times2286 = 2057.4 \text{ kN} > N^* = 450 \text{ kN, OK}$$

3. Section moment capacities

3.1. Section moment capacities without axial forces

For bending about the major principal x-axis,

$$M_{sx} = Z_{ex} f_y = 656\times10^3\times300\times10^{-6} = 196.8 \text{ kN m}$$

$$\phi M_{sx} = 0.9\times196.8 = 177.12 \text{ kN m} > M_x^* = 85 \text{ kN m, OK}$$

For bending about the major principal y-axis,

$$M_{sy} = Z_{ey} f_y = 299 \times 10^3 \times 300 \times 10^{-6} = 89.7 \text{ kN m}$$

$$\phi M_{sy} = 0.9 \times 89.7 = 80.7 \text{ kN m} > M_y^* = 15 \text{ kN m, OK}$$

3.2. Reduced section moment capacities due to axial compression

For compact doubly symmetric I-section with k_f = 1.0, the section moment capacity reduced by axial compression for bending about x-axis can be calculated as follows:

$$M_{rx} = 1.18 M_{sx}\left(1 - \frac{N^*}{\phi N_s}\right) \leq M_{sx}$$

$$= 1.18 \times 196.8 \times \left(1 - \frac{450}{2057.4}\right) = 181.4 \text{ kN m} < M_{sx} = 196.8 \text{ kN m}$$

$$\therefore M_{rx} = 181.4 \text{ kN m}$$

$$\phi M_{rx} = 0.9 \times 181.4 = 163.3 \text{ kN m} > M_x^* = 85 \text{ kN m, OI}$$

For bending about the minor principal y-axis, the reduced section moment capacity is

$$M_{ry} = 1.19 M_{sy}\left[1 - \left(\frac{N^*}{\phi N_s}\right)^2\right] \leq M_{sy}$$

$$= 1.19 \times 89.7 \times \left[1 - \left(\frac{450}{2057.4}\right)^2\right] = 101.6 \text{ kN m} > M_{sy} = 89.7 \text{ kN m}$$

$$\therefore M_{ry} = 89.7 \text{ kN m}$$

$$\phi M_{ry} = 0.9 \times 89.7 = 80.73 \text{ kN m} > M_y^* = 15 \text{ kN m, OK}$$

3.3. Section capacities under biaxial bending

For compact doubly symmetric I-section under biaxial bending, the section capacity is determined as follows:

$$\gamma_b = 1.4 + \frac{N^*}{\phi N_s} = 1.4 + \frac{450}{2057.4} = 1.619 < 2$$

$$\left(\frac{M_x^*}{\phi M_{rx}}\right)^{\gamma_b} + \left(\frac{M_y^*}{\phi M_{ry}}\right)^{\gamma_b} \leq 1$$

$$= \left(\frac{85}{163.29}\right)^{1.619} + \left(\frac{15}{80.73}\right)^{1.619} = 0.413 < 1.0, \text{ OK}$$

4. Axial member capacities

4.1. Axial member capacity N_{cx}

Since the effect of end restraints has been taken into account in the second-order elastic analysis, the effective length factor k_e is taken as 1.0 for combined actions. The effective length is $L_e = k_e L = 1.0 \times 8000 = 8000$ mm.

The modified member slenderness is

$$\lambda_{nx} = \frac{L_{ex}}{r_x}\sqrt{k_f}\sqrt{\frac{f_y}{250}} = \frac{8000}{89.7}\sqrt{1.0}\sqrt{\frac{300}{250}} = 97.7$$

For hot-rolled universal column (UC) section with $k_f = 1.0$, $\alpha_b = 0$. The slenderness reduction factor can be obtained from Table 5.3 by linear interpolation as

$$\alpha_c = 0.575 - \frac{(0.575-0.541)(97.7-95)}{(100-95)} = 0.556$$

The design axial capacity is therefore

$$\phi N_{cx} = \phi\alpha_c N_s = 0.9 \times 0.556 \times 2286 = 1144 \text{ kN}$$

4.2. Axial member capacity N_{cy}

The portal frame column is braced by struts at its mid-height, so that the effective length is taken as 1.0. The effective length $L_e = k_e L = 1.0 \times 4000 = 4000$ mm.

The modified member slenderness is calculated as

$$\lambda_{ny} = \frac{L_{ey}}{r_y}\sqrt{k_f}\sqrt{\frac{f_y}{250}} = \frac{4000}{51.7}\sqrt{1.0}\sqrt{\frac{300}{250}} = 84.75$$

For hot-rolled UC section with $k_f = 1.0$, $\alpha_b = 0$. The slenderness reduction factor can be obtained from Table 5.3 by linear interpolation as

$$\alpha_c = 0.681 - \frac{(0.681-0.645)(84.75-80)}{(85-80)} = 0.647$$

The design axial capacity is therefore

$$\phi N_{cy} = \phi\alpha_c N_s = 0.9 \times 0.647 \times 2286 = 1331 \text{ kN}$$

5. In-plane member moment capacities

For bending about the major principal x-axis, the in-plane member moment capacity can be calculated as

$$M_{ix} = M_{sx}\left(1 - \frac{N^*}{\phi N_{cx}}\right) = 196.8 \times \left(1 - \frac{450}{1144}\right) = 119.4 \text{ kN m}$$

$$\phi M_{ix} = 0.9 \times 119.4 = 107.5 \text{ kN m} > M_x^* = 85 \text{ kN m, OK}$$

For bending about the minor principal y-axis, the in-plane member moment capacity can be calculated as

$$M_{iy} = M_{sy}\left(1 - \frac{N^*}{\phi N_{cy}}\right) = 89.7 \times \left(1 - \frac{450}{1331.5}\right) = 59.4 \text{ kNm}$$

$$\phi M_{iy} = 0.9 \times 59.4 = 53.5 \text{ kNm} > M_y^* = 15 \text{ kNm, OK}$$

6. Out-of-plane member capacity

6.1. Member moment capacity without full lateral restraints

The beam–column is braced by struts at its mid-height. The length of the upper beam–column for flexural–torsional buckling is taken as $L = 4000$ mm.

As the segment is fully restrained against twist at both ends, the twist restraint factor is $k_t = 1.0$.

The segment is subjected to only moments; the load height factor is $k_l = 1.0$.

As the segment is restrained against lateral rotation, the lateral rotational restraint factor is taken as $k_r = 1.0$.

The effective length is therefore

$$L_e = k_t k_l k_r L = 1.0 \times 1.0 \times 1.0 \times 4000 = 4000 \text{ mm}$$

The elastic buckling moment is calculated as

$$M_o = \sqrt{\frac{\pi^2 E_s I_y}{L_e^2}\left(GJ + \frac{\pi^2 E_s I_w}{L_e^2}\right)}$$

$$= \sqrt{\frac{\pi^2 \times 200 \times 10^3 \times 20.4 \times 10^6}{4000^2}\left(80 \times 10^3 \times 477 \times 10^3 + \frac{\pi^2 \times 200 \times 10^3 \times 195 \times 10^9}{4000^2}\right)}$$

$$= 396 \text{ kNm}$$

The member slenderness reduction factor is computed as

$$\alpha_s = 0.6\left[\sqrt{\left(\frac{M_s}{M_{oa}}\right)^2 + 3} - \frac{M_s}{M_{oa}}\right] = 0.6\left[\sqrt{\left(\frac{196.8}{396}\right)^2 + 3} - \frac{196.8}{396}\right] = 0.783$$

Assume that the beam–column about the x-axis bending undergoes double curvature bending, having a contraflexure point at the mid-height lateral restraint with zero moment. The moment ratio of the upper beam–column segment is

$$\beta_m = \frac{M_2^*}{M_1^*} = \frac{0}{85} = 0$$

The moment modification factor is determined as

$$\alpha_m = 1.75 + 1.05\beta_m + 0.3\beta_m^2 = 1.75$$

The member moment capacity is

$$M_{bx} = \alpha_m \alpha_s M_{sx} \leq M_{sx}$$

$$= 1.75 \times 0.783 \times 196.8 = 269.7 \text{ kN m} > M_{sx} = 196.8 \text{ kN m}$$

$$\therefore M_{bx} = 196.8 \text{ kN m}$$

6.2. Out-of-plane member capacity

The out-of-plane member moment capacity of the column under axial compression and bending can be calculated as

$$M_{ox} = M_{bx}\left(1 - \frac{N^*}{\phi N_{cy}}\right) \leq M_{rx}$$

$$= 196.8 \times \left(1 - \frac{450}{1331}\right) = 130.3 \text{ kN m} < M_{rx} = 181.4 \text{ kN m}$$

$$\therefore M_{ox} = 130.3 \text{ kN m}$$

$$\phi M_{ox} = 0.9 \times 130.3 = 117.3 \text{ kN m} > M_x^* = 85 \text{ kN m, OK.}$$

7. Member capacities under biaxial bending

The in-plane and out-of-plane member moment capacities have been calculated as

$$\phi M_{ix} = 107.5 \text{ kN m}, \quad \phi M_{ox} = 117.3 \text{ kN m}, \quad \phi M_{iy} = 53.5 \text{ kN m}$$

The critical member moment capacity about the x-axis is taken as

$$\phi M_{cx} = \min(\phi M_{ix}, \ \phi M_{ox}) = \min(107.5, \ 117.3) = 107.5 \text{ kN m}$$

The biaxial member capacities are checked as follows:

$$\left(\frac{M_x^*}{\phi M_{cx}}\right)^{1.4} + \left(\frac{M_y^*}{\phi M_{iy}}\right)^{1.4} \leq 1$$

$$= \left(\frac{85}{107.5}\right)^{1.4} + \left(\frac{15}{53.5}\right)^{1.4} = 0.888 < 1.0, \text{ OK}$$

Therefore, the capacity of the beam–column under axial load and biaxial bending is adequate.

REFERENCES

AS 4100 (1998) *Australian Standard for Steel Structures*, Sydney, New South Wales, Australia: Standards Australia.

Bennetts, I.D., Thomas, I.R., and Hogan, T.J. (1986) Design of statically loaded tension members, *Civil Engineering Transactions*, Institution of Engineers Australia, 28 (4), 318–327.

Bradford, M.A., Bridge, R.Q., Hancock, G.J., Rotter, J.M., and Trahair, N.S. (1987) Australian limit state design rules for the stability of steel structures, Paper presented at *the International Conference on Steel and Aluminium Structures*, Cardiff, UK, pp. 11–23.

Bridge, R.Q. and Trahair, N.S. (1987) Limit state design rules for steel beam-columns, *Steel Construction*, Australian Institute of Steel Construction, 21 (2): 2–11.

Bulson, P.S. (1970) *The Stability of Flat Plates*, London, U.K.: Chatto and Windus.

Cochrane, V.H. (1922) Rules for rivet hole deductions in tension members, *Engineering News-Record*, 89 (16): 847–848.

Cuk, P.E., Bradford, M.A., and Trahair, N.S. (1986) Inelastic lateral buckling of steel beam-columns, *Canadian Journal of Civil Engineering*, 13 (6): 693–699.

Davids, A.J. and Hancock, G.J. (1985) The strength of long-length I-section columns fabricated from slender plates, *Civil Engineering Transactions*, Institution of Engineers Australia, 27 (4): 347–352.

Dhalla, A.K. and Winter, G. (1974a) Steel ductility measurements, *Journal of the Structural Division*, ASCE, 100 (ST2): 427–444.

Dhalla, A.K. and Winter, G. (1974b) Suggested steel ductility requirements, *Journal of the Structural Division*, ASCE, 100 (ST2): 445–462.

Duan, L. and Chen, W.F. (1988) Effective length factor for columns in braced frames, *Journal of Structural Engineering*, ASCE, 114 (10): 2357–2370.

Duan, L. and Chen, W.F. (1989) Effective length factor for columns in unbraced frames, *Journal of Structural Engineering*, ASCE, 115 (1): 149–165.

Galambos, T.V. (ed.) (1988) *Guide to Stability Design Criteria for Metal Structures*, 4th edn., New York: John Wiley & Sons.

Hancock, G.J. (1982) Design methods for interaction buckling in box and I-section columns, *Civil Engineering Transactions*, Institution of Engineers Australia, 24 (2): 183–186.

Hancock, G.J., Davids, A.J., Keys, P.W., and Rasmussen, K. (1987) Strength tests on thin-walled high tensile steel columns, Paper presented at *the International Conference on Steel and Aluminium Structures*, Cardiff, UK, pp. 475–486.

Key, P.W., Hasan, S.W., and Hancock, G.J. (1988) Column behaviour of cold-formed hollow sections, *Journal of Structural Engineering*, ASCE, 114 (ST2): 390–407.

McGuire, W. (1968) *Steel Structures*, Englewood Cliffs, NJ: Prentice Hall.

Rasmussen, K.J.R. and Hancock, G.J. (1989) Compression tests of welded channel section columns, *Journal of Structural Engineering*, ASCE, 115 (ST4): 789–808.

Rasmussen, K.J.R., Hancock, G.J., and Davids, A.J. (1989) Limit state design of columns fabricated from slender plates, *Civil Engineering Transactions*, Institution of Engineers Australia, 27 (3): 268–274.

Rotter, J.M. (1982) Multiple column curves by modifying factors, *Journal of the Structural Division*, ASCE, 108 (ST7): 1665–1669.

Timoshenko, S.P. and Gere, J.M. (1961) *Theory of Elastic Stability*, 2nd edn., New York: McGraw-Hill.

Trahair, N.S. (1986) Design strengths of steel beam-columns, *Canadian Journal of Civil Engineering*, 13 (6): 639–646.

Trahair, N.S. and Bradford, M.A. (1998) *The Behaviour and Design of Steel Structures to AS 4100*, 3rd edn., London, U.K.: Taylor & Francis Group.

Woolcock, S.T. and Kitipornchai, S. (1986) Design of single angle web struts in trusses, *Journal of the Structural Division*, ASCE, 112 (6): 1665–1669.

Woolcock, S.T., Kitipornchai, S., and Bradford, M.A. (2003) *Limit State Design of Portal Frame Buildings*, Sydney, New South Wales, Australia: Australian Institute of Steel Construction.

Chapter 6

Steel connections

6.1 INTRODUCTION

Structural connections are used to connect a structural member to another member or to the support so that the forces carried by the structural member can be transferred to the other member or to the support. A steel connection consists of connection components and connectors. Cleats, gusset plates, brackets and connecting plates used in steel connections are called connection components, while bolts, welds and pins are connectors. Members are joined together in a connection that consists of several elements, which results in complex stress distributions within the connection. Connections in a steel structure may become potential weak spots that need careful considerations in the design. Structural connections are important parts of a steel structure that influence the overall performance of a steel structure.

This chapter deals with the behaviour and design of structural steel connections in accordance with AS 4100 (1998). The behaviour and design of bolts and bolt groups under shear, tension and combined shear and tension are discussed. The designs of welds and weld groups under in-plane and out-of-plane design actions are also given. One of the emphases of this chapter is placed on the design of bolted moment end plate connections, which includes beam normal to column connections, knee connections and ridge connections in rigid steel construction. The design principles presented for bolted moment end plate connections can be extended to the design of welded beam-to-column moment connections. Another emphasis is on the behaviour and design of pinned column base plate connections. Design procedures of structural steel connections are illustrated through worked examples.

6.2 TYPES OF CONNECTIONS

Steel connections may be classified by the amount of rotational restraint provided by the connections, which are related to the type of steel frames. Steel connections are usually classified into rigid, simple and semi-rigid connections.

Rigid connections provide full continuity at the connections which hold the angles between intersecting members unchanged after deformations. This requires that the rigid connection needs to have the rotational restraint equal to or greater than 90% of that necessary to prevent any angle change between the intersecting members. It is assumed that the deformations of rigid connections have no significant effects on the distribution of design actions or on the overall deformation of the frame. Rigid connections are used to transfer the design actions of bending moment, shear force and axial force from one member to another in steel rigid frames. Typical examples of rigid connections are welded moment

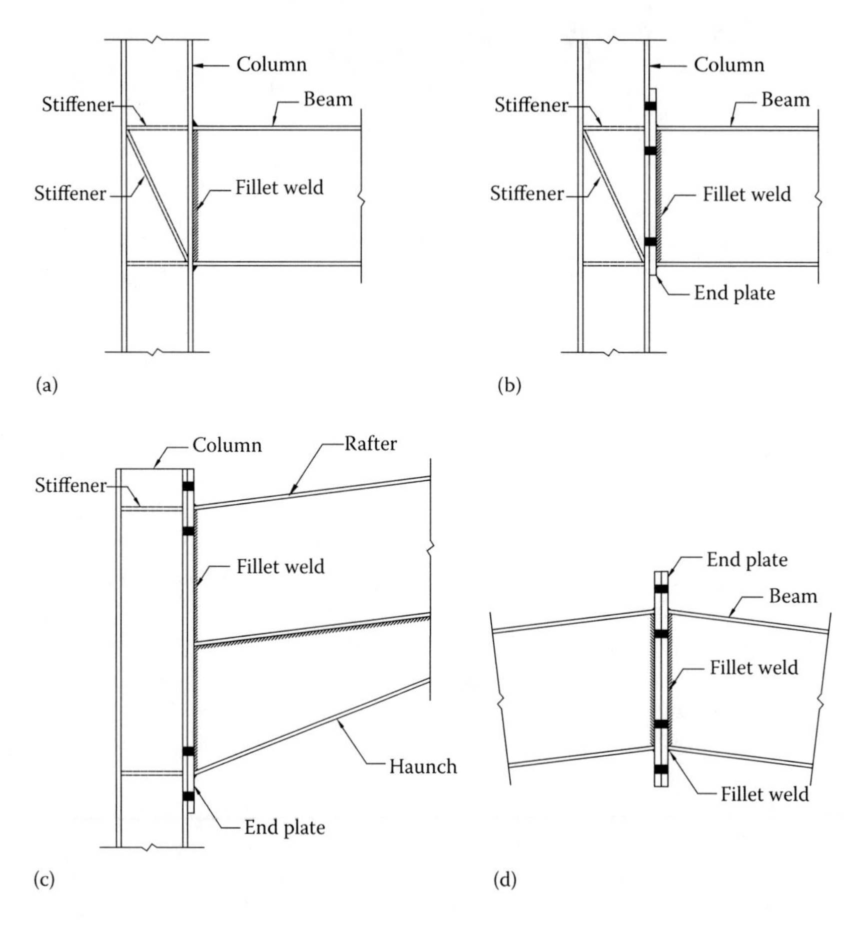

Figure 6.1 Rigid connections: (a) welded moment connection, (b) bolted moment end plate connection, (c) knee joint and (d) ridge connection.

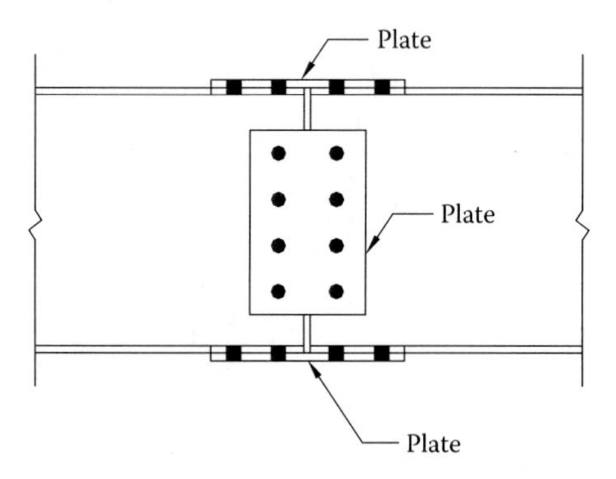

Figure 6.2 Splice connection.

connections and bolted moment end plate connections depicted in Figure 6.1 and bolted splices illustrated in Figure 6.2.

Simple connections provide little rotational restraint at the ends of a member so that the ends of the member can rotate under applied loads. In simple connections, the change in the original angle between intersecting members is 80% or more of that caused by the use of frictionless hinged connections. Simple connections are designed to transfer shear force only from one member to another in a simple framing system. Some standard simple connections are depicted in Figures 6.3 and 6.4, including angle seat, bearing pad, flexible end plate, angle cleat, beam-to-column and beam-to-beam web side plate connections. In simple construction, simple connections must be designed to not only withstand the reactions from

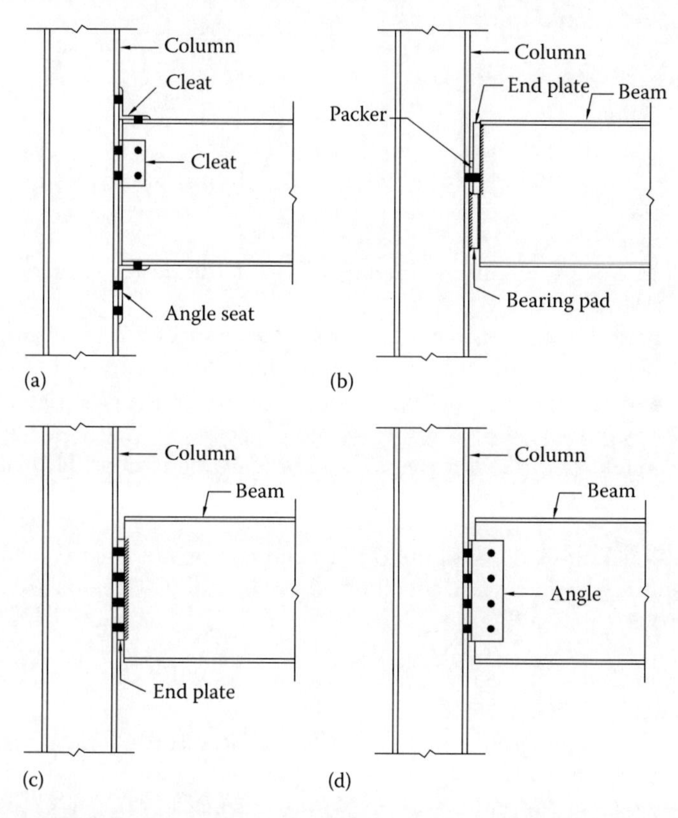

Figure 6.3 Flexible connections: (a) angle seat connection, (b) bearing pad connection, (c) flexible end plate connection and (d) angle cleat connection.

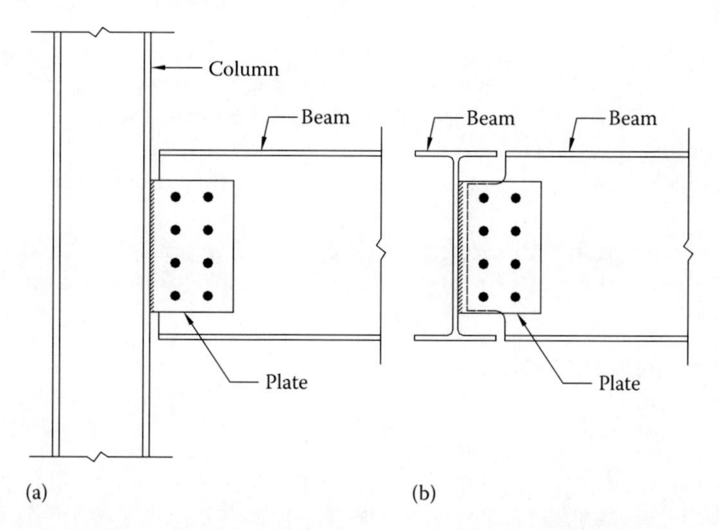

Figure 6.4 Flexible web side plate connections: (a) beam-to-column connection and (b) beam-to-beam connection.

the simply supported beams and the factored lateral loads but also have sufficient inelastic rotation capacity to allow angle changes between intersecting members.

Semi-rigid connections provide some degrees of rotational restraint at the ends of a member so that the connections can transfer bending moments and shear and axial forces in semi-rigid steel frames. The rotational stiffness of semi-rigid connections is between that of rigid connections and simple connections.

6.3 MINIMUM DESIGN ACTIONS

Clause 9.1.4 of AS 4100 (1998) requires that steel connections be designed to transmit the greater of the design action in the member or the minimum design action given as follows:

1. The minimum design bending moment ($M^*_{\min}$) for the design of a rigid connection is taken as $0.5\phi M_b$.
2. The minimum design shear force ($V^*_{\min}$) for simple connections to a beam is taken as the lesser of $0.15\phi V_v$ and 40 kN, where ϕV_v is the member design shear capacity.
3. The minimum design axial force ($N^*_{\min}$) for connections at the ends of tensile or compression members is 0.3 times the member design capacity.
4. The minimum tensile force for threaded rod bracing member with turnbuckles is taken as the member design capacity.

The minimum design actions for designing splice connections in tension members, compression members, flexural members and members under combined actions are also specified in Clause 9.1.4. of AS 4100 as follows:

1. The minimum design force for splices in tension members is taken as $0.3\phi N_t$, where ϕN_t is the member design capacity in axial tension.
2. Splices in axial compression members prepared for full contact at their ends must carry the compressive actions by bearing on contact surfaces.
3. The minimum design force for fasteners in the splices is $0.15\phi N_c$, where ϕN_c is the member design capacity in axial compression.
4. The minimum design forces for splice connections in compression members that are not prepared for full contact is $0.3\phi N_c$.
5. Splice connections between points of effective lateral supports under axial compression must be designed for combined actions of axial compression and bending moment taking as $M^* = \delta_m N^* l_s/1000$, where δ_m is the amplification factor and l_s is the distance between points of effective lateral supports.
6. The minimum design bending moment for splice connections in flexural members is $M^*_{\min} = 0.3\phi M_b$.
7. The splice connections in members under combined actions must satisfy all minimum design action requirements for members under single action as described earlier.

6.4 BOLTED CONNECTIONS

6.4.1 Types of bolts

The types of bolts used in steel connections include Property Class 4.6 commercial bolts, Property Class 8.8 high-strength structural bolts and Property Class 8.8, 10.9 and 12.9 precision bolts. Property Class 4.6 commercial bolts conforming to AS 1111 are made of low-carbon steel. They are used only for snug-tight installation designated as 4.6/S bolts. Property Class 8.8 high-strength structural bolts conforming to AS/NZS 1252 are made of medium carbon steel. Their properties are enhanced by quenching and tempering. Class 8.8 high-strength structural bolts can be highly tensioned and are used for snug-tight installation designated as 8.8/S. These high-strength structural bolts are designated as 8.8/TB when used in bearing mode connections and as 8.8/TF when used in friction mode connections. Property Class 8.8, 10.9 and 12.9 precision bolts are used for mechanical assembly. The minimum yield stress of Property Class 4.6 bolts is 240 MPa, while their minimum tensile strength is 400 MPa. Property Class 8.8 high-strength structural bolts have a minimum yield stress of 660 MPa and a minimum tensile strength of 830 MPa.

6.4.2 Bolts in shear

Bolts in steel connections are subjected to shear and bearing as depicted in Figure 6.5. The shear strengths of bolts can be determined by experiments in which bolts are subjected to double shear caused by plates either in tension or compression. Test data indicated that the average shear strength was about 62% of the tensile strength of the bolt (Kulak et al. 1987). In addition, it was found that the level of initial tension applied to the bolt does not have a significant effect on the ultimate shear strength of the bolt. The shear strength of a bolt also depends on the shear area of the bolt, the number of shear planes and the length of the joint. The total strength of a bolted lap splice connection was found to decrease with an increase in the length of the connection. In AS 4100, a reduction factor is used to account for the effect of the length of the bolted lap connections on the shear strength of the bolts.

The nominal shear strength of a bolt is calculated by the following equation provided in Clause 9.3.2.1 of AS 4100 (1998) as follows:

$$V_f = 0.62 f_{uf} k_{rc} (n_n A_c + n_x A_o) \tag{6.1}$$

where

f_{uf} stands for the minimum tensile strength of the bolt

k_{rc} denotes the reduction factor accounting for the effect of the length of a bolted lap connection

The factor k_{rc} is taken as 1.0 for the connection length (l_j) less than 300 mm, 0.75 for $l_j > 1300$ mm and $(1.075 - l_j/4000)$ for $300 \le l_j \le 1300$ mm (McGuire 1968; Kulak et al. 1987). In Equation 6.1, n_n is the number of shear planes with threads intercepting the

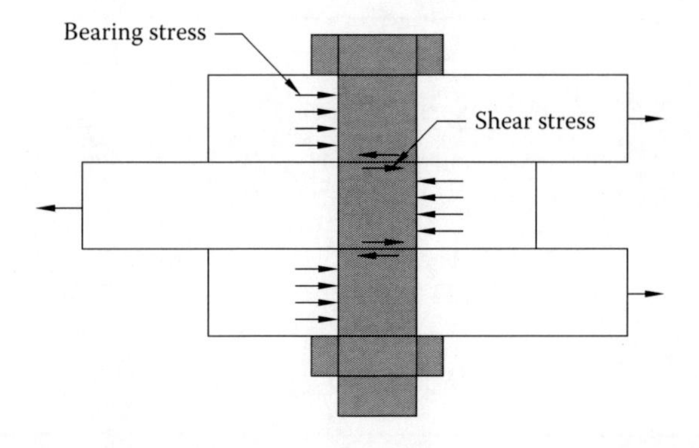

Figure 6.5 Bolt in shear and bearing.

shear plane, A_c is the core area of the bolt, n_x is the number of shear planes without threads intercepting the shear plane, and A_o is the plain shank area of the bolt. The nominal diameters (d_f) of commonly used bolts vary from 12 to 36 mm. The core, shank and tensile stress areas of bolts are given in Table 6.1. Based on AS 1275 (1985), the tensile stress area is calculated as $A_s = \pi(d_f - 0.9382p)^2/4$, where p is the thread pitch. The core area is calculated as $A_c = \pi(d_f - 1.0825p)^2/4$, and the shank area is computed as $A_o = \pi d_f^2/4$.

A bolt under a design shear force (V_f^*) must satisfy the following strength requirement:

$$V_f^* \leq \phi V_f \tag{6.2}$$

where the capacity reduction factor $\phi = 0.8$. The design capacities of 4.6/S bolts and of 8.8/S and 8.8/TB bolts in single shear are given in Tables 6.2 and 6.3, respectively. The value in bracket in Table 6.3 for M20 8.8 bolt is the currently used design value.

Table 6.1 Geometric properties of bolts

Nominal diameter d_f (mm)	*Thread pitch p (mm)*	*Tensile stress area A_s (mm²)*	*Core area A_c (mm²)*	*Shank area A_o (mm²)*
12	1.75	84.3	80.2	113.1
16	2	156.7	150.3	201.1
20	2.5	244.8	234.9	314.2
24	3	352.5	338.2	452.4
30	3.5	560.6	539.6	706.9
36	4	816.7	787.7	1017.9

Table 6.2 Design capacities of 4.6/S bolts

		Single shear ϕV_f (kN)	
Nominal diameter d_f (mm)	*Axial tension ϕN_{tf} (kN)*	*Threads included*	*Threads excluded*
12	27.0	15.9	22.4
16	50.1	29.8	39.9
20	78.3	46.6	62.3
24	112.8	67.1	89.8
30	179.4	107.1	140.2
36	261.3	156.3	202

Table 6.3 Design capacities of 8.8/S and 8.8/TB bolts

		Single shear ϕV_f (kN)	
Nominal diameter d_f (mm)	*Axial tension ϕN_{tf} (kN)*	*Threads included*	*Threads excluded*
16	104.0	61.9	82.8
20	162.5	96.7 (92.6)	129.3
24	234.1	139.2	186.2
30	372.2	222.1	291.0
36	542.3	324.3	419.0

For friction-type connections such as 8.8/TF category bolts, the slip needs to be limited under the serviceability loads. Connections where slip theoretically exceeds 2–3 mm are classified as slip critical and need to be designed for serviceability limit state (Fisher et al. 1978; Galambos et al. 1982; Birkemoe 1983). In Clause 9.3.3.1 of AS 4100, the nominal shear capacity of a bolt under service load is given by

$$V_{sf} = \mu n_{ei} N_{ti} k_h \tag{6.3}$$

where

μ is the slip factor
n_{ei} denotes the number of effective interfaces
N_{ti} is the minimum bolt tension at installation
k_h is the factor accounting for the effect of different hole types and is taken as 1.0 for standard holes, 0.85 for short slotted and oversize holes and 0.7 for long slotted holes

If surfaces in contact are clean as rolled surfaces, the slip factor is taken as 0.35 (Kulak et al. 1987).

The design requirement of bolts subjected to a design shear force for the serviceability limit state is

$$V_{sf}^{*} \le \phi V_{sf} \tag{6.4}$$

6.4.3 Bolts in tension

The strength of a bolt in axial tension is governed by the threaded part of the bolt. Before subjected to the applied axial tensile force, the bolt is usually tightened by turning the nut. However, this does not have a significant effect on the tensile strength of the bolt (Kulak et al. 1987). In addition, it was found that tensioned bolts can withstand direct axial tensile forces without any significant reduction in their tensile strength.

The nominal tensile capacity of a bolt can be determined in accordance with Clause 9.3.2.2 of AS 4100 as follows:

$$N_{tf} = A_s f_{uf} \tag{6.5}$$

where A_s is the tensile stress area of a bolt as given in AS 1275 and Table 6.1.

The design of a bolt in axial tension must satisfy

$$N_{tf}^{*} \le \phi N_{tf} \tag{6.6}$$

where N_{tf}^{*} is the design tension force and the capacity reduction factor $\phi = 0.8$. The design capacities of 4.6/S and 8.8/S and 8.8/TB bolts in axial tension are given in Tables 6.2 and 6.3, respectively.

6.4.4 Bolts in combined shear and tension

For a bolt subject to combined shear and tension, an interaction relationship based on experimental results (Kulak et al. 1987) is used to determine the ultimate strength of the bolt as specified in Clause 9.3.2.3 of AS 4100:

$$\left(\frac{V_f^*}{\phi V_f}\right)^2+\left(\frac{N_{tf}^*}{\phi N_{tf}}\right)^2 \le 1.0 \tag{6.7}$$

where

ϕV_f denotes the design shear capacity of the bolt under shear force alone
ϕN_{tf} is the design tensile capacity of the bolt subject to tension force alone
$\phi = 0.8$

The slip of friction-type connections subjected to combined service loads of shear and tension is required to be limited for the serviceability limit state. For this purpose, a bolt under combined shear and tension must satisfy the following linear interaction equation (Research Council on Structural Connections 1988) given in Clause 9.3.3.3 of AS 4100:

$$\left(\frac{V_{sf}^*}{\phi V_{sf}}\right)+\left(\frac{N_{tf}^*}{\phi N_{tf}}\right) \le 1.0 \tag{6.8}$$

where

V_{sf}^* denotes the design shear force acting on the bolt in the plane of the interface and stands for the design tension force acting on the bolt
V_{sf} is the design shear capacity of the bolt given in Equation 6.1
N_{tf} is the nominal tensile capacity of the bolt and is taken as the minimum bolt tension at installation (N_{ti})
$\phi = 0.7$

6.4.5 Ply in bearing

In a bolted connection under shear force, the connection plate (ply) is subjected to bearing due to bolts in shear as illustrated in Figure 6.6. The local bearing failure of the ply occurs at a bearing stress between $4.5f_{yp}$ and $4.9f_{yp}$ (Hogan and Thomas 1979a; Kulak et al. 1987).

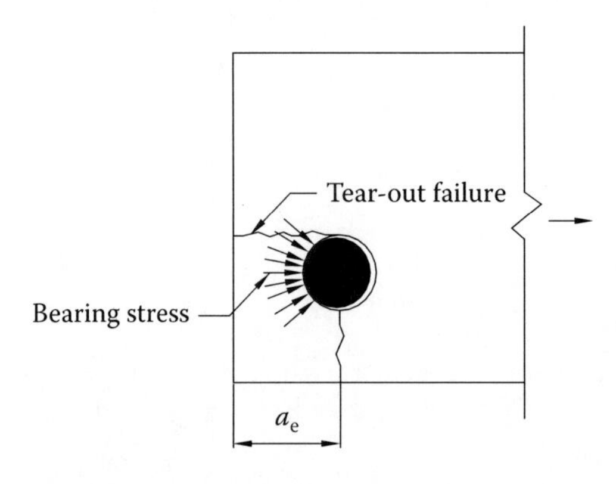

Figure 6.6 Bearing and tear-out of ply.

The design equation given in Clause 9.3.2.4 of AS 4100 for calculating the nominal bearing capacity of a ply due to a bolt in shear is based on the lower bearing stress of $4.5f_{yp}$ and is expressed by

$$V_{bp} = 3.2 d_f t_p f_{up} \tag{6.9}$$

where
t_p is the thickness of the ply
f_{up} is the tensile strength of the ply

For a ply subjected to a force acting towards an edge as shown in Figure 6.6, the bearing or tearing failure may occur. The strength of a ply in bearing may be limited by the bearing or tearing failure. The tearing failure is usually more critical than the bearing failure when the end distance (a_e) measured from the centre of the bolt hole to the edge of the ply in the direction of the force is less than $3.2d_f$. As specified in Clause 9.3.2.4 of AS 4100, the tear-out capacity of the ply is determined as (Kulak et al. 1987)

$$V_{tp} = a_e t_p f_{up} \tag{6.10}$$

The nominal bearing capacity (V_{fb}) of the ply subjected to a force towards an edge should be taken as the lesser of V_{bp} and V_{tp}. A ply subjected to a design bearing force (V_b^*) due to a bolt in shear must satisfy the following condition:

$$V_b^* \le \phi V_{fb} \tag{6.11}$$

where $\phi = 0.9$ is the capacity reduction factor.

6.4.6 Design of bolt groups

6.4.6.1 Bolt groups under in-plane loading

For bolt group subjected to in-plane loading, the elastic analysis can be used to determine the design actions in a bolt group, provided that the assumptions given in Clause 9.4.1 of AS 4100 are satisfied. These assumptions are: (a) the connection plates must be rigid; (b) the connection plates rotate about the instantaneous centre of the bolt group; (c) for a bolt group subjected to a pure couple, instantaneous centre of rotation is located at the centroid of the bolt group; (d) the superposition method can be used; and (e) the design shear force in each bolt acts at right angle to the radius from the instantaneous centre to the bolt.

Assuming the cross-sectional area of each bolt in a group is unity and all bolts have the same size, the second moments of area of a bolt group can be computed by the following:

$$I_x = \sum y_n^2 \tag{6.12}$$

$$I_y = \sum x_n^2 \tag{6.13}$$

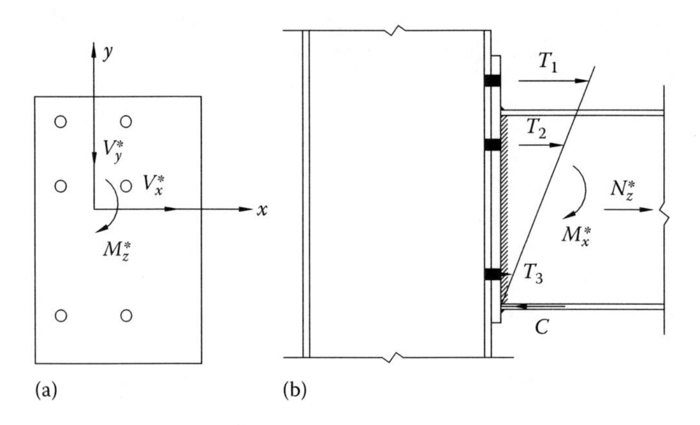

Figure 6.7 Bolt groups: (a) in-plane actions and (b) out-of-plane actions.

$$I_p = I_x + I_y \tag{6.14}$$

where

I_x and I_y are the second moments of area of the bolt group about its centroidal axis
I_p represents the polar second moment of area of the bolts in the group
x_n and y_n are the coordinates of a bolt

It is assumed that the horizontal force V_x^* and vertical force V_y^* applied to a bolt group as presented in Figure 6.7a are equally shared by all bolts in the group. The forces on any bolt in the group can be determined by

$$V_{xb}^* = \frac{V_x^*}{n_b} \tag{6.15}$$

$$V_{yb}^* = \frac{V_y^*}{n_b} \tag{6.16}$$

where n_b is the total number of bolts in the bolt group.

The bolt force due to the design bending moment M_z^* about the centroid of the bolt group is proportional to the distance from the centroid of the bolt group. The maximum bolt forces in x and y directions due to M_z^* occur at the farthest bolt from the centroid of the bolt group (Thomas et al. 1985; Hogan and Thomas 1994) and are determined by

$$V_{xbm}^* = \frac{M_z^* y_{\max}}{I_p} \tag{6.17}$$

$$V_{ybm}^* = \frac{M_z^* x_{\max}}{I_p} \tag{6.18}$$

in which $x_{\max}$ and $y_{\max}$ are the distances from the bolt group centroid to the farthest corner bolt.

The resultant design shear force on the bolt located farthest away from the centre of the bolt group can be determined as

$$V_{res}^* = \sqrt{\left(V_{xb}^* + V_{xbm}^*\right)^2 + \left(V_{yb}^* + V_{ybm}^*\right)^2} \tag{6.19}$$

The resultant design shear force (V_{res}^*) on the farthest bolt must be less than the design shear capacity (ϕV_f) of the bolt and the bearing capacity (ϕV_{fb}) of the ply.

6.4.6.2 Bolt groups under out-of-plane loading

For a bolt group subjected to the out-of-plane actions as depicted in Figure 6.7b, the forces in tension bolts can be determined by assuming a linear distribution of force from the neutral axis to the farthest bolts. The methods of analysis for bolt groups under out-of-plane actions are given by McGuire (1968), Kulak et al. (1987), AISC-LRFD Manual (1994) and Hogan and Thomas (1994). The neutral axis can be assumed to be placed at the $d/6$ from the bottom of the end plate of a depth d (Gorenc et al. 2005). The tension force on a bolt can be calculated by the following equation (Trahair and Bradford 1998):

$$T_i = \frac{N_z^*}{n_b} + \frac{M_x^* y_i}{\sum y_i^2} \tag{6.20}$$

where y_i is the coordinate of the bolt from the centroid of the bolt in the y direction.

The tension force in each of the critically loaded bolt is $N_{tf}^* = T_1/n_{b1}$, where n_{b1} is the number of bolts in the top row. The design shear force (V_o^*) on the bolt group is assumed to be equally shared by all bolts. Therefore, the design shear force on each bolt is $V_f^* = V_o^*/n_b$. The capacity of the bolt under combined shear and tension can be checked using Equation 6.7.

Example 6.1: Capacity of bolted splice connection in tension

Figure 6.8 shows a bolted splice connection in double shear arrangement under a design axial tension force of $N^* = 850$ kN. Grade 300 steel and M20 8.8/S bolts are used. Check the design capacity of this bolted splice connection.

1. Design capacity of steel member

The yield stress and tensile strength of the member and splice plate sections are obtained from Table 2.1 of AS 4100 as $f_y = 300$ MPa and $f_u = 430$ MPa, respectively.

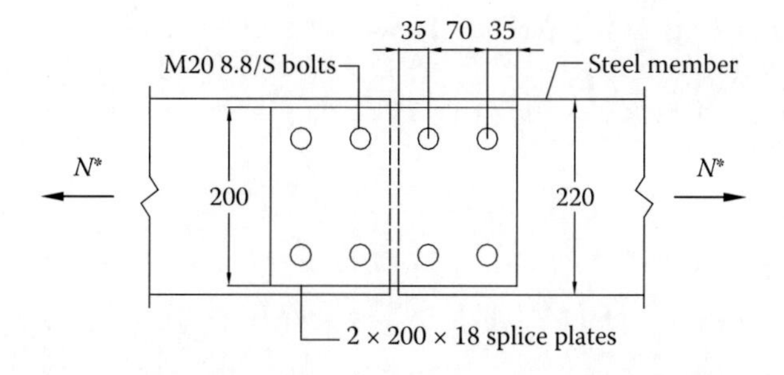

Figure 6.8 Bolted splice connection.

The gross cross-sectional area of the steel member is

$$A_g = 220 \times 20 = 4400 \text{ mm}^2$$

The net cross-sectional area of the splice plate is

$$A_n = 4400 - 2 \times 24 \times 20 = 3440 \text{ mm}^2$$

The connection is symmetric so that $k_{tc} = 1.0$. The fracture capacity of the steel member is

$$\phi N_{ta} = \phi 0.85 k_{tc} A_n f_u = 0.9 \times 0.85 \times 1.0 \times 3440 \times 430 \times 10^{-3} = 1131.6 \text{ kN}$$

The gross yield capacity of the steel member is

$$\phi N_{ty} = \phi A_g f_y = 0.9 \times 4400 \times 300 \times 10^{-3} = 1188 \text{ kN}$$

Thus, $\phi N_t = \min(1131.6;1188) = 1131.6 \text{ kN} > N^* = 850 \text{ kN}$, OK

The minimum design axial tension force is

$$N^*_{\min} = 0.3(\phi N_t) = 0.3 \times 1131.6 = 339.5 \text{ kN} < N^* = 850 \text{ kN}$$

Therefore, the design tension force $N^* = 850$ kN is used in the design of the connection.

2. Design capacity of splice plate

The gross cross-sectional area of the steel member is

$$A_g = 200 \times 18 = 3600 \text{ mm}^2$$

The net cross-sectional area of the splice plate is

$$A_n = 3600 - 2 \times 24 \times 18 = 2736 \text{ mm}^2$$

The connection is symmetric so that $k_{tc} = 1.0$. The fracture capacity of the steel splice plate is

$$\phi N_{ta} = \phi 0.85 k_{tc} A_n f_u = 0.9 \times 0.85 \times 1.0 \times 2736 \times 430 \times 10^{-3} = 900 \text{ kN}$$

The gross yield capacity of the steel member is

$$\phi N_{ty} = \phi A_g f_y = 0.9 \times 3600 \times 300 \times 10^{-3} = 972 \text{ kN}$$

Thus, $\phi N_t = \min(900;972) = 900 \text{ kN} > N^* = 850 \text{ kN}$, OK

3. Shear capacity of bolts

The core and shank areas of a M20 bolt are obtained from Table 6.1 as

$$A_c = 234.9 \text{ mm}^2, \quad A_o = 314.2 \text{ mm}^2$$

The design capacity of a bolt in double shear is computed as

$$\phi V_f = \phi 0.62 f_{uf} k_{rc}(n_n A_c + n_x A_o) = 0.8 \times 0.62 \times 830 \times 1.0 \times (1 \times 234.9 + 1 \times 314.2)\ \text{N} = 226\ \text{kN}$$

The design shear capacity of 4 bolts: 4 × 226 = 904 kN > N^* = 850 kN, OK

4. Bearing capacity of connection plate

The design tear-out capacity of a ply can be calculated as

$$\phi V_{tp} = \phi a_e t_p f_{up} = 0.9 \times 35 \times 20 \times 430 \times 10^{-3} = 270.9\ \text{kN}$$

The total design capacity of 4 bolts in bearing is

$$4\phi V_{tp} = 4 \times 270.9 = 1083.6\ \text{kN} > N^* = 850\ \text{kN, OK}$$

The design bearing capacity of the splice plate due to a bolt in shear is

$$\phi V_{bp} = \phi 3.2 d_f t_p f_{up} = 0.9 \times 3.2 \times 20 \times 20 \times 430 \times 10^{-3} = 495.4\ \text{kN}$$

The total design bearing capacity of the splice plate due to 4 bolts in shear is

$$4\phi V_{bp} = 4 \times 495.4 = 1981.4\ \text{kN} > N^* = 850\ \text{kN, OK}$$

6.5 WELDED CONNECTIONS

6.5.1 Types of welds

Welding is used in the fabrication of steel sections, connections and members and in the attachment of stiffeners. The types of welds used in steel connections include butt, fillet and compound welds. From the strength consideration, butt welds are preferable. Butt welds, however, require careful preparations of the plates for welding and are hence costly. In contrast, fillet welds require only minimal weld preparations involving a straightforward welding process, which makes them less costly. Compound welds consist of butt and fillet welds and are used to provide a smoother transition which reduces the stress concentrations. The weld qualities or categories, which are a measure of the permitted level of defects present on deposited welds, are usually classified into structural purpose (SP) and general purpose (GP). SP weld category is used for highly stressed welds, while GP weld category is for lowly stressed welds and non-structural welds.

6.5.2 Butt welds

Butt welds can be divided into two groups, namely, complete penetration butt welds and incomplete penetration butt welds. A complete penetration butt weld has fusion between the weld and parent metal throughout the complete depth of the joint as depicted in Figure 6.9. An incomplete penetration butt weld has fusion between the weld and parent metal over part of the depth of the joint.

As specified in Clause 9.7.2.7 of AS 4100, the design capacity of a complete penetration butt weld can be taken as the design capacity of the weaker part of the parts joined, where

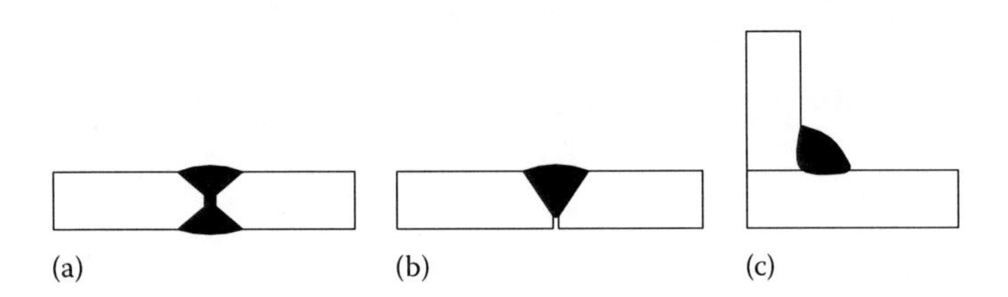

Figure 6.9 Butt and fillet welds: (a) complete penetration butt weld, (b) incomplete penetration butt weld and (c) fillet weld.

the capacity reduction factor (ϕ) is taken as 0.9 for SP category welds and 0.6 for GP category welds. The design capacity of an incomplete penetration butt weld is determined as for a fillet weld.

6.5.3 Fillet welds

The failure plane of a fillet weld may be subjected to resultant forces including shear force parallel to the longitudinal axis of the weld, shear force perpendicular to the longitudinal axis of the weld and normal compression or tensile force to the theoretical plane. It is assumed that the normal or shear stresses on the failure plane are uniformly distributed. The capacity of a fillet weld is determined by the nominal shear capacity across the weld throat or failure plane. The nominal capacity of a fillet weld per unit length is given in Clause 9.7.3.10 of AS 4100 as follows:

$$v_w = 0.6 f_{uw} t_t k_{rw} \tag{6.21}$$

in which f_{uw} is the tensile strength of weld metal, which is 410 MPa for E41XX welds and 480 MPa for E48XX welds. The design throat thickness (t_t) is taken as $0.707D_w$ (D_w is the leg length of the fillet weld). The reduction factor k_{rw}, which accounts for effect of the length (l_w) of a welded lap connection, is taken as follows:

- $k_{rw} = 1.0$ for $l_w \le 1.7\,\text{m}$
- $k_{rw} = 1.10 - 0.06 l_w$ for $1.7 < l_w \le 8.0$
- $k_{rw} = 0.62$ for $l_w > 8\,\text{m}$

The fillet weld subjected to a design force per unit length of weld (v_w^*) must satisfy

$$v_w^* \le \phi v_w \tag{6.22}$$

where the capacity reduction factor ϕ is 0.8 for SP category welds, 0.6 for GP category welds and 0.7 for SP category longitudinal welds to rectangular hollow sections with wall thickness less than 3 mm. The design force (v_w^*) is the vector resultant of all forces acting on the fillet weld. The design capacities of equal-leg fillet welds are given in Table 6.4.

The design capacity of an incomplete butt weld is determined as that of the fillet weld by taking k_{rw} = 1.0. The design of compound weld should satisfy the strength requirement of a butt weld.

Table 6.4 *Design capacities of fillet welds*

	ϕv_w (kN/mm)			
	Category SP		Category GP	
Leg size D_w (mm)	E41XX	E48XX	E41XX	E48XX
4	0.557	0.652	0.417	0.489
5	0.696	0.814	0.522	0.611
6	0.835	0.977	0.626	0.733
8	1.113	1.303	0.835	0.977
10	1.391	1.629	1.044	1.222
12	1.670	1.955	1.252	1.466

6.5.4 Weld groups

6.5.4.1 Weld group under in-plane actions

Welded connections may be subjected to in-plane actions of forces and bending moment. Weld groups in the connection need to be designed to resist these in-plane actions. To simplify the analysis of weld groups, the following assumptions are made: (a) the welds are treated as homogeneous, isotropic and elastic elements, (b) the plate being welded is rigid in the plane of the weld group and (c) the effects of residual stresses and stress concentration are ignored (Swannell 1979; Hogan and Thomas 1994, 1979b).

Figure 6.10a shows the in-plane design forces and bending moment acting on the weld group. The forces act at the centroid of the weld group with the design bending moment about the centroid. The forces per unit length in the weld segment farthest from the centroid of the weld group can be calculated as follows:

$$v_x^* = \frac{V_x^*}{L_w} - \frac{M_z^* y_{\max}}{I_{wp}} \quad (6.23)$$

$$v_y^* = \frac{V_y^*}{L_w} + \frac{M_z^* x_{\max}}{I_{wp}} \quad (6.24)$$

where

$x_{\max}$ and $y_{\max}$ are the coordinates of the weld segment located farthest from the centroid of the weld group

L_w is the total length of the weld in the weld group, and the polar second moment of area of the weld group is given by

$$I_{wp} = \sum \left(x_i^2 l_{iw} + y_i^2 l_{iw} \right) \quad (6.25)$$

where

x_i and y_i are the coordinates of the ith weld segment

l_{iw} is the length of the ith weld segment

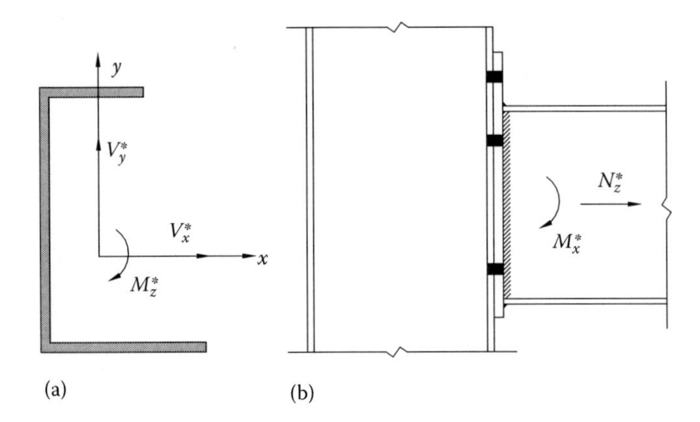

Figure 6.10 Weld groups: (a) in-plane actions and (b) out-of-plane actions.

The resultant force per unit length acting on the most critically loaded part of the weld group must satisfy

$$v_{res}^* = \sqrt{\left(v_x^*\right)^2 + \left(v_y^*\right)^2} \le \phi v_w \tag{6.26}$$

6.5.4.2 Weld group under out-of-plane actions

The out-of-plane design forces and bending moment acting on the weld group are depicted in Figure 6.10b. The same assumptions used for weld groups subjected to in-plane actions are adopted for the analysis of weld groups under out-of-plane actions. The weld group is subjected to a design bending moment M_x^* about the centroid.

The forces per unit length in the weld segment farthest from the centroid of the weld group subjected to the out-of-plane design actions of N_z^* and M_x^* can be calculated as follows:

$$v_z^* = \frac{N_z^*}{L_w} + \frac{M_x^* y_{\max}}{I_{wx}} \tag{6.27}$$

where I_{wx} is the second moment of area of the weld group about the x-axis of the weld group and is expressed by

$$I_{wx} = \sum y_i^2 l_{iw} \tag{6.28}$$

The weld group may also be subjected to an in-plane design shear force V_y^*. The shear per unit length in the weld segment farthest from the centroid of the weld group is given by

$$v_y^* = \frac{V_y^*}{L_w} \tag{6.29}$$

The resultant force per unit length acting on the most critically loaded part of the weld group under combined in-plane and out-of-plane design actions must satisfy

$$v_{res}^* = \sqrt{\left(v_y^*\right)^2 + \left(v_z^*\right)^2} \le \phi v_w \tag{6.30}$$

For a weld group with welds around the perimeter of a steel I-section, the weld group can be divided into subgroups to simplify the analysis of the weld group. It is assumed that the welds around the flanges of the I-section resist the bending moment and the total shear force is resisted by the welds around the web (Gorenc et al. 2005). The force acting at the flange caused by the bending moment is

$$N_f^* = \frac{M_o^*}{d - t_f} \tag{6.31}$$

where
  d is the depth of the steel I-section
  t_f is the thickness of the flanges
The flange fillet welds must satisfy

$$N_f^* \le \phi v_w L_w \tag{6.32}$$

where the length of the weld around each flange is taken as $L_w = 2b_f$ and b_f is the width of the flange.

Similarly, the fillet welds around the web under shear force V_z^* must satisfy

$$V_z^* \le \phi v_w L_w \tag{6.33}$$

where the length of the fillet welds around web is taken as $L_w = 2d_1$ and d_1 is the clear depth of the web.

Example 6.2: Design of welded beam-to-column connection

A welded beam-to-column moment connection is subjected to a vertical design shear force of 35 kN and an out-of-plane design bending moment of 142 kN m. The steel beam with Grade 300 steel 360UB50.7 section shown in Figure 6.11 is fully restrained from lateral buckling. Use the simple method to design this welded connection.

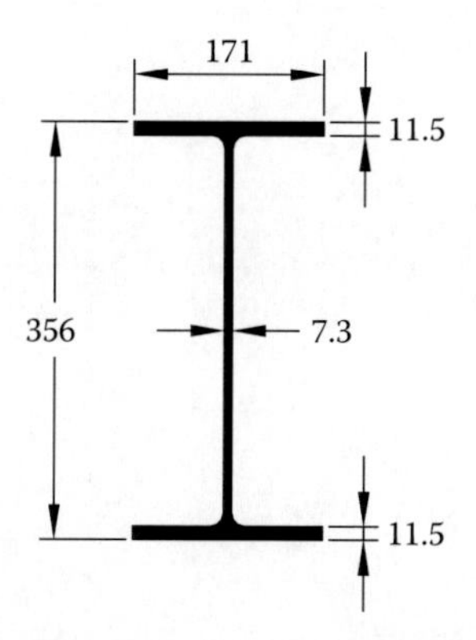

Figure 6.11 Cross section of 360UB50.7.

1. Design actions

The section design capacity of the steel beam is

$$\phi M_{sx} = \phi Z_{ex} f_y = 0.9 \times 897 \times 10^3 \times 300 \times 10^{-6} = 242.2\,\text{kN m}$$

Since the steel beam is fully restrained from lateral buckling, the member design moment capacity of the beam is

$$\phi M_{bx} = \phi M_{sx} = 242.2\,\text{kN m}$$

The minimum design bending moment for the connection is

$$M^*_{\min} = 0.5\phi M_{bx} = 0.5 \times 242.2 = 121\,\text{kN m} < M^* = 142\,\text{kN m}$$

Therefore, the design actions for the design of connections are

$$M^* = 142\,\text{kN m}, \quad V^* = 35\,\text{kN}$$

2. Design of flange welds

The flanges and web of the steel beam section are fillet welded to the flange of the steel column. The flange forces due to the design bending moment M^* are transmitted by flange welds alone and are calculated as

$$N_f^* = \frac{M^*}{d - t_f} = \frac{142 \times 10^3}{356 - 11.5} = 412.2\,\text{kN}$$

The total length of fillet weld on each flange is $L_w = 2b_f = 2 \times 171 = 342$ mm.
The design shear on fillet welds on flange is therefore

$$v_f^* = \frac{N_f^*}{L_w} = \frac{412.2}{342} = 1.2\ \text{kN/mm}$$

Use 8 EXX48 SP fillet welds to the beam flanges; the design shear capacity of the weld per unit length is obtained from Table 6.4 as

$$\phi v_w = 1.303\ \text{kN/mm} > v_f^* = 1.2\ \text{kN/mm, OK}$$

3. Design of web welds

The shear force is assumed to be transmitted by the fillet welds on both sides of the steel beam web. The total length of the web welds is

$$L_w = 2d_1 = 2 \times 333 = 666\ \text{mm}$$

The design shear on fillet welds on the web is therefore

$$v_w^* = \frac{V^*}{L_w} = \frac{35}{666} = 0.053\ \text{kN/mm}$$

Use 6 E48XX SP fillet welds to both sides of the web; the design shear capacity of the weld per unit length is obtained from Table 6.4 as

$$\phi v_w = 0.977 \text{ kN/mm} > v_w^* = 0.053 \text{ kN/mm, OK}$$

6.6 BOLTED MOMENT END PLATE CONNECTIONS

Bolted moment end plate connections are used to transfer design bending moment, shear force and axial force from members to supporting members in steel portal frames or multistorey rigid steel frames. The steel beam is usually shop welded to the end plate which is field bolted to the column flange or supporting element. Typical bolted moment end plate connections are knee and ridge connections in portal frames and beam normal to column connections as shown in Figure 6.1. The behaviour of bolted moment end plate connections is characterised by their moment–rotation curves. The behaviour and design of bolted moment end plate connections are introduced herein. Further information can be found in the book by Hogan and Thomas (1994).

6.6.1 Design actions

A bolted moment end plate connection used in a rigid steel frame is subjected to a design bending moment M^*, a design shear force V^* and a design axial force N^*. These design action effects can be determined by performing either a first-order elastic analysis with moment amplification or a second-order elastic analysis or a plastic analysis. In the connection design, different design actions are calculated for the design of flange and web welds and for the design of bolts, end plates and stiffeners due to the fact that different assumptions are adopted in the design models.

6.6.1.1 Design actions for the design of bolts, end plates and stiffeners

When calculating the design actions for the design of bolts, end plates and stiffeners, it is assumed that the design bending moment M^* is transmitted by the two flanges, the design shear force is transmitted by the web and the design axial force (N^*) is transmitted by the two flanges. The axial force carried by each flange is proportional to its cross-sectional area. The force components of design actions acting on the connection are depicted in Figure 6.12. The design force on the flanges due to the design bending moment is given by

$$N_{tm}^* = N_{cm}^* = \frac{M^*}{d - t_f} \tag{6.34}$$

where d and t_f are the depth and thickness of the I-beam cross section, respectively.

Design actions on the components of a ridge connection with a symmetric cross section under bending moment (M^*), axial force (N^*) and shear force (V^*) can be obtained from Figure 6.12 as follows:

$$N_{ft}^* = \frac{M^*}{d - t_f}\cos\theta + \frac{N^*}{2}\cos\theta - \frac{V^*}{2}\sin\theta \tag{6.35}$$

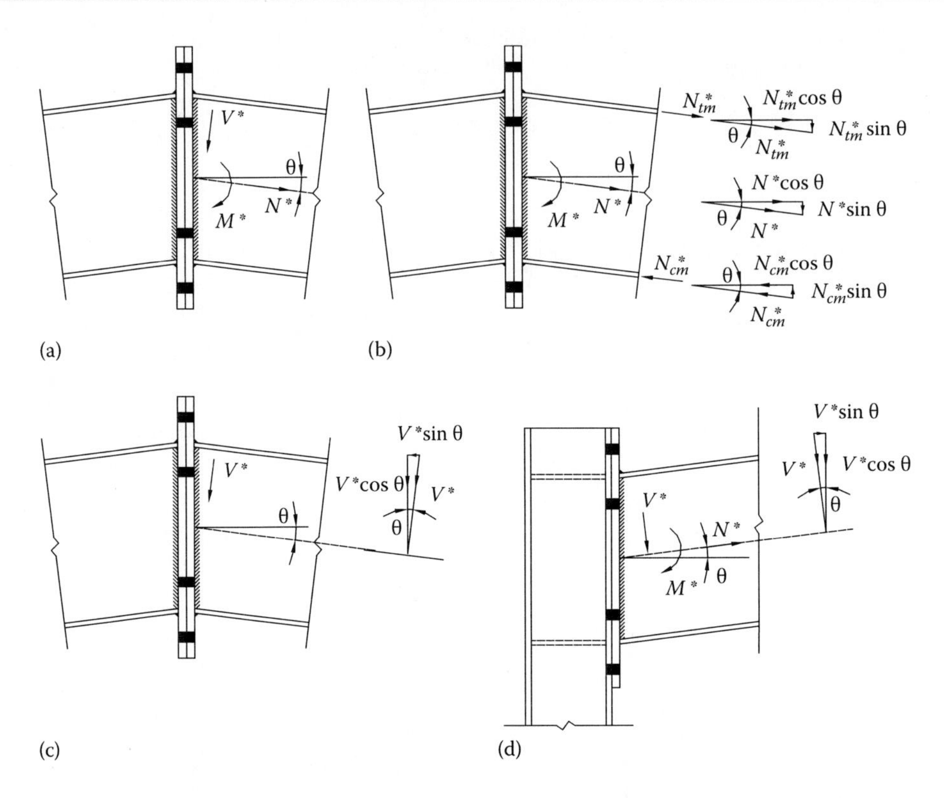

Figure 6.12 Design actions for the design of bolts, end plates and stiffeners: (a) design actions, (b) force components due to moment and axial force, (c) force components of shear force in ridge connection and (d) force components of shear force in knee joint.

$$N_{fc}^{*} = \frac{M^{*}}{d - t_f}\cos\theta - \frac{N^{*}}{2}\cos\theta + \frac{V^{*}}{2}\sin\theta \tag{6.36}$$

$$V_{vc}^{*} = V^{*}\cos\theta + N^{*}\sin\theta \tag{6.37}$$

where

N_{ft}^{*} is the resultant horizontal design force in the tension flange

N_{fc}^{*} is the resultant horizontal design force in the compression flange

V_{vc}^{*} is the resultant vertical design shear force at the end plate and column interface

It is noted that the signs of design actions are positive in the directions shown in Figure 6.12a.

Equations 6.35 through 6.37 can be used to calculate the design actions on the beam normal to column connections by taking $\theta = 0$. For knee connections in portal frames as illustrated in Figure 6.12d, the design forces in the flanges are expressed by

$$N_{ft}^{*} = \frac{M^{*}}{d - t_f}\cos\theta + \frac{N^{*}}{2}\cos\theta + \frac{V^{*}}{2}\sin\theta \tag{6.38}$$

$$N_{fc}^{*} = \frac{M^{*}}{d - t_f}\cos\theta - \frac{N^{*}}{2}\cos\theta - \frac{V^{*}}{2}\sin\theta \tag{6.39}$$

6.6.1.2 Design actions for the design of flange and web welds

For the design of flange and web welds, it is assumed that the design bending moment (M^*) is transmitted by the web and the flanges. The proportion transmitted by each component depends on the second moments of area of the web and flanges. The bending moments carried by the web and two flanges are determined by (Hogan and Thomas 1994)

$$M_w^* = k_{mw} M^* \tag{6.40}$$

$$M_f^* = (1 - k_{mw}) M^* \tag{6.41}$$

where M_w^* and M_f^* are the design bending moments transmitted by the web and flanges, respectively, and k_{mw} is calculated by

$$k_{mw} = \frac{I_{web}}{I_{web} + I_f} \tag{6.42}$$

where I_{web} and I_f are the second moments of area of the web and the two flanges about the principal x-axis, respectively.

The design axial force (N^*) is assumed to be carried by the flanges and web. The proportion of the design axial force carried by each component is proportional to their cross-sectional areas. The design axial forces transmitted by the web and each flange can be determined as follows (Hogan and Thomas 1994):

$$N_w^* = k_w N^* \tag{6.43}$$

$$N_f^* = \frac{(1 - k_w) N^*}{2} \tag{6.44}$$

where the factor k_w is expressed by

$$k_w = \frac{A_w}{A_g} \tag{6.45}$$

where

A_w is the cross-sectional area of the beam web
A_g is the gross cross-sectional area of the beam section

Design actions for the design of flange welds in ridge connections under bending moment (M^*), axial force (N^*) and shear force (V^*) as illustrated in Figure 6.12 can be determined by

$$N_{ft}^* = \frac{M_f^*}{(d - t_f)} \cos\theta + N_f^* \cos\theta - \frac{V^*}{2} \sin\theta \tag{6.46}$$

$$N_{fc}^* = \frac{M_f^*}{(d - t_f)} \cos\theta - N_f^* \cos\theta + \frac{V^*}{2} \sin\theta \tag{6.47}$$

6.6.2 Design of bolts

The bolts in a bolted moment end plate connection are subjected to design tension force (N_{ft}^*) in tension flange and vertical shear force (V_{vc}^*). Therefore, the bolts in the connection (bolt group) must be checked for their design tensile capacity ϕN_{tb} and design shear capacity ϕV_{fn} as follows:

$$N_{ft}^* \le \phi N_{tb} \tag{6.48}$$

$$V_{vc}^* \le \phi V_{fn} \tag{6.49}$$

For a bolted end plate connection with four bolts placed symmetrically about the tension flange, the design capacity of bolts in tension (ϕN_{tb}) can be calculated by (Hogan and Thomas 1994):

$$\phi N_{tb} = \frac{4(\phi N_{tf})}{1 + k_{pr}} \tag{6.50}$$

where the capacity factor $\phi = 0.8$, ϕN_{tf} is the design capacity of a bolt in tension and k_{pr} is the factor that accounts for the effect of additional bolt force due to prying. Prying occurs in bolted connections when bolts are subjected to tension. The edge of the end plate under bending causes bearing stresses on the mating surface. The resulting reaction acting on the end plate must add to the bolt tension. The prying action is found to increase the bolt tension force by 20%–33% (Mann and Morris 1979; Grundy et al. 1980). The factor k_{pr} is between 0.2 and 0.33. A typical value of $k_{pr} = 0.25$ can be used in the design of the connections.

Because the bolts at the tension flange have been utilised to carry the tension force, only those bolts along the web and at the compression flange are assumed to be effective in transmitting the design shear force. The design capacity of bolt group in shear is determined as

$$\phi V_{fn} = n_{cw}(\phi V_{fc}) \tag{6.51}$$

where

n_{cw} is the number of bolts along the web and at the compression flange
ϕV_{fc} is the design capacity of single bolt in shear, which is taken as

$$\phi V_{fc} = \min(\phi V_f; \phi V_{fb}; \phi V_{bc}) \tag{6.52}$$

where

ϕV_f is the design shear capacity of a bolt
ϕV_{fb} is the design capacity of the end plate due to the local bearing or tear-out of the end plate
ϕV_{bc} is the design capacity of the supporting member due to local bearing or tear-out

6.6.3 Design of end plate

The end plate is subjected to bending induced by the tension force at the tension flange, vertical shear and horizontal shear in the bolted end plate connection. The end plate under combined actions must satisfy

$$N_{ft}^* \le \phi N_{pb} \tag{6.53}$$

$$N_{ft}^* \text{ and } N_{fc}^* \leq \phi V_{pb} \tag{6.54}$$

$$V_{vc}^* \leq \phi V_{pv} \tag{6.55}$$

where
ϕN_{pb} is the design capacity of the end plate under bending
ϕV_{pb} is the design capacity of the end plate in horizontal shear
ϕV_{pv} is the design capacity of the end plate in vertical shear

Assuming one dimensional yield line and double curvature bending (Sherbourne 1961; Grundy et al. 1980), the design capacity (ϕN_{pb}) of the end plate under bending can be obtained as

$$\phi N_{pb} = \frac{0.9 f_{yp} b_p t_p^2}{a_{fe}} \tag{6.56}$$

where
f_{yp} is the yield stress of the end plate
b_p and t_p are the width and thickness of the end plate, respectively
a_{fe} effective design value of the distance a_f shown in Figure 6.14

The design capacities of the end plate under horizontal and vertical shear forces are given by (Hogan and Thomas 1994)

$$\phi V_{pb} = 0.9(0.5 f_{yp} b_p t_p) \tag{6.57}$$

$$\phi V_{pv} = 0.9(0.5 f_{yp} d_p t_p) \tag{6.58}$$

where d_p is the depth of the end plate.

6.6.4 Design of beam-to-end-plate welds

In a bolted moment end plate connection, the beam section is welded to the end plate as depicted in Figure 6.13. The flange welds transfer the total horizontal design forces N_{ft}^* and N_{fc}^* which are calculated by Equations 6.46 and 6.47. If fillet weld is used along the flanges, the weld must satisfy the following design requirement:

$$\phi N_w \geq N_{ft}^* \text{ and } N_{fc}^* \tag{6.59}$$

where ϕN_w is the design capacity of fillet weld around a flange of the steel I-section, which is determined as

$$\phi N_w = 2 L_w (\phi v_w) \tag{6.60}$$

in which the weld length L_w across the flange is taken as the width of the beam flange b_f and ϕv_w is the design capacity of fillet weld per unit length of the weld given in Table 6.4.

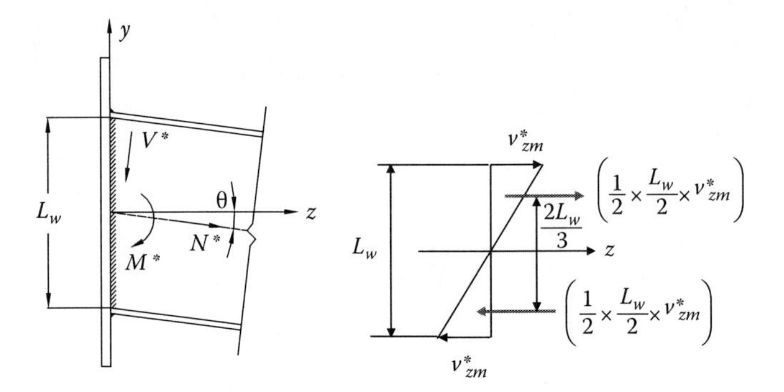

Figure 6.13 Shear in z direction caused by the bending moment.

The web of the steel beam transmits the design actions of M_w^*, N_w^* and V^* as depicted in Figure 6.13. It is assumed that each side of the web of the steel beam is welded to the end plate using fillet weld, which transmits M_w^*, N_w^* and V^*. From the stress distribution shown in Figure 6.13, the moment equilibrium condition gives

$$2 \times \left[\left(\frac{1}{2}\frac{L_w v_{zm}^*}{2}\right) \times \frac{2}{3} L_w\right] = M_w^* \tag{6.61}$$

where v_{zm}^* is the maximum shear stress in the horizontal direction caused by the design bending moment M_w^*. From Equation 6.61, v_{zm}^* can be obtained as

$$v_{zm}^* = \frac{3M_w^*}{L_w^2} \tag{6.62}$$

in which L_w is the weld length along the web, which is taken as $L_w = (d-2t_f)/\cos\theta$ for ridge connection.

The total horizontal design force acting on one web weld is

$$N_{wnv}^* = \frac{N_w^* \cos\theta}{2} - \frac{V^* \sin\theta}{2} \tag{6.63}$$

The shear in the z direction caused by the design force N_{wnv}^* is given by

$$v_{znv}^* = \frac{N_{wnv}^*}{L_w} = \frac{N_w^* \cos\theta - V^* \sin\theta}{2L_w} \tag{6.64}$$

The total shear in the z direction can be determined by

$$v_z^* = \frac{N_w^* \cos\theta - V^* \sin\theta}{2L_w} + \frac{3M_w^*}{L_w^2} \tag{6.65}$$

The shear on one weld caused by the vertical design shear force V_{vc}^* in the y direction is

$$v_y^* = \frac{V_{vc}^*/2}{L_w} = \frac{V_{vc}^*}{2L_w} \tag{6.66}$$

The resultant shear on the weld per unit length is

$$v_{res}^* = \sqrt{\left(v_z^*\right)^2 + \left(v_y^*\right)^2} \tag{6.67}$$

The design requirement for the web weld is

$$\sqrt{\left(v_z^*\right)^2 + \left(v_y^*\right)^2} \le \phi v_w \tag{6.68}$$

6.6.5 Design of column stiffeners

6.6.5.1 Tension stiffeners

The tension flange of a bolted moment end plate connection may be subjected to a large design tension force. This force may cause excessive yielding and distortion of the column flange which is bolted to the end plate. As a result, the column flange or web may fail. Therefore, it is necessary to check the need for the column stiffeners at the tension flange of the beam.

Tension stiffeners are required if the following condition is satisfied:

$$N_{ft}^* > \phi R_t = \min\left(\phi R_{t1}; \phi R_{t2}\right) \tag{6.69}$$

where ϕR_{t1} and ϕR_{t2} are expressed by (Packer and Morris 1977)

$$\phi R_{t1} = 0.9 f_{ycf} t_{fc}^2 \left[\frac{3.14a_d + (2a_c + s_p - d_h)}{a_d}\right] \tag{6.70}$$

$$\phi R_{t2} = 0.9 f_{ycf} t_{fc}^2 \left[\frac{3.14(a_d + a_c) + 0.5s_p}{(a_d + a_p)}\right] + 3.6N_{tf}^* \left[\frac{a_p}{a_d + a_p}\right] \tag{6.71}$$

where

f_{ycf} is the yield stress of the column flange
t_{fc} is the thickness of the column flange depicted in Figure 6.14
s_p is the pitch of bolts
N_{tf}^* is the maximum design tension force acting on a bolt

distances a_c, a_d and a_p are determined as

$$a_c = \frac{b_{fc} - s_g}{2} \tag{6.72}$$

$$a_d = \frac{s_g - t_{wc} - 2b_{rc}}{2} \tag{6.73}$$

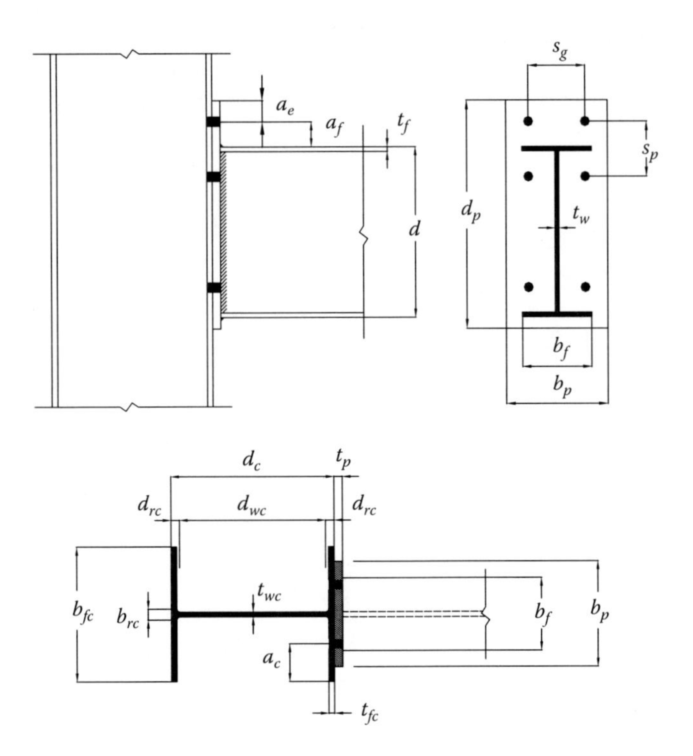

Figure 6.14 Beam-to-column connection details.

$$a_p = \frac{b_p - s_g}{2} \tag{6.74}$$

If column stiffeners are required, column stiffeners need to be designed to carry the excess of the design tension force as follows (Hogan and Thomas 1994):

$$N_{ts}^* = \begin{cases} N_{ft}^* - \phi R_t & \text{(beam on one side)} \\ \max\left[\left(N_{ft1}^* - \phi R_t\right);\left(N_{ft2}^* - \phi R_t\right)\right] & \text{(beams on both sides)} \end{cases} \tag{6.75}$$

Tension stiffeners must satisfy the following design requirement:

$$N_{ts}^* \le \phi N_{ts} \tag{6.76}$$

where the design capacity of the tension stiffeners is given by

$$\phi N_{ts} = 0.9 f_{ys} A_s \tag{6.77}$$

in which A_s is the total cross-sectional area of the stiffeners, taken as $A_s = 2b_{es}t_s$, where b_{es} is the width of the stiffener and t_s is the thickness of the stiffener. The width of the stiffener is taken as $b_{es} \le (15t_s/\sqrt{f_{ys}/250})$ as required by the Clause of 5.14.3 of AS 4100. It is common practice to design the stiffener with $b_{es} \ge b_f/3$ and $t_s \ge t_f/2$.

6.6.5.2 Compression stiffeners

The compression flange of the beam in a bolted moment end plate connection may be subjected to a large design compression force, which may cause the web buckling of the steel column. Therefore, it is necessary to check the need for the column stiffeners at the compression flange of the beam.

Compression stiffeners are required if the following condition is satisfied:

$$N_{fc}^* > \phi R_c = \min(\phi R_{c1};\ \phi R_{c2}) \tag{6.78}$$

where ϕR_{c1} and ϕR_{c2} are the design bearing yield capacity and design bearing buckling capacity of the column web, respectively. The design force N_{fc}^* acting at the compression flange is assumed to be distributed on a 2.5:1 slope to the line at a distance of d_{cr} measured from the top face of the column flange as depicted in Figure 6.14. Expressions for ϕR_{c1} and ϕR_{c2} derived based on test results (Chen and Newlin 1973; Kulak et al. 1987) are given by

$$\phi R_{c1} = 0.9 f_{ycw} t_{wc}(t_f + 5d_{rc} + 2t_p) \tag{6.79}$$

$$\phi R_{c2} = 0.9 \frac{10.8 t_{wc}^2 \sqrt{f_{ycw}}}{d_{wc}} \tag{6.80}$$

where f_{ycw} is the yield stress of the web of the steel column and other symbols are defined in Figure 6.14.

Alternatively, the design bearing yield and buckling capacities of the column web can be determined using the specifications given in AS 4100.

If compression stiffeners are required, column stiffeners need to be designed to carry the excess of the design compression force as follows:

$$N_{ts}^* = \begin{cases} N_{fc}^* - \phi R_c & \text{(beam on one side)} \\ \max\left[\left(N_{fc1}^* - \phi R_c\right);\left(N_{fc2}^* - \phi R_c\right)\right] & \text{(beams on both sides)} \end{cases} \tag{6.81}$$

The design of compression stiffeners is similar to that of the tension stiffeners. If compression stiffeners are provided, the capacity of the stiffened column web needs to be checked.

6.6.5.3 Shear stiffeners

The column web in the connection region is subjected to shear forces composed of a horizontal design force N_{ft}^* or N_{fc}^* on the flange and a design shear force V_c^* in the column as shown in Figure 6.15. V_c^* is taken as positive if it acts in the same direction as the design force in the flange of the beam (Hogan and Thomas 1994). The column web under shear may fail by yielding or shear buckling. Shear stiffeners are required if the resultant horizontal force (V_{res}^*) acting on the flange and column web is greater than the design capacity (ϕV_c) of the column web in shear. The design capacity (ϕV_c) of the column web in shear is determined as

$$\phi V_c = \min(\phi V_w;\ \phi V_b) \tag{6.82}$$

where

ϕV_w is the design shear yield capacity of the column web
ϕV_b is the design shear buckling capacity of the column web as given in Chapter 5

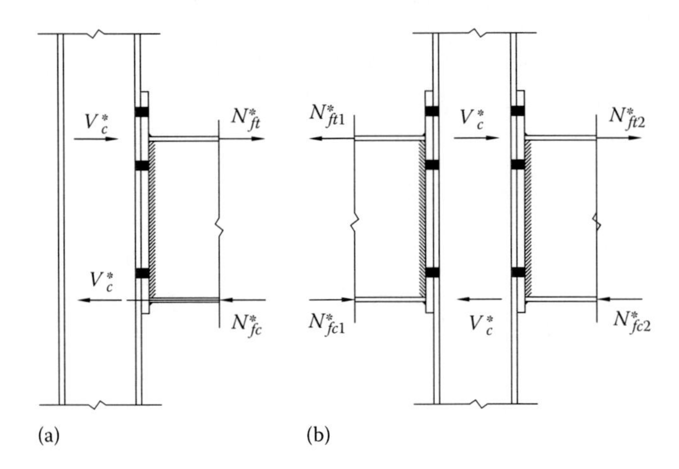

Figure 6.15 Shear forces for the design of column stiffeners: (a) beam on one side of column and (b) beam on both sides of column.

When diagonal stiffeners are used as the web stiffeners of the column with a beam connected on one side, the design force (N_{vs}^*) carried by the diagonal stiffeners is taken as the maximum of ($V_{res}^* - \phi V_c$) on the tension and compression flanges. The diagonal stiffeners must satisfy

$$\frac{N_{vs}^*}{\cos\theta} \le \phi N_{vs} \tag{6.83}$$

where θ is the angle between the diagonal stiffener and the horizontal axis and $\phi N_{vs} = 0.9 f_{ys} A_s$.

6.6.5.4 Stiffened columns in tension flange region

When conventional tension stiffeners are provided, the strength of the stiffened flange of the column needs to be checked. This requires that the design capacity of the stiffened column flange (ϕN_{ts}) must be greater or equal to the design tension force at the tension flange N_{ft}^*. The design capacity of the column flange (Packer and Morris 1977) is calculated as follows:

$$\phi N_{ts} = 0.9 f_{ycf} t_{fc}^2 \left[\frac{2w_1 + 2w_2 - d_h}{a_d} + \left(\frac{1}{w_1} + \frac{1}{w_2} \right) (2a_c + 2a_d - d_h) \right] \tag{6.84}$$

$$w_1 = \sqrt{a_d (a_c + a_d - 0.5 d_h)} \tag{6.85}$$

$$w_2 = (s_p - t_s - 2t_w)/2 \le w_1 \tag{6.86}$$

If $\phi N_{ts} < N_{ft}^*$, a larger section of the column needs to be used or flange doubler plates can be welded to the column flange. The design requirement for the stiffened column flange doubler plates is $N_{ft}^* \le \phi R_{td}$, where the design capacity of stiffened column flange ϕR_{td} is estimated by (Zoetemeijer 1974)

$$\phi R_{td} = 0.9\left(t_{fc}^2 f_{ycf} + 0.5 t_d^2 f_{yd}\right)\left(\frac{s_p + 4a_d + 1.25a_c}{a_d}\right) \tag{6.87}$$

where

t_d is the thickness of doubler plates
f_{yd} is the yield stress of doubler plates

When conventional tension stiffeners are used in addition to doubler plates, ϕR_{td} should be calculated using $(t_{fc} + t_d)$ instead of t_{fc}.

6.6.5.5 Stiffened columns in compression flange region

The stiffened column web in the compression flange region must withstand the design compression force (N_{fc}^*) acting at the compression flange. The design capacity of the stiffened column web (Mann and Morris 1979) can be estimated by

$$\phi R_{cs} = 0.9\left(f_{ys} A_s + 1.63 f_{ycw} t_{fc} \sqrt{b_{fc} t_{wc}}\right) \tag{6.88}$$

The design requirement is $\phi R_{cs} \ge N_{fc}^*$.

6.6.6 Geometric requirements

Bolted moment end plate connections shall be designed to satisfy the geometric restrictions. The symbols used in the connection designs are shown in Figure 6.14. The geometric restrictions are given as follows (Hogan and Thomas 1994):

- $b_p \le b_{fc}$
- $s_g \le b_f - d_f$ and $s_g \le b_{fc} - 2.5d_f$, but $s_g \ge 80$ mm (M20 bolts), $s_g \ge 120$ mm (M24 bolts)
- $30 \le a_e \le 2.5d_f$ mm (M20 bolts), $36 \le a_e \le 2.5d_f$ mm (M24 bolts)
- a_f as small as possible, but $a_f \ge d_f + L_a \cot\phi$, $a_f \ge 0.5d_s + L_s \cot\phi$, and $a_f \ge 0.5d_w + L_w$

The length L_a is taken as $L_a = 2.2d_f$ + grip (actual bolt length), and the distance d_s is the socket diameter taken as $d_s = 50$ mm for M20 bolts and $d_s = 60$ mm for M24 bolts. The socket length L_s is taken as $L_s = 65$ mm for M20 bolts and $L_s = 80$ mm for M24 bolts.

Example 6.3: Design of bolted ridge connection

A bolted ridge connection in a steel portal frame is subjected to a design bending moment $M^* = 160$ kNm, a design axial tension force $N^* = 68$ kN and a design shear force $V^* = -7.5$ kN. The rafter of the ridge connection is a Grade 300 steel section 360UB56.7. The rafter slope is 8°. The design bending moment capacity of the rafter is 250 kN m. Design this bolted ridge connection.

1. Design actions

a. Minimum design actions

The minimum design bending moment is taken as

$$M^*_{\min} = 0.3M_b = 0.3 \times 250 = 75 \text{ kN m} < M^* = 160 \text{ kN m}$$

Thus, M^* = 160 kNm is used in the design of the connection.

The design shear force $|V^*| < 40$ kN; thus, V^* is taken as $V^* = -40$ kN acting in the same direction of the shear force.

b. Design actions for the design of bolts and end plate

The dimensions of 360UB56.7 steel section are:
d = 359 mm, b_f = 172 mm, t_f = 13mm
Design forces at the flanges and shear force are calculated as follows:

$$N^*_{ft} = \frac{M^*}{d - t_f}\cos\theta + \frac{N^*}{2}\cos\theta - \frac{V^*}{2}\sin\theta$$

$$= \frac{160 \times 10^3}{359 - 13}\cos 8^\circ + \frac{68}{2}\cos 8^\circ - \frac{-40}{2}\sin 8^\circ = 494.4 \text{ kN}$$

$$N^*_{ct} = \frac{M^*}{d - t_f}\cos\theta - \frac{N^*}{2}\cos\theta - \frac{V^*}{2}\sin\theta$$

$$= \frac{160 \times 10^3}{359 - 13}\cos 8^\circ - \frac{68}{2}\cos 8^\circ + \frac{-40}{2}\sin 8^\circ = 421.5 \text{ kN}$$

$$V^*_{vc} = V^* \cos\theta + N^* \sin\theta = -40 \times \cos 8^\circ + 68 \times \sin 8^\circ = -30 \text{ kN}$$

c. Design actions for the design of web and flange welds

The load sharing factors are calculated as $k_{mw} = 0.155$ and $k_w = 0.368$.

The design bending moments transmitted by the web and flanges are

$$M^*_w = k_{mw}M^* = 0.155 \times 160 = 24.8 \text{ kN m}$$

$$M^*_f = (1 - k_{mw})M^* = (1 - 0.155) \times 160 = 135.2 \text{ kN m}$$

The design axial forces transmitted by the web and flange are

$$N^*_w = k_w N^* = 0.368 \times 68 = 25 \text{ kN}$$

$$N^*_f = \frac{(1 - k_w)N^*}{2} = \frac{(1 - 0.368) \times 68}{2} = 21.5 \text{ kN}$$

The design actions for the design of flange welds are calculated as

$$N^*_{ft} = \frac{M^*_f}{(d - t_f)}\cos\theta + N^*_f \cos\theta - \frac{V^*}{2}\sin\theta$$

$$= \frac{135.2 \times 10^3}{359 - 13}\cos 8^\circ + 21.5\cos 8^\circ - \frac{-40}{2}\sin 8^\circ = 411 \text{ kN}$$

$$N_{fc}^* = \frac{M_f^*}{(d - t_f)} \cos\theta - N_f^* \cos\theta + \frac{V^*}{2} \sin\theta$$

$$= \frac{135.2 \times 10^3}{359 - 13} \cos 8° - 21.5 \cos 8° + \frac{-40}{2} \sin 8° = 363 \text{ kN}$$

2. Design of bolts

Use 4 M20 8.8/TB bolts at each flange of the rafter section; the capacities of a single bolt are $\phi N_{tf} = 163$ kN (tension) and $\phi V_f = 92.6$ kN (shear) (Table 6.3).

Taking $k_{pr} = 0.25$, the design capacity of bolts at the tension flange can be computed as

$$\phi N_{tb} = \frac{4(\phi N_{tf})}{1 + k_{pr}} = \frac{4 \times 163}{1 + 0.25} = 521.6 \text{ kN} > N_{ft}^* = 494.4 \text{ kN, OK}$$

There are four bolts at the compression flange, $n_w = 4$. The design capacity of bolts in shear is determined as

$$\phi V_{fn} = n_{cw}(\phi V_f) = 4 \times 92.6 = 370.4 \text{ kN} > V_{vc}^* = 30 \text{ kN, OK}$$

Adopt total 8 M20 8.8/TB bolts.

3. Design of end plate

Use Grade 250 steel $b_p \times t_p = 200 \times 25$ mm end plate. The yield stress of the end plate is $f_{yp} = 250$ MPa.

The pitch of bolts is chosen as 140 mm. By placing the two bolts symmetrically about the centroid of the top flange, the distance a_f is $a_f = (140 - 13)/2 = 63.5$ mm.

The effective value of a_f is $a_{fe} = a_f - d_h/2 = 63.5 - 24/2 = 51.5$ mm

The design capacity of end plate in flexure can be computed as

$$\phi N_{pb} = \frac{0.9 f_{yp} b_p t_p^2}{a_{fe}} = \frac{0.9 \times 250 \times 200 \times 25^2 \times 10^{-3}}{51.5} = 546 \text{ kN} > N_{ft}^* = 494.4 \text{ kN, OK}$$

The design capacity of end plate under horizontal shear with double shear planes is calculated as

$$\phi N_{ph} = 2\phi(0.5 f_{yp} b_p t_p) = 2 \times 0.9 \times (0.5 \times 250 \times 200 \times 25) \times 10^{-3}$$

$$= 1125 \text{ kN} > N_{ft}^* = 494.4 \text{ kN, OK}$$

Assuming 35 mm edge distance, the total depth of the end plate is determined as

$$d_p = 2a_e + 2a_f - t_f + d = 2 \times 35 + 2 \times 63.5 - 13 + 359 = 543 \text{ mm}$$

The design capacity of end plate under vertical shear with double shear planes is therefore

$$\phi N_{ph} = 2\phi(0.5 f_{yp} d_p t_p) = 2 \times 0.9 \times (0.5 \times 250 \times 543 \times 25) \times 10^{-3}$$

$$= 3054.4 \text{ kN} > V_{vc}^* = 30 \text{ kN, OK}$$

Thus, adopt 200 × 25 mm end plate.

4. Design of flange welds

Use 8 E48XX SP fillet weld to flanges.
The design capacity of fillet weld per unit length is ϕv_w = 1.303 kN/m Table 6.4.
The design capacity of the fillet weld to each flange is

$$\phi N_w = 2L_w(\phi v_w) = 2\times 172\times 1.303 = 448.2\,\text{kN} > N_{ft}^* = 411\text{ kN, OK}$$

$$> N_{fc}^* = 362\text{ kN, OK}$$

Adopt 8 E48XX SP fillet welds to two flanges.

5. Design of web welds

The length of the weld on one side of the web is

$$L_w = (d - 2t_f)/\cos\theta = (359 - 2\times 13)/\cos 8° = 336.3\text{ mm}$$

The horizontal shear on web weld is computed as

$$v_z^* = \frac{N_w^*\cos\theta - V^*\sin\theta}{2L_w} + \frac{3M_w^*}{L_w^2} = \frac{25\times\cos 8° - (-40\sin 8°)}{2\times 336.3} + \frac{3\times 24.8\times 10^3}{336.3^2} = 0.701\text{ kN/mm}$$

The vertical shear on web weld is

$$v_y^* = \frac{V_{vc}^*}{2L_w} = \frac{30}{2\times 336.3} = 0.045\text{ kN/mm}$$

The resultant shear is determined as

$$v_{res}^* = \sqrt{(v_z^*)^2 + (v_y^*)^2} = \sqrt{0.701^2 + 0.045^2} = 0.702\,\text{kN/mm}$$

Use six E48XX SP fillet welds to both sides of the web; from Table 6.4, we obtain

$$\phi v_w = 0.978\text{ kN/mm} > v_{res}^* = 0.702\text{ kN/mm, OK}$$

Therefore, the bolted ridge connection is specified as follows:

8 M20 8.8/TB bolts, 90 mm gauge, 140 mm pitch
200 × 543 × 25 mm steel end plate
8 E48XX SP fillet welds to flanges
6 E48XX SP fillet welds to both sides of web

6.7 PINNED COLUMN BASE PLATE CONNECTIONS

Pinned column base plate connections are used to transmit the design actions from the steel columns to the foundations. The components of a pinned column base plate connection include concrete foundation, steel base plate, fillet welds and anchor bolts. Pinned column base plates may be subjected to an axial design force N^* (either compression N_c^* or tension N_t^*)

and a design shear V^* acting in the direction of principal axis or both (V_x^*, V_y^*). The design of pinned column base plate connections must check for the strengths of the connection components under axial compression/tension and shear forces. The behaviour and design of pinned column base plate connections are introduced herein. Further information can be found in the book by Hogan and Thomas (1994).

6.7.1 Connections under compression and shear

6.7.1.1 Concrete bearing strength

The large axial compression force transmitted from the steel column to the base plate results in high bearing stresses on the concrete footing. This bearing stress may reach the compressive strength of the concrete, which causes the crushing of the concrete. The bearing strength of the concrete depends on the bearing area of the base plate, the supporting surface area of the footing and the compressive strength of the concrete. Clause 12.3 of AS 3600 (2001) gives specifications on the design bearing strength of the concrete as follows:

$$\phi N_{bc} = \phi A_1 0.85 f'_c \sqrt{\frac{A_2}{A_1}} \leq \phi A_1 2 f'_c \tag{6.89}$$

where

$\phi = 0.6$ is the capacity reduction factor
f'_c is the compressive strength of concrete
A_1 is the bearing area
A_2 is the largest area of the supporting surface that is geometrically similar to and concentric with A_1

The anchor bolt holes of the base plate are ignored in the calculation of the bearing area A_1.

6.7.1.2 Base plates due to axial compression in columns

It is assumed that the base plate is rigid and the axial compression force is concentrated over an area of $0.8b_{fc} \times 0.95d_c$ for the steel I-section column base plate connection as shown in Figure 6.16. The base plate under bearing stresses can be treated as a cantilever plate bending about the edges of this area (Stockwell 1975; DeWolf 1978, 1990). The maximum value (a_{max}) of distances a_m and a_n is used to calculate the bending moment of the cantilever plate under bearing stress of $\phi N_{sc1}/A_1$. The bending moment per unit width at the edges of this area is equal to the moment capacity of the plate:

$$\frac{\phi N_{sc1}}{A_1} \times \frac{a_{max}^2}{2} = \frac{0.9 f_{yp} t_p^2}{4} \tag{6.90}$$

where

f_{yp} is the yield stress of the base plate
t_p is the thickness of the base plate and the design capacity of the base plate under compression ϕN_{sc1} can be obtained as

$$\phi N_{sc1} = \frac{0.9 f_{yp} t_p^2 A_1}{2 a_{max}^2} \tag{6.91}$$

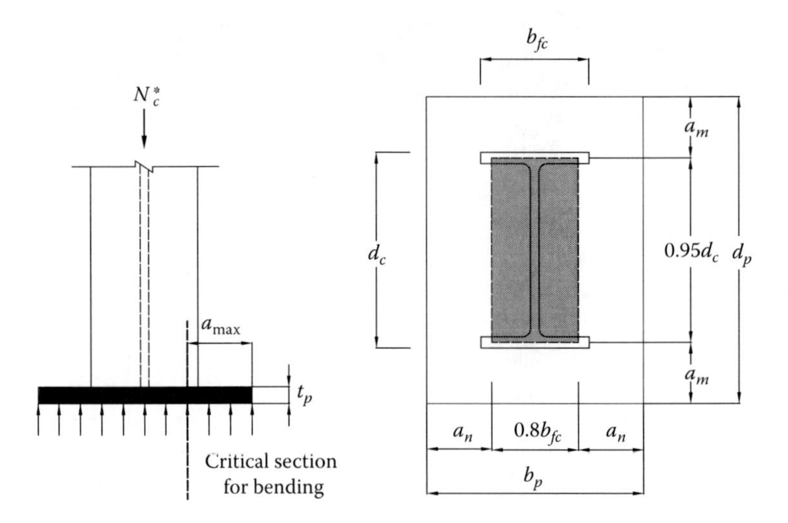

Figure 6.16 Critical section for bending of the cantilever plate.

The actual bearing stress distribution under the base plate may not be uniform but rather is confined to an H-shaped area characterised by the dimension a_o (Stockwell 1975; DeWolf 1978; Murry 1983) as depicted in Figure 6.17. Equation 6.91 can be modified as

$$\phi N_{sc2} = \frac{0.9 f_{yp} t_p^2 A_H}{2 a_o^2} \tag{6.92}$$

where the H-shaped area A_H is taken as the lesser of the values calculated by the following equations (Stockwell 1975; DeWolf 1978; Murry 1983):

$$A_H = \frac{N_c^*}{\phi 0.85 f'_c \sqrt{A_2/(b_{fc} d_c)}} \quad (\phi = 0.6) \tag{6.93}$$

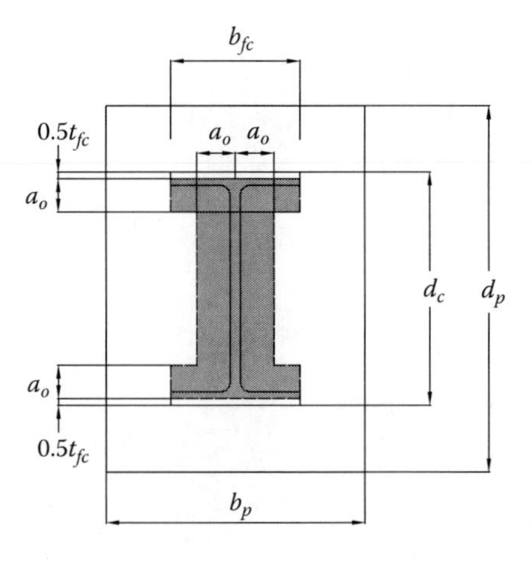

Figure 6.17 H-shaped bearing area A_H.

$$A_H = \frac{N_c^*}{\phi 2 f_c'} \quad (\phi = 0.6) \tag{6.94}$$

The dimension a_o can be calculated as

$$a_o = \frac{1}{4}\left[(b_{fc} + d_c) - \sqrt{(b_{fc} + d_c)^2 - 4A_H}\right] \tag{6.95}$$

The design capacity (ϕN_{sc}) of the base plate under compression should be taken as the lesser of ϕN_{sc1} and ϕN_{sc2}.

6.7.1.3 Column to base plate welds

If the column end is not prepared for full contact with the base plate, the fillet weld at the base of the column under axial compression must satisfy the following requirement:

$$N_c^* \le \phi N_w \tag{6.96}$$

where ϕN_w is the design capacity of the fillet weld at the base of column and is calculated as $\phi N_w = (\phi v_w) L_w$, where L_w is the total length of fillet weld.

The column end is fillet welded to the base plate to transmit the axial compression force and design shear forces (V_x^*, V_y^*) acting in both principal axes. Under the combined actions of axial compression and shear, the fillet weld must satisfy

$$v_{res}^* = \sqrt{\left(v_x^*\right)^2 + \left(v_y^*\right)^2 + \left(v_z^*\right)^2} \le \phi v_w \tag{6.97}$$

where

$v_x^* = V_x^*/L_w$, $v_y^* = V_y^*/L_w$, $v_z^* = N_c^*/L_w$ and L_w is the total length of fillet weld around the column section profile

ϕv_w is the design capacity of fillet weld per unit length

6.7.1.4 Transfer of shear force

In pinned column base plate connections, the horizontal shear force may be resisted by (a) the anchor bolts, (b) friction between the base plate and the concrete foundation, (c) shear key welded to the underside of the base plate and (d) recessing the base plate into the concrete foundation or a combination of these (Hogan and Thomas 1994). It is not recommended that shear be resisted by the anchor bolts alone. The reason for this is that the shear induces bending of the anchor bolt that has a low bending capacity. Under axial compression, the shear should be resisted by friction. However, if friction is not sufficient to resist shear, anchor bolts can be designed to resist part of the shear, or a combination of friction and a shear key may be used. Under axial tension, the shear can be resisted by anchor bolts or a shear key.

When shear at the base plate is resisted by friction alone, the design shear capacity (ϕV_{d1}) based on friction must be greater than the resultant shear. This can be expressed as

$$V_{res}^* = \sqrt{\left(V_x^*\right)^2 + \left(V_y^*\right)^2} \leq \phi V_{d1} \tag{6.98}$$

where the design shear capacity is $\phi V_{d1} = 0.8\mu N_c^*$. The coefficient of friction μ is taken as 0.55 for contact plane between the grout and the rolled steel column above the concrete surface, 0.7 for contact plane at the concrete surface and 0.9 for the contact plane of the base plate thickness below the concrete surface (DeWolf 1990).

6.7.1.5 *Anchor bolts in shear*

Shear force on anchor bolt is transferred by bearing on the surrounding concrete and bending the bolt. The possible failure modes for anchor bolt under shear force (Ueda et al. 1988) include (a) concrete failure with wedge cone, (b) concrete failure without wedge cone, (c) concrete failure with pull-out cone and (d) shear failure of the anchor bolt. The failure mode (a) can be prevented by sufficient edge distance, while failure mode (c) can be prevented by providing sufficient embedment of the anchor bolt.

The strength of bolt in shear and the distance between the plane of the applied shear force and the concrete surface have influences on the shear capacity of anchor bolt. If the shear acts towards an edge of the concrete footing, the edge distance may govern the shear capacity of the anchor bolt. The concrete failure surface is assumed to be a semi-cone of height equal to the edge distance and an inclination of 45° with respect to the concrete edge. The design capacity of the embedded anchor bolt under shear force can be estimated by using the tensile strength of concrete over the projected area of the semi-cone surface (ACI 349 1976) as follows:

$$\phi V_{us} = \phi 0.32 a_e^2 \sqrt{f_c'} \tag{6.99}$$

where

$\phi = 0.8$ is the capacity reduction factor
a_e is the distance measured from the centre of an anchor bolt to the concrete edge

The minimum distance a_e is taken as

$$a_e > d_f \sqrt{\frac{f_{uf}}{0.83\sqrt{f_c'}}} \tag{6.100}$$

The distance a_e should be greater than $12d_f$ for Grade 250 rod or Grade 4.6 bolts and $17d_f$ for Grade 8.8 bolts.

An anchor bolt subjected to design shear force in a principal axis or in both directions must satisfy (Hogan and Thomas 1994)

$$V_f^* \leq \phi V_{fe} \tag{6.101}$$

where $V_f^* = V_x^*/n_b$, $V_f^* = V_y^*/n_b$ or $V_f^* = V_{res}^*/n_b$ and the design capacity of the anchor bolt in shear $\phi V_{fe} = \min(\phi V_f; \phi V_{us})$. The design capacity of a single bolt in shear (ϕV_f) is given in Tables 6.2 and 6.3.

6.7.2 Connections under tension and shear

6.7.2.1 Base plates due to axial tension in columns

The steel base plate due to axial tension in column is subjected to uplift force but held down by the anchor bolts. The failure mechanism of the base plate welded to an I-section column is characterised by three yield lines radiating from the centre of the column web (Murry 1983). Based on the yield line theory, the design capacity of steel base plate due to axial tension in the column can be estimated by (Murry 1983)

$$\phi N_{st} = \frac{\phi 4 b_{fo} f_{yp} t_p^2}{\sqrt{2} s_g}\left(\frac{n_b}{2}\right) \quad \text{for } b_{fo} \le \frac{d_c}{\sqrt{2}} \tag{6.102}$$

$$\phi N_{st} = \frac{\phi\left(2b_{fo}^2 + d_c^2\right) f_{yp} t_p^2}{s_g d_c}\left(\frac{n_b}{2}\right) \quad \text{for } b_{fo} > \frac{d_c}{\sqrt{2}} \tag{6.103}$$

where

$\phi = 0.9$ is the capacity factor
n_b is the total number of bolts in the connection
b_{fo} is length of yield line defined in Figure 6.18
s_g is the gauge of anchor bolts

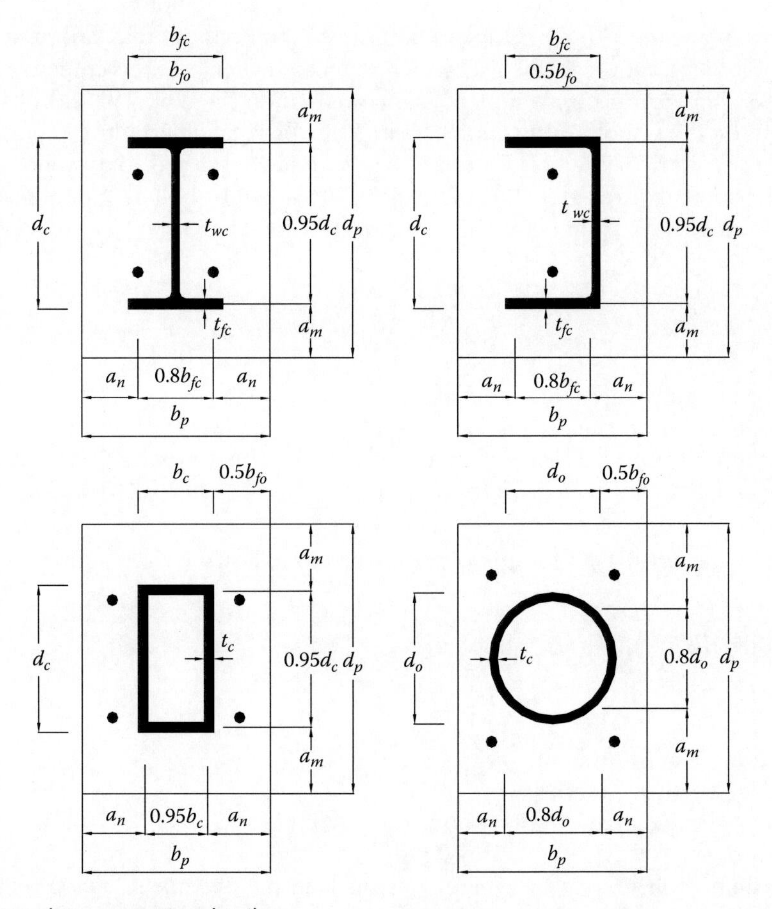

Figure 6.18 Base plate connection details.

These equations can be used for the design of base plate welded to channel sections, RHS and CHS columns with two pars of anchor bolts. However, the length of yield lines must be similar to that for I-sections as defined in Figure 6.18.

6.7.2.2 Column to base plate welds

For the column to base plate welds subjected to axial tension and shear force, the design capacity of the fillet weld in shear needs to be checked as follows:

$$v_{res}^{*} = \sqrt{\left(v_{x}^{*}\right)^{2} + \left(v_{y}^{*}\right)^{2} + \left(v_{z}^{*}\right)^{2}} \leq \phi v_{w} \tag{6.104}$$

where

$v_x^* = V_x^*/L_w$, $v_y^* = V_y^*/L_w$, $v_z^* = N_t^*/L_w$ and L_w is the total length of fillet weld around the column section profile

ϕv_w is the design capacity of fillet weld per unit length

6.7.2.3 Anchor bolts under axial tension

Anchor bolts used in column base plate connections are classified into cast-in-place bolts and drilled-in bolts. Cast-in-place bolts include hooked bolts, bolts with head, bolts with nut, bolts with plate and U-bolts. Hooked bolts are often used but may fail by straightening and pulling out of the concrete when subjected to tension. They are recommended to be used in column base plate connections under axial compression (DeWolf 1990). Bolts with head, nut and plate or the U-bolts offer more positive anchorage. The failure modes of anchor bolts are (a) the failure of the bolt group in tension and (b) the pull-out failure of a cone of concrete radiating outwards at 45° from the head of the nut or bolt as shown in Figure 6.19. To prevent these failures from occurring, anchor bolts must satisfy

$$N_t^* \leq \phi N_t = \min(\phi N_{tb}; \phi N_{cc}) \tag{6.105}$$

where

ϕN_t is the design capacity of embedded bolts

ϕN_{tb} is the design capacity of the bolt group calculated as $\phi N_{tb} = n_b(\phi N_{tf})$

ϕN_{cc} is the pull-out resistance of concrete (Marsh and Burdette 1985; DeWolf 1990) given by

$$\phi N_{cc} = 0.8\left(0.33\sqrt{f_c'}\right)A_{ps} \quad \text{for all bolt types but hook bolts} \tag{6.106}$$

$$\phi N_{cc} = 0.8 n_b \left(0.7 f_c'\right) d_f L_h \quad \text{for hook bolts} \tag{6.107}$$

where

d_f is the diameter of the hook bolt

L_h is the length of the hook

A_{ps} is the projected area of failure cone of concrete

For isolated single bolt, $A_{ps} = \pi L_d^2$, where L_d is the length embedment. The projected area of failure cone of concrete for bolt group is given by Marsh and Burdette (1985).

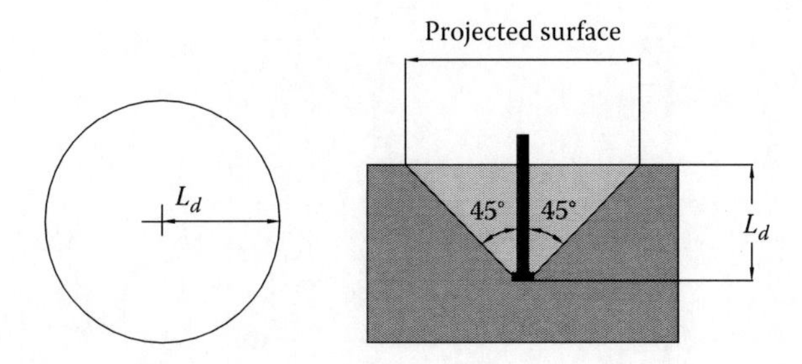

Figure 6.19 Failure cone of embedded bolt in tension.

The requirement on the edge distance is

$$a_e \geq d_f \sqrt{\frac{f_{uf}}{6\sqrt{f'_c}}} \tag{6.108}$$

The edge distance (a_e) should be greater than $5d_f$ for Grade 250 rod or Grade 4.6 bolts and $7d_f$ for Grade 8.8 bolts and 100 mm.

6.7.2.4 Anchor bolts under tension and shear

For an anchor bolt subject to combined tension and shear forces, the bolt must satisfy the following additional requirement (Hogan and Thomas 1994):

$$\frac{V_f^*}{\phi V_f} + \frac{N_{tf}^*}{\phi N_{tf}} \leq 1.0 \tag{6.109}$$

where $N_{tf}^* = N_t^*/n_b$.

Example 6.4: Design of column base plate connection

Design a pinned base plate connection for the steel column of 460UB74.6 subjected to axial forces and shear forces. The end of the steel column is cold sawn. The steel column is supported on a 850 mm diameter concrete pier foundation. The compressive strength of concrete is $f'_c = 25$ MPa. The column is subjected to the following design actions: (a) $N_c^* = 87$ kN, $V_y^* = 30$ kN and (b) $N_t^* = 105$ kN, $V_y^* = 70$ kN.

a. Connection under axial compression and shear

1. Connection geometry

Based on standard base plate connections, the initial sizing of the base plate connection is selected as 200 × 490 × 20 mm end plate, 4 M20 4.6/S bolts with 300 mm pitch and 100 mm gauge as schematically depicted in Figure 6.20.

Connection geometry and material properties:

$$d_c = 457 \text{ mm}, \quad b_{fc} = 190 \text{ mm}, \quad t_{fc} = 14.5 \text{ mm}, \quad t_{wc} = 9.1 \text{ mm}$$

$$d_p = 490 \text{ mm}, \quad b_p = 200 \text{ mm}, \quad t_p = 20 \text{ mm}, \quad f_{yp} = 250 \text{ MPa}$$

$$s_g = 100 \text{ mm}, \quad s_p = 300 \text{ mm}$$

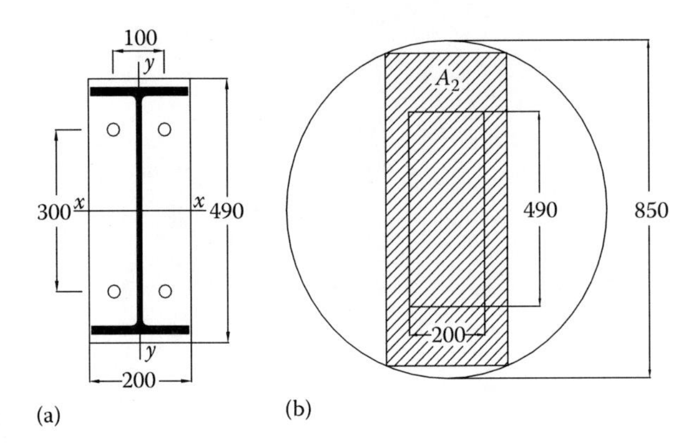

Figure 6.20 Base plate connection: (a) base plate and (b) area A_2.

2. Concrete bearing strength

The area of the bearing base plate is

$$A_1 = b_p d_p = 200 \times 490 = 98{,}000 \text{ mm}^2$$

The supporting surface area A_2 that is geometrically similar to A_1 can be calculated as

$$A_2 = 320 \times 784 = 250{,}880 \text{ mm}^2$$

The design bearing strength of concrete is

$$\phi N_{bc} = \phi A_1 0.85 f'_c \sqrt{\frac{A_2}{A_1}} = 0.6 \times 98{,}000 \times 0.85 \times 25 \times \sqrt{\frac{250{,}880}{98{,}000}} = 1999.2 \text{ kN}$$

$$\phi N_{bc} = \phi A_1 2 f'_c = 0.6 \times 98{,}000 \times 2 \times 25 \times 10^{-3} = 2{,}940 \text{ kN}$$

Thus, $\phi N_{bc} = \min(1999.2; 2940) = 1999.2 \text{ kN} > N_c^* = 87 \text{ kN}$, OK.

3. Base plate due to axial compression in column

The distance $a_{\max}$ is calculated as follows:

$$a_m = (d_p - 0.95 d_c)/2 = (490 - 0.95 \times 457)/2 = 28 \text{ mm}$$

$$a_n = (b_p - 0.8 b_{fc})/2 = (200 - 0.8 \times 190)/2 = 24 \text{ mm}$$

$$a_{\max} = \max(28; 24) = 28 \text{ mm}$$

The design capacity of base plate ϕN_{sc1} is

$$\phi N_{sc1} = \frac{0.9 f_{yp} t_p^2 A_1}{2 a_{\max}^2} = \frac{0.9 \times 250 \times 20^2 \times 98{,}000 \times 10^{-3}}{2 \times 28^2} = 5{,}625 \text{ kN}$$

The H-shaped area is calculated as follows:

$$A_H = \frac{N_c^*}{\phi 0.85 f'_c \sqrt{A_2/(b_{fc} d_c)}} = \frac{87 \times 10^3}{0.6 \times 0.85 \times 25 \sqrt{250{,}880/(190 \times 457)}} = 4{,}014 \text{ mm}^2$$

$$A_H = \frac{N_c^*}{\phi 2 f_c'} = \frac{87 \times 10^3}{0.6 \times 2 \times 25} = 2900 \text{ mm}^2$$

Thus, $A_H = \max(4014;2900) = 4014 \text{ mm}^2$.

The distance a_o of the H-shaped area is

$$a_o = \frac{1}{4}\left[(b_{fc} + d_c) - \sqrt{(b_{fc} + d_c)^2 - 4A_H}\right] = \frac{1}{4}\left[(190 + 457) - \sqrt{(190 + 457)^2 - 4 \times 4014}\right]$$

$$= 3.13 \text{ mm}$$

The design capacity of base plate ϕN_{sc2} is

$$\phi N_{sc2} = \frac{0.9 f_{yp} t_p^2 A_H}{2a_o^2} = \frac{0.9 \times 250 \times 20^2 \times 4014 \times 10^{-3}}{2 \times 3.13^2} = 18437.5 \text{ kN}$$

The design capacity of base plate is therefore

$$\phi N_{sc} = \min(\phi N_{sc1}; \phi N_{sc2}) = \min(5625; 18437.5) = 5625 \text{ kN} > N_c^* = 87 \text{ kN, OK}$$

4. Column to base plate welds

The total length of fillet weld around the column section profile is

$$L_w = 2b_{fc} + 2(b_{fc} - t_{wc}) + 2(d_c - 2t_{fc}) = 2 \times 190 + 2 \times (190 - 9.1) + 2 \times (457 - 2 \times 14.5)$$

$$= 1598 \text{ mm}$$

The shears per unit length under shear and axial compression are

$$v_y^* = \frac{V_y^*}{L_w} = \frac{30}{1598} = 0.019 \text{ kN/mm}$$

$$v_z^* = \frac{N_c^*}{L_w} = \frac{87}{1598} = 0.054 \text{ kN/mm}$$

The resultant shear $v_{res}^* = \sqrt{(v_y^*)^2 + (v_z^*)^2} = \sqrt{0.019^2 + 0.054^2} = 0.057$ kN/mm.
Use 5 EXX48 GP fillet weld; $\phi v_w = 0.522 \text{ kN/mm} > v_{res}^* = 0.057$ kN/mm, OK.

5. Transfer of shear force

The shear force is assumed to be resisted by friction alone. The base plate is supported on a grout pad on the top of the concrete pier foundation so that the coefficient of friction is $\mu = 0.55$.

The design shear capacity is calculated as

$$\phi V_{d1} = 0.8 \mu N_c^* = 0.8 \times 0.55 \times 87 = 38.3 \text{ kN} > V_y^* = 30 \text{ kN, OK}$$

6. Anchor bolts in shear

The minimum edge distance is

$$a_e > d_f\sqrt{\frac{f_{uf}}{0.83\sqrt{f_c'}}} = 20\times\sqrt{\frac{400}{0.83\sqrt{25}}} = 196.4 \text{ mm}$$

$$> 12d_f = 12\times 20 = 240 \text{ mm}$$

Adopt a_e = 250 mm.

The design capacity of embedded bolt under horizontal shear is

$$\phi V_{us} = \phi 0.32 a_e^2\sqrt{f_c'} = 0.8\times 0.32\times 250^2\times\sqrt{25} \text{ N} = 80 \text{ kN}$$

The design shear capacity of a single bolt with threads included in the shear plane is obtained from Table 6.2 as ϕV_f = 46.6 kN.

Thus, $\phi V_{fe} = \min(\phi V_{us};\phi V_f)$ = min(80;46.6) = 46.6 kN.

The design shear force on a bolt is

$$V_f^* = V_y^*/4 = 30/4 = 7.5 \text{ kN} < \phi V_{fe} = 46.6 \text{ kN, OK}$$

b. Connection under axial tension and shear

1. Base plate due to axial tension on column

For I-section column, the length of yield line is

$$b_{fo} = 190 \text{ mm} < \frac{d_c}{\sqrt{2}} = \frac{457}{\sqrt{2}} = 323 \text{ mm}$$

The design capacity of the base plate is

$$\phi N_{st} = \frac{\phi 4 b_{fo} f_{yp} t_p^2}{\sqrt{2}s_g}\left(\frac{n_b}{2}\right) = \frac{0.9\times 4\times 190\times 250\times 20^2}{\sqrt{2}\times 100}\left(\frac{4}{2}\right) \text{N} = 967.3 \text{ kN} > N_t^* = 105 \text{ kN, OK}$$

2. Column to base plate welds

The total length of fillet weld around the column section profile is

$$L_w = 1598 \text{ mm}$$

The shears per unit length under shear and axial tensions are

$$v_y^* = \frac{V_y^*}{L_w} = \frac{70}{1598} = 0.044 \text{ kN/mm}$$

$$v_z^* = \frac{N_t^*}{L_w} = \frac{105}{1598} = 0.066 \text{ kN/mm}$$

The resultant shear $v_{res}^* = \sqrt{(v_y^*)^2 + (v_z^*)^2} = \sqrt{0.044^2 + 0.066^2} = 0.079$ kN/mm.

Use 5 EXX48 GP fillet weld; $\phi v_w = 0.522 \text{ kN/mm} > v_{res}^* = 0.079 \text{ kN/mm}$, OK.

3. Anchor bolts under axial tension

The minimum length of bolt embedment is

$$L_d = 12d_f = 12 \times 20 = 240 \text{ mm, adopt } L_d = 250 \text{ mm}$$

The projected area of failure cone for a single bolt is

$$A_{ps} = \pi L_d^2 = 3.14 \times 250^2 = 196{,}250 \text{ mm}^2$$

The capacity of a single bolt in tension is $\phi N_{tf} = 78.3$ kN (Table 6.2).

The pull-out resistance of concrete can be determined as

$$\phi N_{cc} = 0.8\left(0.33\sqrt{f_c'}\right)A_{ps} = 0.8 \times 0.33\sqrt{25} \times 196{,}250 \times 10^{-3} = 259 \text{ kN} > \phi N_{tf} = 78.3 \text{ kN, OK}$$

The projected area of failure cones for bolt group is illustrated in Figure 6.21. The area A_{p1} is calculated as

$$A_{p1} = \pi L_d^2 + 2L_d(s_g + s_p) + s_g s_p$$
$$= 3.14 \times 250^2 + 2 \times 250 \times (100 + 300) + 100 \times 300 = 426{,}250 \text{ mm}^2$$

The shaped area A_{p2} as shown in Figure 6.21 is calculated as follows (Marsh and Burdette 1985):

$$A_{p2} = \left[2L_d - \sqrt{L_d^2 - \frac{s_p^2}{4}}\right]\frac{s_p}{2} - \frac{\sin^{-1}(s_p/2L_d)\pi L_d^2}{180^\circ}$$

$$= \left[2 \times 250 - \sqrt{250^2 - \frac{300^2}{4}}\right]\frac{300}{2} - \frac{\sin^{-1}(300/(2 \times 250))\pi \times 250^2}{180^\circ}$$

$$= 45{,}000 - 40{,}219 = 4{,}781 \text{ mm}^2$$

$$A_{ps} = A_{p1} - 2A_{p2} = 426{,}250 - 2 \times 4{,}781 = 416{,}688 \text{ mm}^2$$

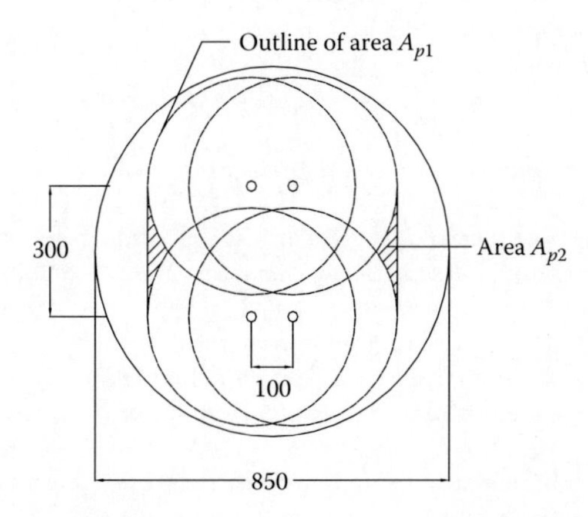

Figure 6.21 Projected area of failure cone.

The pull-out resistance of concrete for bolt group is

$$\phi N_{cc} = 0.8\left(0.33\sqrt{f'_c}\right)A_{ps} = 0.8 \times 0.33\sqrt{25} \times 416{,}688 \times 10^{-3} = 550 \text{ kN}$$

The capacity of the bolt group in tension is

$$\phi N_{tb} = n_b(\phi N_{tf}) = 4 \times 78.4 = 313.6 \text{ kN} < \phi N_{cc} = 550 \text{ kN}$$

Thus, $\phi N_t = 313.6 \text{ kN} > N_t^* = 105 \text{ kN}$, OK.

The required minimum edge distance for the bolt is

$$a_e > d_f\sqrt{\frac{f_{uf}}{6\sqrt{f'_c}}} = 20 \times \sqrt{\frac{400}{6\sqrt{25}}} = 73 \text{ mm} < 5d_f = 5 \times 20 = 100 \text{ mm, adopt } 100 \text{ mm}$$

4. Anchor bolts under tension and shear

The forces on a single bolt under combined tension and shear forces are

$$V_f^* = \frac{V_y^*}{n_b} = \frac{70}{4} = 17.5 \text{ kN}$$

$$N_{tf}^* = \frac{N_t^*}{n_b} = \frac{105}{4} = 26.25 \text{ kN}$$

$$\frac{V_f^*}{\phi V_f} + \frac{N_{tf}^*}{\phi N_{tf}} = \frac{17.5}{44.6} + \frac{26.25}{78.4} = 0.73 < 1.0, \text{ OK}$$

REFERENCES

ACI 349. (1976) Code requirements for nuclear safety related structures, Manual of Concrete Practice, American Concrete Institute, Detroit, Michigan.

AISC-LRFD Manual. (1994) *Load and Resistance Factor Design*, Vol. II, *Connections*, *Manual of Steel Construction*, Chicago, IL: American Institute of Steel Construction.

AS 1275. (1985) Australian standard for metric screw threads for fasteners, Sydney, New South Wales, Australia: Standards Australia.

AS 3600 (2001) *Australian Standard for Concrete Structures*, Sydney, New South Wales, Australia: Standards Australia.

AS 4100 (1998) *Australian Standard for Steel Structures*, Sydney, New South Wales, Australia: Standards Australia.

Birkemoe, P.C. (1983) High-strength bolting: Recent research and design practice, Paper presented at *the W.H. Munse Symposium on Behaviour of Metal Structures*, ASCE, 103–127, Philadelphia, PA, May 1983.

Chen, W.F. and Newlin, D.E. (1973) Column web strength in beam-to-column connections, technical notes, *Journal of the Structural Division*, ASCE, 99 (ST9): 1978–1984.

DeWolf, J.T. (1978) Axially loaded column base plates, *Journal of the Structural Division*, ASCE, 104 (ST5): 781–794.

DeWolf, J.T. (1990) Column base plates, Design guide series no. 1, Chicago, IL: American Institute of Steel Construction.

Fisher, J.W., Galambos, T.V., Kulak, G.L., and Ravindra, M.K. (1978) Load and resistance factor design criteria for connectors, *Journal of the Structural Division*, ASCE, 104 (ST9): 1427–1441.

Galambos, T.V., Reinhold, T.A., and Ellingwood, B. (1982) Serviceability limit states: Connection slip, *Journal of the Structural Division*, ASCE, 108 (ST12): 2668–2680.

Gorenc, B.E., Tinyou, R., and Syam, A.A. (2005) *Steel Designers' Handbook*, 7th edn., Sydney, New South Wales, Australia: UNSW Press.

Grundy, P., Thomas, I.R., and Bennetts, I.D. (1980) Beam-to-column moment connections, *Journal of the Structural Division*, ASCE, 106 (ST1): 313–330.

Hogan, T.J. and Thomas, I.R. (1979a) Bearing stress and edge distance requirements for bolted steelwork connections, *Steel Construction*, Australian Institute of Steel Construction, 13 (3).

Hogan, T.J. and Thomas I.R. (1979b) Fillet weld design in the AISC Standardized Structural Connections, *Steel Construction*, Australian Institute of Steel Construction, 13 (1): 16–29.

Hogan, T.J. and Thomas, I.R. (1994) *Design of Structural Connections*, 4th edn., Sydney, New South Wales, Australia: Australian Institution of Steel Construction.

Kulak, G.L., Fisher, J.W., and Struik, J.H.A. (1987) *Guide to Design Criteria for Bolted and Riveted Joints*, 2nd edn., New York: John Wiley & Sons.

Mann, A.P. and Morris, J.L. (1979) Limit design of extended end-plate connections, *Journal of the Structural Division*, ASCE, 105 (ST3): 511–526.

Marsh, M.L. and Burdette, E.G. (1985) Anchorage of steel building components to concrete, *Engineering Journal*, American Institute of Steel Construction, 22 (1), 33–39.

McGuire, W. (1968) *Steel Structures*, Englewood Cliffs, NJ: Prentice Hall.

Murry, T.M. (1983) Design of lightly loaded steel column base plates, *Engineering Journal*, American Institute of Steel Construction, 20 (4), 143–152.

Packer, J.A. and Morris, L.J. (1977) A limit state design method for the tension region of bolted beam-column connections, *The Structural Engineer*, 55 (10): 446–458.

Research Council on Structural Connections. (1988) Specification for structural joints using ASTM A325 or A490 bolts, AISC.

Sherbourne, A.N. (1961) Bolted beam to column connections, *The Structural Engineer*, 39 (6): 203–210.

Stockwell, F.W. (1975) Preliminary base plate selection, *Engineering Journal*, American Institute of Steel Construction, 12 (3), 92–93.

Swannell, P. (1979) Design of fillet weld groups, *Steel Construction*, Australian Institute of Steel Construction, 13 (1): 2–15.

Thomas, I.R., Bennetts, I.D., and Elward, S.J. (1985) Eccentrically loaded bolted connections, Paper presented at the *Third Conference on Steel Developments*, Australian Institute of Steel Construction, Melbourne, Victoria, Australia, May 1985, pp. 37–43.

Trahair, N.S. and Bradford, M.A. (1998) *The Behaviour and Design of Steel Structures to AS 4100*, 3rd edn. (Australian), London, U.K.: Taylor & Francis Group.

Ueda, T., Kitipornchi, S., and Link, K. (1988) An experimental investigation of anchor bolts under shear, Research report no. CE93, Brisbane, Queensland, Australia: Department of Civil Engineering, University of Queensland.

Zoetemeijer, P. (1974) A design method for the tension side of statically loaded bolted beam-to-column connections, *Heron*, STEVIN Laboratory and I.B.B.C. Institute TNO, the Netherlands, 20 (1): 1–59.

Chapter 7

Plastic analysis of steel beams and frames

7.1 INTRODUCTION

The plastic analysis methods are widely used in the design of simply supported steel beams, continuous steel beams, steel portal frames and multistorey rectangular steel frames. The goal of the plastic analysis is to determine the ultimate loads of a steel structure at which the structure will fail due to the development of excessive deflections (Neal 1977). The plastic methods of structural analysis provide economical designs of steel structures and have the advantage of simplicity compared to the elastic methods of structural analysis. The plastic analysis assumes that (1) the behaviour of the steel structure being analysed is ductile, (2) the deflections of the structure are not the critical design criteria, and (3) the local and overall buckling of the structure will not occur before the collapse load is reached.

This chapter gives an introduction to the plastic methods of structural analysis. The simple plastic theory is described, providing insight into the plastic hinge, full plastic moment, plastic section modulus, shape factor and the effects of axial and shear forces on the full plastic moment. The plastic analysis of simply supported and continuous steel beams is presented by introducing the collapse mechanism, the work equation and the mechanism method. The method of combined mechanisms is provided to deal with the plastic analysis of steel frames. The plastic design to AS 4100 is also discussed.

7.2 SIMPLE PLASTIC THEORY

7.2.1 Plastic hinge

The basic concepts of the simple plastic theory can be demonstrated by investigating the actual behaviour of a simply supported steel beam under uniformly distributed load. The typical load–deflection curve of the steel beam is shown in Figure 7.1a (Baker and Heyman 1969). The behaviour from O to A on the load–deflection curve is elastic. When the load is increased from A to B, the beam develops some permanent deformations which cannot be recovered after removing the load. In addition to this, the deflections increase more rapidly with increasing the loading. It can be observed that further increase of the loading from B to C leads to rapid increase of large deflections. The beam is considered to have collapsed when the load has reached the loading at point B. It should be noted that the strain hardening of the steel material results in the raising characteristics of the load–deflection curve beyond point B. The idealised load–deflection curve is given in Figure 7.1b, which shows that under the constant load (W_c), the deflection increases without limit. This load W_c is called the collapse load of the steel beam.

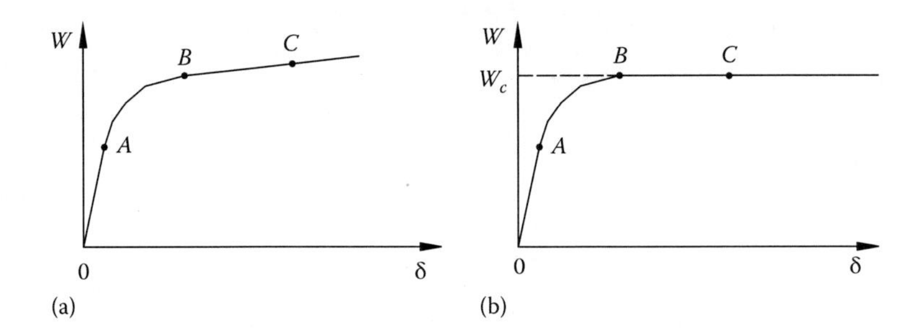

Figure 7.1 Load–deflection curves for beam: (a) typical and (b) idealised.

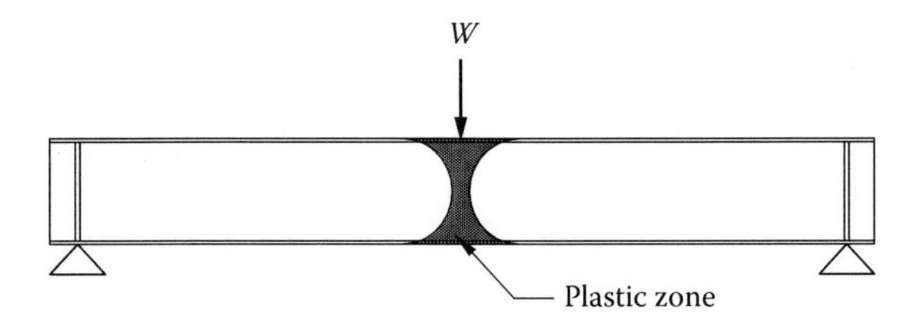

Figure 7.2 Plastic zone in simply supported beam under a concentrated load.

In the collapsed state, large deflections occur at the central kink in the simply supported steel beam due to the rotation of the hinge. This hinge is known as a plastic hinge that forms at the section of maximum bending moment in the beam. When a plastic high forms in a steel member, yielding starts at a local section of the greatest bending moment. The gradual spread of yielding towards the neutral axis and locally along the member takes place when the moment capacity is increased as depicted in Figure 7.2. This results in the plastic zone at the plastic hinge. In the simple plastic theory, however, the spread of plasticity along the member is usually ignored and the plastic hinge is assumed to be confined at the cross section of maximum bending moment.

7.2.2 Full plastic moment

The relationship between the bending moment and curvature can be derived from the stress–strain relation based on the simple beam theory. Figure 7.3 schematically depicts the stress distributions in a rectangular cross section of a beam. The section is assumed to remain plane after deformation, which results in a linear strain distribution through the depth of the section. As shown in Figure 7.3d, the yield strain (ε_y) is attained at a distance h from the neutral axis. The compression and tension forces shown in Figure 7.3c are determined as $C_1 = T_1 = (1/2)bdf_y$. The forces shown in Figure 7.3d are $C_2 = T_2 = b(d-h)f_y$ and $C_3 = T_3 = (1/2)bhf_y$. The bending moment can be determined from the stress distribution given in Figure 7.3d as follows:

$$M = \left(\frac{1}{2}f_y bh\right)\left(\frac{4}{3}h\right) + f_y b(d-h)(d+h) = b\left(d^2 - \frac{h^2}{3}\right)f_y \tag{7.1}$$

The curvature is determined as $\phi = \varepsilon_y/h$. When $h = d$, the yielded zones disappear and the extreme fibre attains the yield stress as shown in Figure 7.3c. The corresponding moment is

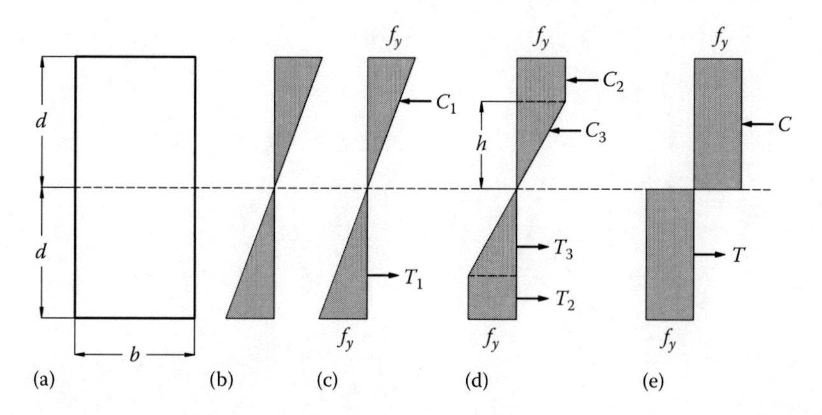

Figure 7.3 Stress distributions in rectangular section: (a) cross section, (b) strain, (c) at first yield, (d) partially plastic and (e) fully plastic.

called the first yield moment (M_y), which is the greatest moment that the section can withstand before yielding. The first yield moment (M_y) of the rectangular section can be obtained from Equation 7.1 as

$$M_y = \left(\frac{2bd^2}{3}\right) f_y \tag{7.2}$$

This equation can be written as

$$M_y = Zf_y \tag{7.3}$$

where Z is the elastic section modulus. The curvature corresponding to M_y is $\phi_y = \varepsilon_y/d$.

The bending moment–curvature relationship of the rectangular cross section can be obtained by combining Equations 7.1 and 7.2 as follows (Neal 1977):

$$\frac{M}{M_y} = 1.5 - 0.5\left(\frac{\phi_y}{\phi}\right)^2 \tag{7.4}$$

Figure 7.4 shows the moment–curvature curve for the rectangular section. It appears that when the curvature is very large, the moment M approaches to $1.5M_y$.

When $h = 0$, the state of full plasticity of the rectangular steel cross section is achieved as shown in Figure 7.3e. From the full plastic stress distribution illustrated in Figure 7.3e, the full plastic moment can be calculated by taking moments about the plastic neutral axis (PNA). It is noted that the PNA is a zero stress axis that divides the section into two equal areas. The full plastic moment of the rectangular section can also be obtained from Equation 7.1 as

$$M_p = (bd^2)f_y \tag{7.5}$$

Equation 7.5 can be rewritten as

$$M_p = Z_p f_y \tag{7.6}$$

where $Z_p = bd^2$ is the plastic section modulus of the rectangular cross section shown in Figure 7.3.

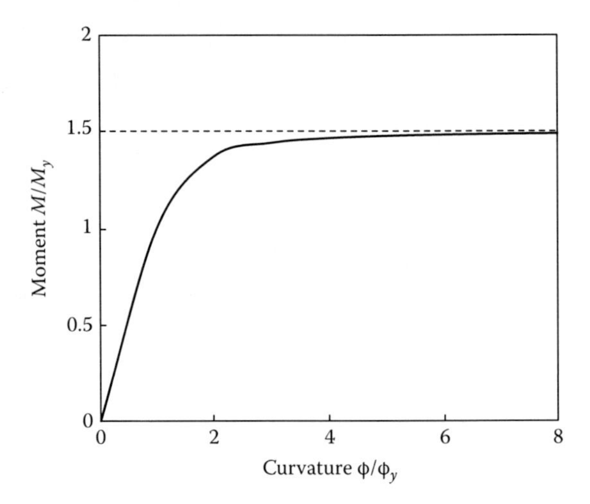

Figure 7.4 Typical moment–curvature curve of beam.

In general, the plastic section modulus of a cross section composed of elements can be computed by summing the first moment of area of each element about the PNA of the section as

$$Z_p = \sum_{i=1}^{m} A_i y_i + \sum_{j=1}^{n} A_j y_j \tag{7.7}$$

where

A_i is the area of the ith element above the PNA
y_i is the distance from the centroid of the ith element to the PNA
A_j is the area of the jth element below the PNA
y_j is the distance from the centroid of the jth element to the PNA
m and n are the total number of elements above and below the PNA, respectively

The shape factor is defined as the ratio of the plastic to elastic section modulus (Neal 1977):

$$\nu = \frac{Z_p}{Z} \tag{7.8}$$

The shape factor indicates the additional moment capacity that a section can support beyond its first yield moment.

Example 7.1: Calculation of full plastic moment of T-section

Figure 7.5 shows a Grade 300 steel T-section bending about its principal x-axis. The yield stress of the steel section is 300 MPa. Calculate (a) the first yield moment of the section, (b) the full plastic moment and (c) the shape factor of the section.

a. First yield moment

The centroid location of the section measured from the top fibre is computed as

$$y_c = \frac{\sum A_n y_n}{\sum A_n} = \frac{200 \times 20 \times (20/2) + 18 \times 250 \times (250/2 + 20)}{200 \times 20 + 18 \times 250} = 81.5 \text{ mm}$$

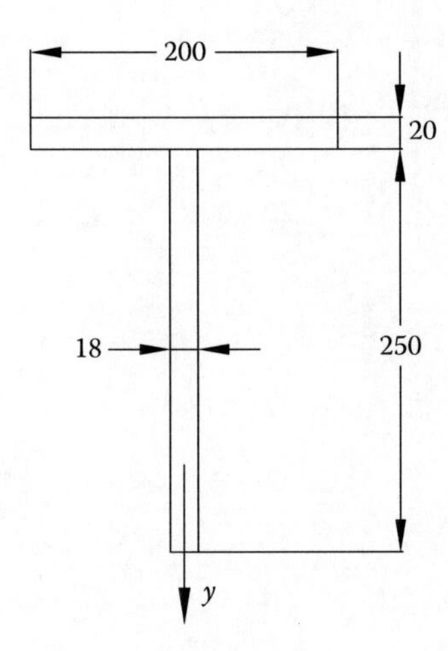

Figure 7.5 Steel T-section.

The second moment of area of the section about the x-axis is

$$I_x = \left[\frac{200 \times 20^3}{12} + 200 \times 20 \times \left(81.5 - \frac{20}{2}\right)^2\right]$$

$$+ \left[\frac{18 \times 250^3}{12} + 18 \times 250 \times \left(\frac{250}{2} + 20 - 81.5\right)^2\right] = 62.16 \times 10^6 \text{ mm}^4$$

The elastic section modulus is

$$Z = \frac{I_x}{y_{max}} = \frac{62.16 \times 10^6}{250 + 20 - 81.5} = 329{,}736 \text{ mm}^3$$

The first yield moment is calculated as

$$M_y = Zf_y = 329{,}736 \times 300 \times 10^{-6} = 98.9 \text{ kN m}$$

b. Full plastic moment

The cross-sectional area of the flange is

$$A_f = 200 \times 20 = 4000 \text{ mm}^2$$

The cross-sectional area of the web is

$$A_w = 18 \times 250 = 4500 \text{ mm}^2 > A_f = 4000 \text{ mm}^2$$

The PNA is located in the web. The depth of the PNA can be determined as

$$200 \times 20 + 18 \times (d_n - 20) = 18 \times (250 + 20 - d_n)$$

Hence, d_n = 33.9 mm.

The plastic modulus of the section is computed as

$$Z_p = 200\times 20\times\left(33.9-\frac{20}{2}\right)+18\times(33.9-20)\times\frac{(33.9-20)}{2}$$

$$+18\times(250+20-33.9)\times\frac{(250+20-33.9)}{2} = 599{,}028\text{ mm}^3$$

The full plastic moment of the section is therefore

$$M_p = Z_p f_y = 599{,}028\times 300\times 10^{-6} = 179.71\text{ kN m}$$

c. Shape factor

The shape factor is

$$\nu = \frac{Z_p}{Z} = \frac{599{,}028}{329{,}736} = 1.82$$

7.2.3 Effect of axial force

For a steel short column subject to axial load and bending, the full plastic moment of the column section is reduced by the axial load. Figure 7.6 depicts the plastic stress distribution of a rectangular column section under combined axial load and bending. In the full plastic state, the axial force (P) and full plastic moment (M_p) in the section can be determined as the stress resultants:

$$P = b(\alpha d)(2f_y) = \alpha P_o \tag{7.9}$$

$$M_p = M_o - P\left(\frac{\alpha d}{2}\right) = (1-\alpha^2)M_o \tag{7.10}$$

where

P_o is the ultimate axial load of the cross section in the absence of bending moment
M_o is the full plastic moment in the absence of the axial load

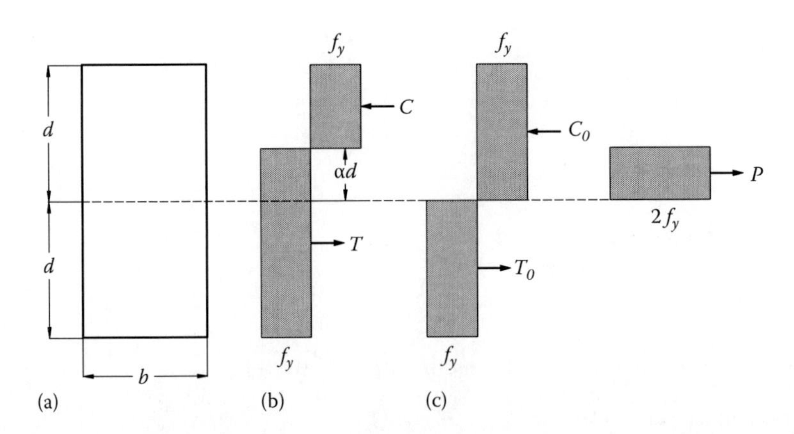

Figure 7.6 Plastic stress distributions in a rectangular column section under axial load and bending: (a) cross-section; (b) actual plastic stress distribution; (c) equivalent plastic stress distribution.

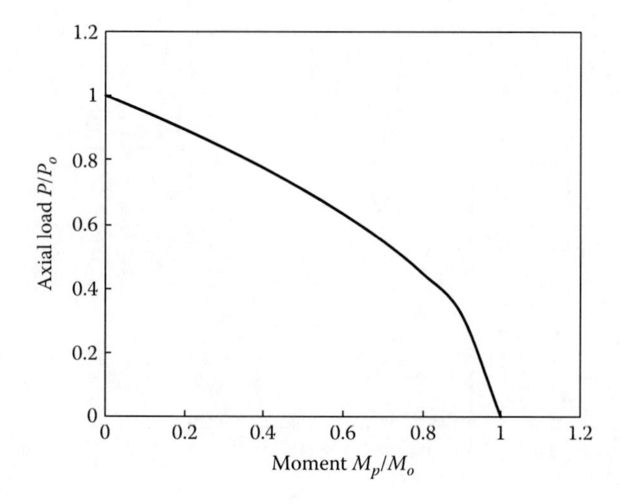

Figure 7.7 Yield surface.

The axial load–moment interaction equation can be obtained by combining Equations 7.9 and 7.10 as

$$\left(\frac{P}{P_o}\right)^2 + \left(\frac{M_p}{M_o}\right) = 1 \tag{7.11}$$

The interaction curve for a steel short column under axial compression and bending moment is given in Figure 7.7. The interaction curve represents a yield surface which is an important concept in the plastic theory (Baker and Heyman 1969). If a point lies within the boundary of the yield surface, the section can carry the combination of axial load and bending moment. A point on the boundary of the yield surface just causes the section to become fully plastic. A point outside the boundary of the yield surface represents an impossible state.

7.2.4 Effect of shear force

The cross section of a steel member under combined shear force and bending is subjected to a 2D stress state. The bending stresses act in the longitudinal direction, while the shear stresses act in the transverse direction. It is assumed that the flanges of a steel I-section do not carry shear stresses and shear stresses are uniformly distributed over the web. In the fully plastic state, the longitudinal stress (σ) in the web for resisting the plastic moment will be less than the yield stress (f_y) due to the presence of the shear stresses (τ). This can be expressed by the von Mises yield criteria as follows:

$$\sigma^2 + 3\tau^2 = f_y^2 \tag{7.12}$$

The longitudinal bending stress on the web of the steel I-section can be obtained from the aforementioned equation as

$$\sigma = \sqrt{f_y^2 - 3\tau^2} \le f_y \tag{7.13}$$

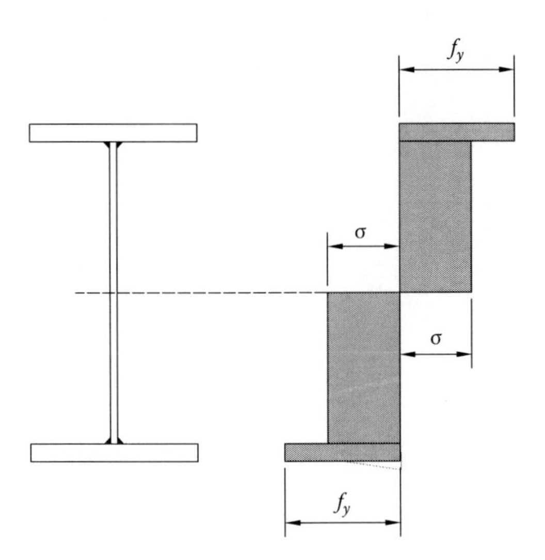

Figure 7.8 Effect of shear on the stress distribution in I-section.

Figure 7.8 depicts the stress distribution over the cross section subjected to combined bending and shear. It can be seen that the contribution from web to the full plastic moment is reduced by shear stresses. Further details on the effects of shear on the full plastic moment were given by Baker and Heyman (1969).

7.3 PLASTIC ANALYSIS OF STEEL BEAMS

7.3.1 Plastic collapse mechanisms

The fixed ended beam depicted in Figure 7.9 is used to demonstrate the development of plastic collapse mechanism (Baker and Heyman 1969). The beam of uniform cross section is subjected to slowly increasing point load W until it collapses. The elastic bending moment diagram is shown in Figure 7.9b. The bending moments at points A, B and C are $M_A = 6WL/27$, $M_B = 8WL/27$ and $M_C = 12WL/27$. As the load W is slowly increased, the bending moment at point C approaches the full plastic moment M_p and the first plastic hinge forms at point C as illustrated in Figure 7.9c. The formation of the plastic hinge causes a redistribution of moments. As the load is continuously increased, a second plastic hinge forms at point B. The two plastic hinges have turned the redundant beam into a statically determinate structure. Further increase in the loading causes the final plastic hinge to form at point A. The formation of the third plastic hinge turns the beam into a mechanism of plastic collapse. The plastic moment distribution of the beam is given in Figure 7.9d. It should be noted that the collapse load does not depend on the order of formation of the plastic hinges. The plastic analysis is concerned with only the collapse state of a structure.

7.3.2 Work equation

The virtual work equation can be used to describe the energy balance for a structure in the collapse state (Baker and Heyman 1969; Neal 1977). The formation of plastic hinges in a structure turns the structure into a collapse state. This implies that a small deformation

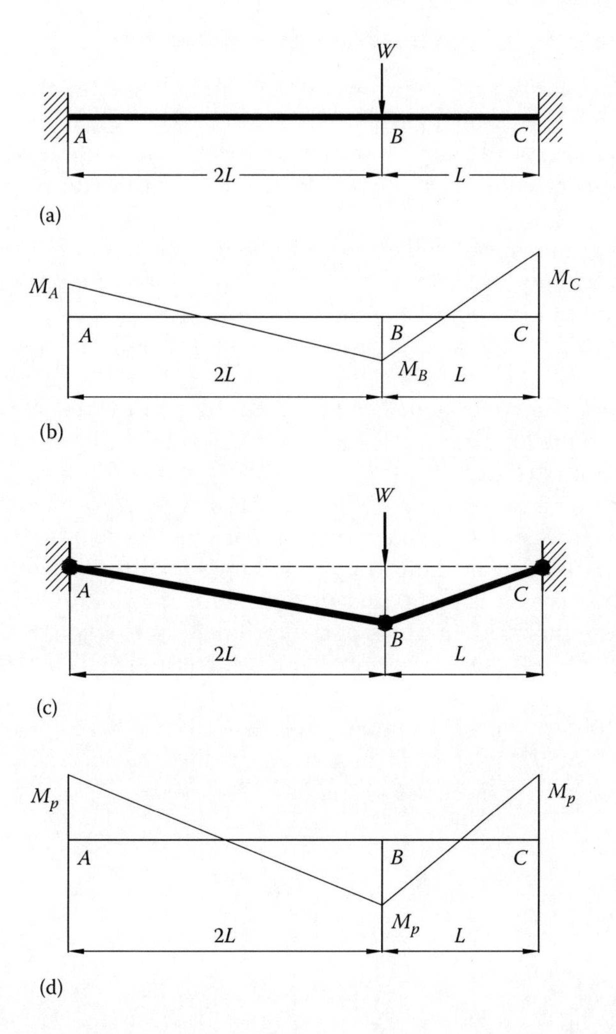

Figure 7.9 Development of plastic collapse mechanism: (a) steel beam with fixed ends, (b) elastic bending moment diagram, (c) plastic collapse mechanism and (d) plastic bending moment diagram.

of the collapse mechanism at constant values of the applied loads can occur. The work done by the applied load W under a small deformation δ is $W\delta$. The total work done by all applied loads on the structure is $\sum W\delta$. The plastic hinges will absorb the work done by external loads by rotating certain angles θ under the constant plastic moment M_p. The work absorbed in all plastic hinges is $\sum M_p\theta$. The work equation based on the simple energy balance theorem is expressed by

$$\sum W\delta = \sum M_p\theta \tag{7.14}$$

The work dissipated at a plastic hinge is always positive. Therefore, the signs of hinge rotations (θ) must be taken as the same as the sign of the corresponding plastic moment (M_p). All collapse mechanisms are usually drawn with straight members between plastic hinges. The use of the work equation is called the mechanism method.

7.3.3 Plastic analysis using the mechanism method

The plastic design of beams is to determine all possible collapse mechanisms and the corresponding values of full plastic moments and then design the beams based on the mechanism which provides the largest full plastic moment. In the plastic analysis using the mechanism method, the following considerations should be taken into account:

- All possible mechanisms of collapse should be investigated.
- Plastic hinges tend to form at the ends of members, at positions of concentrated loads and at the point of maximum bending moment.
- The mechanism and M_p of each span in a continuous beam should be investigated individually.
- At each support of a continuous beam, the plastic hinge forms in the weaker member with a smaller value of M_p.

The propped cantilever beam shown in Figure 7.10a is used to illustrate the mechanism method (Baker and Heyman 1969; Horne and Morris 1981). The propped cantilever of span L is subjected to slowly increase uniformly distributed load w. The collapse mechanism is given in Figure 7.10b, which is composed of two rigid rinks. The *central* hinge is located some distance x from the right-hand support. The angle of rotation at the left-hand end of the beam is assumed to be θ_1. Other rotations can be determined from the geometry in terms of θ_1. This gives $\theta_2 = (L-x)\theta_1/x$ and $\theta_3 = L\theta_1/x$. The resultant force acting on each rigid rink is shown in Figure 7.10b. Under the resultant force on each rigid rink, the rigid rink undergoes a mean displacement of $\delta/2$, where δ is the displacement at the point of *central* plastic hinge. The work equation can be written as

$$[w(L-x)+wx]\times\frac{(L-x)\theta_1}{2} = M_p(\theta_1)+M_p\left(\frac{L}{x}\theta_1\right) \tag{7.15}$$

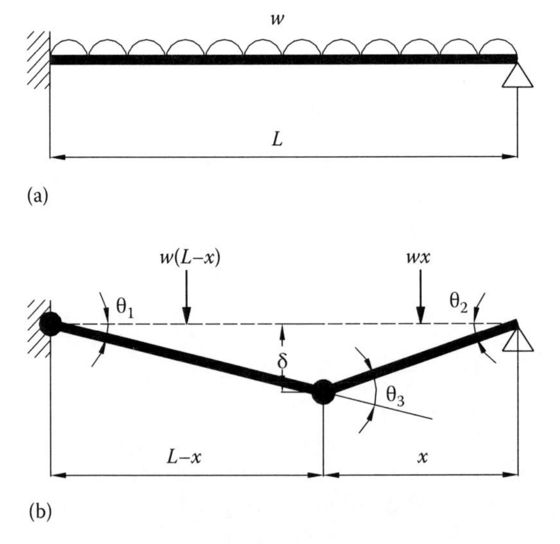

Figure 7.10 (a) Propped cantilever and (b) collapse mechanism.

The full plastic moment can be obtained from the preceding equation as

$$M_p = \frac{wLx}{2}\left(\frac{L-x}{L+x}\right) \tag{7.16}$$

The maximum full plastic moment is $M_p = wL^2/11.66$ when $x = 0.414L$ (Horne and Morris 1981).

Example 7.2: Largest plastic moment of two-span continuous beam

A two-span continuous steel beam with different uniform cross sections under factored concentrated loads is schematically depicted in Figure 7.11a. Determine the largest full plastic moment of the continuous beam.

1. Mechanism 1

Mechanism 1 is shown in Figure 7.11b. At the support, the plastic hinge is correctly located in the weaker member. Since no mechanism has been assumed in the second span,

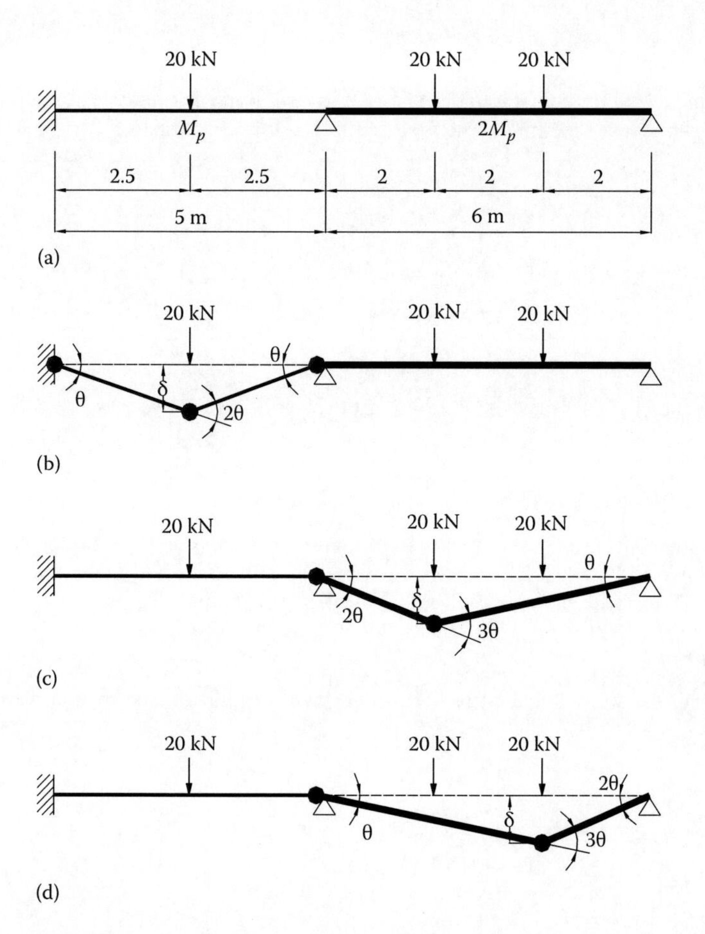

Figure 7.11 Mechanisms of two-span continuous beam: (a) continuous beam, (b) mechanism 1, (c) mechanism 2 and (d) mechanism 3.

there is no displacement and the work done by the loading in that span is zero. The work equation for mechanism 1 can be written as

$$\sum W\delta = \sum M_p\theta$$

$$20\times(2.5\times\theta) = M_p(\theta) + M_p(2\theta) + M_p(\theta)$$

$$\therefore M_p = 12.5 \text{ kNm}$$

2. Mechanism 2

Mechanism 2 is shown in Figure 7.11c. At the support, the plastic hinge is correctly located in the weaker member. The work equation for mechanism 2 can be written as

$$\sum W\delta = \sum M_p\theta$$

$$20\times(2\times 2\theta) + 20\times(2\times\theta) = M_p(2\theta) + 2M_p(3\theta)$$

$$\therefore M_p = 15 \text{ kNm}$$

3. Mechanism 3

Mechanism 3 is shown in Figure 7.11d. At the support, the plastic hinge is correctly located in the weaker member. The work equation for mechanism 3 can be written as

$$\sum W\delta = \sum M_p\theta$$

$$20\times(2\times\theta) + 20\times(4\times\theta) = M_p(\theta) + 2M_p(3\theta)$$

$$\therefore M_p = 17.1 \text{ kNm}$$

Therefore, the greatest full plastic moment of the continuous beam is 17.1 kN m.

Example 7.3: Collapse load of three-span continuous beam

A three-span continuous steel beam with a uniform cross section under concentrated loads is shown in Figure 7.12a. The full plastic moment of the beam cross section is M_p = 450 kN m. Determine the collapse load W of the continuous beam.

1. Mechanism 1

Mechanism 1 is shown in Figure 7.12b. The work equation for mechanism 1 can be written as

$$\sum W\delta = \sum M_p\theta$$

$$1.5W\times(3\times\theta) = M_p(2\theta) + M_p(\theta)$$

$$\therefore W = \frac{3M_p}{4.5} = \frac{3\times 450}{4.5} = 300 \text{ kN}$$

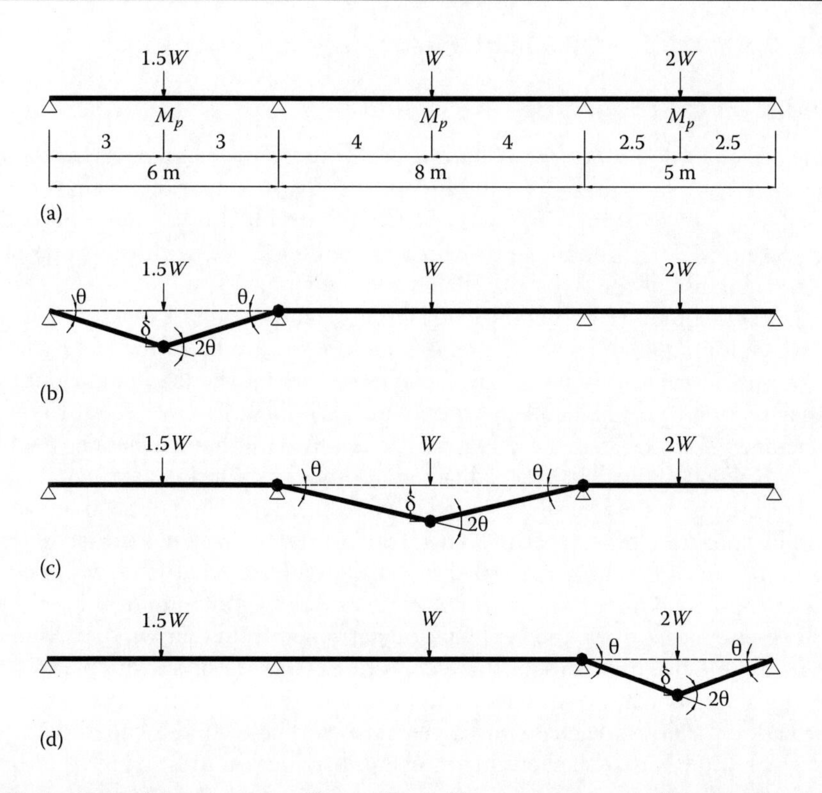

Figure 7.12 Mechanisms of three-span continuous beam: (a) continuous beam, (b) mechanism 1, (c) mechanism 2 and (d) mechanism 3.

2. Mechanism 2

Mechanism 2 is shown in Figure 7.12c. The work equation for mechanism 2 can be written as

$$\sum W\delta = \sum M_p\theta$$

$$W \times (4 \times \theta) = M_p(\theta) + M_p(2\theta) + M_p(\theta)$$

$$\therefore W = \frac{4M_p}{4} = M_p = 450 \text{ kN}$$

3. Mechanism 3

Mechanism 3 is shown in Figure 7.12d. The work equation for mechanism 3 can be written as

$$\sum W\delta = \sum M_p\theta$$

$$2W \times (2.5 \times \theta) = M_p(\theta) + M_p(2\theta)$$

$$\therefore W = \frac{3M_p}{5} = \frac{3 \times 450}{5} = 270 \text{ kN}$$

Therefore, the minimum collapse load W of the continuous beam is 270 kN.

7.4 PLASTIC ANALYSIS OF STEEL FRAMES

7.4.1 Fundamental theorems

In the plastic design, only proportional loading is allowed. This means that the loads applied to a structure will not vary randomly and independently. It is considered that the structure is initially subjected to working loads which can be multiplied by a common load factor λ as the load increases. The fundamental theorems are concerned with the value of the load factor λ_c at the collapse of the structure (Baker and Heyman 1969).

The uniqueness theorem states that the load factor (λ_c) at the collapse of a structure has a definite value, which is unique for the structure. As the loads are gradually increased, the structure collapses at a certain value λ_c. The unsafe theorem states that the load factor (λ) determined from the analysis of an assumed collapse mechanism will be greater or equal to the true collapse load factor λ_c. This theorem means that if the assumed mechanism happens to be correct, the load factor is equal to the collapse load factor λ_c; otherwise, the load factor calculated from the assumed mechanism is greater than λ_c and is overestimated. The safe theorem is concerned with the equilibrium state of a structure. This theorem states that the load factor determined from the equilibrium of bending moment distribution with external loads will be less the or equal to the collapse load factor λ_c. This implies that if the bending moment distribution does not cause a collapse mechanism, the load factor determined from that will be less than λ_c.

A structure at collapse must satisfy three conditions (Baker and Heyman 1969). The first is called the mechanism condition, which requires that a sufficient number of plastic hinges must be formed to turn the structure into a mechanism. The second is called the equilibrium condition, which implies that the bending moment distribution must be in equilibrium with external loads at all loading stages. The third condition is called the yield condition, which means that the bending moment at any section must not exceed the full plastic moment M_p. If these three conditions are satisfied simultaneously, the structure is at the state of collapse and the load factor determined is equal to the collapse factor λ_c which is unique.

7.4.2 Method of combined mechanism

The work equation can be written for any mechanism which satisfies the equilibrium condition of a structure. However, there is a limit to the number of independent equations of equilibrium for a structure. For a multistorey and multi-bay frame, the number of independent mechanisms can be calculated as (Horne and Morris 1981)

$$n_m = k(j+1) \tag{7.17}$$

where

k is the total number of storeys
j is the total number of bays

Beam and sway mechanisms are independent mechanisms as depicted in Figure 7.13b and c for a portal frame. All other mechanisms can be deduced from these independent mechanisms. The combined mechanism is obtained by combining the beam and sway mechanisms into one. Some of the plastic hinges in the two mechanisms are cancelled in order to lock together in an equilibrium state. Figure 7.13d shows a combined mechanism. The plastic analysis of frames using the combined mechanism method is demonstrated in Examples 7.4 and 7.5. Further details on the plastic analysis of frames can be found in the book by Horne and Morris (1981).

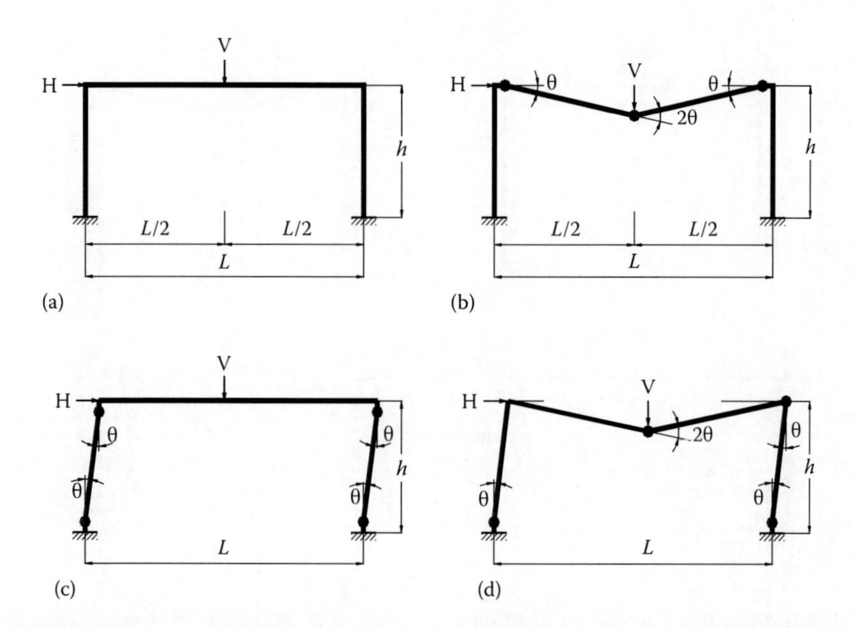

Figure 7.13 Beam, sway and combined mechanisms: (a) portal frame, (b) beam mechanism, (c) sway mechanism and (d) combined mechanism.

Example 7.4: Collapse load factor of steel portal frame

Figure 7.14a shows a steel portal frame under working loads. The portal frame has a uniform full plastic moment of 150 kN m. Determine the collapse load factor of this portal frame.

1. Beam mechanism

The beam mechanism is shown in Figure 7.14b. The work equation for the beam mechanism can be written as

$$\sum W\delta = \sum M_p\theta$$

$$100\lambda \times (4 \times \theta) = M_p(\theta) + M_p(2\theta) + M_p(\theta)$$

$$\therefore \lambda = \frac{4M_p}{400} = \frac{M_p}{100} = \frac{150}{100} = 1.5$$

2. Sway mechanism

The sway mechanism is shown in Figure 7.14c. The work equation for the sway mechanism can be written as

$$\sum W\delta = \sum M_p\theta$$

$$60\lambda \times (3.5 \times \theta) = M_p(\theta) + M_p(\theta) + M_p(\theta) + M_p(\theta)$$

$$\therefore \lambda = \frac{4M_p}{210} = \frac{4 \times 150}{210} = 2.86$$

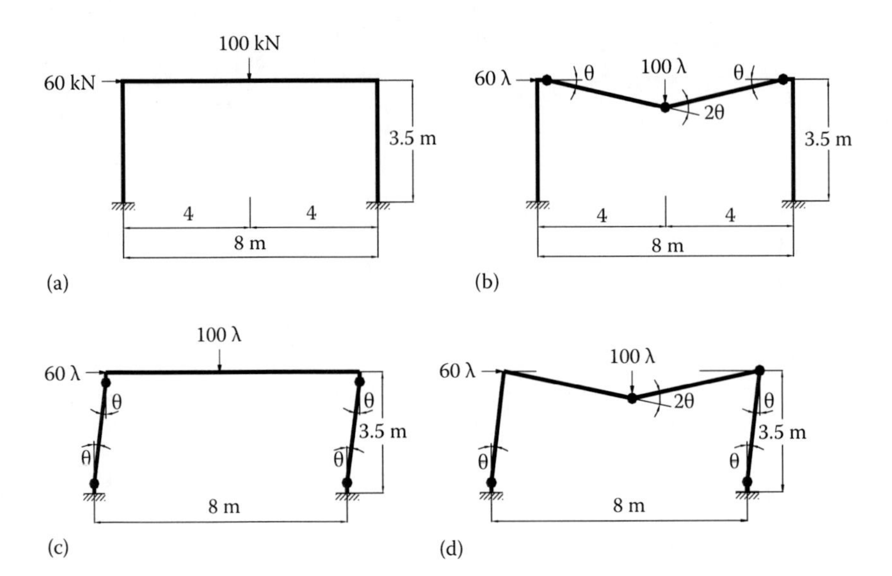

Figure 7.14 Mechanisms of steel portal frame: (a) portal frame, (b) beam mechanism, (c) sway mechanism and (d) combined mechanism.

3. Combined mechanism

The combined mechanism 3 is shown in Figure 7.14d. The work equation for the combined mechanism can be written as

$$\sum W\delta = \sum M_p\theta$$

$$100\lambda \times (4 \times \theta) + 60\lambda \times (3.5 \times \theta) = M_p(\theta) + M_p(2\theta) + M_p(2\theta) + M_p(\theta)$$

$$\therefore \lambda = \frac{6M_p}{610} = \frac{6 \times 150}{610} = 1.48$$

Therefore, the collapse load factor is $\lambda_c = 1.48$.

Example 7.5: Collapse load factor of two-storey frame

Figure 7.15 shows a two-storey steel frame under working loads. The full plastic moments M_p (kN m) of the frame members are shown in the figure. Determine the collapse load factor of this two-storey frame.

1. Number of independent mechanisms

This is a two-storey and one-bay frame; thus, $k = 2$ and $j = 1$. The number of independent mechanisms is

$$n_m = k(j + 1) = 2 \times (1 + 1) = 4$$

There are four independent mechanisms, which include two beam and two sway mechanisms.

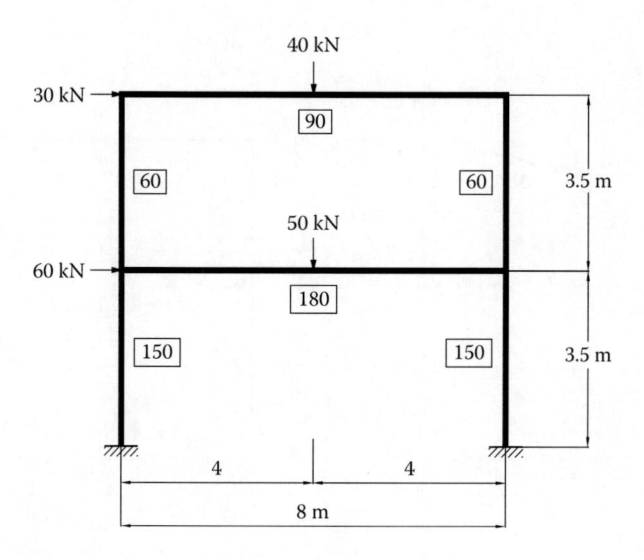

Figure 7.15 Two-storey frame.

2. Mechanism 1

Mechanism 1 is a beam mechanism shown in Figure 7.16a. The work equation for the sway mechanism can be written as

$$\sum W\delta = \sum M_p\theta$$

$$40\lambda \times (4 \times \theta) = 60 \times (\theta) + 90 \times (2\theta) + 60 \times (\theta)$$

$$\therefore \lambda = \frac{300}{160} = 1.875$$

3. Mechanism 2

Mechanism 2 is also a beam mechanism shown in Figure 7.16b. The work equation for this mechanism can be written as

$$\sum W\delta = \sum M_p\theta$$

$$50\lambda \times (4 \times \theta) = 180 \times (\theta) + 180 \times (2\theta) + 180 \times (\theta)$$

$$\therefore \lambda = \frac{720}{200} = 3.6$$

4. Mechanism 3

Mechanism 3 is a sway mechanism shown in Figure 7.16c. The work equation for this mechanism can be written as

$$\sum W\delta = \sum M_p\theta$$

$$30\lambda \times (3.5 \times \theta) = 60 \times (\theta) + 60 \times (\theta) + 60 \times (\theta) + 60 \times (\theta)$$

$$\therefore \lambda = \frac{240}{200} = 2.286$$

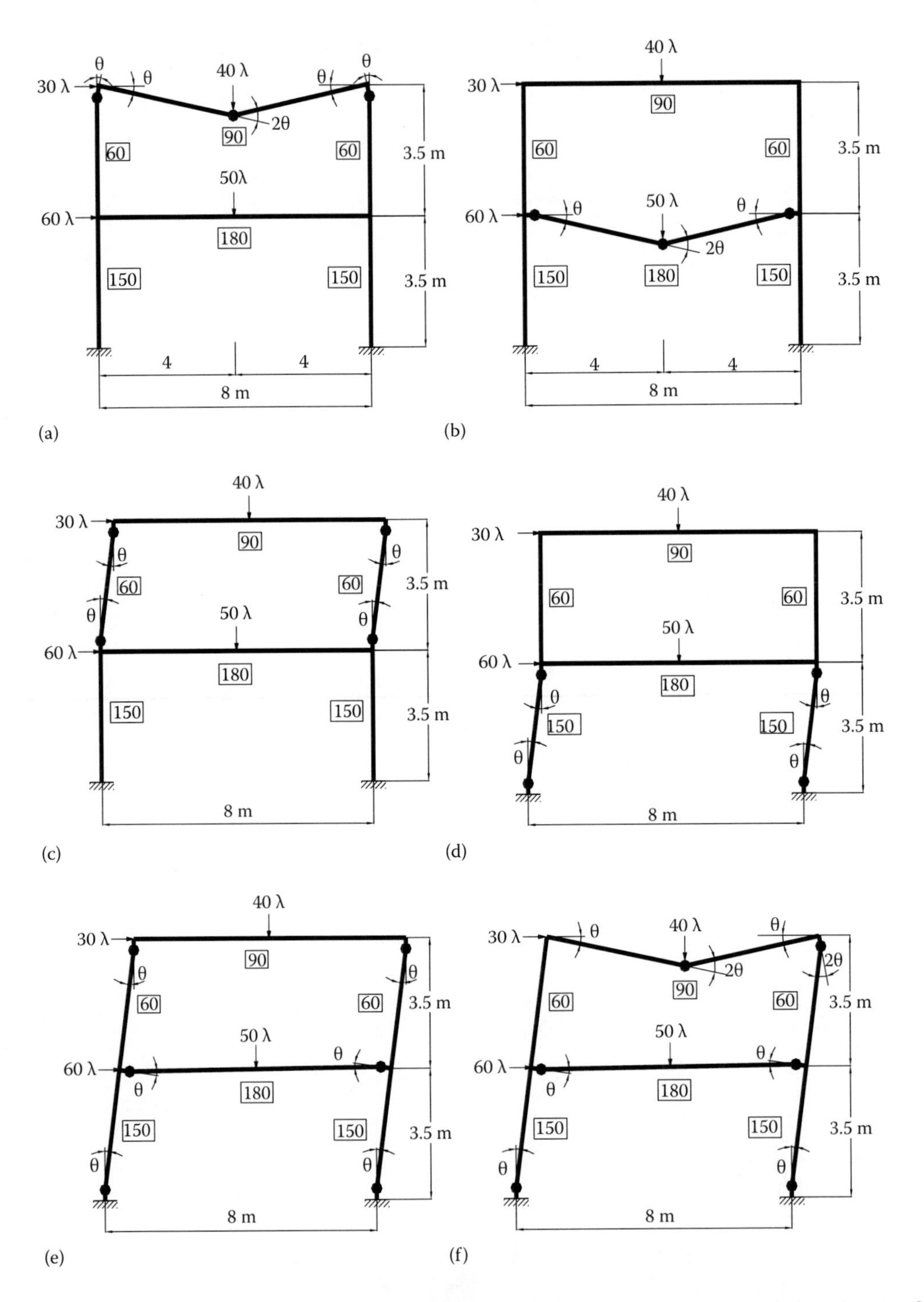

Figure 7.16 Mechanisms of two-storey frame: (a) mechanism 1, (b) mechanism 2, (c) mechanism 3, (d) mechanism 4, (e) mechanism 5 and (f) mechanism 6.

5. Mechanism 4

Mechanism 4 is also a sway mechanism shown in Figure 7.16d. The work equation for this mechanism can be written as

$$\sum W\delta = \sum M_p\theta$$

$$(60+30)\lambda \times (3.5\times\theta) = 150\times(\theta)+150\times(\theta)+150\times(\theta)+150\times(\theta)$$

$$\therefore \lambda = \frac{600}{315} = 1.9$$

6. Mechanism 5

Mechanism 5 is a combined mechanism shown in Figure 7.16e. The work equation for this mechanism can be written as

$$\sum W\delta = \sum M_p\theta$$

$$30\lambda\times(7\times\theta)+60\lambda\times(3.5\times\theta) = 2\times150\times(\theta)+2\times60\times(\theta)+2\times180\times(\theta)$$

$$\therefore \lambda = \frac{780}{420} = 1.857$$

7. Mechanism 6

Mechanism 6 is a combined mechanism shown in Figure 7.16f. The work equation for this mechanism can be written as

$$\sum W\delta = \sum M_p\theta$$

$$30\lambda\times(7\times\theta)+60\lambda\times(3.5\times\theta)+40\lambda\times(4\times\theta) = 2\times150\times(\theta)+60\times(2\theta)+2\times180\times(\theta)+90\times(2\theta)$$

$$\therefore \lambda = \frac{960}{580} = 1.655$$

Therefore, the collapse load factor is $\lambda_c = 1.655$.

7.5 PLASTIC DESIGN TO AS 4100

7.5.1 Limitations on plastic design

Clause 4.5 of AS 4100 requires that if the plastic method of structural analysis is used, all of the following conditions shall be satisfied:

- The members used shall be hot-formed, doubly symmetric, compact I-sections.
- The minimum yield stress of the steel shall not exceed 450 MPa.
- The stress–strain characteristics of the steel shall not be significantly different from those of AS/NZS 3678 or AS/NZS 3679.1.
- The stress–strain curve of the steel shall have a yield plateau extending for at least six times the yield strain.
- The ratio of f_u/f_y is not less than 1.2.
- The elongation of the steel is not less than 15% and it exhibits strain-hardening characteristics.

- No impact loading or fluctuating loading that requires a fatigue assessment is applied to the members.
- The connections shall have the capacity to cope with the formation of the plastic hinges and do not suppress the formation of plastic hinges.

7.5.2 Section capacity under axial load and bending

AS 4100 gives specifications on the plastic design of in-plane beams, beam–columns and frames. However, the biaxial bending is not considered in AS 4100 owing to the complexity of biaxial interaction behaviour.

The design moment capacity of the section reduced by axial force for bending about the major principal axis is given in Clause 8.4.3.4 of AS 4100 (1998) as follows:

$$\phi M_{prx} = \phi 1.18 M_{sx}\left[1 - \frac{N^*}{\phi N_s}\right] \le \phi M_{sx} \tag{7.18}$$

where

$\phi = 0.9$, the capacity reduction factor
ϕM_{sx} is the design section moment capacity for bending about the major principal x-axis
N^* is the design axial force
ϕN_s is the design axial section capacity

For a section bent about the minor principal axis, AS 4100 provides the following equation for calculating the reduced design moment capacity of the section:

$$\phi M_{pry} = \phi 1.19 M_{sy}\left[1 - \left(\frac{N^*}{\phi N_s}\right)^2\right] \le \phi M_{sy} \tag{7.19}$$

where

$\phi = 0.9$, the capacity reduction factor
ϕM_{sy} is the design section moment capacity for bending about the minor principal y-axis

7.5.3 Slenderness limits

Clause 8.4.3.2 of AS 4100 gives limits on the slenderness of members which contain plastic hinges in terms of the design axial compressive force. The design axial compressive force N^* in a member containing a plastic hinge shall satisfy the following conditions:

$$\frac{N^*}{\phi N_s} \le \left[\frac{0.6 + 0.4\beta_m}{\sqrt{N_s/N_{cr}}}\right]^2 \quad \text{when } \frac{N^*}{\phi N_s} \le 0.15 \tag{7.20}$$

$$\frac{N^*}{\phi N_s} \le \frac{\left[1 + \beta_m - \sqrt{N_s/N_{cr}}\right]}{\left[1 + \beta_m + \sqrt{N_s/N_{cr}}\right]} \quad \text{when } \frac{N^*}{\phi N_s} > 0.15 \tag{7.21}$$

where

N_s is the nominal axial section capacity of the member
N_{cr} is the elastic buckling load of the member
β_m is the ratio of the smaller to the larger end bending moments

The member, which does not have a plastic hinge, should be designed based on the elastic method if the following condition is satisfied:

$$\frac{N^*}{\phi N_s} > \frac{\left[1+\beta_m - \sqrt{N_s/N_{cr}}\right]}{\left[1+\beta_m + \sqrt{N_s/N_{cr}}\right]} \quad \text{and} \quad \frac{N^*}{\phi N_s} > 0.15 \tag{7.22}$$

Clause 8.4.3.3 of AS 4100 also gives limits on the webs of members containing plastic hinges in terms of the design axial compression force. In members containing plastic hinges, the design axial compressive forces should satisfy the following conditions:

$$\frac{N^*}{\phi N_s} \le 0.6 - \frac{\lambda_n}{137} \quad \text{for } 45 \le \lambda_n \le 82 \tag{7.23}$$

$$\frac{N^*}{\phi N_s} \le 1.91 - \frac{\lambda_n}{24.7} \quad \text{for } 25 < \lambda_n < 45 \tag{7.24}$$

$$\frac{N^*}{\phi N_s} \le 1.0 \quad \text{for } 0 \le \lambda_n \le 25 \tag{7.25}$$

When $\lambda_n > 82$, the web of the member is slender so that it must not contain any plastic hinge. The member must be designed based on the elastic method or the frame should be redesigned.

REFERENCES

AS 4100 (1998) Australian standard for steel structures, Sydney, New South Wales, Australia: Standards Australia.

Baker, J. and Heyman, J. (1969) *Plastic Design of Frames*, London, U.K.: Cambridge University Press.

Horne, M.R. and Morris, L.J. (1981) *Plastic Design of Low-Rise Frames*, London, U.K.: Collins.

Neal, B.G. (1977) *The Plastic Methods of Structural Analysis*, London, U.K.: Chapman and Hall.

Chapter 8

Composite slabs

8.1 INTRODUCTION

Composite floor systems are formed by connecting floor slabs to the top flanges of structural steel beams, girders or trusses using mechanical shear connectors. The concrete floor slab can be a conventional reinforced concrete slab or a composite slab with profiled steel sheeting supporting the concrete. Composite slabs have been widely used in multistorey composite buildings in many countries. This composite slab system utilises the best load-resisting characteristics of steel and concrete materials. Structural steel has the properties of high strength, high ductility and high speed of erection, while structural concrete has the properties of excellent fire resistance, inherent mass and low material cost. Composite slabs can be designed as either simply supported one-way slabs or continuous slabs.

Currently, there are no Australian Standards available for the design of composite slabs. This chapter presents the behaviour and design of composite slabs for strength and serviceability to Eurocode 4 (2004) and to Australian practice. The concept of shear connection is introduced first. The design of simply supported composite slabs with complete and partial shear connections to Eurocode 4 is then described. This is followed by the presentations of the design of continuous composite slabs for positive moment and negative moment regions in terms of flexural and vertical shear strengths in accordance with Australian practice. The longitudinal shear and punching shear are also covered. The design of composite slabs for serviceability is given.

8.2 COMPONENTS OF COMPOSITE SLABS

The components of a composite slab include the profiled steel sheeting, cast in situ concrete and reinforcement in the form of welded-wire mesh or deformed bars as schematically depicted in Figure 8.1.

The profiled steel sheeting is very thin with basis metal thickness between 0.6 and 1.0 mm for Australian products. The steel sheeting is pressed or cold rolled and is designed to span in the longitudinal direction only. In the construction stage, before casting the concrete, the profiled steel sheeting acts as a platform for construction. After casting the slab concrete, the sheeting supports the wet concrete and acts as permanent formwork for the concrete. After the concrete has hardened and composite action between the sheeting and the concrete has been developed, the steel sheeting acts as bottom face tensile reinforcement for the concrete slab.

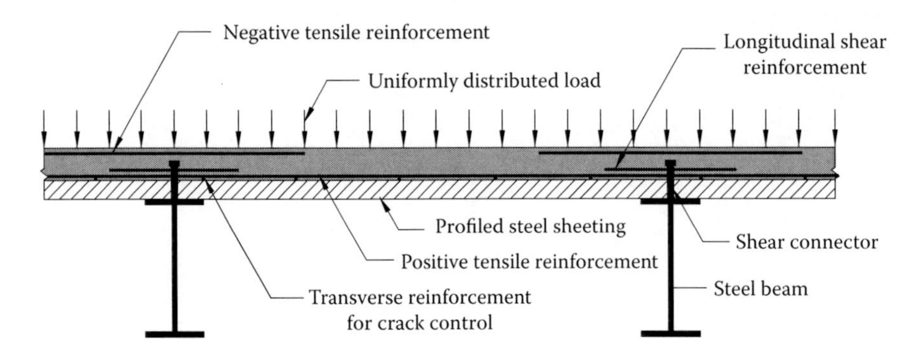

Figure 8.1 Components of composite slab.

The profiled steel sheeting used in composite beam construction must satisfy the geometric requirements given in the Clause 1.2.4 of AS 2327.1 (2003) as illustrated in Figure 8.2:

- The height of the steel rib (h_r) should not be greater than 80 mm.
- The concrete cover slab thickness ($h_c = D_c - h_r$) should not be less than 65 mm.
- The opening width of the steel rib at its base should not be greater than 20 mm.
- The area of the voids due to the opening of the rib should not be greater than 20% of the area of the concrete within the depth of the ribs.
- The width of concrete between the mid-heights of adjacent ribs should not be less than 150 mm.

The profiled steel sheeting usually provides more than adequate bottom reinforcement for the composite slab so that it can be designed as simply supported to utilise the strength of the profiled steel sheeting. However, top longitudinal reinforcement at the supports is still needed to control cracks if the slabs are treated as simply supported. In Australia, it is common practice to design continuous composite slabs with negative tensile reinforcement over the supports for bending and crack control. Positive tensile reinforcement may be provided to increase the moment capacity of composite slabs. Transverse reinforcement must be provided in composite slabs for crack control due to shrinkage and temperature effects.

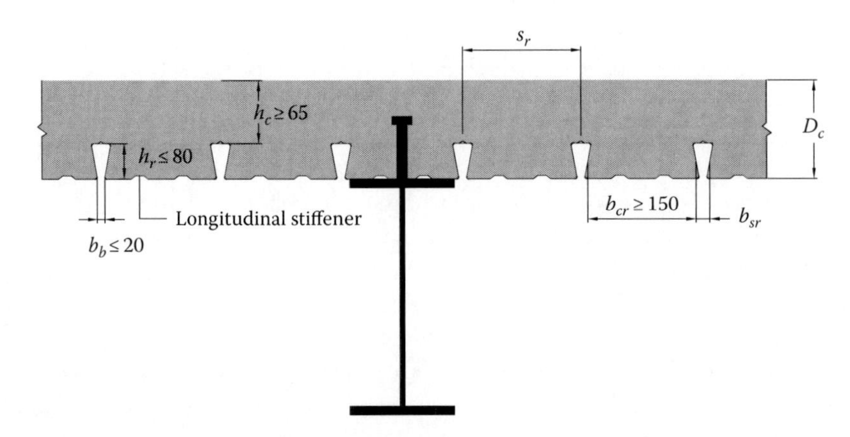

Figure 8.2 Profiled steel sheeting geometric restrictions.

8.3 BEHAVIOUR OF COMPOSITE SLABS

The behaviour of composite slabs can be determined by either experiments or numerical analysis such as the finite element analysis. There are three possible failure modes associated with a simply supported composite slab in a two-point load test (Johnson 2004). The failure mode depends on the ratio of the shear span to the effective depth of the slab (L_s/D_e). When the L_s/D_e ratio is high, the composite slab fails by flexure in the region of maximum positive bending moment. When the L_s/D_e ratio is low, the composite slab fails by the vertical shear near the supports. At intermediate values of L_s/D_e, longitudinal shear failure occurs at the interface of the sheeting ribs and the concrete cover slab. The longitudinal shear failure is initialised by the crack in the concrete under one of the load points, which associates with the loss of bond along the shear span and slip at the end of the slab. The shear connection between the concrete and sheeting is brittle if longitudinal shear failure occurs. For continuous composite slabs, flexural and vertical shear failures may occur in the negative moment regions. The design of composite slabs is to ensure that the failure modes mentioned earlier will not occur. For this purpose, continuous composite slabs need to be designed for positive and negative bending moments and vertical shear forces.

8.4 SHEAR CONNECTION OF COMPOSITE SLABS

8.4.1 Basic concepts

The shear connection of a composite slab is the interconnection between the profiled steel sheeting and the concrete, which enables the two components to act as a single structural member. The shear connection resists the longitudinal slip at the interface of the steel sheeting and concrete. There are three mechanisms that contribute to the shear connection of a composite slab. The first mechanism is the chemical bond between the two components. The second mechanism is the mechanical interlock provided by the dimples which are pressed into the surface of the steel sheeting. The third mechanism is the end anchorage which may be provided by pins, welding studs through the sheeting to the top flange of the steel beam or friction between the sheeting and the supports.

When no shear connection between the sheeting and concrete is provided, there is no bond between these two components so that they act separately. If the slip and slip strain in a composite slab are everywhere zero, this condition is called full interaction of a composite slab. This implies that plane sections remain plane after deformation. The full interaction of a composite slab is a stiffness criterion. When the slip at the interface of the sheeting and concrete occurs along the length of a composite slab, this condition is called partial interaction, which is a stiffness criterion. Complete/full shear connection of a composite slab is the condition for which its section moment capacity is governed by the strength of the steel sheeting or concrete cover slab above the steel ribs. In contrast, the partial shear connection of composite slab is the condition for which its section moment capacity is governed by the strength of the shear connection. It is noted that the complete or partial shear connections is concerned with the strength of composite slabs so that it is a strength criterion.

8.4.2 Strength of shear connection

The shear connection strength of a composite slab depends on the mechanical resistance which includes the contributions of chemical bond and mechanical interlock along the slab and on the frictional resistance at its supports. The steel sheeting in a simply supported composite slab under bending is subjected to a resultant tensile force (T_p), while the top part

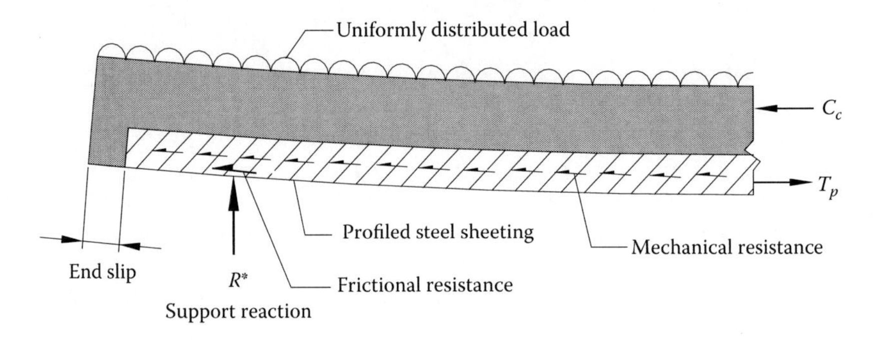

Figure 8.3 Mechanical and frictional resistance in composite slab.

of the concrete slab is in compression as schematically depicted in Figure 8.3. The resultant tensile force (T_p) at a critical cross section is resisted by the mechanical resistance (H_m) and the frictional resistance (H_f). It is assumed that the mechanical resistance (H_m) is developed uniformly across the full width of the composite slab and is expressed as force per unit plan area (kPa). The mechanical resistance of a composite slab is usually determined by either full-scale slab tests or small-scale slip-block tests (Patrick 1990; Patrick and Bridge 1994). Test results showed that the mechanical resistance (H_m) depends on the profile geometry, the sheeting thickness and the compressive strength of concrete. The mechanical resistance (H_m) is determined experimentally by Bridge (1998) as $88\sqrt{t_{bm}f'_c}$, 235 and 210 kPa for profiled steel sheeting Bondek II, Comform and Condeck HP, respectively, and are given in Table 8.1.

At a cross section with complete shear connection, the resultant tensile force in the steel sheeting (T_{pcs}) can be determined from the force equilibrium condition using the rectangular stress block theory. The strength of complete shear connection is governed by either the strength of the steel sheeting or the strength of the concrete cover slab including the contribution of longitudinal tensile reinforcement in the concrete. For a composite slab reinforced with conventional tensile reinforcement in the bottom face, the strength of the reinforced concrete cover slab can be expressed by

$$F_{cst} = 0.85f'_c b(D_c - h_r) - T_{yr} \tag{8.1}$$

where

$T_{yr} = A_r f_{yr}$ is the yield capacity of the steel reinforcement in the bottom face of the composite slab
A_r is the cross-sectional area of the reinforcement
f_{yr} is the yield stress of the reinforcement

Table 8.1 Properties of profiled steel sheeting

Profiled steel sheeting	h_r *(mm)*	b_{cr} *(mm)*	s_r *(mm)*	H_m *(kPa)*	A_p *(mm²)*	M_{up} *(kN m/m)*	φ_b
Bondek II	54	187	200	$88\sqrt{t_{bm}f'_c}$	$1678t_{bm}$	$13.8t_{bm}$	$1-\beta_{sc}^2$
Comform	58	300	300	235	$1563t_{bm}$	$10.7t_{bm}$	$1-\beta_{sc}^3$
Condeck HP	55	300	300	210	$1620t_{bm}$	$11.6t_{bm}$	$1-\beta_{sc}^3$

Source: Adapted from Goh, C.C. et al., Design of composite slabs for strength, composite structures design manual – Design booklet DB3.1, BHP Integrated Steel, Melbourne, Victoria, Australia, 1998.

The resultant tensile force in the steel sheeting (T_{pcs}) with complete shear connection is taken as

$$T_{pcs} = \min(F_{cst}, T_{yp}) \tag{8.2}$$

where
$T_{yp} = A_p f_{yp}$ is the yield capacity of the steel sheeting
A_p is the cross-sectional area of the sheeting
f_{yp} is the yield stress of the sheeting

8.4.3 Degree of shear connection

The degree of shear connection at a cross section in a composite slab is defined as the ratio of the resultant tensile force (T_p) to the resultant tensile force (T_{pcs}) in the steel sheeting with complete shear connection (Goh et al. 1998), which is expressed by

$$\beta_{sc} = \frac{T_p}{T_{pcs}} \quad 0 \le \beta_{sc} \le 1.0 \tag{8.3}$$

If the degree of shear connection at a cross section is known, the strength of the shear connection governing the moment capacity of the composite slab with partial shear connection is obtained from Equation 8.3 as $T_p = \beta_{sc} T_{pcs}$.

8.5 MOMENT CAPACITY BASED ON EUROCODE 4

At a cross section of a composite slab with complete shear connection and under bending, the plastic neutral axis of the cross section is usually located in the concrete cover slab (above the steel sheeting), except where the sheeting is very deep that the plastic neutral axis may lie in the sheeting. However, there are two neutral axes in a cross section with partial shear connection. The first plastic neutral axis lies in the concrete cover slab, while the second falls in the sheeting. The ultimate moment capacity of a composite slab with any degree of shear connection depends on the location of the plastic neutral axis. The calculation of the ultimate moment capacity of composite slabs based on Eurocode 4 (2004) is given in detail in the following sections.

8.5.1 Complete shear connection with neutral axis above sheeting

The longitudinal bending stress distribution through the depth of the cross section of a composite slab with complete shear connection is schematically depicted in Figure 8.4. For clarity, only part of the cross section of the composite slab is shown in Figure 8.4. The rectangular stress block theory is assumed for concrete in compression. The plastic neutral axis is assumed to be above the sheeting. The effective area of width (b) of sheeting and the height (h_p) of the centre of area above the bottom of the sheeting are determined by tests. The compressive force in the concrete cover slab can be calculated by

$$N_{cc} = 0.85 f_c' b \gamma d_n \tag{8.4}$$

where
d_n is the neutral axis depth
γ is given in AS3600 (2001) as

$$\gamma = 0.85 - 0.007(f_c' - 28) \quad 0.65 \le \gamma \le 0.85 \tag{8.5}$$

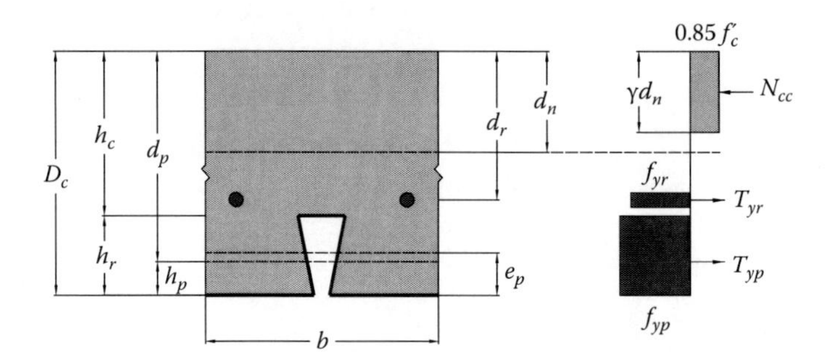

Figure 8.4 Stress distributions in section with complete shear connection: PNA above sheeting.

Assuming both reinforcing steel and profiled steel sheeting are at yield at the ultimate limit state, the compressive force in concrete with complete shear connection is

$$N_{cc} = T_{yp} + T_{yr} \tag{8.6}$$

The neutral axis depth d_n can be determined from the force equilibrium as

$$d_n = \frac{N_{cc}}{0.85 f_c' b \gamma} \tag{8.7}$$

The nominal ultimate moment capacity of the composite slab can be calculated by taking moments about the top fibre as

$$M_u = T_{yp} d_p + T_{yr} d_r - N_{cc}(0.5\gamma d_n) \tag{8.8}$$

where
- d_p is the distance from the top fibre to the elastic centroid of the sheeting
- d_r is the distance from the top fibre to the centroid of steel reinforcement

8.5.2 Complete shear connection with neutral axis within sheeting

When the plastic neutral axis is located within the sheeting as shown in Figure 8.5, the compressive force in the concrete with complete shear connection ignoring the compressive concrete in the ribs is given by

$$N_{cc} = 0.85 f_c' b h_c \tag{8.9}$$

where $h_c = (D_c - h_r)$ is the height of the concrete cover slab above the ribs.

As depicted in Figure 8.5, there is a compressive force N_{ac} in the steel sheeting below the plastic neutral axis. There is no simple method for determining the plastic neutral axis depth (d_n) and N_{ac} due to the complex properties of profiled steel sheeting. In Eurocode 4, the approximate method is used (Johnson 2004). The tensile force in steel sheeting is decomposed into a force at the bottom equal to N_{ac} and a force $N_p = N_{cc}$. The moment capacity M_{pr}

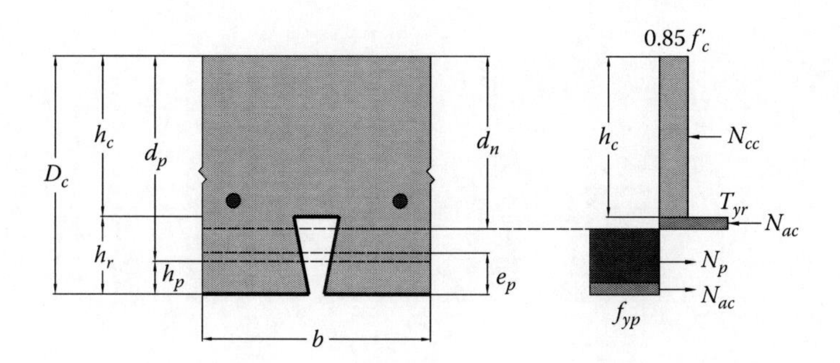

Figure 8.5 Stress distributions in section with complete shear connection: PNA in sheeting.

due to the couple forces (N_{ac}) is determined as the moment capacity of the steel sheeting (M_{pa}) reduced by the axial force N_{cc}. In Eurocode 4 (2004), M_{pr} is approximately determined by

$$M_{pr} = 1.25 M_{pa} \left[1 - \frac{N_{cc}}{N_p} \right] \tag{8.10}$$

The moment capacity of the composite slab is

$$M_u = N_{cc} z + M_{pr} \tag{8.11}$$

where the level arm z is given by

$$z = D_c - 0.5 h_c - e_p + (e_p - h_p) \frac{N_{cc}}{N_p} \tag{8.12}$$

where

e_p is the distance of plastic neutral axis above the base of steel sheeting
h_p is the distance of elastic centroid above the base of steel sheeting

8.5.3 Partial shear connection

The stress distribution in section with partial shear connection is presented in Figure 8.6. When the cross section of a composite slab is in partial shear connection, the compressive force in the concrete (N_{cp}) is less than N_{cc} and is determined by the strength of the shear connection. The depth (d_n) of the neutral axis in the concrete cover slab is

$$d_n = \frac{N_{cp}}{0.85 f'_c b \gamma} \tag{8.13}$$

As shown in Figure 8.6, the second neutral axis falls in the sheeting and the stress distribution is similar to that shown in Figure 8.5. In Eurocode 4 (2004), the moment capacity (M_{pr}) due to couple forces (N_{ac}) is approximately determined by

$$M_{pr} = 1.25 M_{pa} \left[1 - \frac{N_{cp}}{N_p} \right] \le M_{pa} \tag{8.14}$$

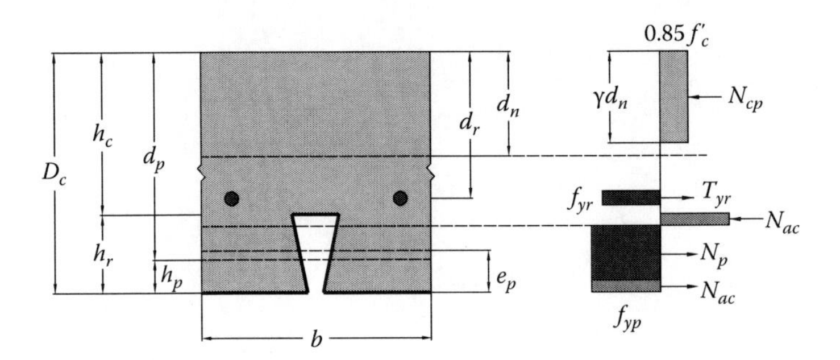

Figure 8.6 Stress distributions in section with partial shear connection.

The moment capacity of the composite slab can be calculated as

$$M_u = N_{cp}z + M_{pr} \tag{8.15}$$

where the level arm z is given by

$$z = D_c - 0.5\gamma d_n - e_p + (e_p - h_p)\frac{N_{cp}}{N_p} \tag{8.16}$$

8.6 MOMENT CAPACITY BASED ON AUSTRALIAN PRACTICE

8.6.1 Positive moment capacity with complete shear connection

In Australian practice of composite slab design, the simple plastic rectangular stress block theory is used in the calculation of the moment capacity of a composite slab. It is assumed that conventional reinforcement located on the tensile side of the neutral axis yields at the ultimate moment, otherwise it is ignored. The sheeting is lumped at the height of its centroid above the bottom of the composite slab. The height (y_p) of the sheeting centroid varies with the degree of shear connection, which is given as follows (Goh et al. 1998):

a. Bondek II: $y_p = \begin{cases} 18\beta_{sc}^2 & \text{for } 0 < \beta_{sc} \le 0.75 \\ 21.6\beta_{sc} - 6.1 & \text{for } 0.75 < \beta_{sc} \le 1.0 \end{cases}$

b. Comform: $y_p = \begin{cases} 18\beta_{sc}^3 & \text{for } 0 < \beta_{sc} \le 0.75 \\ 23.1\beta_{sc} - 9.7 & \text{for } 0.75 < \beta_{sc} \le 1.0 \end{cases}$

c. Condeck HP: $y_p = \begin{cases} 16\beta_{sc}^2 & \text{for } 0 < \beta_{sc} \le 0.75 \\ 24.1\beta_{sc} - 11.3 & \text{for } 0.75 < \beta_{sc} \le 1.0 \end{cases}$

Figure 8.7 gives the stress distribution in the section with complete shear connection. For a cross section with complete shear connection, the neutral axis depth that lies above the concrete cover slab can be calculated using Equation 8.7, providing that both steel reinforcement and sheeting are at yield. The strain in the steel reinforcement is given by

$$\varepsilon_r = 0.003 \times \frac{d_r - d_n}{d_n} \tag{8.17}$$

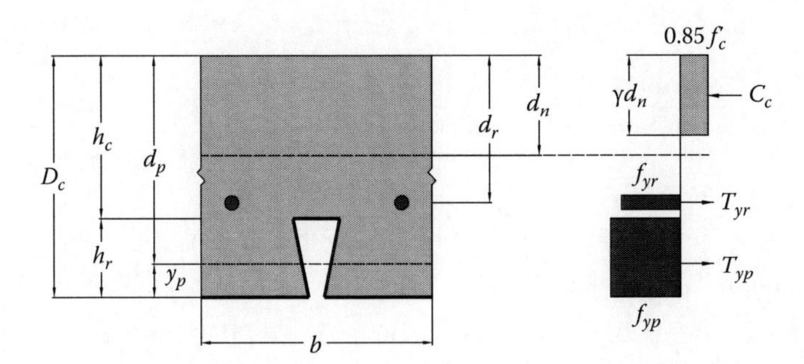

Figure 8.7 Stress distributions in section with complete shear connection: PNA above sheeting.

If the strain in steel reinforcement is greater than the yield strain of the steel reinforcement, its contribution to the moment capacity of the composite slab is considered, otherwise it is ignored. The nominal moment capacity of the composite slab can be calculated using Equation 8.8.

Example 8.1: Moment capacity of section with complete shear connection

The cross section of a composite slab incorporating Bondek II profiled steel sheeting has complete shear connection. The overall depth of the slab (D_c) is 150 mm. The compressive strength of concrete (f'_c) is 32 MPa. The thickness of the sheeting (t_{bm}) is 1.0 mm. The yield stress of the sheeting is 550 MPa. The cross-sectional area of bottom face tensile reinforcement (A_r) in the composite slab is 393 mm²/m. Determine the design positive moment capacity of the section with complete shear connection.

1. Resultant tensile force in sheeting

The area of Bondek II sheeting is obtained from Table 8.1 as

$$A_p = 1678t_{bm} = 1678 \times 1.0 = 1678 \text{ mm}^2/\text{m}$$

For complete shear connection, the resultant tensile force in sheeting is equal to its yield capacity, which is computed as

$$T_{yp} = A_p f_{yp} = 1678 \times 550 \times 10^{-3} = 922.9 \text{ kN/m}$$

2. Neutral axis depth

Assume the plastic neutral axis is located in the concrete cover slab and the steel reinforcement is at yield at the ultimate moment. The yield force in reinforcement is

$$T_{yr} = A_r f_{yr} = 393 \times 400 \times 10^{-3} = 157.2 \text{ kN/m}$$

The compressive force in the concrete cover slab is

$$C_c = 0.85 f'_c b \gamma d_n$$

$$\gamma = 0.85 - 0.007(f'_c - 28) = 0.85 - 0.007 \times (32 - 28) = 0.822$$

From the force equilibrium $C_c = T_{yp} + T_{yr}$, the neutral axis depth d_n is computed as

$$d_n = \frac{T_{yp} + T_{yr}}{0.85 f'_c b \gamma} = \frac{(922.9 + 157.2) \times 10^3}{0.85 \times 32 \times 1000 \times 0.822} = 48.3 \text{ mm}$$

$$h_c = D_c - h_r = 150 - 54 = 96 \text{ mm} > d_n = 48.3 \text{ mm, OK}$$

3. Check strain in reinforcement

Using Y10 bars in both directions in the composite slab, the depth of the longitudinal reinforcement from the top fibre of the slab is

$$d_r = 150 - 54 - 10 - \frac{10}{2} = 81 \text{ mm}$$

The strain in steel reinforcement can be calculated as

$$\varepsilon_r = 0.003 \times \frac{d_r - d_n}{d_n} = 0.003 \times \frac{81 - 48.3}{48.3} = 0.00203 > \varepsilon_y = \frac{400}{200{,}000} = 0.002$$

The steel reinforcement yields at the ultimate limit state.

4. Design moment capacity

The height y_p for which T_{yp} acts for section with complete shear connection is determined as

$$y_p = 21.6\beta_{sc} - 6.1 = 21.6 \times 1.0 - 6.1 = 15.5 \text{ mm}$$

$$d_p = 150 - 15.5 = 134.5 \text{ mm}$$

The compressive force in the concrete cover slab is computed as

$$C_c = 0.85 f'_c b \gamma d_n = 0.85 \times 32 \times 1000 \times 0.822 \times 48.3 \times 10^{-3} = 1080 \text{ kN/m}$$

The nominal moment capacity of the section is

$$M_u = T_{yp} d_p + T_{yr} d_r - C_c (0.5 \gamma d_n)$$

$$= 922.9 \times 134.5 + 157.2 \times 81 - 1080 \times (0.5 \times 0.822 \times 48.3) \text{ kN mm} = 115.4 \text{ kN m/m}$$

The design moment capacity of the composite slab section is therefore

$$\phi M_u = 0.8 \times 115.4 = 92.3 \text{ kN m/m}$$

8.6.2 Positive moment capacity with partial shear connection

The resultant tensile force (T_p) developed in the steel sheeting depends on the degree of shear connection at the cross section and is resisted by the mechanical resistance force $H_m x$ and the frictional force μR^* (Goh et al. 1998). The resultant tensile force in sheeting at the critical section with a distance x from one end of the sheeting in the composite slab with partial shear connection can be determined by

$$T_p = (H_m x + \mu R^*) \le T_{pcs} \tag{8.18}$$

where μ is the friction coefficient, taken as 0.5. It is noted that the tensile force in the sheeting varies with the distance from the end of the steel sheeting and is affected by the support reaction. If the steel sheeting does not extend over the full width of the support, the frictional resistance is taken as zero. The resultant tensile force (T_p) in sheeting should be taken as the lesser values of $T_{p \cdot L}$ and $T_{p \cdot R}$ calculated using Equation 8.18 for the critical section with the distance from the left and right ends of the sheeting. By ignoring the frictional resistance force, the distance measured from the end of the sheeting to the cross section where the complete shear connection is attained can be computed from Equation 8.18 as

$$x_{cs} = \frac{T_{yp}}{H_m} \tag{8.19}$$

Cross sections located at a distance from the end of the sheeting less than x_{cs} are in partial shear connection and shall be designed based on the partial shear connection strength theory. For the cross section with partial shear connection, the first neutral axis is located in the concrete cover slab as shown in Figure 8.8. The compressive force in concrete is given by

$$C_c = 0.85 f_c' b \gamma d_n \tag{8.20}$$

It is assumed that the steel reinforcement yields at the ultimate moment and the resultant tensile force (T_p) in the sheeting is less than T_{pcs}. This neutral axis depth (d_n) in the concrete cover slab can be calculated by

$$d_n = \frac{T_p + T_{yr}}{0.85 f_c' b \gamma} \tag{8.21}$$

It should be noted that the strain in the conventional steel reinforcement needs to be checked against its yield strain. If the reinforcement is not at yield, it can be ignored in the calculation.

The moment capacity due to the couple forces N_{ac} is represented by $M_{up}\varphi_b$, which depends on the axial force N_{ac} and the section properties of the profiled steel sheeting. The nominal ultimate moment capacity of the composite slab can be determined by taking moments about the top fibre of the section as

$$M_u = T_p d_p + T_{yr} d_r - C_c(0.5\gamma d_n) + M_{up}\varphi_b \tag{8.22}$$

where

- M_{up} is the nominal moment capacity of the sheeting alone
- φ_b is the bending factor of the sheeting which is a function of the degree of shear connection given in Table 8.1 (Goh et al. 1998)

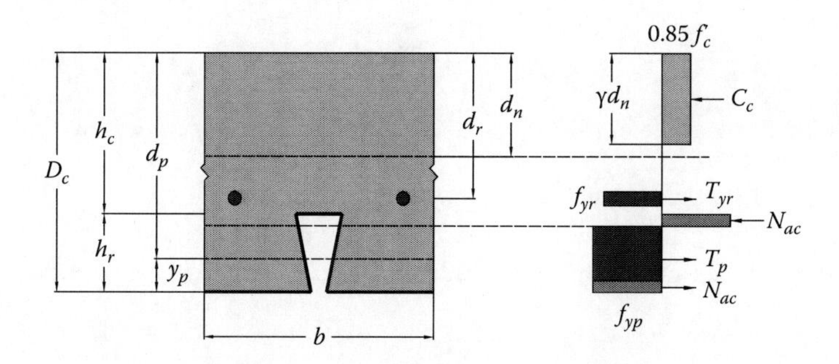

Figure 8.8 Stress distributions in section with partial shear connection.

8.6.3 Minimum bending strength

To prevent the sudden collapse of composite slabs that exhibit brittle failure, the moment capacity at each cross section in the positive moment regions must satisfy the following minimum bending strength requirement (AS 3600 2001):

$$M_u \geq M_{u,\min} = 0.12 b D_c^2 \sqrt{f_c'} \tag{8.23}$$

In the positive moment regions, the minimum bending strength requirement is satisfied if the mechanical resistance (H_m) is greater than 100 MPa, the composite slabs are subjected to uniformly distributed loads and the slabs meet the deflection limits and have a span to depth ratio of $L/D_c \geq 15$ (Goh et al. 1998).

Example 8.2: Moment capacity of section with partial shear connection

The cross section of a composite slab incorporating Comform profiled steel sheeting has partial shear connection of β_{sc} = 0.6. The overall depth of the slab (D_c) is 160 mm. The compressive strength of concrete (f_c') is 40 MPa. The thickness of the sheeting (t_{bm}) is 1.0 mm. The yield stress of the sheeting is 550 MPa. The cross-sectional area of bottom face tensile reinforcement (A_{st}) in the composite slab is 393 mm²/m. Determine the design positive moment capacity of the section with partial shear connection.

1. Resultant tensile force in sheeting

The area of the Comform sheeting is calculated as

$$A_p = 1563 t_{bm} = 1563 \times 1.0 = 1563 \text{ mm}^2/\text{m}$$

Assume the reinforcement is at yield at the ultimate moment. The yield capacity of steel reinforcement is

$$T_{yr} = A_r f_{yr} = 393 \times 400 \times 10^{-3} = 157.2 \text{ kN/m}$$

The strength of reinforced concrete cover slab is

$$F_{cst} = 0.85 f_c' b (D_c - h_r) - T_{yr}$$

$$= 0.85 \times 40 \times 1000 \times (160 - 58) \times 10^{-3} - 157.2 = 3310.8 \text{ kN/m}$$

The yield capacity of sheeting is

$$T_{yp} = A_p f_{yp} = 1563 \times 550 \times 10^{-3} = 859.65 \text{ kN/m}$$

The resultant tensile force in sheeting with complete shear connection is taken as

$$T_{pcs} = \min(F_{cst}, T_{yp}) = \min(3310.8, 859.65) = 859.65 \text{ kN/m}$$

The resultant tensile force in sheeting with partial shear connection is given by

$$T_p = \beta_{sc} T_{p\cdot cs} = 0.6 \times 859.65 = 515.79 \text{ kN/m}$$

2. Neutral axis depth

For section with partial shear connection, the first plastic neutral axis is located in the concrete cover slab. The compressive force in the concrete cover slab is

$$C_c = 0.85 f_c' b \gamma d_n$$

$$\gamma = 0.85 - 0.007(f_c' - 28) = 0.85 - 0.007 \times (40 - 28) = 0.766$$

From the force equilibrium $C_c = T_p + T_{yr}$, the neutral axis depth d_n is

$$d_n = \frac{T_p + T_{yr}}{0.85 f_c' b \gamma} = \frac{(515.79 + 157.2) \times 10^3}{0.85 \times 40 \times 1000 \times 0.766} = 25.8 \text{ mm}$$

$$h_c = D_c - h_r = 160 - 58 = 102 \text{ mm} > d_n = 25.8 \text{ mm, OK}$$

3. Check reinforcement strain

Using Y10 bars in both directions in the composite slab, the depth of the longitudinal reinforcement from the top fibre of the slab is

$$d_r = 160 - 58 - 10 - \frac{10}{2} = 87 \text{ mm}$$

The strain in steel reinforcement can be calculated as

$$\varepsilon_r = 0.003 \times \frac{d_r - d_n}{d_n} = 0.003 \times \frac{87 - 25.8}{25.8} = 0.007 > \varepsilon_y = \frac{400}{200{,}000} = 0.002$$

The steel reinforcement is at yield.

4. Design moment capacity

The height of sheeting y_p for section with $\beta_{sc} = 0.6$ is calculated as

$$y_p = 18\beta_{sc}^3 = 18 \times 0.6^3 = 3.9 \text{ mm}$$

$$d_p = 160 - 3.9 = 156.1 \text{ mm}$$

The compressive force in concrete cover slab is computed as

$$C_c = 0.85 f_c' b \gamma d_n = 0.85 \times 40 \times 1000 \times 0.766 \times 25.8 \times 10^{-3} = 617.9 \text{ kN/m}$$

The nominal moment capacity of the bare sheeting is obtained from Table 8.1 as

$$M_{up} = 10.7 t_{bm} = 10.7 \times 1.0 = 10.7 \text{ kN m/m}$$

The bending factor of the sheeting is

$$\varphi_b = 1 - \beta_{sc}^3 = 1 - 0.6^3 = 0.784$$

The nominal moment capacity of the section is

$$M_u = T_p d_p + T_{yr} d_r - C_c(0.5\gamma d_n) + M_{up}\varphi_b$$
$$= 515.79 \times 0.1561 + 157.2 \times 0.087 - 617.9 \times (0.5 \times 0.766 \times 0.0258) + 10.7 \times 0.784$$
$$= 96.5 \text{ kN m/m}$$

The design moment capacity of the composite slab section is therefore

$$\phi M_u = 0.8 \times 96.5 = 77.2 \text{ kN m/m}$$

8.6.4 Design for negative moments

Continuous composite slabs at the interior supports are subjected to negative bending moments. Negative tensile reinforcement needs to be provided at the top face of the continuous composite slab over the supports. The contribution of steel sheeting to the negative moment capacity of composite slabs is usually ignored. Therefore, the design negative moment capacity of the composite slab is calculated as

$$\phi M_u = \phi(0.85 f_c' b \gamma k_u d)(1 - 0.5\gamma k_u)d \tag{8.24}$$

where

$\phi = 0.8$ is the capacity reduction factor
$k_u = d_n / d$
d is the effective depth of the composite slab measured from the centroid of top face reinforcement to the extreme fibre of compression

To achieve ductile designs, the neutral axis parameter (k_u) must not exceed 0.4 as required by AS 3600. The required neutral axis parameter corresponding to the minimum amount of top face reinforcement can be determined from Equation 8.24 as

$$k_u = \frac{q_1 - \sqrt{q_1^2 - q_2}}{\gamma} \tag{8.25}$$

where

q_1 is taken as 1.0
q_2 is given by

$$q_2 = \frac{2M_-^*}{\phi 0.85 f_c' b d^2} \tag{8.26}$$

The major Australian products of profiled steel sheeting have a high yield stress of 550 MPa so that they provide the composite slab with a large positive moment capacity. To achieve economical designs of continuous composite slabs, it is desirable to redistribute the bending moments from the negative moment regions to the positive moment regions. The moment redistribution in continuous composite slabs should be in accordance with the Clause 7.6.8 of AS 3600 (2001). If the moment redistribution is used in the design, Class N conventional reinforcement must be used as negative tensile reinforcement. The negative design bending moment after redistribution is given by

$$M_-^{*R} = (1-\xi_m)M_-^* \quad (8.27)$$

where

M_-^* is the negative design bending moment at the support obtained by elastic analysis

ξ_m is the moment redistribution parameter, which is taken as 0.3 for the neutral axis parameter $k_u \le 0.2$ and $(0.3-0.75k_u)$ for $0.2 < k_u \le 0.4$ (Goh et al. 1998)

For design incorporating moment redistribution from negative moment regions to positive moment regions, the parameters q_1 and q_2 in Equation 8.25 are given by (Goh et al. 1998)

$$q_1 = \begin{cases} 1 & \text{for } k_u \le 0.2 \\ 1 - \dfrac{0.75M_-^*}{\phi 0.85 f_c' \gamma b d^2} & \text{for } 0.2 < k_u \le 0.4 \end{cases} \quad (8.28)$$

$$q_2 = \frac{1.4M_-^*}{\phi 0.85 f_c' b d^2} \quad (8.29)$$

The minimum cross-sectional area of negative reinforcement can be determined from the force equilibrium of the section as follows:

$$A_{st} = \frac{0.85 f_c' b \gamma k_u d}{f_{yr}} \quad (8.30)$$

Example 8.3: Design of composite slab for negative moments

The interior support of a continuous composite slab supported on steel beams is subjected to a negative design bending moment of 27 kNm. The depth of the composite slab is 140 mm. The compressive strength of concrete is 32 MPa. The concrete cover is 25 mm. The depth of negative tensile reinforcement from the top fibre is assumed to be 30 mm. The yield stress of the reinforcement is 400 MPa. (a) Determine the amount of negative tensile reinforcement required at the interior support for design not incorporating moment distribution. (b) Determine the amount of negative tensile reinforcement required at the interior support for design incorporating moment distribution.

a. Design not incorporating moment redistribution

The effective depth of the composite slab under negative moment is

$$d = D_c - d_{ct} = 140 - 30 = 110 \text{ mm}$$

For the design of composite slab without moment redistribution, $q_1 = 1$ and q_2 is calculated as follows:

$$q_2 = \frac{2M_-^*}{\phi 0.85 f_c' b d^2} = \frac{2 \times 27 \times 10^6}{0.8 \times 0.85 \times 32 \times 1000 \times 110^2} = 0.205$$

$$\gamma = 0.85 - 0.007(f_c' - 28) = 0.85 - 0.007 \times (32 - 28) = 0.822$$

The neutral axis parameter k_u is computed as

$$k_u = \frac{q_1 - \sqrt{q_1^2 - q_2}}{\gamma} = \frac{1 - \sqrt{1^2 - 0.205}}{0.822} = 0.132 < 0.4, \text{OK}$$

The required minimum cross-sectional area of negative tensile reinforcement is

$$A_{st} = \frac{0.85 f'_c b \gamma k_u d}{f_{yr}} = \frac{0.85 \times 32 \times 1000 \times 0.822 \times 0.132 \times 110}{400} = 812 \text{ mm}^2/\text{m}$$

b. Design incorporating moment redistribution

Assume the neutral axis parameter $k_u \leq 0.2$. For the design of composite slab incorporating moment redistribution, $q_1 = 1$ and q_2 is calculated as follows:

$$q_2 = \frac{1.4 M_-^*}{\phi 0.85 f'_c b d^2} = \frac{1.4 \times 27 \times 10^6}{0.8 \times 0.85 \times 32 \times 1000 \times 110^2} = 0.144$$

The neutral axis parameter k_u is computed as

$$k_u = \frac{q_1 - \sqrt{q_1^2 - q_2}}{\gamma} = \frac{1 - \sqrt{1^2 - 0.144}}{0.822} = 0.091 \leq 0.2, \text{OK}$$

The required minimum cross-sectional area of negative reinforcement is

$$A_{st} = \frac{0.85 f'_c b \gamma k_u d}{f_{yr}} = \frac{0.85 \times 32 \times 1000 \times 0.822 \times 0.091 \times 110}{400} = 560 \text{ mm}^2/\text{m}$$

8.7 VERTICAL SHEAR CAPACITY OF COMPOSITE SLABS

8.7.1 Positive vertical shear capacity

Experiments have been conducted on simply supported composite slabs incorporating profiled steel sheeting under a vertical line load placed at a distance of $1.5D_c$ from the support (Patrick 1993). Test results indicated that the composite slab did not fail by vertical shear before the ultimate load corresponding to its moment capacity was attained. This implies that the positive vertical shear capacity (ϕV_{uc}) of a simply supported composite slab can be calculated by its positive moment capacity (ϕM_u) at the cross section with a distance of $1.5D_c$ from the support. The sheeting and fully anchored reinforcement contribute to the vertical shear capacity of the composite slab in the positive moment regions. A hypothetical line load is assumed to be placed at a distance of $1.5D_c$ from the face of the hypothetical support as depicted in Figure 8.9. For continuous composite slabs under uniformly distributed load on all spans, a hypothetical support can be placed at each point of contraflexure. The design vertical shear capacity of a composite slab in the positive moment regions can be calculated by (Goh et al. 1998)

$$\phi V_u = \frac{\phi M_u}{1.5(D_c - y_p)} \tag{8.31}$$

where

$\phi = 0.8$ is the capacity reduction factor
ϕM_u is the design moment capacity of the composite slab
y_p is the height of the profiled steel sheeting at which the tensile force T_p acts

It should be noted that ϕM_u and y_p are calculated at the location of the hypothetical line load.

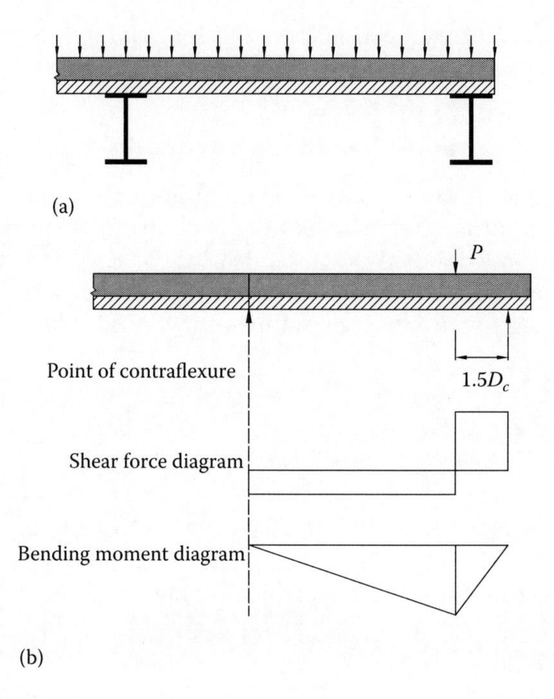

Figure 8.9 Model for positive vertical shear capacity: (a) actual continuous composite slab and (b) hypothetical simply supported composite slab.

For simple spans and the edge support regions of end spans, the design vertical shear capacity considering the contribution of sheeting but ignoring the contribution of reinforcement can be approximately computed by (Goh et al. 1998)

$$\phi V_u = \frac{\phi\left(1.5H_m + (M_{up}/bD_c^2)\right)bD_c}{1.5 - \mu} \tag{8.32}$$

where the capacity reduction factor $\phi = 0.8$.

8.7.2 Negative vertical shear capacity

The design for vertical shear of composite slabs in negative moment regions is treated as the same as that of reinforced concrete slabs. Composite slabs in negative moment regions are treated as solid reinforced concrete slabs. As specified in AS 3600 (2001), the design negative vertical shear capacity of a composite slab is calculated by

$$\phi V_u = \phi\beta_1\beta_2\beta_3 b_v d_o \left(\frac{f_c' A_{st}}{b_v d_o}\right)^{1/3} \tag{8.33}$$

where

b_v is the effective width of the slab for vertical shear
$\beta_2 = 1$
$\beta_3 = 1$
β_1 is given by

$$\beta_1 = 1.1\left(1.6 - \frac{d_o}{1000}\right) \geq 1.1 \tag{8.34}$$

In Equation 8.33, A_{st} is the cross-sectional area of longitudinal negative tensile reinforcement which is fully anchored.

8.7.3 Vertical shear capacity based on Eurocode 4

In Eurocode 4, the vertical shear capacity of a composite slab is assumed to be provided by the concrete ribs. The reinforcement that is fully anchored beyond the shear critical cross section is considered to contribute to the vertical shear capacity. However, the contribution of steel sheeting is ignored. The resistance of a composite slab to vertical shear (design vertical shear capacity) per unit width is given in Eurocode 4 (2004) as

$$V_u = \left(\frac{b_{cr}}{s_r}\right) d_p v_{\min} \tag{8.35}$$

where

b_{cr} is the width of concrete rib at the mid-height of the steel ribs in the composite slab
s_r is the spacing of steel ribs
$v_{\min}$ is the shear strength of the concrete, which is expressed by

$$v_{\min} = 0.035\left[1 + \sqrt{\frac{200}{d_p}}\right]^{3/2} \sqrt{f_{ck}} \tag{8.36}$$

where

$d_p \geq 200$ mm
$v_{\min}$ and f_{ck} are in MPa

8.8 LONGITUDINAL SHEAR

As described in Section 8.4.1, three mechanisms contribute to the transfer of longitudinal shear in composite slabs incorporating profiled steel sheeting. Shear-bond tests were usually performed to determine the resistance of composite slabs to longitudinal shear. The m–k method is used in the design of longitudinal shear in composite slabs in Eurocode 4. As specified in Eurocode 4 (2004), the design longitudinal shear capacity of a composite slab must satisfy

$$\phi V_l = \phi b d_p \left[\frac{m A_p}{b L_s} + k\right] \geq V^* \tag{8.37}$$

where

$\phi = 0.8$ the capacity reduction factor
b is the width of slab
m and k are constants that are determined by experiments
V^* is the vertical shear at an end support where the longitudinal shear failure occurs in a shear span of L_s (Johnson 2004)

The shear span L_s is taken as $L/4$ for a composite slab with span of L and under uniformly distributed load.

The m–k method is shown to be adequate for designing composite slabs with short spans (Johnson 2004). However, this method is not based on a mechanical model and does not account for the effects of end anchorage and friction above the supports.

8.9 PUNCHING SHEAR

Punching shear failure may occur in thin composite slabs under concentrated loads. The punching shear capacity of thin composite slabs that support point loads needs to be checked. It is assumed that punching shear occur on a critical perimeter of length u_{ps}. The loaded area $a_p \times b_p$ of the concentrated load is assumed to spread through a screed of thickness h_f at 45°. The effective depth of the composite slab is taken as h_c. The critical perimeter length is determined as (Johnson 2004)

$$u_{ps} = 2\pi h_c + 2(b_p + 2h_r) + 2(a_p + 2h_f + 2d_p - 2h_c) \tag{8.38}$$

It is assumed that the areas of reinforcing mesh per unit width above the steel sheeting ribs are A_{sx} and A_{sy} in x and y directions, respectively. The reinforcement ratios are $\rho_x = A_{sx}/h_c$ and $\rho_y = A_{sy}/h_c$. The effective reinforcement ratio is given in EN 1992-1-1 as $\rho_s = \sqrt{\rho_x \rho_y} \le 0.02$. The design punching shear stress is given by (Eurocode 4 2004)

$$v_{ps} = 0.12(100\rho_s f_{ck})^{1/3}\left[1 + \sqrt{\frac{200}{d_{om}}}\right] \ge v_{\min} \tag{8.39}$$

where $d_{om} \ge 200$ mm is the average effective depth of the two layers of reinforcement and $v_{\min}$ is given by Equation 8.36.

The punching shear capacity of the composite slab is

$$\phi V_{ps} = v_{ps} u_{ps} d_{om} \tag{8.40}$$

8.10 DESIGN CONSIDERATIONS

8.10.1 Effective span

The effective span of a composite slab depends on its support conditions. When a composite slab is supported on steel beams, its effective span is taken as the distance between the centre lines of adjacent steel beams. When a composite slab is supported on masonry walls, its effective span is taken as the lesser of $[L_n + (b_{s1} + b_{s2})/2]$ and $(L_n + D_c)$, where L_n denotes the clear distance between the support faces and b_{s1} and b_{s2} are the widths of the adjacent masonry supports. For a composite slab where the steel ribs are not oriented perpendicular to the support lines, the slab should be designed as a series of parallel strips. The effective span of each design strip is taken as the distance between the centre lines of the strip.

8.10.2 Potentially critical cross sections

The potentially critical cross section of a composite slab is a cross section that may govern the flexural and shear strengths of the slab. Design check for strengths should be undertaken at the potentially critical cross sections of a composite slab. For design for bending and shear, potentially critical cross sections are as follows:

- Sections subject to the maximum design positive bending moment
- Section subject to maximum negative bending moment
- Sections subject to maximum design shear force
- Sections with a distance equal to the tensile development length away from the terminated end of the reinforcement
- For a composite slab under uniformly distributed load, sections at one-third and two-thirds of the distance measured from the maximum positive moment to the ends of the span or adjacent contraflexure points

8.10.3 Effects of propping

The construction of composite slabs is classified into unpropped and propped. In unpropped construction, the profiled steel sheeting must support its self-weight, the weight of wet concrete and reinforcement and any construction loads before the hardening of the concrete. The span of composite slabs which are unpropped in construction is usually 2–3 m. It is assumed that the composite action between the interface of the steel sheeting and the concrete is achieved when the concrete compressive strength reaches 15 MPa as specified in AS 2327.1 (2003). In propped construction, the steel sheeting spans can be chosen to avoid large deflections. The positive moment capacity of a composite slab is not affected by the construction method, namely, unpropped or propped construction. As a result of this, the construction sequence is not considered in the strength design of a composite slab.

Example 8.4: Design of continuous composite slab for strength

A two-span continuous composite slab supported on steel beams is shown in Figure 8.10. The slab is subjected to a live load of 4 kPa and a superimposed dead load of 1.0 kPa. The concrete compressive strength (f'_c) is 25 MPa. The cross-sectional area of the bottom face tensile reinforcement is A_{st} = 500 mm^2/m. The centroid height of the bottom face reinforcement from the slab soffit is 60 mm. The yield stress of the reinforcement is 400 MPa. The Condeck HP profiled steel sheeting with t_{bm} = 0.75 mm is used. The yield stress of the sheeting is 550 MPa. Calculate the amount of negative tensile reinforcement at support B for design not incorporating moment distribution, check the positive moment capacity of the section with a distance x = 1401 mm measured from the end of the sheeting as depicted in Figure 8.5 and check the positive and negative vertical shear capacities of the composite slab.

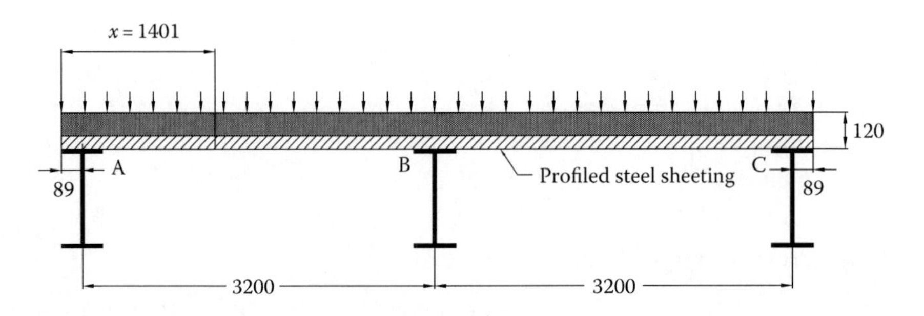

Figure 8.10 Two-span continuous composite slab.

1. Design actions

The design width of the slab is taken as 1 m and the unit weight of composite slab with reinforcement is taken as 25 kN/m³.

Dead load: $G = (0.12 \times 25 + 1.0) \times 1 = 4$ kN/m
Live load: $Q = 4 \times 1 = 4$ kN/m
The design load: $w^* = 1.2G + 1.5Q = 1.2 \times 4 + 1.5 \times 4 = 10.8$ kN/m

The maximum positive design bending moment occurs at $x = 1401$ mm from the end of the sheeting when live load is on the first span only:

$$M_+^* = 9.45 \text{ kN m/m}$$

The positive shear force at support A is $V_A^* = 14.3$ kN/m.
When live load is on both spans, the maximum design negative bending moment at support B is obtained as

$$M_-^* = 13.3 \text{ kN m/m}$$

The negative shear force at support B is $V_B^* = 21.4$ kN/m.
The reaction at support B is

$$R_A^* = 14.3 + 10.8 \times 0.089 = 15.3 \text{ kN/m}$$

2. Negative tensile reinforcement

The effective depth of the composite slab in the negative moment region is

$$d = D_c - d_{ct} = 120 - 30 = 90 \text{ mm}$$

The moment redistribution is not considered in the design of this composite slab. The parameters $q_1 = 1$ and q_2 is calculated as

$$q_2 = \frac{2M_-^*}{\phi 0.85 f_c' b d^2} = \frac{2 \times 13.3 \times 10^6}{0.8 \times 0.85 \times 25 \times 1000 \times 90^2} = 0.193$$

$$\gamma = 0.85 - 0.007(f_c' - 28) = 0.85 - 0.007 \times (25 - 28) = 0.871 > 0.85$$

$$\therefore \gamma = 0.85$$

The natural axis parameter k_u is computed as

$$k_u = \frac{q_1 - \sqrt{q_1^2 - q_2}}{\gamma} = \frac{1 - \sqrt{1^2 - 0.193}}{0.85} = 0.12 < 0.4, \text{OK}$$

The required minimum cross-sectional area of negative reinforcement is

$$A_{st} = \frac{0.85 f_c' b \gamma k_u d}{f_{sy}} = \frac{0.85 \times 25 \times 1000 \times 0.85 \times 0.12 \times 90}{400} = 488 \text{ mm}^2\text{/m}$$

3. Positive moment capacity

3.1. Resultant tensile force in sheeting

The cross-sectional area of bare steel sheeting is (Table 8.1)

$$A_p = 1620 t_{bm} = 1620 \times 0.75 = 1215 \text{ mm}^2/\text{m}$$

The yield capacity of steel sheeting is computed as

$$T_{yp} = A_p f_{py} = 1215 \times 550 \times 10^{-3} = 668.25 \text{ kN/m}$$

The mechanical resistance of Condeck HP is $H_m = 210$ kPa.

The distance x_{cs} from the end of sheeting to the section with complete shear connection is given by

$$x_{cs} = \frac{T_{yp}}{H_m} = \frac{668.25}{210} = 3.182 \text{ m}$$

Since $x = 1.401 \text{ m} < x_{cs} = 3.182$ m, the section at $x = 1.401$ m is in partial shear connection.

The yield capacity of bottom reinforcement is

$$T_{yr} = A_r f_{yr} = 500 \times 400 \times 10^{-3} = 200 \text{kN}$$

The strength of the reinforced concrete cover slab is computed as

$$F_{cst} = 0.85 f_c' b (D_c - h_r) - T_{yr}$$

$$= 0.85 \times 25 \times 1000 \times (120 - 55) \times 10^{-3} - 200 = 1181.25 \text{ kN/m}$$

The resultant tensile force developed in sheeting with complete shear connection is

$$T_{pcs} = \min(F_{cst}, T_{pcs}) = \min(1181.25, 668.25) = 668.25 \text{ kN}$$

The tensile force in sheeting at section with distance $x = 1.401$ m from the left end of the sheeting is determined as

$$T_{p \cdot L} = H_m x + \mu R_A^* = 210 \times 1.401 + 0.5 \times 15.3 = 301.86 \text{kN/m} < T_{pcs} = 668.25 \text{ kN/m}$$

Hence, $T_p = 301.86$ kN/m.

3.2. Neutral axis depth

The neutral axis depth d_n in the concrete cover slab is calculated as

$$d_n = \frac{T_p + T_{yr}}{0.85 f_c' b \gamma} = \frac{(301.86 + 200) \times 10^3}{0.85 \times 25 \times 1000 \times 0.85} = 27.78 \text{ mm}$$

3.3. Check reinforcement strain

The strain in the steel reinforcement is

$$\varepsilon_r = 0.003 \times \frac{d_r - d_n}{d_n} = 0.003 \times \frac{(120 - 60 - 27.78)}{27.78} = 0.0035$$

The yield stain of steel reinforcement is

$$\varepsilon_{sy} = \frac{f_{sy}}{E_s} = \frac{400}{200{,}000} = 0.002 < \varepsilon_r = 0.0035$$

Hence, the steel reinforcement yields at ultimate moment capacity.

3.4. Design moment capacity

The degree of shear connection at the section of x = 1.401 m is given by

$$\beta_{sc} = \frac{T_p}{T_{yp}} = \frac{301.86}{668.25} = 0.45$$

The height of centroid of sheeting for $0 < \beta_{sc} = 0.45 \le 0.75$ is obtained from Table 8.1 as

$$y_p = 16\beta_{sc}^3 = 16 \times 0.45^3 = 1.46 \text{ mm}$$

Hence,

$$d_p = 120 - 1.46 = 118.54 \text{ mm}, \quad d_r = 120 - 60 = 60 \text{ mm}$$

The bending factor of the sheeting is

$$\varphi_b = 1 - \beta_{sc}^3 = 1 - 0.45^3 = 0.909$$

The compressive force in the concrete cover slab is

$$C_c = 0.85 f_c' b \gamma d_n = 0.85 \times 25 \times 1000 \times 0.85 \times 27.78 \times 10^{-3} = 501.8 \text{ kN/m}$$

The nominal moment capacity of the bare steel sheeting is

$$M_{up} = 11.6 t_{bm} = 11.6 \times 0.75 = 8.7 \text{ kN m/m}$$

The minimal positive moment capacity of the composite slab at x = 1.401 m is calculated as

$$\begin{aligned} M_u &= T_p d_p + T_{yr} d_r - C_c(0.5\gamma d_n) + M_{up}\varphi_b \\ &= 301.86 \times 0.11854 + 200 \times 0.060 - 501.8 \times (0.5 \times 0.85 \times 0.02778) + 8.7 \times 0.909 \\ &= 49.8 \text{ kN m/m} \end{aligned}$$

The design positive moment capacity is

$$\phi M_u = 0.8 \times 49.8 = 39.84 \text{ kN m/m} > M_+^* = 9.45 \text{ kN m/m, OK}$$

4. Positive vertical shear capacity

The design positive shear force at a distance of D_c from the support A is

$$V^* = 14.3 - 10.8 \times 0.12 = 13 \text{ kN/m}$$

The design vertical shear capacity of the composite slab is calculated as

$$\phi V_u = \frac{\phi\left(1.5H_m + (M_{up}/bD_c^2)\right)bD_c}{1.5-\mu}$$

$$= \frac{0.8\times\left(1.5\times 210+(8.7/1\times 0.12^2)\right)\times 1\times 0.12}{1.5-0.5} = 88.24 \text{ kN/m} > V^* = 13 \text{ kN/m, OK}$$

5. Negative vertical shear capacity

The design negative shear force at a distance of D_c from the support B is

$$V^* = 21.4 - 10.8\times 0.12 = 20.1 \text{ kN/m}$$

$$\beta_1 = 1.1\left(1.6-\frac{d_o}{1000}\right) = 1.1\left(1.6-\frac{120-30}{1000}\right) = 1.66 > 1.1$$

The design negative vertical shear capacity of a composite slab is therefore

$$\phi V_u = \phi\beta_1\beta_2\beta_3 b_v d_o\left(\frac{f_c' A_{st}}{b_v d_o}\right)^{1/3}$$

$$= 0.8\times 1.66\times 1\times 1\times 1000\times 90\left(\frac{25\times 488}{1000\times 90}\right)^{1/3} = 47.4 \text{ kN/m} > V^* = 20.1 \text{ kN/m, OK.}$$

8.11 DESIGN FOR SERVICEABILITY

8.11.1 Crack control of composite slabs

Crack control is an important design consideration of composite slabs. If the composite slab is continuous over the internal support, cracking will occur in the top face of the slab over the support. Each span of the slab may be designed as simply supported to use the beneficial effect of high-strength steel sheeting material. However, this will lead to more severe cracking in the top face of the slab over the support. To control cracking, longitudinal reinforcement must be provided above internal supports. In Eurocode 4 (2004), the minimum cross-sectional area of this reinforcement is taken as follows: 0.2% of the cross-sectional area of the concrete cover slab above the ribs should be provided for unpropped construction and 0.4% for propped construction.

As specified in Clause 9.1.1 of AS 3600, for reinforced concrete slabs supported on beams or walls, the minimum tensile reinforcement ratio of A_{st}/bd should not be less than $0.8/f_{sy}$. For composite slabs supported on beams or walls, the minimum tensile steel area including the areas of steel sheeting and conventional reinforcement should be taken as not less than $0.002bh_c$. To control flexural cracking in composite slabs, the centre-to-centre spacing of bars in primary direction should not exceed the lesser of $2.5D$ or 500 mm. The area of steels required to control cracking due to shrinkage and temperature effects is influenced by the flexure action, the degree of restraint against in-plane movement and exposure classification and should be determined in accordance with Clause 9.4.3 of AS 3600 (2001). The steel sheeting is considered to contribute to the control of cracking due to shrinkage and temperature effects.

8.11.2 Short-term deflections of composite slabs

In AS 3600, the deflection of one-way reinforced concrete slabs under uniformly distributed load is calculated using a prismatic beam of unit width. A simplified method is given in AS 3600 for calculating the deflections of reinforced concrete beams. This simplified method is adopted here for calculating the deflections of composite slabs with profiled steel sheeting. The immediate deflections of composite slab under short-term service loads can be calculated using Young's modulus of concrete (E_{cj}) and the effective second moment of area of the composite slab (I_{ef}). The effective second moment of area (I_{ef}) of a section is between the second moment of area of the cracked section (I_{cr}) and the second moment of area of the uncracked gross section (I_g). The second moment of area of the cracked section (I_{cr}) in a composite slab can be computed using the transformed section method of elastic analysis. In this method, the areas of steel sheeting and conventional reinforcements are transformed to equivalent concrete areas using the modulus ratio ($n = E_s/E_c$) as depicted in Figure 8.11. The neutral axis depth d_n can be determined by equating the first moments of area of the compressive and tensile areas about the neutral axis as follows:

$$\frac{1}{2}bd_n^2 = nA_p(d_p - d_n) + nA_r(d_r - d_n) \tag{8.41}$$

The second moment of area of the cracked section can be obtained by taking the second moments of areas about the neutral axis as

$$I_{cr} = \frac{1}{3}bd_n^3 + nA_p(d_p - d_n)^2 + nA_r(d_r - d_n)^2 \tag{8.42}$$

The effective second moment of area of the section considered is evaluated by (Branson 1963)

$$I_{ef} = I_{cr} + (I_g - I_{cr})\left(\frac{M_{cr}}{M_{se}}\right)^3 \leq I_g \tag{8.43}$$

where

M_{se} is the bending moment at the section under short-term service load
M_{cr} is the cracking moment at the section

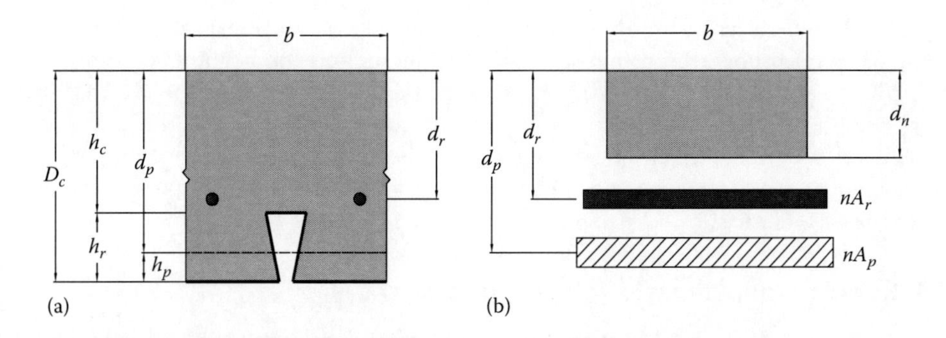

Figure 8.11 Transformed cracked section: (a) cross section and (b) transformed section.

The concrete cracks when the tensile stress of the concrete reaches its tensile strength f'_{ct}. By setting the concrete tensile stress at the extreme fibre of the cross section equal to f'_{ct}, the cracking moment at the section can be determined as

$$M_{cr} = f'_{ct}\frac{I_g}{y_t} \tag{8.44}$$

where

$f'_{ct} = 0.6\sqrt{f'_c}$

y_t is the distance from the centroidal axis of the cross section to the extreme tensile fibre

For a composite slab with several regions of peak moments, the short-term deflection can be calculated using the average value ($I_{ef\cdot av}$) of the effective second moments of area I_{ef} at nominated cross sections as follows:

- For simply supported composite slab, $I_{ef\cdot av} = I_{ef}$ at the mid-span.
- For an end span of a continuous composite slab, $I_{ef\cdot av} = 0.5(I_{ef\cdot M} + I_{ef\cdot S})$, where $I_{ef\cdot M}$ and $I_{ef\cdot S}$ are the effective second moments of area at mid-span and at the continuous support, respectively.
- For an interior span of a continuous composite slab, $I_{ef\cdot av} = 0.5[I_{ef\cdot M} + 0.5(I_{ef\cdot L} + I_{ef\cdot R})]$, where $I_{ef\cdot L}$ and $I_{ef\cdot R}$ are the effective second moments of area at the left support and at the right support, respectively.
- For a cantilever composite slab, $I_{ef\cdot av} = I_{ef\cdot S}$ at the support.

8.11.3 Long-term deflections of composite slabs

The long-term deflections of composite slabs under long-term service loads are induced by the shrinkage and creep of concrete. The deflection due to shrinkage should be estimated using the shrinkage properties of the concrete. The deflections caused by creep of concrete can be calculated by multiplying the short-term deflections by the final creep coefficients. In AS 3600 (2001), a simplified multiplier method is used to determine the long-term deflections induced by shrinkage and creep. In this method, the additional long-term deflection is computed by multiplying the short-term deflection caused by the sustained loads by a multiplier given by

$$k_{cs} = \left[2 - 1.2\left(\frac{A_{sc}}{A_r}\right)\right] \geq 0.8 \tag{8.45}$$

where A_{sc} is the cross-sectional area of compressive reinforcement in the top face, A_r is the cross-sectional area of tensile reinforcement in the bottom, the steel ratio A_{sc}/A_r is taken at the mid-span for simply supported composite slab or at the support for a cantilever composite slab.

In Eurocode 4, the second moment of area of the composite slab for internal spans is taken as the mean value of the second moments of area of the cracked and uncracked sections. Deflection calculation can be omitted if the shear connection of the composite slab is so strong that the end slip does not occur under service loads and the span to the effective depth ratio is less than 20.

8.11.4 Span-to-depth ratio for composite slabs

The Clause 9.3.4 of AS 3600 (2001) provides the span-to-depth ratio method as an alternative to checking the deflections of reinforced concrete slabs. If the slabs satisfy the span-to-depth limits, the calculation of deflections can be avoided. This method is adopted for composite slabs with uniform depth and subjected to uniformly distributed loads and where the live load does not exceed the dead load. The composite slab satisfies deflection limits if the span-to-depth of the composite slab satisfies the following condition:

$$\frac{L_{ef}}{d} \leq k_3 k_4 \left[\frac{(\Delta / L_{ef}) E_c}{F_{d.ef}} \right] \tag{8.46}$$

where

- L_{ef} is the effective span
- d is the effective depth of the composite slab
- Δ/L_{ef} is the deflection limit
- $k_3 = 1$
- k_4 is the deflection constant which is 1.6 for simply supported slabs, 2.0 in an end span and 2.4 in interior spans of a continuous composite slab where in adjoining spans, ratio of longer span to shorter span does not exceed 1.2 and where no end span is longer than an interior span

The effective design load per unit length ($F_{d\cdot ef}$) in Equation 8.46 for calculating the total deflection is taken as

$$F_{d\cdot ef} = (1.0 + k_{cs})g + (\psi_s + k_{cs}\psi_l)q \tag{8.47}$$

For calculating the deflection which occurs after the addition or attachment of the partitions, $F_{d\cdot ef}$ is taken as

$$F_{d\cdot ef} = k_{cs}g + (\psi_s + k_{cs}\psi_l)q \tag{8.48}$$

where

- ψ_s is the short-term load factor
- ψ_l is the long-term load factor

Example 8.5: Design of simply supported composite slab

A simply supported composite slab supported on steel beams is shown in Figure 8.12. The composite slab is to be constructed unpropped and is subjected to a live load of 7.5 kPa and a superimposed dead load of 1.0 kPa in addition to its own weight. In the construction stage 1, the load from stacked materials is 4.0 kPa and live load is 1.0 kPa. The concrete compressive strength (f'_c) is 25 MPa. The Bondek II profiled steel sheeting with t_{bm} = 0.75 mm is used. The yield stress of the sheeting is 550 MPa. (1) Check the deflection and strengths of the steel sheeting during construction; (2) check the flexural and shear strength and deflections of the composite slab.

a. Design of formwork

1. Design for serviceability

Self-weight of sheeting: G_p = 0.1 kPa
Self-weight of concrete and reinforcement (0.1 kPa for reinforcement):

$$G_c = 0.12 \times 2400 \times \left(\frac{9.8}{1000} \right) + 0.1 = 2.92 \text{ kPa}$$

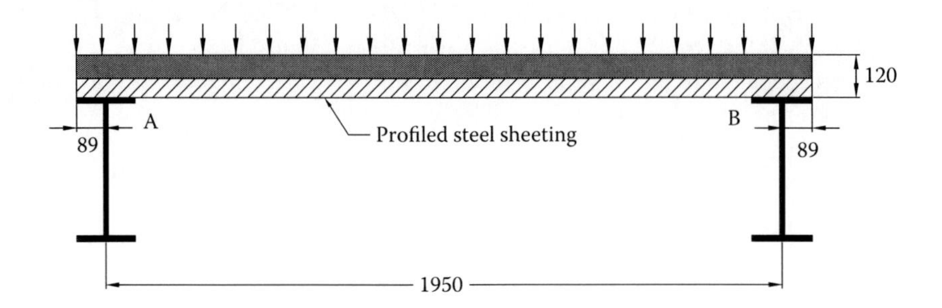

Figure 8.12 Simply supported composite slab.

Taking the design width of the slab as 1 m, the design service load is

$$w = (G_p + G_c)\times 1.0 = (0.1 + 2.92\ )\times 1.0 = 3.02 \text{ kN/m} \quad \left(\text{AS 3601-1995}\right)$$

The second moment of area of the Bondek II profiled steel sheeting is $I_p = 0.4798 \times 10^6$ mm^4/m and $E_s = 200\times10^3$ MPa. The deflection at the mid-span of the sheeting is

$$\delta_{C1.3} = \frac{5wL^4}{384E_sI_p} = \frac{5\times 3.02\times 1950^4}{384\times 200\times 10^3\times 0.4798\times 10^6} = 5.9 \text{ mm}$$

The deflection limit is

$$\Delta_{\text{limit}} = \frac{L}{250} = \frac{1950}{250} = 7.8 \text{ mm} > \delta_{C1.3} = 5.9 \text{ mm, OK}$$

1.2. Design for strength

At stage 1, before placing concrete, the design load is

$$w_1^* = 1.2G_p + 1.5Q_{uv} + 1.5Q_m = 1.2\times 0.1 + 1.5\times 1 + 1.5\times 4 = 7.62 \text{ kN/m}$$

At stage 2, after placing concrete, the design load is

$$w_2^* = 1.2G_p + 1.2G_c + 1.5Q_{uv} = 1.2\times 0.1 + 1.2\times 2.92 + 1.5\times 1 = 5.12 \text{ kN/m} < w_1^*$$

Therefore,

$$w^* = 7.62 \text{ kN/m}$$

The design maximum bending moment at mid-span of the slab at stage 1 is

$$M^* = \frac{w^*L^2}{8} = \frac{7.62\times 1.95^2}{8} = 3.62 \text{ kN m/m}$$

The nominal moment capacity of the bare steel sheeting is (Table 8.1)

$$M_u = 13.8t_{bm} = 13.8\times 0.75 = 10.35 \text{ kN m/m}$$

The design moment capacity of the sheeting is

$$\phi M_u = 0.8 \times 10.35 = 8.28 \text{ kN m/m} > M^* = 3.62 \text{ kN m/m, OK}$$

b. Design of the composite slab

1. Design actions

Dead load: $G = (0.12 \times 25 + 1.0) \times 1 = 4$ kN/m
Live load: $Q = 7.5 \times 1 = 7.5$ kN/m
The design load: $w^* = 1.2G + 1.5Q = 1.2 \times 4 + 1.5 \times 7.5 = 16.05$ kN/m

The maximum design bending moment is calculated as

$$M^* = \frac{w^*L^2}{8} = \frac{16.05 \times 1.95^2}{8} = 7.6 \text{ kN m/m}$$

The vertical shear force: $V^* = \frac{w^*L}{2} = \frac{16.05 \times 1.95}{2} = 15.6$ kN/m

The reaction at support A: $R_A^* = 15.6 + 0.089 \times 16.05 = 17.1$ kN

2. Design moment capacity

2.1. Resultant tensile force in sheeting

The cross-sectional area and capacity of bare steel sheeting are (Table 8.1)

$$A_p = 1678 t_{bm} = 1678 \times 0.75 = 1258.5 \text{ mm}^2/\text{m}$$

The yield capacity of steel sheeting is computed as

$$T_{yp} = A_p f_{py} = 1258.5 \times 550 \times 10^{-3} = 692.2 \text{ kN/m}$$

The mechanical resistance of Bondek II is

$$H_m = 88\sqrt{t_{bm} f_c'} = 88 \times \sqrt{0.75 \times 25} = 381 \text{ kPa}$$

The distance x_{cs} from the end of sheeting to the section with complete shear connection is given by

$$x_{cs} = \frac{T_{yp}}{H_m} = \frac{692.2}{381} = 1.816 \text{ m}$$

The distance from the end of sheeting to the mid-span of the composite slab is $x = 1.95/2 + 0.089 = 1.064$ m $< x_{cs} = 1.816$ m; therefore, the section at is in partial shear connection. The strength of the reinforced concrete cover slab is computed as

$$F_{cst} = 0.85 f_c' b(D_c - h_r) - T_{yr}$$

$$= 0.85 \times 25 \times 1000 \times (120 - 54) \times 10^{-3} - 0 = 1402.5 \text{ kN/m}$$

The resultant tensile force developed in sheeting with complete shear connection is

$$T_{pcs} = \min(F_{cst}, T_{pcs}) = \min(1402.5, 692.2) = 692.2 \text{ kN/m}$$

The tensile force in sheeting at section with distance x = 1.064 m from the left end of the sheeting is determined as

$$T_{p\cdot L} = H_m x + \mu R_A^* = 381 \times 1.064 + 0.5 \times 17.1 = 414 \text{ kN/m} < T_{pcs} = 692.2 \text{ kN/m}$$

Hence,

$$T_p = 414 \text{ kN/m}$$

2.2. Neutral axis depth

The neutral axis depth d_n in the concrete cover slab is

$$d_n = \frac{T_p + T_{yr}}{0.85 f_c' b \gamma} = \frac{(414+0)\times 10^3}{0.85 \times 25 \times 1000 \times 0.85} = 22.92 \text{ mm}$$

2.3. Design moment capacity

The degree of shear connection at the section of x = 1.064 m is given by

$$\beta_{sc} = \frac{T_p}{T_{yp}} = \frac{414}{692.2} = 0.6$$

The height of sheeting where T_p acts for $0 < \beta_{sc} = 0.6 \le 0.75$ is calculated as

$$y_p = 18\beta_{sc}^2 = 18 \times 0.6^2 = 6.48 \text{ mm}$$

Hence,

$$d_p = 120 - 6.48 = 113.52 \text{ mm}$$

The bending factor of the sheeting is

$$\varphi_b = 1 - \beta_{sc}^2 = 1 - 0.6^2 = 0.64$$

The compressive force in the concrete cover slab is

$$C_c = 0.85 f_c' b \gamma d_n = 0.85 \times 25 \times 1000 \times 0.85 \times 22.92 \times 10^{-3} = 414 \text{ kN}$$

The nominal positive moment capacity of the composite slab at x = 1.308 m is calculated as

$$\begin{aligned} M_u &= T_p d_p + T_{yr} d_r - C_c(0.5\gamma d_n) + M_{up}\varphi_b \\ &= 414 \times 0.11352 + 0 - 414 \times (0.5 \times 0.85 \times 0.02292) + 10.35 \times 0.64 \\ &= 49.6 \text{ kN m/m} \end{aligned}$$

The design positive moment capacity is

$$\phi M_u = 0.8 \times 49.6 = 39.7 \text{ kN m/m} > M^* = 7.6 \text{ kN m/m, OK}$$

3. Positive vertical shear capacity

The design positive shear force at a distance of D_c from the support A is

$$V^* = 15.6 - 16.05 \times 0.12 = 13.7 \text{ kN}$$

The design vertical shear capacity of the composite slab is calculated as

$$\phi V_u = \frac{\phi\left(1.5H_m + (M_{up}/bD_c^2)\right)bD_c}{1.5 - \mu}$$

$$= \frac{0.8 \times \left(1.5 \times 381 + (10.35/1 \times 0.12^2)\right) \times 1 \times 0.12}{1.5 - 0.5} = 123.9 \text{ kN} > V^* = 13.7 \text{ kN, OK}$$

4. Deflection check

4.1. Second moment of area of cracked section

Young's modulus of concrete is computed as

$$E_c = 3{,}320\sqrt{f_c'} + 6{,}900 = 3{,}320 \times \sqrt{25} + 6{,}900 = 23{,}500 \text{ MPa}$$

The modulus ratio is

$$n = \frac{E_s}{E_c} = \frac{200{,}000}{23{,}500} = 8.5$$

The height of the elastic centroid from the sheeting bottom is 15.6 mm. Assume the neutral axis is in the concrete cover slab. The neutral axis depth d_n can be determined by taking the first moment of area about the neutral axis as

$$\frac{1}{2}bd_n^2 = nA_p(d_p - d_n)$$

$$\frac{1}{2} \times 1000 \times d_n^2 = 8.5 \times 1258.5(120 - 15.6 - d_n)$$

$d_n = 37.8 \text{ mm} < h_c = 120 - 54 = 66 \text{ mm}$; thus, the neutral axis is in the concrete cover slab.

The second moment of area of the cracked section is

$$I_{cr} = \frac{1000 \times 37.8^3}{3} + 8.5 \times 1259 \times (120 - 15.6 - 37.8)^2$$

$$= 65.53 \times 10^6 \text{ mm}^4/\text{m}$$

4.2. Effective second moment of area

The second moment of area of the cross section ignoring the sheeting is

$$I_g = \frac{1000 \times 120^3}{12} = 144 \times 10^6 \text{ mm}^4/\text{m}$$

The cracking moment at the section is calculated as

$$M_{cr} = f_c' \frac{I_g}{y_t} = 0.6\sqrt{25} \times \frac{144 \times 10^6}{120/2} = 7.2 \text{ kN m/m}$$

The short-term service load is

$$w_s = G + \psi_s Q = (3.22 + 1.0) + 1.0 \times 7.5 = 11.52 \text{ kN/m}$$

The bending moment at the mid-span under short-term service load is

$$M_{se} = \frac{w_s L^2}{8} = \frac{11.52 \times 1.95^2}{8} = 5.5 \text{ kN m/m}$$

The effective second moment of area at mid-span section is computed as

$$I_{ef} = I_{cr} + (I_g - I_{cr})\left(\frac{M_{cr}}{M_{se}}\right)^3 \le I_g$$

$$= 65.53 \times 10^6 + (144 \times 10^6 - 65.53 \times 10^6)\left(\frac{7.2}{5.5}\right)^3 = 241.57 \times 10^6 \text{ mm}^4\text{/m} > I_g$$

Hence,

$$I_{ef} = I_g = 144 \times 10^6 \text{ mm}^4\text{/m}$$

4.3. Short-term deflection

The short-term deflection due to short-term service load is

$$\delta_s = \frac{5wL^4}{384E_c I_{ef}} = \frac{5 \times 11.52 \times 1{,}950^4}{384 \times 23{,}500 \times 144 \times 10^6} = 0.64 \text{ mm}$$

4.4. Long-term deflection

The long-term service load is

$$w = G + \psi_l Q = 4 + 0.6 \times 7.5 = 8.5 \text{ kN/m}$$

The deflection due to the sustained load is

$$\delta_{sus} = \frac{8.5}{11.5} \times 0.64 = 0.47 \text{ mm}$$

There is no compressive reinforcement in the composite slab, $A_{sc} = 0$:

$$k_{cs} = \left[2 - 1.2\left(\frac{A_{sc}}{A_r}\right)\right] = 2$$

The long-term deflection due to shrinkage and creep is therefore

$$\delta_l = k_{cs}\delta_{sus} = 2 \times 0.47 = 0.94 \text{ mm}$$

4.5. Total deflection

The total deflection is

$$\delta_{tot} = \delta_{C1.3} + \delta_s + \delta_l = 5.9 + 0.64 + 0.94 = 7.5 \text{ mm} < \delta_{\text{limit}} = 7.8 \text{ mm, OK}$$

REFERENCES

AS 2327.1 (2003) Australian standard for composite structures, Part 1: Simply supported beams, Sydney, New South Wales, Australia: Standards Australia.

AS 3600 (2001) Australian standard for concrete structures, Sydney, New South Wales, Australia: Standards Australia.

Branson, D.E. (1963) Instantaneous and time-dependent deflection of simple and continuous reinforced concrete beams, HPR Report No. 7. Birmingham, AL: Alabama Highway Department, US Bureau of Public Roads.

Bridge, R.Q. (1998) *Shear Connection Parameters for Bondek II, Comform and Condeck HP*, Sydney, New South Wales, Australia: University of Western Sydney.

Eurocode 4 (2004) Design of composite steel and concrete structures, Part 1.1: General rules and rules for buildings, Brussels, Belgium: European Committee for Standardization.

Goh, C.C., Patrick, M., Proe, D., and Wilkie, R. (1998) Design of composite slabs for strength, composite structures design manual – Design booklet DB3.1, Melbourne, Victoria, Australia: BHP Integrated Steel.

Johnson, R.P. (2004) *Composite Structures of Steel and Concrete: Beams, Slabs, Columns, and Frames for Buildings*, Oxford, U.K.: Blackwell Publishing.

Liang, Q.Q. and Patrick, M. (2001) Design of the shear connection of simply-supported composite beams to Australian standards AS 2327.1-1996, Composite structures design manual – Design booklet DB1.2, Sydney, New South Wales, Australia: OneSteel Manufacturing Limited.

Patrick, M. (1990) A new partial shear connection strength model for composite slabs, *Steel Construction Journal*, Australian Institute of Steel Construction, 24 (3): 2–17.

Patrick, M. (1993) Testing and design of Bondek II composite slabs for vertical shear, *Steel Construction Journal*, Australian Institute of Steel Construction, 27 (2): 2–26.

Patrick, M. and Bridge, R.Q. (1994) Partial shear connection design of composite slabs, *Engineering Structures*, 16 (5): 348–362.

Chapter 9

Composite beams

9.1 INTRODUCTION

A steel–concrete composite beam is constructed by connecting the concrete slab to the top flange of a steel beam by shear connectors. In a simply supported composite beam, the concrete slab is subjected to compression, while part or whole of the steel beam is in tension. The best properties of both steel and concrete materials are utilised in composite beam construction. Shear connectors not only transfer the longitudinal shear at the interface of the concrete slab and the steel beam but also resist the longitudinal slip and vertical separation of these two components. The strength of composite beams depends on the degree of shear connection between the concrete slab and the steel beam. Continuous composite beams have the advantages of reduced steel quantity and improved flexural stiffness compared to simply supported composite beams. However, additional slab reinforcement needs to be placed in the negative moment regions. The use of partial shear connection leads to economical designs of simply supported composite beams while continuous composite beams are usually designed with complete shear connection.

This chapter presents the behaviour and design of simply supported composite beams for strength and serviceability to AS 2327.1 (2003). The design of continuous composite beams is also introduced. The method for determining the effective sections of concrete slabs and steel beams is given first. The basic concepts and design of the shear connection of composite beams is introduced. The veridical shear capacity of composite beams is then described. This is followed by the introduction of the design of composite beams for positive and negative moment regions. The design of longitudinal shear reinforcement is presented. The design of composite beams for serviceability is discussed.

9.2 COMPONENTS OF COMPOSITE BEAMS

The main components of a composite beam consist of the steel beam, concrete slab and shear connectors as schematically depicted in Figure 9.1. The most common types of steel beams include hot-rolled I-sections, welded I-sections, rectangular cold-formed hollow sections, fabricated I-sections and any of these mentioned sections with an additional plate welded to the bottom flange, as shown in Figure 9.2. In general, AS 2327.1 requires that the cross section of the steel beam must be symmetrical about its vertical axis.

The concrete slab can be either a solid slab or a composite slab incorporating profiled steel sheeting. The concrete slab must be reinforced with deformed bars or mesh to carry tensile forces and longitudinal shear in the slab arising from direct loading, shrinkage and temperature effects or fire. The design of solid reinforced concrete slabs must be in accordance with

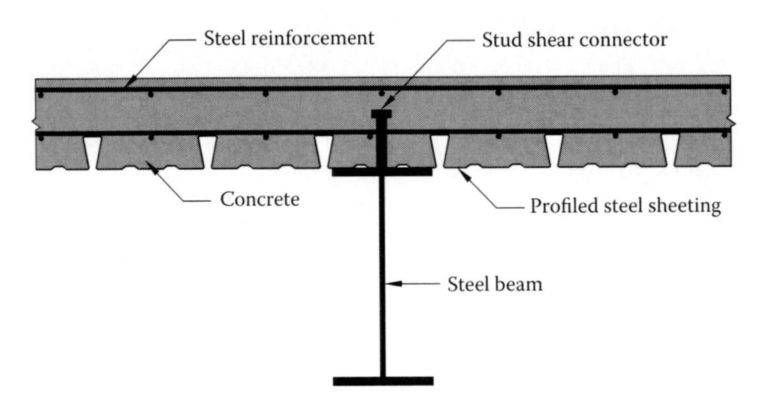

Figure 9.1 Components of a composite beam.

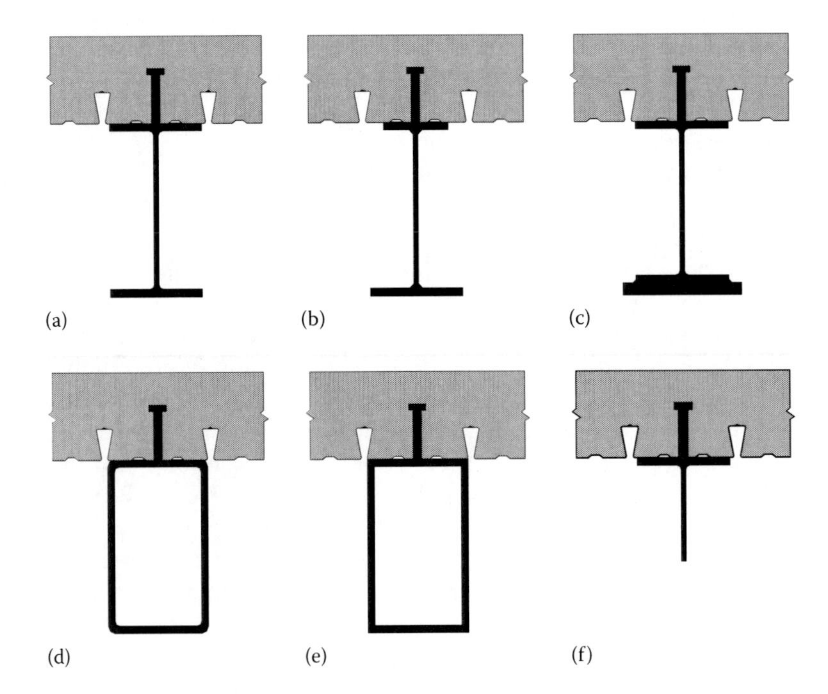

Figure 9.2 Typical composite beams incorporating profiled steel sheeting: (a) composite beam with hot-rolled steel I-section; (b) composite beam with welded steel I-section; (c) composite beam with hot-rolled steel I-section welded with bottom plate; (d) composite beam with cold-formed rectangular hollow steel section; (e) composite beam with welded rectangular hollow steel section; (f) composite beam with steel T-section.

AS 3600. The design of composite slabs is given in Chapter 8. The profiled steel sheeting incorporated in a composite slab must satisfy the geometric requirements given in Clause 1.2.4 of AS 2327.1. The major Australian profiled steel sheeting products such as Bondek II, Comform and Condeck HP satisfy these geometric requirements.

The shear connectors are attached to the top flange of the steel beam to resist the longitudinal slip at the interface and the vertical separation between the steel beam and the concrete slab. The commonly used shear connectors are headed studs, channels and high-strength structural bolts as shown in Figure 9.3. The headed studs are the most widely used shear connectors in composite beam construction.

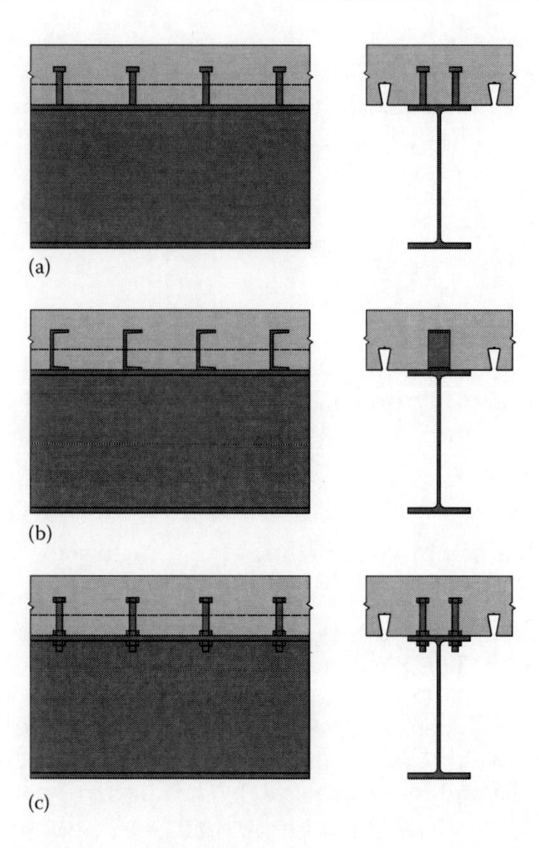

Figure 9.3 Types of shear connectors: (a) headed studs, (b) channels and (c) high-strength structural bolts.

9.3 BEHAVIOUR OF COMPOSITE BEAMS

The behaviour of composite beams can be determined by either experiments (Chapman and Balakrishnan 1964; Ansourian 1981) or numerical analysis such as the finite element analysis (Liang et al. 2004, 2005; Pi et al. 2006a,b; Ranzi 2008; Zona and Ranzi 2011). It depends on the shear connection between the concrete slab and the steel beam. Either push-out tests or full-scale composite beam tests can be used to determine the load–slip characteristics and ultimate shear capacity of shear connectors. Push-out tests indicate that the shear connection may fail by crushing of the concrete or by shearing off the shear connectors (Chapman and Balakrishnan 1964). The extent of cracking and crushing in the concrete slab depends on the type and diameter of the studs. During the tests of simply supported composite beams, a distinct bond-breaking sound may occur, which signifies that extensive slip has occurred between the concrete slab and the steel beam. However, in some cases, the bond may be gradually destroyed so that no bond-breaking sound can be heard. Simply supported composite beams under a concentrated load applied at the mid-span may fail by crushing of the concrete in the top face and yielding of the steel section at the mid-span. The connection may fail suddenly by shearing off the connectors in one half of the beam, which significantly reduces the load-carrying capacity of the beam. The failure mode of shear connection is brittle. Test results demonstrate that the end slips and slips at the mid-span occur. The pull-out failure of shear connectors leads to a rapid increase in slip and uplift and in deflections.

Tests on two-span continuous composite beams indicate that the top of the concrete slab at mid-span may crush and spall, while the entire steel section may yield in tension at the mid-span (Ansourian 1981). In addition, the bottom flange and web in the interior support may buckle locally. Test results show that the concrete slab and composite action contribute significantly to the vertical shear strength of composite beams (Clawson and Darwin 1982). This was confirmed by the finite element analyses undertaken by Liang et al. (2004, 2005) on simply supported and continuous composite beams.

9.4 EFFECTIVE SECTIONS

The section moment capacity of a composite beam is calculated using its effective cross section, which is composed of the effective width of concrete flange and the effective portion of the steel beam section.

9.4.1 Effective width of concrete flange

The in-plane shear strain in the concrete slab of a composite section under bending causes the longitudinal displacements in the parts of the slab remote from the steel web to lag behind those near the web. This phenomenon is called shear lag, which affects longitudinal displacements and stresses in the composite section (Moffatt and Dowling 1978). The distribution of elastic strains between the concrete slab and the steel beam is not uniform. The strains are large above the steel beam and decrease with the distance from the beam (Adekola 1968; Vallenilla and Bjorhovde 1985). The effective width concept is employed as a simplified method for determining the strength and stiffness of composite beams, which indirectly accounts for shear lag effects. This concept assumes that the effective concrete flange carries the maximum uniform stress over the steel beam. The effective width (b_{cf}) of concrete flange in a composite beam depends on the effective span (L_{ef}) of the composite beam, centre-to-centre spacing (b_1,b_2) of adjacent beams and the overall thickness of the slab (D_c).

For an internal composite beam shown in Figure 9.4a, the effective widths b_{e1} and b_{e2} of the concrete flange in a solid slab are given in Clause 5.2.2 of AS 2327.1 (2003) as follows:

$$b_{e1} = \min\left[\left(\frac{L_{ef}}{8}\right), \left(\frac{b_1}{2}\right), \left(\frac{b_{f1}}{2} + 8D_c\right)\right] \quad (9.1)$$

$$b_{e2} = \min\left[\left(\frac{L_{ef}}{8}\right), \left(\frac{b_2}{2}\right), \left(\frac{b_{f1}}{2} + 8D_c\right)\right] \quad (9.2)$$

where b_{f1} is the width of the top flange of the steel section in the composite beam. The effective span (L_{ef}) of a composite beam is the distance between points of zero bending moment. For simply supported beams, it should be determined in accordance with Appendix H of AS 2327.1. In Eurocode 4 (2004), for continuous composite beams, the effective span for positive bending is taken as $0.8L_o$ for an end span and $0.7L_o$ for an interior span, where L_o is the centre-to-centre spacing of the supports. For negative bending, L_{ef} is taken as $(L_1 + L_2)/4$, where L_1 and L_2 are adjacent spans.

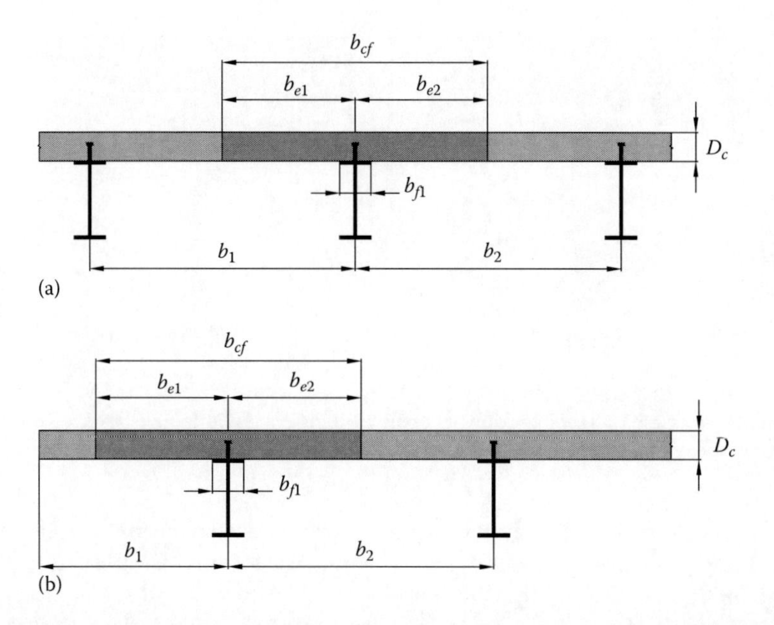

Figure 9.4 Effective width of concrete flange in composite beams with solid slabs: (a) internal beam and (b) edge beam.

For an edge composite beam schematically depicted in Figure 9.4b, Clause 5.2.2 of AS 2327.1 (2003) suggests that the effective widths b_{e1} and b_{e2} of the concrete flange in a solid slab are calculated by

$$b_{e1} = \min\left[\left(\frac{L_{ef}}{8}\right),(b_1),\left(\frac{b_{f1}}{2}+6D_c\right)\right] \tag{9.3}$$

$$b_{e2} = \min\left[\left(\frac{L_{ef}}{8}\right),\left(\frac{b_2}{2}\right),\left(\frac{b_{f1}}{2}+8D_c\right)\right] \tag{9.4}$$

The effective width of the concrete flange in a composite beam where the slab is a composite slab is illustrated in Figure 9.5. For the portion of the concrete cover slab above the ribs, the

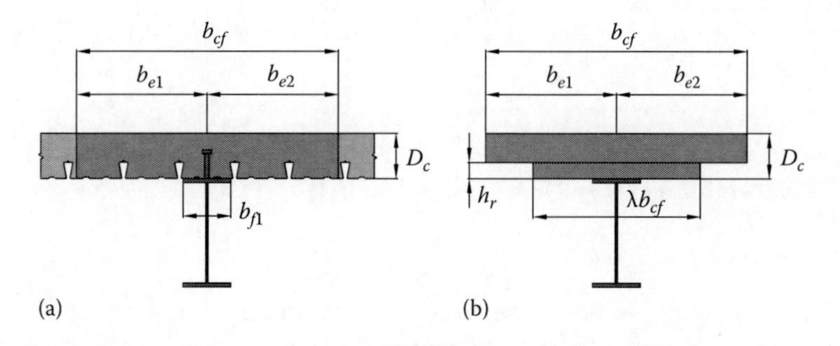

Figure 9.5 Effective width of concrete flange in composite beams with composite slabs: (a) ribs orientated parallel to steel beam and (b) ribs orientated with an angle to steel beam.

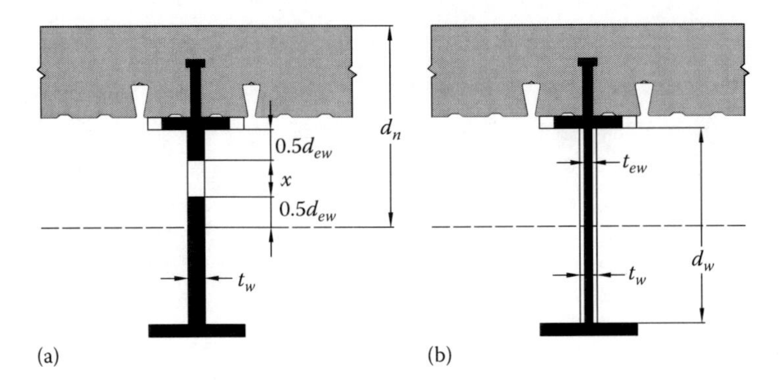

Figure 9.6 Effective portion of steel section in positive bending: (a) effective area of steel section and (b) simplified effective steel section.

effective width is calculated using Equations 9.1 and 9.2 for an internal composite beam and Equations 9.3 and 9.4 for an edge composite beam, respectively. Clause 5.2.2 of AS 2327.1 specifies that for the portion of the slab within the depth of the ribs, the effective width is taken as λb_{cf}. The multiplier λ depends on the orientation angle (θ) of sheeting ribs with respect to the longitudinal axis of the steel beam and is taken as 1.0 for $0 < \theta \le 15°$, $(b_{cr}\cos^2\theta)/s_r$ for $15 < \theta \le 60°$ and 0.0 for $\theta > 60°$.

9.4.2 Effective portion of steel beam section

When part of the flange or part or the entire web of the steel beam cross section is in compression, local buckling of these plate elements may occur. AS 2327.1 does not allow steel beams with slender plate elements to be used in composite beams. If a steel beam in a composite beam has a compact section, the entire steel section is assumed to be effective. The effective width concept can be used to determine the effective portion of the steel beam with non-compact section.

Figure 9.6a shows the effective portion of steel beam with the non-compact top flange and web. The effective portion of the non-compact steel web can be determined by calculating the ineffective length x, which is given as $x = d_w - 30t_w\sqrt{250/f_y}$. The Clause 5.2.3.3 of AS 2327.1 provides a simplified method for determining the effective portion of a non-compact steel web as illustrated in Figure 9.6b. In the simplified method, the effective thickness (t_{ew}) of the steel web is calculated by ignoring the ineffective portion of the web in the compression zone as $t_{ew} = t_w(d_w - x)/d_w$. The plate element plasticity and yield slenderness limits are given in AS 2327.1.

9.5 SHEAR CONNECTION OF COMPOSITE BEAMS

9.5.1 Basic concepts

The shear connection of a composite beam is the interconnection between the concrete slab and the steel beam, which enables the two components to act together as a single structural member. This is achieved by mechanical shear connectors which are attached to the top flange of the steel beam. The shear connection of a composite beam is composed of five components, including shear connectors, concrete slab, top flange of the steel beam, slab

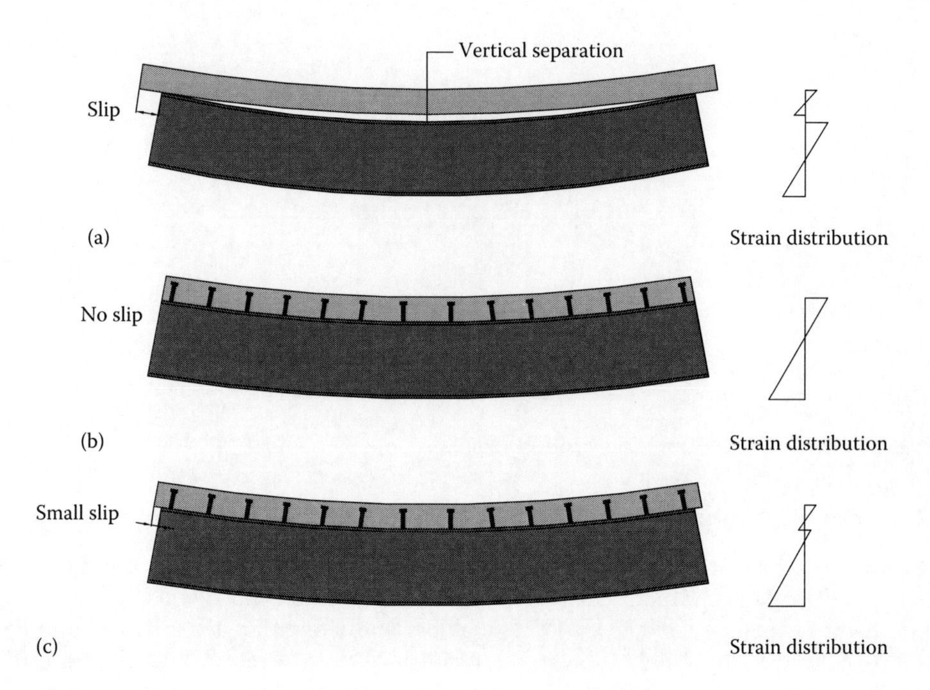

Figure 9.7 Effect of shear connection on the behaviour of composite beams: (a) no shear connection, (b) complete shear connection and (c) partial shear connection.

reinforcement and profiled steel sheeting as schematically depicted in Figure 9.1 (Liang and Patrick 2001). The behaviour of shear connection is influenced by these components.

When no shear connection is provided at the interface between the concrete slab and the steel beam, the two components will work independently to resist the loading as shown in Figure 9.7a. The end of the slab is free to slip and there is a vertical separation between these two components. The ultimate strength of the non-composite beam is conservatively determined as the plastic capacity of the steel beam alone and the contribution from the concrete slab is ignored. Perfect connection requires a connection with infinite shear, bending and axial stiffness. It is difficult to achieve perfect connection since no mechanical shear connectors can provide this degree of shear connection. In practice, the shear connectors of a simply supported composite beam are designed to transfer the longitudinal shear force, which is the smaller of either the tensile capacity of the steel beam or the effective compressive capacity of the concrete slab. The connection so designed is called complete shear connection or full shear connection as depicted in Figure 9.7b, which results in the maximum possible capacity of a composite section (Liang 2005).

The incomplete interaction or partial shear connection is between no connection and complete shear connection as illustrated in Figure 9.7c. In partial shear connection, the total shear transferred by the shear connectors in a simply supported composite beam is less than the smaller of the tensile capacity of the steel beam and the effective compressive capacity of the concrete slab. This implies that the section moment capacity of the composite beam is governed by the strength of shear connection. The partial shear connection offers economical designs of simply supported composite beams. The partial shear connection theory has been adopted in AS 2327.1 (2003), Eurocode 4 (2004) and AISC-LRFD Specification (1994) for the design of simply supported composite beams. On the other hand, the codes allow only complete shear connection to be considered in the design of composite beams in negative moment regions.

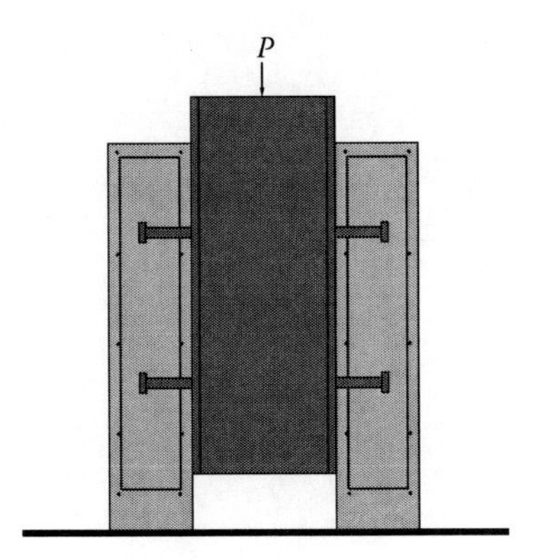

Figure 9.8 Standard push-out test in Eurocode 4 (2004).

9.5.2 Load–slip behaviour of shear connectors

The behaviour of shear connectors embedded in the concrete slab of a composite beam under applied loads is characterised by the load–slip relationship, which can be obtained from push-out tests (Ollgaard et al. 1971; Oehlers and Coughlan 1986; Liang and Patrick 2001; Patrick and Liang 2002). The standard push-out test given in Eurocode 4 (2004) is schematically depicted in Figure 9.8. The slip capacity of the specimen is taken as the larger slip measured at the characteristic load and can be determined by statistical analysis of the push-out test results. The load–slip relationship of stud shear connectors developed based on experimental results (Ollgaard et al. 1971) is expressed by

$$Q_n = f_{vs}(1 - e^{-18\delta})^{0.4} \tag{9.5}$$

where
- Q_n is the longitudinal shear force acting on a shear connector
- f_{vs} is the nominal shear capacity of a welded headed stud
- δ is the longitudinal slip

Figure 9.9 shows a typical load–slip curve calculated using Equation 9.5 for a 19 mm diameter stud shear connector embedded in 25 MPa concrete. It becomes apparent that this headed stud shear connector exhibits a ductile behaviour. In AS 2327.1, it is required that the shear connection of a composite beam must be ductile because the design methods for composite beams given in the codes are based on the ductile behaviour of shear connection. Shear connectors with a slip capacity of 6 mm are regarded as ductile in Eurocode 4.

9.5.3 Strength of shear connectors

Clause 8.2.2 of AS 2327.1 (2003) gives some geometric requirements on headed studs, channels and high-strength structural bolts. Standard headed studs are 15.9 or 19 mm diameter studs. The overall height of studs after welding should not be less than four times the nominal shank diameter (d_{bs}) and 40 mm above the top of ribs in composite slabs. The length of channel shear connectors should be greater than 50 mm and less than 60 mm. M20 high-strength structural bolts are usually used in composite beams. AS 2327.1 requires that the overall height of the bolts measured from the top face of the steel flange to the top of the bolt

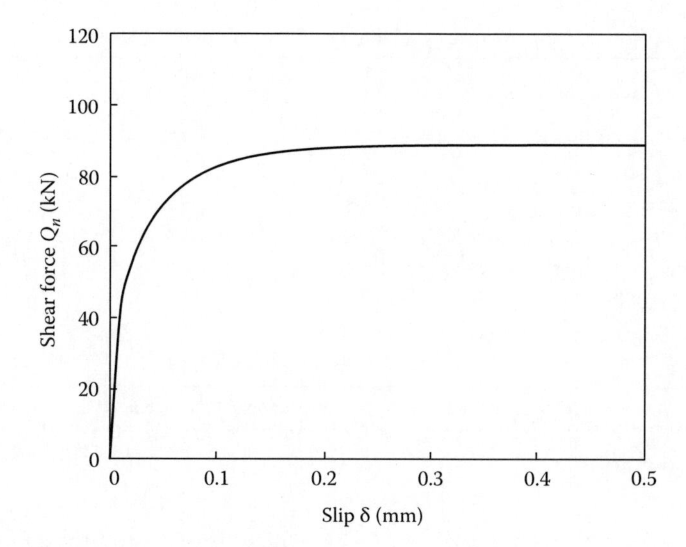

Figure 9.9 Typical load–slip curve for headed stud shear connector.

should not be less than 100 mm. Only automatically welded headed studs are allowed to be attached directly to the steel top flange through profiled steel sheeting.

A shear connector in a concrete slab under shear force may fail by either shearing off the shear connector in stronger concrete or crushing of the concrete when the concrete is weak. The shear capacity of a shear connector embedded in a concrete slab is governed by either the stud strength or the concrete strength. In Clause 8.3.2.1 of AS 2327.1 (2003), the nominal shear capacity (f_{vs}) of a welded headed stud is taken as the lesser value calculated by the following equations:

$$f_{vs} = 0.63 d_{bs}^2 f_{uc} \tag{9.6}$$

$$f_{vs} = 0.31 d_{bs}^2 \sqrt{f'_{cj} E_c} \tag{9.7}$$

where

d_{bs} denotes the diameter of the shank of the stud
f_{uc} is the tensile strength of shear connector material ($f_{uc} \leq 500$ MPa)
f'_{cj} is the estimated characteristic compressive strength of concrete at j days
E_c can be calculated as $E_c = 0.043 \rho_c^{1.5} \sqrt{f'_{cj}}$ for normal-weight and lightweight concrete

The nominal shear capacity (f_{vs}) of a channel shear connector embedded in a solid concrete slab is given in AISC-LRFD Specification (1994) as

$$f_{vs} = 0.3(t_{cf} + 0.5 t_{cw}) L_c \sqrt{f'_{cj} E_c} \tag{9.8}$$

where t_{cf}, t_{cw} and L_c are the flange thickness, web thickness and length of the channel shear connector.

The nominal shear capacities of shear connectors in normal-weight concrete are given in Table 9.1. These values are calculated using the concrete density of ρ_c = 2400 kg/m^3 and

Table 9.1 Nominal shear capacity of shear connectors in normal-weight concrete

	f_{vs} (kN)		
Type of shear connector	$f'_c = 25$ MPa	$f'_c = 32$ MPa	$f'_c = 40$ MPa
Headed stud			
$d_{bs} = 19$ mm	89	93	93
$d_{bs} = 15.9$ mm	62	65	65
Channel (l = 50 mm)			
100TFC/100PFC	100	110	125
High-strength structural bolt			
M20/8.8	98	118	126

Source: Adapted from AS 2327.1, Australian standard for composite structures, Part 1: Simply supported beams, Standards Australia, Sydney, New South Wales, Australia, 2003.

the minimum tensile steel strength of $f_{uc} = 410$ MPa for headed studs and $f_{uc} = 500$ MPa for high-strength structural bolt shear connectors. For channels and high-strength structural bolts in lightweight concrete, f_{vs} shall be taken as 80% of the values determined for normal-weight concrete of the same grade.

In AS 2327.1, the strength of shear connectors located in the ribs of profiled steel sheeting that satisfies the geometry requirements specified in Clause of 1.2.4 is taken as the same as that of shear connectors in solid slabs. Profiled steel sheeting that does not satisfy these geometry requirements may reduce the strength of shear connectors welded to the steel flange through the sheeting (Grant et al. 1977; Liang and Patrick 2001). For ribs oriented perpendicular to the steel beam, the strength reduction factor (Grant et al. 1977) for the stud is given by

$$\varphi_{pe} = \frac{0.85}{\sqrt{n_x}}\left(\frac{b_{cr}}{h_r}\right)\left(\frac{h_s}{h_r} - 1\right) \leq 1.0 \tag{9.9}$$

where

h_s is the height of the stud after welding
b_{cr} is the width of concrete rib at the mid-height of steel ribs
n_x is the number of shear connectors at a cross section of the composite beam

For ribs oriented parallel to the steel beam, the strength reduction factor (Grant et al. 1977) for the stud is expressed by

$$\varphi_{pa} = 0.6\left(\frac{b_{cr}}{h_r}\right)\left(\frac{h_s}{h_r} - 1\right) \leq 1.0 \tag{9.10}$$

In a real composite beam, shear connectors are distributed along the beam. The longitudinal shear force is shared by shear connectors in the composite beam. It is assumed that all shear connectors are ductile and have the same design shear capacity, which is influenced by the number of shear connectors in the group. In Clause 8.3.4 of AS 2327.1 (2003), the design shear capacity of a shear connector in a group of shear connectors is given by

$$f_{ds} = \phi k_n f_{vs} \tag{9.11}$$

where

$\phi = 0.85$ is the capacity reduction factor
k_n is the load-sharing factor, which is determined as

$$k_n = 1.18 - \frac{0.18}{\sqrt{n_c}} \tag{9.12}$$

The number of shear connectors (n_c) is taken as the lesser number of shear connectors between each end of the beam and the cross section being considered.

9.5.4 Degree of shear connection

The moment capacity of a composite beam cross section with complete shear completion is governed by either the tensile capacity (F_{st}) of the steel beam or the effective compressive capacity (F_{cc}) of the concrete slab as depicted in Figure 9.10. This means that the strength of shear connection (F_{sh}) is greater than F_{st} and F_{cc}. In contrast, the moment capacity of a composite beam cross section with partial shear connection is governed by the strength of shear connection, which implies that $F_{st} > F_{sh}$ and $F_{cc} > F_{sh}$.

The degree of shear connection of composite beams is defined in Clause 1.4.3 of AS 2327.1 as

$$\beta = \frac{F_{cp}}{F_{cc}} \quad 0 \le \beta \le 1.0 \tag{9.13}$$

in which F_{cc} and F_{cp} are the compressive forces in the concrete slab with complete shear connection and with partial shear connection, respectively. If the degree of shear connection is known, the compressive force in the concrete with partial shear connection is calculated

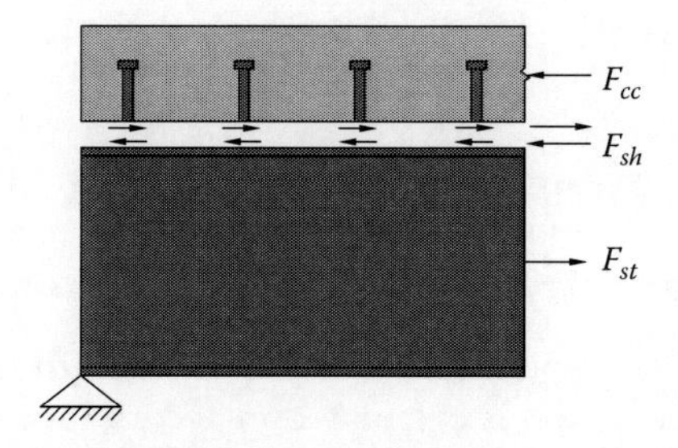

Figure 9.10 Strength of the components of a composite beam.

as $F_{cp} = \beta F_{cc}$. If the distribution of shear connectors along the composite beam is known, the compressive force in the concrete slab at the potentially critical cross section is taken as

$$F_{cp} = \min[n_A f_{ds \cdot A}; n_B f_{ds \cdot B}] \leq F_{cc} \quad (9.14)$$

where

n_A and $f_{ds \cdot A}$ are the number of shear connectors between the left end of the beam to the section considered and their corresponding design shear capacity, respectively

n_B and $f_{ds \cdot B}$ are the number of shear connectors between the right end of the beam to the section considered and their corresponding design shear capacity, respectively

9.5.5 Detailing of shear connectors

Clause 8.4 of AS 2327.1 provides detailing requirements for shear connector distributions in longitudinal and transverse directions. The shear connectors should be detailed along the length of the beam as follows:

- Shear connectors should be uniformly distributed between potentially critical cross sections and the ends of the beam.
- The maximum longitudinal spacing of shear connectors is taken as the lesser of $4D_c$ or 600 mm.
- The minimum centre-to-centre spacing of headed studs or high-strength structural bolts in solid slabs and in composite slabs with sheeting oriented parallel to the steel beam is $5d_{bs}$.
- For channel shear connectors, the minimum clear distance between the adjacent edges is 100 mm.
- The minimum distance between adjacent faces of a headed stud and sheeting rib measured parallel to the longitudinal axis of the beam is 60 mm.

The shear connectors should be detailed along the transverse cross section of the beam as follows:

- The maximum number of headed stud shear connectors per transverse cross section is three for solid slabs and two for composite slabs, while it is two for high-strength structural bolts and headed studs in composite slabs.
- The minimum transverse spacing of headed studs and high-strength structural bolts between their heads is $1.5d_{bs}$.
- The minimum clearance between the shear connector and the nearest part of sheeting rib or end of an opened rib profiled is 30 mm.

9.6 VERTICAL SHEAR CAPACITY OF COMPOSITE BEAMS

9.6.1 Vertical shear capacity ignoring concrete contribution

In AS 2327.1, the vertical shear capacity of a composite beam is assumed to be resisted by the web of the steel beam alone and is calculated in accordance with AS 4100. This implies that the contribution from the concrete slab to the vertical shear capacity of composite beam is ignored. The design requirement is expressed by $V^* \leq \phi V_u$, where $\phi = 0.9$ is the capacity reduction factor, and the nominal shear capacity of the steel web V_u is given in Section 4.5.

In AS 2327.1, the shear ratio is defined as the ratio of the design vertical shear force to the design vertical shear capacity of the steel web, which is expressed by

$$\gamma = \frac{V^*}{\phi V_u} \tag{9.15}$$

The design section moment capacity of a composite beam may be influenced by the design shear force acting on the section. There is a strength interaction between the moment capacity and the vertical shear capacity. The design section moment capacity of a composite beam depends on the shear ratio. If $\gamma \le 0.5$, the design shear force is small so that it does not reduce the moment capacity of the composite beam. However, if $0.5 < \gamma \le 1.0$, the design shear force reduces the section moment capacity of the composite beam and its effect must be taken into account in the evaluation of the flexural strength.

9.6.2 Vertical shear capacity considering concrete contribution

Composite beams under applied loads are often subjected to combined actions of bending and vertical shear. Despite experimental evidence, the contributions of the concrete slab and composite action to the vertical shear strength of composite beams are ignored in current design codes, such as AS 2327.1 (2003), Eurocode 4 (2004) and AISC-LRFD Specification (1994). The design codes assume that the web of the steel section resists the entire vertical shear. This assumption obviously leads to conservative designs of composite beams. The effects of the concrete slab and composite action on the flexural and vertical shear strengths of simply supported and continuous composite beams have been investigated by Liang et al. (2004, 2005) using the finite element analysis method. Their investigations indicate that the concrete slab and composite action contribute significantly to the flexural and vertical shear strengths of composite beams.

When no shear connection is provided between the steel beam and the concrete slab, the vertical shear capacity of the non-composite section can be determined by (Liang et al. 2004, 2005)

$$V_o = V_c + V_s \tag{9.16}$$

where

V_c is the contribution of the concrete slab

V_s is the shear capacity of the web of the steel beam

Tests indicated that the pull-out failure of stud shear connectors in composite beams may occur. This failure mode limits the vertical shear capacity of the concrete slab. As a result, the contribution of the concrete slab V_c should be taken as the lesser of the shear strength of the concrete slab V_{slab} and the pull-out capacity of stud shear connectors T_{po}. The shear strength of the concrete slab proposed by Liang et al. (2004, 2005) is expressed by

$$V_{slab} = \varphi_1 \left(f_c'\right)^{1/3} A_{ec} \tag{9.17}$$

where

φ_1 is equal to 1.16 for simply supported composite beams and 1.31 for continuous composite beams

f_c' is the compressive strength of the concrete (MPa)

A_{ec} is the effective shear area of the concrete slab

The effective shear area of a solid slab can be taken as $A_{ec} = (b_{f1} + D_c)D_c$, in which b_{f1} is the width of the top flange of the steel beam and D_c is the total depth of the concrete slab. For a composite slab incorporating profiled steel sheeting placed perpendicular to the steel beam, A_{ec} can be taken as $(b_{f1} + h_r + D_c)(D_c - h_r)$, in which h_r is the rib height of the profiled steel sheeting.

The pull-out capacity of stud shear connectors in a composite beam comprising a solid slab can be calculated by

$$T_{po} = [\pi(d_s + h_s) + 2(n_x - 1)s_x]h_s f_{ct} \tag{9.18}$$

where

- d_s is the head diameter of the headed stud
- h_s is the total height of the stud
- n_x is the number of studs per cross section
- s_x is the transverse spacing of studs
- f_{ct} is the tensile strength of concrete (MPa)

The pull-out capacity of stud shear connectors in composite slabs incorporating profiled steel sheeting should be determined using the effective pull-out failure surfaces in the aforementioned equations. It should be noted that the transverse spacing of stud shear connectors should not be greater than two times the stud height.

The shear capacity of the web of the steel beam can be calculated by

$$V_s = 0.6\alpha_w f_{yw} d_w t_w \tag{9.19}$$

where

- f_{yw} is the yield strength of the steel web (MPa)
- d_w is the depth of the steel web
- t_w is the thickness of the steel web
- α_w is the reduction factor for slender webs in shear buckling

For stocky steel webs without shear buckling, the reduction factor α_w is equal to 1.0.

Equation 9.16 can be used to determine the vertical shear capacity of non-composite sections. To take advantage of composite actions, design models for the vertical shear strength of composite beams with any degree of shear connection were proposed by Liang et al. (2004, 2005) as

$$V_{uo} = V_o(1 + \varphi_2\sqrt{\beta}) \quad 0 \le \beta \le 1 \tag{9.20}$$

where

- V_{uo} is the ultimate shear strength of the composite section in pure shear
- φ_2 is 0.295 for simply supported composite beams and sagging moment regions in continuous composite beams and 0.092 for hogging moment regions in continuous composite beams
- β is the degree of shear connection

It should also be noted that the pull-out failure of stud shear connectors leads to the damage of composite action. If this occurs, the ultimate shear strength of the damaged composite beam (V_{uo}) should be taken as V_o for safety.

Interaction equations are used in AS 2327.1 and Eurocode 4 to account for the effect of vertical shear on the ultimate moment capacity of composite beams. However, the design codes allow only the shear strength of the steel web to be considered in the interaction equations. Strength interaction equations accounting for the effects of the concrete slab and composite action were given by Liang et al. (2004, 2005) as

$$\left(\frac{M_u}{M_{uo}}\right)^{e_m} + \left(\frac{V_u}{V_{uo}}\right)^{e_v} = 1 \tag{9.21}$$

where

M_u and V_u are the ultimate moment and shear capacities of the composite beam in combined bending and shear, respectively

M_{uo} is the ultimate moment capacity of the composite section in pure bending

The exponents e_m and e_v are equal to 6.0 for simply supported composite beams and 5.0 for sagging moment regions in continuous composite beams. For hogging moment regions in continuous composite beams, e_m and e_v are equal to 0.6 and 6.0, respectively

The moment–shear interaction diagrams for composite beams under sagging and hogging are shown in Figure 9.11. The ultimate moment capacity of the composite section (M_{uo}) can be determined using the rigid plastic analysis method in accordance with the codes of practice such as AS 237.1 and Eurocode 4. It should be noted that the ultimate moment-to-shear ratio is equal to the applied moment-to-shear ratio. If the applied moment and vertical shear are known, the ultimate strengths of a composite beam in combined actions of bending and shear can be determined using Equation 9.21.

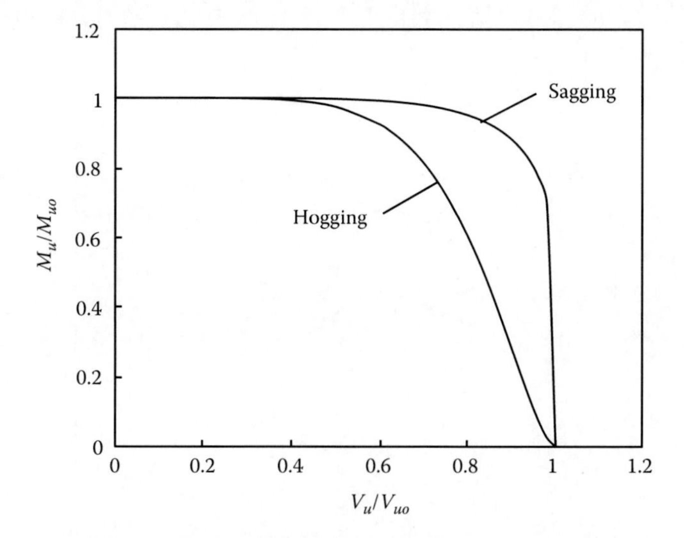

Figure 9.11 Moment–shear interaction of composite beams.

9.7 DESIGN MOMENT CAPACITY FOR POSITIVE BENDING

9.7.1 Assumptions

In the analysis of the cross section of a composite beam for determining its ultimate moment capacity, the main assumptions are as follows:

1. Each of the plane cross sections of steel beam and concrete flange remains plane after deformation, resulting in linear distribution of strain on the cross section of each component.
2. The effective portion of steel section is stressed to its yield strength in compression or in tension.
3. The rectangular stress block from the extreme compressive fibre of concrete to the plastic neutral axis (PNA) has a compressive stress of $0.85f'_c$.
4. The tensile strength of concrete is ignored.
5. Shear connectors are ductile.

9.7.2 Cross sections with $\gamma \leq 0.5$ and complete shear connection

9.7.2.1 *Nominal moment capacity M_{bc}*

The design moment capacity (ϕM_{bv}) of the cross section of a composite beam under positive bending is a function of the degree of shear connection (β) and shear ratio (γ) at the section. Figure 9.12 shows the dimensionless moment capacities of a typical composite beam with various degrees of shear connection and shear ratios. For cross sections where $\gamma \leq 0.5$, the vertical shear force does not affect the design moment capacity of the cross sections. The design moment capacity of a composite beam with any degree shear connection can be determined from the plastic stress distributions in the cross section (AS 2327.1 2003). The equivalent plastic stress distribution in the composite beam cross section with $\gamma \leq 0.5$ and complete shear connection is schematically presented in Figure 9.13, where the PNA is shown to lie in the web of the steel beam. However, it should be noted that the PNA can be located in the concrete cover slab, the steel ribs, the top flange or the web of the steel beam.

From the equivalent plastic stress distribution given in Figure 9.13, the nominal moment capacity (M_{bc}) of the cross section with $\gamma \leq 0.5$ and complete shear connection can be

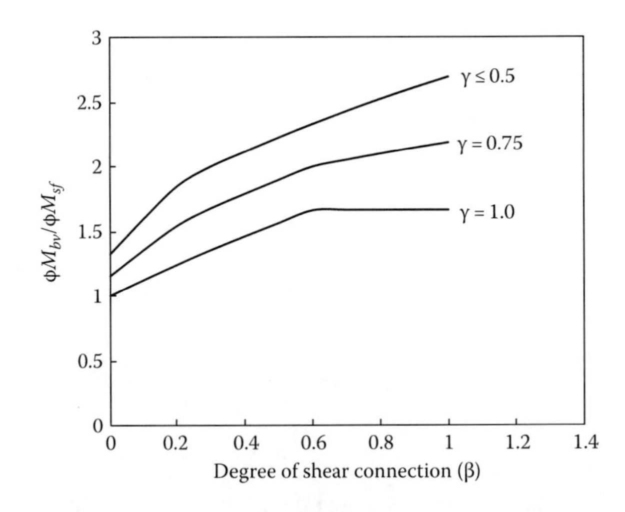

Figure 9.12 Design moment capacity as a function of degree of shear connection.

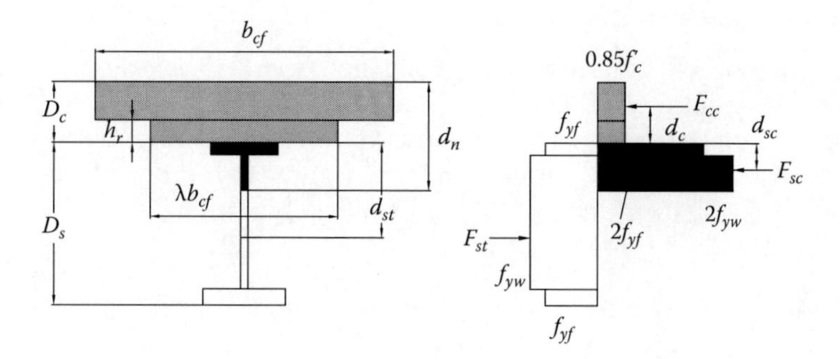

Figure 9.13 Plastic stress distributions in composite section with $\gamma \leq 0.5$ and $\beta = 1.0$.

obtained by taking moments about the line of action of the resultant compressive force (F_{sc}) in the steel section as follows:

$$M_{bc} = F_{cc}(d_c + d_{sc}) + F_{st}(d_{st} - d_{sc}) \tag{9.22}$$

where

d_c is the distance from the centroid of the compressive force F_{cc} in the concrete slab to the top face of the steel section

d_{sc} is the distance from the centroid of the resultant compressive force F_{sc} in the steel section to the top face of the steel section

For the case of no compression in the steel section, $d_{sc} = 0$. The design moment capacity of the composite cross section is therefore ϕM_{bc}, where the capacity reduction factor $\phi = 0.9$.

The tensile capacity of the steel section is calculated as

$$F_{st} = (b_{f1}t_{f1} + b_{f2}t_{f2})f_{yf} + d_w t_w f_{yw} \tag{9.23}$$

where

subscript 1 refers to the top flange
subscript 2 refers to the bottom flange
subscript w refers to the web of the steel section
f_{yf} and f_{yw} are the yield stress of the flange and web, respectively

The distance from the line of action of F_{st} to the top face of the steel section is

$$d_{st} = \frac{F_{f1}(t_{f1}/2) + F_w(d_w/2 + t_{f1}) + F_{f2}(D_s - t_{f2}/2)}{F_{st}} \tag{9.24}$$

where

$F_{f1} = b_{f1}t_{f1}f_{yf}$
$F_{f2} = b_{f2}t_{f2}f_{yf}$
$F_w = d_w t_w f_{yw}$

The compressive capacity of the concrete cover slab (F_{c1}) and concrete between the ribs (F_{c2}) are calculated by

$$F_{c1} = 0.85 f'_c b_{cf}(D_c - h_r) \tag{9.25}$$

$$F_{c2} = 0.85 f'_c \lambda b_{cf} h_r \tag{9.26}$$

9.7.2.2 *Plastic neutral axis depth*

Case 1: If the compressive capacity of the concrete cover slab is greater than the tensile capacity of the steel section, such as $F_{c1} \geq F_{st}$, the PNA falls in the concrete cover slab above the steel ribs. This gives $d_n \leq h_c$. The compressive force in concrete with complete shear connection is $F_{cc} = 0.85 f_c' b_{cf} d_n$ as can be seen from Figure 9.13. From the force equilibrium condition of $F_{cc} = F_{st}$, the depth of the PNA (d_n) can be obtained as

$$d_n = \frac{F_{st}}{0.85 f_c' b_{cf}} \tag{9.27}$$

Case 2: If $F_{c1} < F_{st} \leq (F_{c1} + F_{c2})$, the PNA is located in the steel ribs so that $h_c < d_n \leq D_c$. The compressive force in concrete with complete shear connection is determined as $F_{cc} = F_{c1} + 0.85 f_c' \lambda b_{cf} (d_n - h_c)$. From the force equilibrium condition of $F_{cc} = F_{st}$, the depth of the PNA (d_n) can be determined by

$$d_n = h_c + \frac{F_{st} - F_{c1}}{0.85 f_c' \lambda b_{cf}} \tag{9.28}$$

Case 3: If $(F_{c1} + F_{c2}) < F_{st} \leq (F_{c1} + F_{c2} + 2F_{f1})$, the PNA lies in the top flange of the steel section so that $D_c < d_n \leq (D_c + t_{f1})$. The compressive force in concrete with complete shear connection becomes $F_{cc} = (F_{c1} + F_{c2})$. The compressive force in the top steel flange can be calculated as $F_{sc} = b_{f1}(d_n - D_c)(2f_{yf})$. The force equilibrium condition requires that $(F_{cc} + F_{sc}) = F_{st}$. The depth of the PNA ($d_n$) is given by

$$d_n = D_c + \frac{F_{st} - F_{cc}}{b_{f1}(2f_{yf})} \tag{9.29}$$

Case 4: If $(F_{c1} + F_{c2} + 2F_{f1}) < F_{st}$, the PNA is located in the web of the steel section as illustrated in Figure 9.13. This implies that $(D_c + t_{f1}) < d_n \leq (D_c + d_w)$. The compressive force in concrete with complete shear connection is $F_{cc} = (F_{c1} + F_{c2})$. The compressive force in the steel section is computed as $F_{sc} = 2F_{f1} + t_w(d_n - D_c - t_{f1})(2f_{yw})$. The depth of the PNA ($d_n$) can be determined from the force equilibrium condition of $(F_{cc} + F_{sc}) = F_{st}$ as

$$d_n = D_c + t_{f1} + \frac{F_{st} - F_{cc} - 2F_{f1}}{t_w(2f_{yw})} \tag{9.30}$$

Example 9.1: Moment capacity of composite beam with complete shear connection

Figure 9.14 shows the cross section of a simply supported composite beam with complete shear connection. The profiled steel sheeting is orientated $\theta = 30°$ to the longitudinal axis of the steel beam. The geometric parameters of the steel sheeting are $h_r = 55$ mm, $s_r = b_{cr} = 300$ mm. The steel I-section is 410UB53.7 of Grade 300 steel with $f_{yf} = f_{yw} = 320$ MPa. The design strength of the concrete flange is $f_c' = 25$ MPa. The design shear force at the section considered is 200 kN. Determine the design moment capacity of this composite beam cross section.

1. Vertical shear capacity

The slenderness of the steel web is

$$\lambda_w = \frac{d_w}{t_w}\sqrt{\frac{f_y}{250}} = \frac{403 - 2 \times 10.9}{7.6}\sqrt{\frac{320}{250}} = 56.7 < \lambda_{yw} = 82$$

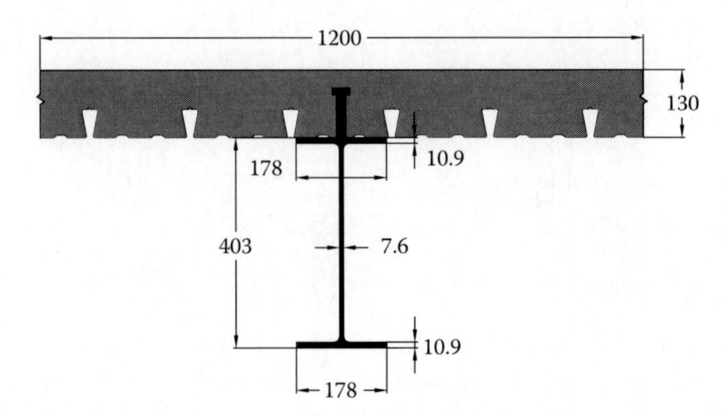

Figure 9.14 Cross section of composite beam under positive bending.

The web is not slender. The shear yield capacity of the web is calculated as

$$\phi V_u = \phi(0.6A_w f_{yw}) = 0.9\times0.6\times403\times7.6\times320 \text{ N} = 529.3 \text{ kN} > V^* = 200 \text{ kN, OK}$$

The shear ratio is

$$\gamma = \frac{V^*}{\phi V_u} = \frac{200}{529.3} = 0.38 < 0.5$$

Therefore, the design moment capacity of the composite beam is not affected by the vertical shear.

2. Plastic neutral axis depth

The tensile capacity of the steel section is computed as

$$F_{st} = (b_{f1}t_{f1} + b_{f2}t_{f2})f_{yf} + d_w t_w f_{yw}$$

$$= (178\times10.9\times2)\times320 + 381.2\times7.6\times320 \text{ N} = 2168.8 \text{ kN}$$

For complete shear connection, $F_{cc} = F_{st} = 2168.8$ kN.

The compressive capacity of the concrete cover slab is

$$F_{c1} = 0.85f'_c b_{cf}(D_c - h_r) = 0.85\times25\times1200\times(130-55)\times10^{-3} = 1912.5 \text{ kN}$$

Since $F_{c1} < F_{st}$, one needs to check if the neutral axis is in the ribs.

The angle between the ribs and the longitudinal axis of the steel beam is $\theta = 30°$. The parameter λ is calculated as

$$\lambda = \frac{b_{cr}\cos^2\theta}{s_r} = \frac{300\times\cos^2 30°}{300} = 0.75$$

The compressive capacity of concrete in the steel ribs is

$$F_{c2} = 0.85f'_c\lambda b_{cf}h_r = 0.85\times25\times0.75\times1200\times55\times10^{-3} = 1051.9 \text{ kN}$$

$$F_{c1} + F_{c2} = 1912.5 + 1051.9 = 2964.4 \text{ kN} > F_{st} = 2168.8 \text{ kN} > F_{c1}$$

Hence, the PNA is located within the ribs.

The PNA depth is calculated as

$$d_n = h_c + \frac{F_{st} - F_{c1}}{0.85 f'_c \lambda b_{cf}} = 75 + \frac{(2168.8 - 1912.5) \times 10^3}{0.85 \times 25 \times 0.75 \times 1200} = 88.4 \text{ mm}$$

3. Distances to centroid of forces

The compressive force in the concrete within the ribs is

$$F_{cn} = 0.85 f'_c \lambda b_{cf} (d_n - h_c) = 0.85 \times 25 \times 0.75 \times 1200 \times (88.4 - 75) \text{ N} = 256.3 \text{ kN}$$

The distance from the centroid of F_{cn} to the top face of the steel section is

$$d_{cn} = h_r - \frac{d_n - h_c}{2} = 55 - \frac{88.4 - 75}{2} = 48.3 \text{ mm}$$

The distance from the centroid of F_{cc} to the top face of the steel section is determined as

$$d_c = \frac{F_{c1} d_{c1} + F_{cn} d_{cn}}{F_{cc}} = \frac{1912.5 \times (130 - 75/2) + 256.3 \times 48.3}{2168.8} = 87.3 \text{ mm}$$

410UB53.7 is a doubly symmetric section. The distance from the centroid of F_{st} to the top face of the steel section is given as

$$d_{st} = \frac{D_s}{2} = \frac{403}{2} = 201.5 \text{ mm}$$

The compressive force in the steel section is $F_{sc} = 0$ and the distance from the centroid of F_{sc} to the top fibre of the steel section is $d_{sc} = 0$.

4. Design moment capacity

Taking moments about the line of action of the compressive force in steel section F_{sc}, the nominal moment capacity is calculated as

$$M_{bc} = F_{cc}(d_c + d_{sc}) + F_{st}(d_{st} - d_{sc})$$

$$= 2168.8 \times (87.3 + 0) + 2168.8 \times (201.5 - 0) \text{ kN mm} = 626.3 \text{ kN m}$$

The design moment capacity is therefore

$$\phi M_{bc} = 0.9 \times 626.3 = 563.7 \text{ kN m}$$

9.7.3 Cross sections with $\gamma \leq 0.5$ and partial shear connection

9.7.3.1 *Nominal moment capacity M_b*

For a cross section with partial shear connection ($0 < \beta < 1.0$), its moment capacity is governed by the strength of shear connection. The compressive force in the concrete slab with partial shear connection can be determined by one of the following expressions:

$$F_{cp} = n_i f_{ds} \leq F_{cc} \tag{9.31}$$

$$F_{cp} = \beta F_{cc} \leq F_{cc} \tag{9.32}$$

where n_i is the number of shear connectors between the potentially critical cross section i and the end of the beam.

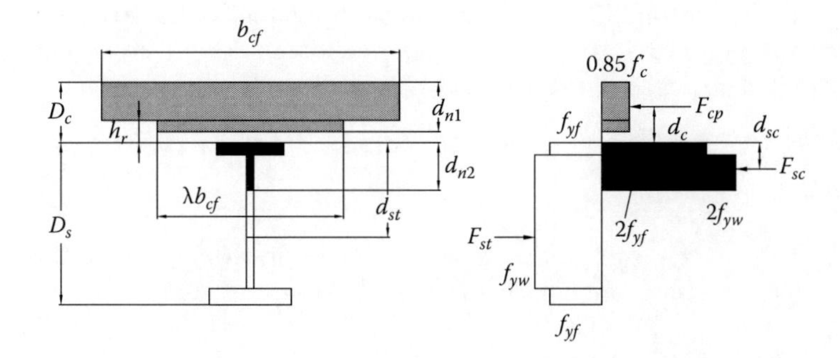

Figure 9.15 Plastic stress distributions in composite section with $\gamma \leq 0.5$ and $0 < \beta < 1.0$.

The equivalent plastic stress distribution in the cross section with $\gamma \leq 0.5$ and partial shear connection is illustrated in Figure 9.15. For a composite beam cross section with partial shear connection, there are two plastic neutral axes in the section as shown in Figure 9.15. The first PNA is located in the concrete slab either in the concrete cover slab or in the steel ribs. The second one falls in the steel section, which can be located in the top flange, web or the bottom flange.

The nominal moment capacity (M_b) of the cross section with $\gamma \leq 0.5$ and partial shear connection is determined by taking moments about the line of action of the resultant compressive force (F_{sc}) in the steel section as follows:

$$M_b = F_{cp}(d_c + d_{sc}) + F_{st}(d_{st} - d_{sc}) \tag{9.33}$$

9.7.3.2 Depth of the first plastic neutral axis

Case 1: If $F_{cp} \leq F_{c1}$, the first PNA lies in the concrete cover slab above the steel ribs so that $d_{n1} \leq h_c$. The compressive force in concrete with partial shear connection is determined as $F_{cp} = 0.85f_c'b_{cf}d_{n1}$. The force F_{cp} depends on the degree of shear connection and is taken as $F_{cp} = \beta F_{cc}$. Consequently, the depth of the first PNA (d_{n1}) can be expressed by

$$d_{n1} = \frac{F_{cp}}{0.85f_c'b_{cf}} \tag{9.34}$$

Case 2: If $F_{cp} > F_{c1}$, the first PNA is located in the steel ribs. The depth of the neutral axis is in the range of $h_c < d_{n1} \leq D_c$. The compressive force in concrete with partial shear connection is determined as $F_{cp} = F_{c1} + 0.85f_c'\lambda b_{cf}(d_{n1} - h_c)$. This compressive force (F_{cp}) must be in equilibrium with the strength (F_{sh}) of the shear connection. The depth of the first PNA (d_{n1}) is derived as follows:

$$d_{n1} = h_c + \frac{F_{cp} - F_{c1}}{0.85f_c'\lambda b_{cf}} \tag{9.35}$$

9.7.3.3 Depth of the second plastic neutral axis

It can be seen from Figure 9.15 that part of the steel section is subjected to compression. The equilibrium condition requires that the resultant force in the steel section must be equal to the strength of the shear connection: $F_{st} - F_{sc} = F_{sh} = F_{cp}$. The resultant compressive force (F_{sc}) in the steel section is determined as $F_{sc} = F_{st} - F_{cp}$.

Case 1: If $F_{sc} \le 2F_{f1}$, the second PNA is located in the top flange of the steel section so that $d_{n2} \le t_{f1}$. It is seen from Figure 9.15 that the compressive force in the steel section is determined as $F_{sc} = b_{f1}d_{n2}(2f_{yf})$. The depth of the second PNA (d_{n2}) can be computed as

$$d_{n2} = \frac{F_{st} - F_{cp}}{b_{f1}(2f_{yf})} \tag{9.36}$$

Case 2: If $2F_{f1} < F_{sc} \le (2F_{f1} + 2F_w)$, the second PNA lies in the steel web. This implies that $t_{f1} < d_{n2} \le (t_{f1} + d_w)$. It is seen from Figure 9.15 that the resultant compressive force in the steel section is obtained as $F_{sc} = 2F_{f1} + t_w(d_{n2} - t_{f1})(2f_{yw})$. The depth of the second PNA is given by

$$d_{n2} = t_{f1} + \frac{F_{sc} - 2F_{f1}}{t_w(2f_{yw})} \tag{9.37}$$

Case 3: If $F_{sc} > (2F_{f1} + 2F_w)$, the second PNA falls in the bottom flange of the steel section. This condition leads to $(t_{f1} + d_w) < d_{n2} \le D_s$. For this case, the resultant compressive force in the steel section is calculated as $F_{sc} = 2F_{f1} + 2F_w + b_{f2}(d_{n2} - d_w - t_{f1})(2f_{yf})$. The depth of the second PNA (d_{n2}) is derived as

$$d_{n2} = t_{f1} + d_w + \frac{F_{sc} - 2F_{f1} - 2F_w}{b_{f2}(2f_{yf})} \tag{9.38}$$

9.7.4 Cross sections with γ = 1.0 and complete shear connection

9.7.4.1 *Nominal moment capacity M_{bfc}*

When the shear ratio (γ) at the cross section of a composite beam under positive bending is equal to unity, the contribution of the steel web to the moment capacity is ignored. The steel web is assumed to resist the entire vertical design shear force. The plastic stress distribution in the composite beam cross section with γ = 1.0 and complete shear connection is schematically depicted in Figure 9.16, where the stresses on the steel web are not drawn because the web is completely ignored in the calculation of the moment capacity. It is noted that the figure shows only the typical case for which the PNA is located in the top flange of the steel section.

In AS 2327.1, the degree of shear connection at the cross section with γ = 1.0 and complete shear connection is calculated as

$$\psi = \frac{F_{ccf}}{F_{cc}} \tag{9.39}$$

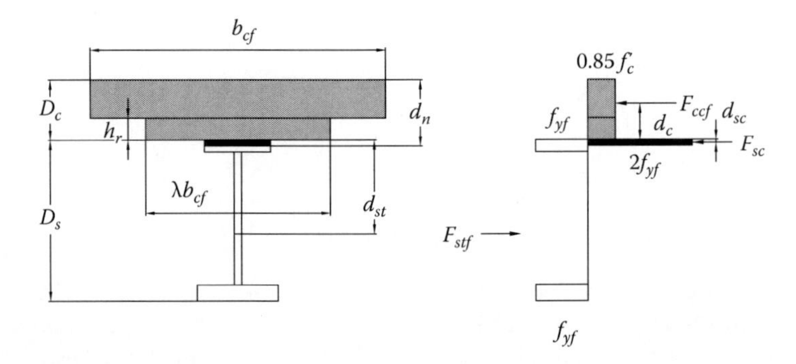

Figure 9.16 Plastic stress distributions in composite section with γ = 1.0 and complete shear connection.

in which F_{ccf} is the compressive force in the concrete slab with $\beta = 1.0$ when the steel web is ignored. It is worth to noting that F_{cc} is the compressive force in the concrete slab with $\beta = 1.0$ when the whole effective steel section is taken into account. For cross sections with $\gamma = 1.0$, the complete shear connection is defined as the condition of $\psi \le \beta \le 1.0$.

The nominal moment capacity (M_{bfc}) of the cross section with $\gamma = 1.0$ and complete shear connection is determined by taking moments about the line of action of the resultant compressive force (F_{sc}) in the steel section as follows:

$$M_{bfc} = F_{ccf}(d_c + d_{sc}) + F_{stf}(d_{st} - d_{sc}) \tag{9.40}$$

where F_{stf} is the tensile capacity of the two flanges of the steel section.

9.7.4.2 Plastic neutral axis depth

Case 1: If the compressive capacity of the concrete cover slab is greater than the tensile capacity of the steel two flanges (F_{stf}), such as $F_{stf} \le F_{c1}$, the PNA lies in the concrete cover slab above the steel ribs so that $d_n \le h_c$. The compressive force in concrete with complete shear connection is $F_{ccf} = 0.85 f'_c b_{cf} d_n$. The force equilibrium condition of $F_{ccf} = F_{stf}$ gives the depth of the PNA (d_n) as follows:

$$d_n = \frac{F_{stf}}{0.85 f'_c b_{cf}} \tag{9.41}$$

Case 2: If $F_{c1} < F_{stf} \le (F_{c1} + F_{c2})$, the PNA is located in the steel ribs. The neutral axis depth satisfies the condition of $h_c < d_n \le D_c$. The compressive force in concrete with complete shear connection is computed as $F_{ccf} = F_{c1} + 0.85 f'_c \lambda b_{cf}(d_n - h_c)$. From the force equilibrium condition of $F_{ccf} = F_{stf}$, the depth of the PNA (d_n) can be determined as

$$d_n = h_c + \frac{F_{stf} - F_{c1}}{0.85 f'_c \lambda b_{cf}} \tag{9.42}$$

Case 3: If $(F_{c1} + F_{c2}) < F_{stf} \le (F_{c1} + F_{c2} + 2F_{f1})$, the PNA lies in the top flange of the steel section and it means that $D_c < d_n \le (D_c + t_{f1})$. The compressive force in concrete becomes $F_{ccf} = (F_{c1} + F_{c2})$. The compressive force in the top flange is $F_{sc} = b_{f1}(d_n - D_c)(2f_{yf})$. The force equilibrium condition is expressed as $(F_{ccf} + F_{sc}) = F_{stf}$. The depth of the PNA ($d_n$) is derived as

$$d_n = D_c + \frac{F_{stf} - F_{ccf}}{b_{f1}(2f_{yf})} \tag{9.43}$$

9.7.5 Cross sections with γ = 1.0 and partial shear connection

9.7.5.1 Nominal moment capacity M_{bf}

For a cross section with partial shear connection ($0 < \beta < \psi$), its moment capacity is governed by the strength of shear connection. The compressive force (F_{cpf}) in the concrete slab at a cross section with $\gamma = 1.0$ and partial shear connection can be determined by one of the following expressions:

$$F_{cpf} = n_i f_{ds} \le F_{ccf} \tag{9.44}$$

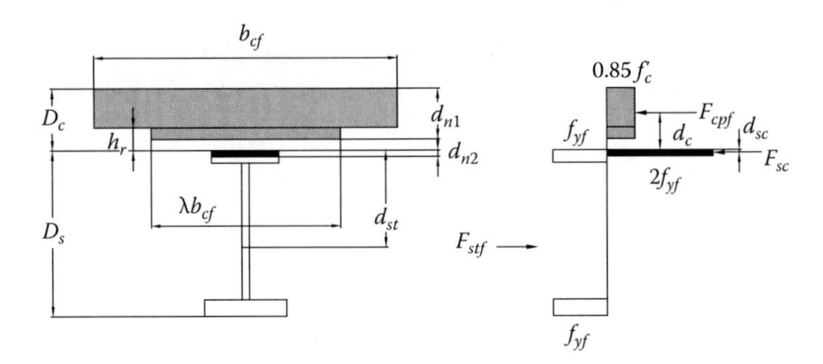

Figure 9.17 Plastic stress distributions in composite section with γ = 1.0 and partial shear connection.

$$F_{cpf} = \beta F_{cc} \leq F_{ccf} \tag{9.45}$$

where F_{ccf} is the compressive force in the concrete slab at a cross section with γ = 1.0 and complete shear connection.

The equivalent plastic stress distribution in the cross section with γ = 1.0 and partial shear connection is illustrated in Figure 9.17, where shows that there are two plastic neutral axes in the cross section. The first PNA is located in the concrete slab either in the concrete cover slab or in the steel ribs. The second one falls in the top or the bottom flange of the steel section.

The nominal moment capacity (M_{bf}) of the cross section with γ = 1.0 and partial shear connection is obtained by taking moments about the line of action of the compressive force (F_{sc}) in the steel flange as follows:

$$M_{bf} = F_{cpf}(d_c + d_{sc}) + F_{stf}(d_{st} - d_{sc}) \tag{9.46}$$

9.7.5.2 Depth of the first plastic neutral axis

Case 1: If $F_{cpf} \leq F_{c1}$, the first PNA lies in the concrete cover slab above the steel ribs so that $d_{n1} \leq h_c$. The compressive force in concrete with partial shear connection is given as $F_{cpf} = 0.85 f'_c b_{cf} d_{n1}$. The force F_{cp} depends on the degree of shear connection and is taken as $F_{cpf} = \beta F_{cc}$. The depth of the first PNA (d_{n1}) is given by

$$d_{n1} = \frac{F_{cpf}}{0.85 f'_c b_{cf}} \tag{9.47}$$

Case 2: If $F_{cpf} > F_{c1}$, the first PNA lies in the steel ribs. The depth of the neutral axis satisfies the condition of $h_c < d_{n1} \leq D_c$. The compressive force in concrete with partial shear connection is determined as $F_{cpf} = F_{c1} + 0.85 f'_c \lambda b_{cf}(d_{n1} - h_c)$. This compressive force (F_{cp}) must be in equilibrium with the strength (F_{sh}) of the shear connection. The depth of the first PNA (d_{n1}) is

$$d_{n1} = h_c + \frac{F_{cpf} - F_{c1}}{0.85 f'_c \lambda b_{cf}} \tag{9.48}$$

9.7.5.3 Depth of the second plastic neutral axis

It can be seen from Figure 9.17 that part of the steel section is subjected to compression. The equilibrium condition requires that the resultant force in the steel section must be equal to the strength of the shear connection, such as $F_{stf}-F_{sc} = F_{sh} = F_{cpf}$. The resultant compressive force (F_{sc}) in the steel section can be obtained as $F_{sc} = F_{stf}-F_{cpf}$.

Case 1: If $F_{sc} \leq 2F_{f1}$, the second PNA is located in the top flange of the steel section so that $d_{n2} \leq t_{f1}$. It is seen from Figure 9.17 that the compressive force in the steel section is determined as $F_{sc} = b_{f1}d_{n2}(2f_{yf})$. The depth of the second PNA (d_{n2}) can be computed as

$$d_{n2} = \frac{F_{stf} - F_{cpf}}{b_{f1}(2f_{yf})} \tag{9.49}$$

Case 2: If $2F_{f1} < F_{sc}$, the second PNA lies in the bottom flange of the steel section. This implies that $(t_{f1} + d_w) < d_{n2} \leq D_s$. The compressive force in the steel flange is obtained as $F_{sc} = 2F_{f1} + b_{f2}(d_{n2}-d_w-t_{f1})(2f_{yf})$. The depth of the second PNA is derived as

$$d_{n2} = t_{f1} + d_w + \frac{F_{sc} - 2F_{f1}}{b_{f2}(2f_{yf})} \tag{9.50}$$

9.7.6 Cross sections with 0.5 < γ ≤ 1.0

For beam cross sections with $0.5 < \gamma \leq 1.0$, the design moment capacity ϕM_{bv} depends on the shear ratio γ and the degree of shear connection β. This means that the design vertical shear force acting at the section reduces the design moment capacity ϕM_{bv}. AS 2327.1 allows a linear interaction equation to be used to determine the design moment capacity ϕM_{bv} of a composite beam with $0.5 < \gamma \leq 1.0$. The moment–shear interaction diagram is presented in Figure 9.18. The design moment capacity (ϕM_{bv}) of the cross section with $0.5 < \gamma \leq 1.0$ is calculated by linear interpolation as follows:

$$\phi M_{bv} = \phi M_{bf} + (\phi M_b - \phi M_{bf})(2 - 2\gamma) \quad \text{for } \beta < 1.0 \tag{9.51}$$

$$\phi M_{bv} = \phi M_{bfc} + (\phi M_{bc} - \phi M_{bfc})(2 - 2\gamma) \quad \text{for } \beta = 1.0 \tag{9.52}$$

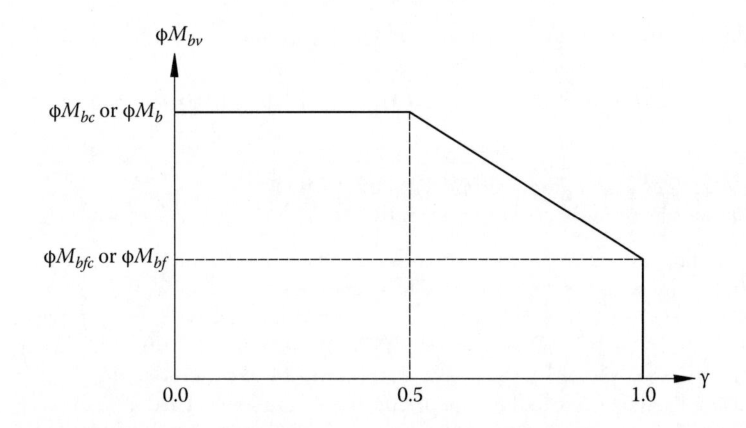

Figure 9.18 Moment–shear interaction diagram for composite sections.

9.7.7 Minimum degree of shear connection

In order to satisfy the strength requirement $M^* \le \phi M_{bv}$ at a potentially critical cross section, the minimum degree of shear connection at that section needs to be determined. For cross sections with $\gamma \le 0.5$, the minimum degree of shear connection β_i is given in Clause 6.5.2 of AS 2327.1 (2003) as follows:

$$\beta_i = \frac{M^* - \phi M_s}{2(\phi M_{b \cdot 5} - \phi M_s)} \ge 0 \quad \text{for } \phi M_s < M^* \le \phi M_{b \cdot 5} \tag{9.53}$$

$$\beta_i = \frac{M^* + \phi M_{bc} - 2\phi M_{b \cdot 5}}{2(\phi M_{bc} - \phi M_{b \cdot 5})} \ge 0 \quad \text{for } \phi M_{b \cdot 5} < M^* \le \phi M_{bc}, \tag{9.54}$$

where

ϕM_s is the design moment capacity of the steel section
$\phi M_{b \cdot 5}$ is the design moment capacity of the cross section by setting $\beta = 0.5$

For cross sections with $0.5 < \gamma \le 1.0$, the minimum degree of shear connection β_i can be calculated in accordance with Clause 6.5.3 of AS 2327.1 (2003) as follows:
For $0 < \beta_i \le \psi$,

$$\beta_i = \frac{[M^* - (2\gamma - 1)\phi M_{sf} - 2(1-\gamma)\phi M_s]\psi}{(1-2\gamma)\phi M_{sf} + (2\gamma - 1)\phi M_{bfc} - 2(1-\gamma)\phi M_s + 2(1-\gamma)\phi M_{b \cdot \psi}} \ge 0 \tag{9.55}$$

For $\psi < \beta_i$,

$$\beta_i = \psi + \frac{(1-\psi)[M^* - 2(1-\gamma)\phi M_{b \cdot \psi} - (2\gamma - 1)\phi M_{bfc}]}{2(1-\gamma)(\phi M_{bc} - \phi M_{b \cdot \psi})} \ge 0 \tag{9.56}$$

where

ϕM_{sf} is the design moment capacity of the steel section neglecting the contribution of the web
$\phi M_{b \cdot \psi}$ is the design moment capacity of the composite cross section by setting $\beta = \psi$

Example 9.2: Design of simply supported composite beam with complete shear connection for strength

Figure 9.19 shows the cross section of an internal secondary simply supported composite beam with complete shear connection. The spacing of the secondary beams is 3.2 m. The effective span of the composite beam is 8 m. The profiled steel sheeting is placed perpendicular to the steel beam. The steel section 360UB50.7 of Grade 300 steel is used with f_{yf} = 300 MPa and f_{yw} = 320 MPa. The design strength of the concrete flange is f'_c = 32 MPa. The composite slab is subjected to a superimposed dead load of 1.0 kPa and a live load of 4 kPa. Check the strengths of the composite beam and provide adequate stud shear connectors to the beam.

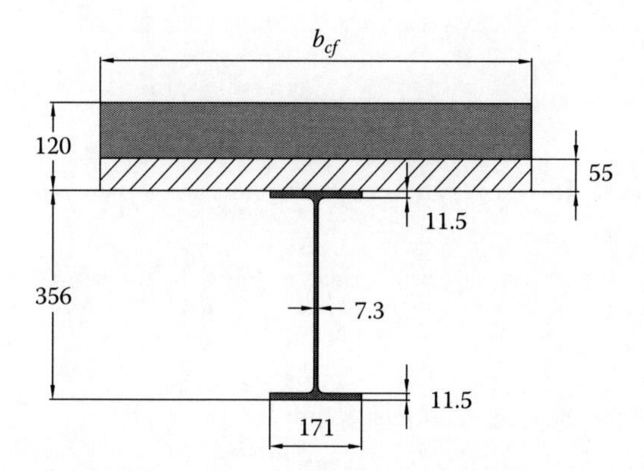

Figure 9.19 Cross section of composite beam under positive bending.

1. Effective width of concrete flange

The profiled steel sheeting is placed perpendicular to the steel beam, $\lambda = 0$. The effective width of the concrete flange is calculated as follows:

$$b_{e1} = \min\left[\left(\frac{8000}{8}\right), \left(\frac{3200}{2}\right), \left(\frac{171}{2} + 8 \times 120\right)\right] = 1000 \text{ mm}$$

$$b_{cf} = 2 \times 1000 = 2000 \text{ mm}$$

2. Design action effects

The self-weight of the steel beam: $50.7 \times 9.81 \times 10^{-3} = 0.497$ kN/m
The self-weight of the slab: $0.12 \times 25 \times 3.2 = 9.6$ kN/m

Superimposed dead load: $1.0 \times 3.2 = 3.2$ kN/m
Total dead load: $G = 0.497 + 9.6 + 3.2 = 13.3$ kN/m
Live load: $Q = 4 \times 3.2 = 12.8$ kN/m
The design load: $w^* = 1.2G + 1.5Q = 1.2 \times 13.3 + 1.5 \times 12.8 = 35.16$ kN/m

The maximum design bending moment at mid-span of the composite beam is

$$M^* = \frac{w^* L_{ef}^2}{8} = \frac{35.16 \times 8^2}{8} = 281.3 \text{ kN m}$$

The design vertical shear at support is

$$V^* = \frac{w^* L_{ef}}{2} = \frac{35.16 \times 8}{2} = 140.6 \text{ kN}$$

3. Vertical shear capacity

The slenderness of the steel web is

$$\lambda_w = \frac{d_w}{t_w}\sqrt{\frac{f_y}{250}} = \frac{333}{7.3}\sqrt{\frac{320}{250}} = 51.6 < \lambda_{yw} = 82$$

The web is not slender. The shear yield capacity of the web is calculated as

$$\phi V_u = \phi(0.6A_w f_{yw}) = 0.9\times 0.6\times 356\times 7.3\times 320\,\text{N} = 449\text{ kN} > V^* = 140.6\text{ kN, OK}$$

$$0.5\phi V_u = 0.5\times 449\text{ kN} = 224.5\text{ kN} > V^* = 140.6\text{ kN}$$

Therefore, the design moment capacity of the composite beam is not affected by the vertical shear.

4. Plastic neutral axis depth

The tensile capacity of the steel section is computed as

$$F_{st} = (b_{f1}t_{f1} + b_{f2}t_{f2})f_{yf} + d_w t_w f_{yw}$$
$$= (171\times 11.5\times 2)\times 300 + 333\times 7.3\times 320\text{ N} = 1957.8\text{ kN}$$

The compressive capacity of the concrete cover slab is computed as

$$F_{c1} = 0.85f'_c b_{cf}(D_c - h_r) = 0.85\times 32\times 2000\times(120-55)\times 10^{-3} = 3536\text{ kN}$$

Since $F_{c1} > F_{st}$, the PNA is located in the concrete cover slab.
For complete shear connection, $F_{cc} = F_{st}$.

The PNA depth is calculated as

$$d_n = \frac{F_{st}}{0.85f'_c b_{cf}} = \frac{1957.8\times 10^3}{0.85\times 32\times 2000} = 36\text{ mm} < h_c = 120-55 = 65\text{ mm}$$

5. Distances to centroid of forces

The distance from the centroid of F_{cc} to the top face of the steel section is

$$d_c = D_c - \frac{d_n}{2} = 120 - \frac{36}{2} = 102\text{ mm}$$

360UB50.7 is a doubly symmetric section. The distance from the centroid of F_{st} to the top face of the steel section is given as

$$d_{st} = \frac{D_s}{2} = \frac{356}{2} = 178\text{ mm}$$

The compressive force in the steel section is $F_{sc} = 0$, and the distance from the centroid of F_{sc} to the top fibre of the steel section is $d_{sc} = 0$.

6. Design moment capacity

Taking moments about the line of action of the compressive force in steel section F_{sc}, the nominal moment capacity is calculated as

$$M_{bc} = F_{cc}(d_c + d_{sc}) + F_{st}(d_{st} - d_{sc})$$
$$= 1957.8\times(102+0)\times 10^{-3} + 1957.8\times(178-0)\times 10^{-3} = 548.2\text{ kN m}$$

The design moment capacity is therefore

$$\phi M_{bc} = 0.9\times 548.2 = 493.4\text{ kN m} > M^* = 281.3\text{ kN m, OK}$$

7. Required number of shear connectors

The nominal shear capacity of 19 mm diameter headed stud embedded in 32 MPa concrete is obtained from Table 9.1 as f_{vs} = 93 kN. Taking $f_{ds} = f_{vs}$ = 93 kN, the required number of shear connectors from the end of the composite beam to its mid-span can be determined as

$$n_c = \frac{F_{cc}}{f_{ds}} = \frac{1957.8}{93} = 21$$

Taking n_c = 22, the load-sharing factor is

$$k_n = 1.18 - \frac{0.18}{\sqrt{n_c}} = 1.18 - \frac{0.18}{\sqrt{22}} = 1.14$$

The design shear capacity of shear connectors in a group is computed as

$$f_{ds} = \phi k_n f_{vs} = 0.85 \times 1.14 \times 93 = 90 \text{ kN}$$

The required number of stud shear connectors is finalized as

$$n_c = \frac{F_{cc}}{f_{ds}} = \frac{1957.8}{90} = 21.8 \approx 22$$

The design strength of the shear connection is determined as

$$F_{sh} = n_c f_{ds} = 22 \times 90 = 1980 \text{ kN} > F_{cc} = 1957.8 \text{ kN,OK}$$

The total number of stud shear connectors in the whole composite beam is 44.

Example 9.3: Design of simply supported composite beam with partial shear connection for strength

As shown in Example 9.1, only 57% of the design moment capacity of the composite beam with complete shear connection is utilised. Redesign this composite beam with partial shear connection of β = 0.6.

1. Plastic neutral axis depth

The composite beam is designed with β = 0.6 at the mid-span section.

The tensile capacity of the steel section is computed as

$$F_{st} = (b_{f1}t_{f1} + b_{f2}t_{f2})f_{yf} + d_w t_w f_{yw}$$

$$= (171 \times 11.5 \times 2) \times 300 + 333 \times 7.3 \times 320 \text{ N} = 1957.8 \text{ kN}$$

The compressive force in the concrete slab with partial shear connection is

$$F_{cp} = \beta F_{cc} = 0.6 \times 1957.8 = 1174.68 \text{ kN}$$

The compressive capacity of the concrete cover slab is computed as

$$F_{c1} = 0.85 f'_c b_{cf}(D_c - h_r) = 0.85 \times 32 \times 2000 \times (120 - 55) \times 10^{-3} = 3536 \text{ kN}$$

Since $F_{cp} < F_{c1}$, the first PNA lies in the concrete cover slab.

The depth of the first PNA in the concrete slab is calculated as

$$d_{n1} = \frac{F_{cp}}{0.85 f'_c b_{cf}} = \frac{1174.68 \times 10^3}{0.85 \times 32 \times 2000} = 22 \text{ mm} < h_c = 120 - 55 = 65 \text{ mm}$$

The compressive force in steel section is computed as

$$F_{sc} = F_{st} - F_{cp} = 1957.8 - 1174.67 = 783 \text{ kN}$$

The slenderness of the top flange in compression is

$$\lambda_{ef} = \frac{b}{t}\sqrt{\frac{f_y}{250}} = \frac{(171 - 7.3)/2}{11.5}\sqrt{\frac{300}{250}} = 7.8 < \lambda_{ey} = 9 \qquad \text{Table 5.1 of AS 2327.1}$$

Hence, the top flange of the steel section is compact.

The capacity of the steel top flange is

$$2F_{f1} = 2 \times b_{f1} t_{f1} f_{yf} = 2 \times 171 \times 11.5 \times 300 \text{ N} = 1180 \text{ kN}$$

If $F_{sc} < 2F_{f1}$, the second neutral axis lies in the top flange of the steel section.

The depth of the second neutral axis is computed as

$$d_{n2} = \frac{F_{sc}}{b_{f1}(2 f_{yf})} = \frac{783 \times 10^3}{171 \times (2 \times 300)} = 7.6 \text{ mm} < t_{f1} = 11.5 \text{ mm}$$

2. Distances to centroid of forces

The distance from the centroid of F_{cp} to the top face of the steel section is

$$d_c = D_c - \frac{d_{n1}}{2} = 120 - \frac{22}{2} = 109 \text{ mm}$$

360UB50.7 is a doubly symmetric section. The distance from the centroid of F_{st} to the top face of the steel section is given as

$$d_{st} = \frac{D_s}{2} = \frac{356}{2} = 178 \text{ mm}$$

The distance from the centroid of F_{sc} to the top fibre of the steel section is

$$d_{sc} = \frac{d_{n2}}{2} = \frac{7.6}{2} = 3.8 \text{ mm.}$$

3. Design moment capacity

Taking moments about the line of action of the compressive force in steel section F_{sc}, the nominal moment capacity is calculated as

$$M_b = F_{cp}(d_c + d_{sc}) + F_{st}(d_{st} - d_{sc})$$

$$= 1174.68 \times (109 + 3.8) \times 10^{-3} + 1957.8 \times (178 - 3.8) \times 10^{-3} = 473.6 \text{ kN m}$$

The design moment capacity is therefore

$$\phi M_{bc} = 0.9 \times 473.6 = 426.24 \text{ kN m} > M^* = 281.3 \text{ kN m, OK}$$

4. Required number of shear connectors

Taking $f_{ds} = f_{vs} = 93$ kN, the required number of shear connectors from the end of the composite beam to its mid-span can be determined as

$$n_c = \frac{F_{cp}}{f_{ds}} = \frac{1174.68}{93} = 12.63$$

Taking $n_c = 14$, the load-sharing factor is

$$k_n = 1.18 - \frac{0.18}{\sqrt{n_c}} = 1.18 - \frac{0.18}{\sqrt{14}} = 1.132$$

The design shear capacity of shear connectors in a group is computed as

$$f_{ds} = \phi k_n f_{vs} = 0.85 \times 1.132 \times 93 = 89.5 \text{ kN}$$

The required number of stud shear connectors is finalized as

$$n_c = \frac{F_{cp}}{f_{ds}} = \frac{1174.67}{89.5} = 13.13$$

The design strength of the shear connection is determined as

$$F_{sh} = n_c f_{ds} = 14 \times 89.5 = 1253 \text{ kN} > F_{cp} = 1174.68 \text{ kN, OK}$$

The total number of stud shear connectors in the whole composite beam is 28.

9.8 DESIGN MOMENT CAPACITY FOR NEGATIVE BENDING

9.8.1 Design concepts

The cross sections of peak negative moments in a continuous composite beam must be designed for complete shear connection to prevent catastrophic failure in negative moment regions. The maximum design shear force usually occurs at the supports of a continuous composite beam. As a result, its effect on the design moment capacity of cross sections in negative bending is more critical than on that of cross sections in positive bending. When $\gamma \le 0.5$, the design moment capacity of cross section is not affected by vertical shear so that the effective portion of the steel web contributes to the resistance to bending. When $\gamma = 1.0$, the web of the steel section is used to resist vertical shear and is ignored in the calculation of the design moment capacity of cross section.

Figure 9.20 presents the plastic stress distribution in a general composite cross section with $\gamma \le 0.5$ and in negative bending. The moment capacity of the cross section depends on the area of longitudinal tensile reinforcement (A_r) in the concrete slab. Any level of reinforcement leads to at least part of the steel section in compression. The maximum area of longitudinal reinforcement corresponds to the condition in which the entire steel section is in compression. The local buckling of the flanges and web of the steel section in the composite beam in negative

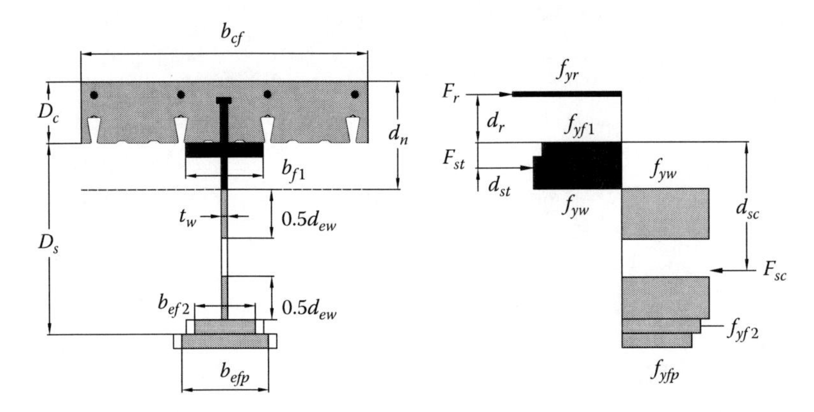

Figure 9.20 Plastic stress distribution in composite section under negative bending with $\gamma \leq 0.5$.

moment regions may occur. The strength and ductility of the composite section in negative bending may be limited by local buckling of the steel section or fracture of the reinforcement. Slender plate elements are not allowed to be used in composite cross sections according to AS 2327.1, which is also applied to the design for negative moment regions. Longitudinal web stiffeners may be welded to the web to reduce its slenderness, and additional plate may be welded to the bottom flange to lower the PNA to place less of the web in compression.

9.8.2 Key levels of longitudinal reinforcement

In the analysis of a composite cross section in negative bending to compute its moment capacity, the location of the PNA needs to be determined. The location of PNA depends on the area of longitudinal tensile reinforcement in the concrete slab and effective steel cross section. The calculation of key levels of reinforcement that defines the key locations of PNA gives a simple direct solution to the problem (Berry et al. 2001a,b). The method for calculating the key level longitudinal tensile reinforcement in the concrete slab is presented herein. Further information on the method is given by Berry et al. (2001a,b).

9.8.2.1 Maximum area of reinforcement

The maximum area of longitudinal tensile reinforcement in the concrete slab, which makes contributions to the negative moment capacity of a composite cross section, is limited by the compressive capacity of the effective steel section. The PNA is located between the top face of the steel section and the bottom of the reinforcement. The force equilibrium condition is expressed by

$$F_{rm} = F_{ef1} + F_{ew} + F_{ef2} + F_{efp} \tag{9.57}$$

where

$F_{rm} = A_{rm} f_{yr}$

$F_{ef1} = b_{ef1} t_{f1} f_{yf1}$

$F_{ew} = d_{ew} t_w f_{yw}$

$F_{ef2} = b_{ef2} t_{f2} f_{yf2}$

$F_{efp} = b_{efp} t_p f_{yfp}$

The subscript *e* represents the effective width of a plate element or effective force.

The maximum area of longitudinal tensile reinforcement in the concrete slab is therefore

$$A_{rm} = \frac{F_{ef1} + F_{ew} + F_{ef2} + F_{efp}}{f_{yr}} \tag{9.58}$$

9.8.2.2 PNA located at the junction of the top flange and web

When the PNA is located at the junction of the top flange and the web of the steel section, the top flange is in tension, while the web and the bottom flange and plate are subjected to compression. From the force equilibrium, the required area of reinforcement is calculated by

$$A_{rfw} = \frac{F_{ew} + F_{ef2} + F_{efp} - F_{f1}}{f_{yr}} \tag{9.59}$$

9.8.2.3 PNA located in the web

When the PNA lies in the web, it divides the steel web into tension and compression zones. If the depth of the web in compression is greater than $d_{ew} = 30t_w\sqrt{250/f_{yw}}$, the local buckling of the steel web occurs and a hole will develop in the web as shown in Figure 9.20. If the depth of the web in compression is equal to d_{ew}, the compressive force (F_{wc}) in the web is $F_{wc} = d_{ew}t_w f_{yw}$. The tensile force in the web is computed as $F_{wt} = F_w - F_{wc}$. The required area of reinforcement is determined from the force equilibrium as

$$A_{rho} = \frac{F_{wc} + F_{ef2} + F_{efp} - F_{f1} - F_{wt}}{f_{yr}} \tag{9.60}$$

9.8.2.4 PNA located at the junction of the web and bottom flange

When the PNA is located at the junction of the steel web and the bottom flange, the top flange and the web are in tension, while the bottom flange and additional flange plate are in compression. The force equilibrium gives

$$A_{rwf} = \frac{F_{ef2} + F_{efp} - F_{f1} - F_w}{f_{yr}} \tag{9.61}$$

9.8.2.5 PNA located at the junction of the bottom flange and plate

When the PNA is located at the junction of the steel bottom flange and the additional flange plate, the entire steel I-section is in tension and the additional flange plate is in compression. For this case, the area of longitudinal reinforcement can be calculated from the force equilibrium as

$$A_{rfp} = \frac{F_{efp} - F_{f1} - F_w - F_{f2}}{f_{yr}} \tag{9.62}$$

9.8.3 Plastic neutral axis depth

The PNA of a composite cross section under negative bending depends on the area of longitudinal tensile reinforcement in the concrete slab. It may be located between the bottom of the reinforcement and top face of the top flange, in the top flange, web, bottom flange and additional bottom flange plate.

Case 1: If $A_{rm} \le A_r$, the PNA is located between the bottom of the longitudinal reinforcement and the top face of the steel top flange. Since the entire steel section is in compression, the effective portion of the steel section should be used to calculate the negative moment capacity.
Case 2: If $A_{rfw} \le A_r < A_{rm}$, the PNA lies in the top flange of the steel section. For this case, the portion of the top flange below the PNA is in compression and the effective width of the steel top flange is used in compression and tension. The depth of the PNA is determined using linear interpolation as

$$d_n = D_c + \left(\frac{A_{rm} - A_r}{A_{rm} - A_{rfw}} \right) t_{f1} \tag{9.63}$$

Case 3: If $A_{rbo} \le A_r < A_{rfw}$, the PNA falls in the web of the steel section. A hole forms in the compressive portion of the web. On the onset of local buckling of the web in compression, the effective depth of the web in compression is d_{ew}, while the depth of the web in tension is equal to $d_{wt} = d_w - d_{ew}$. The PNA varies within the depth of d_{wt}. The depth of the PNA is given by

$$d_n = D_c + t_{f1} + \left(\frac{A_{rfw} - A_r}{A_{rfw} - A_{rbo}} \right) d_{wt} \tag{9.64}$$

Case 4: If $A_{rwf} \le A_r < A_{rbo}$, the PNA is located within the depth d_{ew} of the web measured from the junction of the web and the bottom flange. The depth of the PNA is obtained using linear interpolation as

$$d_n = D_c + t_{f1} + d_{wt} + \left(\frac{A_{rbo} - A_r}{A_{rbo} - A_{rwf}} \right) d_{ew} \tag{9.65}$$

Case 5: If $A_{rfp} \le A_r < A_{rwf}$, the PNA lies in the bottom flange. For this case, the portion of the bottom flange below the PNA is in compression and the effective width of the steel bottom flange is used in compression and tension. The depth of the PNA is expressed by

$$d_n = D_c + t_{f1} + d_w + \left(\frac{A_{rwf} - A_r}{A_{rwf} - A_{rfp}} \right) t_{f2} \tag{9.66}$$

9.8.4 Design negative moment capacity

Once the depth of the PNA has been determined, the nominal negative moment capacity (M_{bc}) of the composite cross section with $\gamma \le 0.5$ can be calculated based on the stress distributions depicted in Figure 9.20 by taking moments about the centroid of the resultant tensile force (F_{st}) in the steel section as

$$M_{bc} = F_r(d_r + d_{st}) + F_{sc}(d_{sc} - d_{st}) \tag{9.67}$$

where

d_r is the distance from the centroid of the longitudinal reinforcement in the concrete slab to the top face of the steel section

d_{st} is the distance from the centroid of the resultant tensile force F_{st} in the steel section to the top face of the steel section

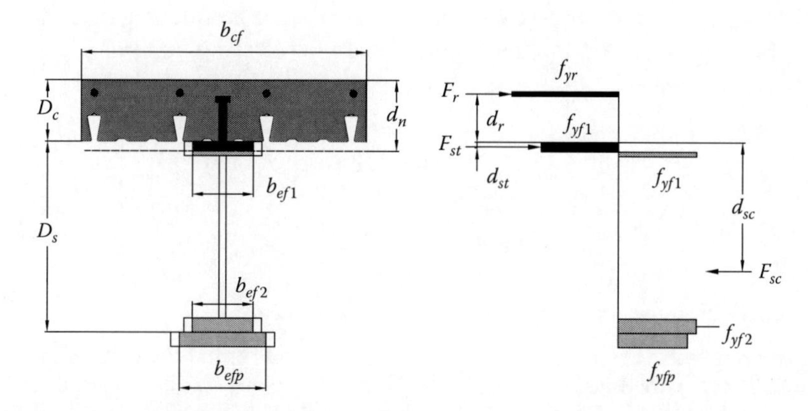

Figure 9.21 Plastic stress distribution in composite section under negative bending with γ = 1.0.

For the case of no tension in the steel section, $d_{st} = 0$. The distance d_{sc} is the distance from the centroid of the resultant compressive force F_{sc} in the steel section to the top face of the steel section. It should be noted that F_{sc} is calculated using the effective areas of steel plate elements which lies below the PNA.

For a cross section with $\gamma = 1.0$, the steel web is ignored in the determination of its nominal negative moment capacity (M_{bfc}). The plastic stress distribution in the composite section with $\gamma = 1.0$ is presented in Figure 9.21. For this situation, the areas of key level longitudinal reinforcement in the concrete slab that need to be calculated are A_{rm}, A_{rfw}, A_{rwf} and A_{rfp}. The depth of the PNA can be determined using the equations given in the preceding section.

For a cross section with $0.5 < \gamma \leq 1.0$, the design vertical shear reduces the negative moment capacity of the section. The moment–shear interaction diagram for composite cross sections under combined negative bending and vertical shear is presented in Figure 9.18. The design negative moment capacity (ϕM_{bv}) of cross sections with $0.5 < \gamma \leq 1.0$ can be calculated using Equation 9.52. Figure 9.22 presents the design negative moment capacity of a typical composite section as a function of the area of reinforcement and shear ratio.

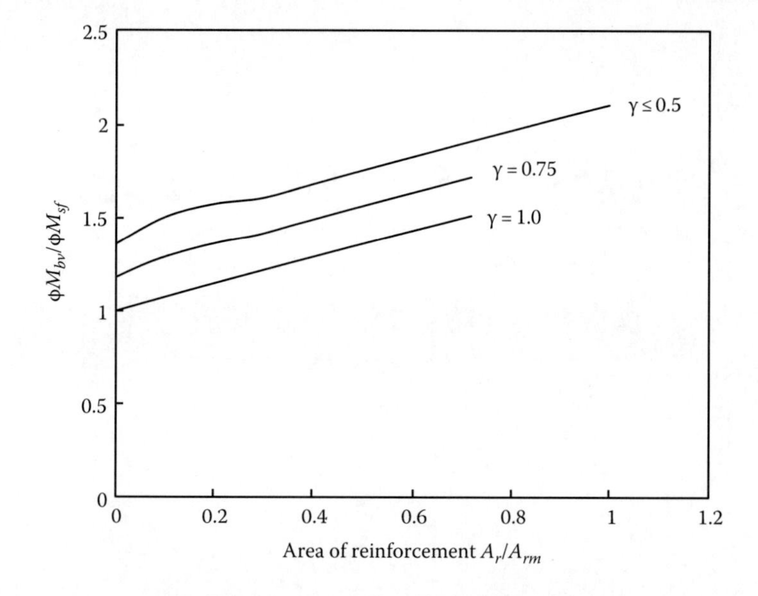

Figure 9.22 Design negative moment capacity as a function of area of reinforcement and shear ratio.

The required number of shear connectors between the maximum negative moment at the support and the adjacent section of zero moment can be determined by

$$n_c = \frac{F_r}{f_{ds}} \tag{9.68}$$

where $F_r = A_r f_{yr}$.

Example 9.4: Negative moment capacity of composite beam

The cross section of a composite beam under negative bending and a design vertical shear force of 320 kN is presented in Figure 9.23. The profiled steel sheeting is placed perpendicular to the steel beam. The hot-rolled steel section 460UB74.6 of Grade 300 steel with $f_{yf} = 300$ Mpa and $f_{yw} = 320$ Mpa is used. The compressive strength of the concrete flange is $f'_c = 32\,\text{MPa}$. The cross-sectional area of longitudinal tensile reinforcement in the concrete flange is 1100 mm² and the distance from the centroid of the reinforcement to the top face of the slab is 35 mm. The yield stress of the reinforcement is 500 MPa. Calculate the design negative moment capacity of the composite beam section.

1. Vertical shear capacity

The slenderness of the steel web under vertical shear is

$$\lambda_{ew} = \frac{b}{t}\sqrt{\frac{f_y}{250}} = \frac{457 - 2 \times 14.5}{9.1}\sqrt{\frac{320}{250}} = 53.2 < \lambda_{ey} = 82$$

The web is not slender. The shear yield capacity of the web is calculated as

$$\phi V_u = \phi(0.6 A_w f_{yw}) = 0.9 \times 0.6 \times 457 \times 9.1 \times 320 \text{ N} = 718.6 \text{ kN} > V^* = 320 \text{ kN, OK}$$

The shear ratio at the section is

$$\gamma = \frac{V^*}{\phi V_u} = \frac{320}{718.6} = 0.45 < 0.5$$

Therefore, the design negative moment capacity of the composite beam is not affected by the vertical shear.

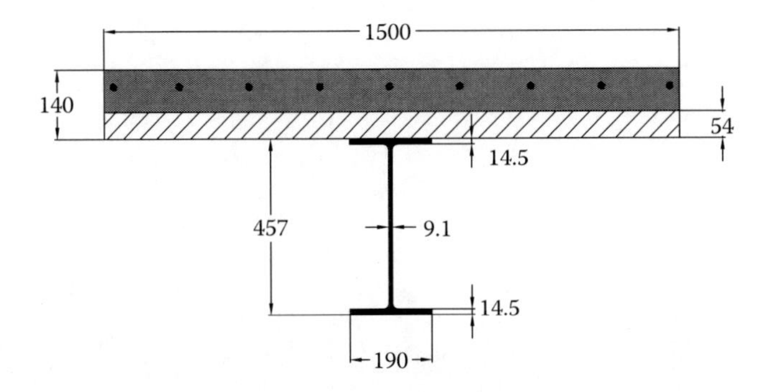

Figure 9.23 Cross section of composite beam under negative bending.

2. Key levels of longitudinal reinforcement

The maximum area of longitudinal reinforcement can be calculated by taking $d_n = D_c$. For this case, the whole steel section is in compression.

The slenderness of the flanges is

$$\lambda_{ef} = \frac{b}{t}\sqrt{\frac{f_y}{250}} = \frac{(190-9.1)/2}{14.5}\sqrt{\frac{300}{250}} = 6.8 < \lambda_{ey} = 9 \quad \text{Table 5.1 of AS 2327.1}$$

The two flanges are compact.

The effective width of the web in compression is calculated as

$$d_{ew} = 30t_w\sqrt{\frac{250}{f_y}} = 30\times 9.1\times\sqrt{\frac{250}{320}} = 241.3 \text{ mm} < d_w = 457-2\times 14.5 = 428 \text{ mm}$$

Hence, local buckling occurs and a hole forms in the steel web.

The capacities of the effective steel flanges and web are calculated as follows:

$$F_{ef1} = b_{ef1}t_{f1}f_{yf} = 190\times 14.5\times 300\ \times 10^{-3} = 826.5 \text{ kN}$$

$$F_{ew} = d_{ew}t_w f_{yw} = 241.3\times 9.1\times 320\ \times 10^{-3} = 702.7 \text{ kN}$$

$$F_{ef2} = b_{ef2}t_{f2}f_{yf} = 190\times 14.5\times 300\ \times 10^{-3} = 826.5 \text{ kN}$$

The capacities of the web in tension are computed as

$$F_w = d_w t_w f_{yw} = 428\times 9.1\times 320\ \times 10^{-3} = 1246.3 \text{ kN}$$

$$F_{wt} = F_w - F_{ew} = 1246.3 - 702.7 = 543.6 \text{ kN}$$

The areas of key level longitudinal reinforcement in the concrete slab are calculated as follows:

$$A_{rm} = \frac{F_{ef1}+F_{ew}+F_{ef2}+F_{efp}}{f_{yr}} = \frac{(826.5+702.7+826.5+0)\times 10^3}{500} = 4711 \text{ mm}^2$$

$$A_{rfw} = \frac{F_{ew}+F_{ef2}+F_{efp}-F_{f1}}{f_{yr}} = \frac{(702.7+826.5+0-826.5)\times 10^3}{500} = 1405 \text{ mm}^2$$

$$A_{rbo} = \frac{F_{ew}+F_{ef2}+F_{efp}-F_{f1}-F_{wt}}{f_{yr}} = \frac{(702.7+826.5+0-826.5-543.6)\times 10^3}{500} = 318 \text{ mm}^2$$

$A_r = 1100\,\text{mm}^2$, hence $A_{rbo} < A_r < A_{rfw}$

3. Depth of the plastic neutral axis

Since $A_{rbo} < A_r < A_{rfw}$, the PNA is located in the web of the steel section. The depth of the PNA is calculated as

$$d_n = D_c + t_{f1} + \left(\frac{A_{rfw}-A_r}{A_{rfw}-A_{rbo}}\right)d_{wt}$$

$$= 140 + 14.5 + \left(\frac{1405-1100}{1405-318}\right)\times(428-241.3) = 206.9 \text{ mm}$$

4. Forces and distances to centroid of forces

The tensile force in reinforcement is calculated as

$$F_r = A_r f_{yr} = 1100 \times 500 \times 10^{-3} = 550 \text{ kN}$$

The distance from the centroid of F_r to the top of the steel section is

$$d_r = D_c - d_t = 140 - 35 = 105 \text{ mm}$$

The tensile force in the top steel flange is

$$F_{ef1} = 826.5 \text{ kN}$$

The distance from the centroid of F_{ef1} to the top of steel section is

$$d_{f1} = \frac{t_{f1}}{2} = \frac{14.5}{2} = 7.25 \text{ mm}$$

The tensile force in the web for d_n = 206.9 mm is computed as

$$F_{wt} = (d_n - D_c - t_{f1}) t_w f_{yw} = (206.9 - 140 - 14.5) \times 9.1 \times 320 \text{ N} = 152.6 \text{ kN}$$

The distance from the centroid of F_{wt} to the top of steel section is

$$d_{wt} = \frac{d_n - D_c - t_{f1}}{2} + t_{f1} = \frac{206.9 - 140 - 14.5}{2} + 14.5 = 40.7 \text{ mm}$$

The resultant tensile force in the steel section is computed as

$$F_{st} = F_{ef1} + F_{wt} = 826.5 + 152.6 = 979.1 \text{ kN}$$

The distance from the centroid of F_{st} to the top of the steel section is computed as

$$d_{st} = \frac{F_{ef1} d_{f1} + F_{wt} d_{wt}}{F_{st}} = \frac{826.5 \times 7.25 + 152.6 \times 40.7}{979.1} = 12.5 \text{ mm}$$

The compressive force in the web is

$$F_{wc} = F_{ew} = 702.7 \text{ kN}$$

The distance from the centroid of F_{wc} to the top of steel section is

$$d_{wc} = D_s - t_{f2} - \frac{D_s - t_{f2} + D_c - d_n}{2} = 457 - 14.5 - \frac{457 - 14.5 + 140 - 206.9}{2}$$
$$= 254.7 \text{ mm}$$

The resultant compressive force in the steel section is

$$F_{sc} = F_{wc} + F_{ef2} = 702.7 + 826.5 = 1529.2 \text{ kN}$$

The distance from the centroid of F_{sc} to the top of the steel section is computed as

$$d_{sc} = \frac{F_{ew}d_{wc} + F_{ef2}(D_s - t_{f2}/2)}{F_{sc}}$$

$$= \frac{702.7 \times 254.7 + 826.5 \times (457 - 14.5/2)}{1529.2} = 360 \text{ mm}$$

5. Design negative moment capacity

The nominal negative moment capacity of the composite section is computed as

$$M_{bc} = F_r(d_r + d_{st}) + F_{sc}(d_{sc} - d_{st})$$

$$= 550 \times (105 + 12.5) + 1529.2 \times (360 - 12.5) \text{ kN mm} = 596 \text{ kN m}$$

The design negative moment capacity of the composite section is

$$\phi M_{bc} = 0.9 \times 596 = 536.4 \text{ kN m}$$

Example 9.5: Design negative moment capacity of continuous composite beam

Figure 9.24 shows the cross section of a continuous composite beam under negative bending and a design vertical shear force of 350 kN. The profiled steel sheeting is placed parallel to the steel beam. The hot-rolled steel section 410UB53.7 of Grade 300 steel with $f_{yf} = f_{yw} = 300$ Mpa is used. The compressive strength of the concrete flange is $f'_c = 25$ MPa. The cross-sectional area of longitudinal tensile reinforcement in the concrete flange is 1600 mm^2 and the distance from the centroid of the reinforcement to the top face of the slab is 35 mm. The yield stress of the reinforcement is 500 MPa. Calculate the design negative moment capacity of the composite beam section and the required number of stud shear connectors in the negative moment region to achieve complete shear connection.

1. Vertical shear capacity

The slenderness of the steel web under vertical shear is

$$\lambda_{ew} = \frac{b}{t}\sqrt{\frac{f_y}{250}} = \frac{403 - 2 \times 10.9}{7.6}\sqrt{\frac{320}{250}} = 56.7 < \lambda_{ey} = 82$$

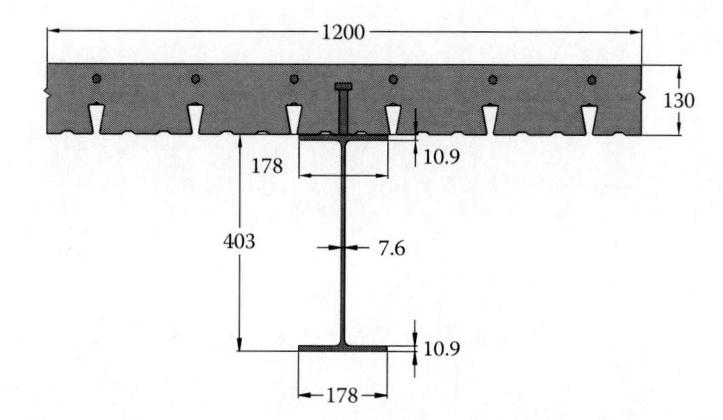

Figure 9.24 Cross section of continuous composite beam under negative bending.

The web is not slender. The shear yield capacity of the web is calculated as

$$\phi V_u = \phi(0.6A_w f_{yw}) = 0.9\times0.6\times403\times7.6\times320 \text{ N} = 529.25 \text{ kN} > V^* = 350 \text{ kN, OK}$$

The shear ratio at the section is

$$\gamma = \frac{V^*}{\phi V_u} = \frac{350}{529.25} = 0.66 > 0.5$$

Therefore, the design negative moment capacity of the composite beam is affected by the vertical shear. It needs to calculate ϕM_{bc} with $\gamma = 0.5$ and ϕM_{bfc} with $\gamma = 1.0$, respectively.

2. Design negative moment capacity with $\gamma = 0.5$

2.1. Key levels of longitudinal reinforcement

The maximum area of longitudinal reinforcement can be calculated by taking $d_n = D_c$. For this case, the whole steel section is in compression.

The slenderness of the flanges is

$$\lambda_{ef} = \frac{b}{t}\sqrt{\frac{f_y}{250}} = \frac{(178-7.6)/2}{10.9}\sqrt{\frac{320}{250}} = 8.8 < \lambda_{ey} = 9 \quad \text{Table 5.1 of AS 2327.1}$$

The two flanges are compact.

The effective width of the web in compression is calculated as

$$d_{ew} = 30t_w\sqrt{\frac{250}{f_y}} = 30\times7.6\times\sqrt{\frac{250}{320}} = 201.5 \text{ mm} < d_w = 381 \text{ mm}$$

Hence, local buckling occurs and a hole forms in the steel web.

The capacities of the effective steel flanges and web are calculated as follows:

$$F_{ef1} = b_{ef1}t_{f1}f_{yf} = 178\times10.9\times320\times10^{-3} = 620.86 \text{ kN}$$

$$F_{ew} = d_{ew}t_w f_{yw} = 201.5\times7.6\times320\times10^{-3} = 490 \text{ kN}$$

$$F_{ef2} = b_{ef2}t_{f2}f_{yf} = 178\times10.9\times320\times10^{-3} = 620.86 \text{ kN}$$

The areas of key level longitudinal reinforcement in the concrete slab are calculated as follows:

$$A_{rm} = \frac{F_{ef1} + F_{ew} + F_{ef2} + F_{efp}}{f_{yr}} = \frac{(620.86+490+620.86+0)\times10^3}{500} = 3463 \text{ mm}^2$$

$$A_{rfw} = \frac{F_{ew} + F_{ef2} + F_{efp} - F_{f1}}{f_{yr}} = \frac{(490+620.86+0-620.86)\times10^3}{500} = 980 \text{ mm}^2$$

$A_r = 1600\,\text{mm}^2$, hence $A_{rfw} < A_r < A_{rm}$.

2.2. Depth of the plastic neutral axis

Since $A_{rfw} < A_r < A_{rm}$, the PNA lies in the top flange of the steel section. The depth of the neutral axis is calculated as

$$d_n = D_c + \left(\frac{A_{rm} - A_r}{A_{rm} - A_{rfw}}\right)t_{f1} = 130 + \left(\frac{3463 - 1600}{3463 - 980}\right) \times 10.9 = 138.2 \text{ mm}$$

2.3. Forces and distances to centroid of forces

The tensile force in reinforcement is

$$F_r = A_r f_{yr} = 1600 \times 500 \times 10^{-3} = 800 \text{ kN}$$

The distance from the centroid of F_r to the top of the steel section is

$$d_r = D_c - d_t = 130 - 35 = 95 \text{ mm}$$

The compressive force in the top steel flange is calculated as

$$F_{ef1\cdot c} = b_{ef1}(D_c + t_{f1} - d_n) f_{yf} = 178 \times (130 + 10.9 - 138.2) \times 320 \times 10^{-3} = 153.8 \text{ kN}$$

The distance from the centroid of $F_{ef1\cdot c}$ to the top of steel section is

$$d_{f1\cdot c} = t_{f1} - \frac{D_c + t_{f1} - d_n}{2} = 10.9 - \frac{130 + 10.9 - 138.2}{2} = 9.55 \text{ mm}$$

The resultant compressive force in the steel section is computed as

$$F_{sc} = F_{ef1\cdot c} + F_{ew} + F_{ef2} = 153.8 + 490 + 620.86 = 1264.7 \text{ kN}$$

The distance from the centroid of F_{sc} to the top of the steel section is computed as

$$d_{sc} = \frac{F_{ef1\cdot c} d_{f1c} + F_{ew}(D_s/2) + F_{ef2}(D_s - t_{f2}/2)}{F_{sc}}$$

$$= \frac{153.8 \times 9.55 + 490 \times (403/2) + 620.86 \times (403 - 10.9/2)}{1264.7} = 274.4 \text{ mm}$$

The distance from the centroid of steel flange in tension to the top of steel section is $d_{st} = \dfrac{d_n - D_c}{2} = \dfrac{138.2 - 130}{2} = 4.1 \text{ mm}$

2.4. Design negative moment capacity

The nominal negative moment capacity of the composite section is computed as

$$M_{bc} = F_r(d_r + d_{st}) + F_{sc}(d_{sc} - d_{st})$$

$$= 800 \times (95 + 4.1) + 1264.8 \times (274.4 - 4.1) \text{ kN mm} = 421.13 \text{ kN m}$$

The design negative moment capacity of the composite section with $\gamma = 0.5$ is

$$\phi M_{bc} = 0.9 \times 421.13 = 379 \text{ kN m}$$

3. Design negative moment capacity with γ = 1.0

3.1. Key levels of longitudinal reinforcement

At the cross section with $\gamma = 1.0$, the steel web is ignored. The areas of key level longitudinal reinforcement are computed as

$$A_{rm} = \frac{F_{ef1} + F_{ew} + F_{ef2} + F_{efp}}{f_{yr}} = \frac{(620.86 + 0 + 620.86 + 0) \times 10^3}{500} = 2483 \text{ mm}^2$$

$$A_{rfw} = \frac{F_{ew} + F_{ef2} + F_{efp} - F_{f1}}{f_{yr}} = \frac{(0 + 620.86 + 0 - 620.86) \times 10^3}{500} = 0 \text{ mm}^2$$

$$A_r = 1600 \text{ mm}^2, \text{hence } A_{rfw} < A_r < A_{rm}$$

3.2. Depth of the plastic neutral axis

Since $A_{rfw} < A_r < A_{rm}$, the PNA lies in the top flange of the steel section. The depth of the neutral axis is calculated as

$$d_n = D_c + \left(\frac{A_{rm} - A_r}{A_{rm} - A_{rfw}} \right) t_{f1} = 130 + \left(\frac{2483 - 1600}{2483 - 0} \right) \times 10.9 = 133.9 \text{ mm}$$

3.3. Forces and distances to centroid of forces

The distance from the centroid of F_r to the top of the steel section is

$$d_r = D_c - d_t = 130 - 35 = 95 \text{ mm}$$

The compressive force in the top steel flange is computed as

$$F_{ef1\cdot c} = b_{ef1}(D_c + t_{f1} - d_n) f_{yf} = 178 \times (130 + 10.9 - 133.9) \times 320 \times 10^{-3} = 398.7 \text{ kN}$$

The distance from the centroid of $F_{ef1\cdot c}$ to the top of the steel section is

$$d_{f1\cdot c} = t_{f1} - \frac{D_c + t_{f1} - d_n}{2} = 10.9 - \frac{130 + 10.9 - 133.9}{2} = 7.4 \text{ mm}$$

The resultant compressive force in the steel section is computed as

$$F_{sc} = F_{ef1\cdot c} + F_{ef2} = 398.7 + 620.86 = 1019.56 \text{ kN}$$

The distance from the centroid of F_{sc} to the top of the steel section is

$$d_{sc} = \frac{F_{ef1\cdot c} d_{f1\cdot c} + F_{ef2}(D_s - t_{f2}/2)}{F_{sc}}$$

$$= \frac{398.7 \times 7.4 + 620.86 \times (403 - 10.9/2)}{1019.56} = 245 \text{ mm}$$

The distance from the centroid of steel flange in tension to the top of steel section is $d_{st} = (d_n - D_c)/2 = (133.9 - 130)/2 = 1.95$ mm

3.4. Design negative moment capacity

The nominal negative moment capacity of the composite section is computed as

$$M_{bfc} = F_r(d_r + d_{st}) + F_{sc}(d_{sc} - d_{st})$$

$$= 800 \times (95 + 1.95) \times 10^{-3} + 1019.56 \times (245 - 1.95) \times 10^{-3} = 325.4 \text{ kN m}$$

The design negative moment capacity of the composite section with $\gamma = 1.0$ is therefore

$$\phi M_{bfc} = 0.9 \times 325.4 = 292.8 \text{ kN m}$$

4. Design negative moment capacity with $\gamma = 0.66$

For the section with $\gamma = 0.66$, the design negative moment capacity is calculated as

$$\phi M_{bv} = \phi M_{bfc} + (\phi M_{bc} - \phi M_{bfc})(2 - 2\gamma)$$

$$= 292.8 + (379 - 292.8)(2 - 2 \times 0.66) = 351.4 \text{ kN m}$$

5. Required number of shear connectors

From Table 9.1, f_{vs} = 89 kN. Taking $f_{ds} = f_{vs}$ = 89 kN, the required number of shear connectors between the maximum negative moment at the support and the adjacent section of zero moment can be determined as

$$n_c = \frac{F_r}{f_{ds}} = \frac{800}{89} = 8.99$$

Taking n_c = 10, the load-sharing factor is

$$k_n = 1.18 - \frac{0.18}{\sqrt{n_c}} = 1.18 - \frac{0.18}{\sqrt{10}} = 1.123$$

The design shear capacity of shear connectors in a group is computed as

$$f_{ds} = \phi k_n f_{vs} = 0.85 \times 1.123 \times 89 = 85 \text{ kN}$$

The required number of stud shear connectors is finalized as

$$n_c = \frac{F_r}{f_{ds}} = \frac{800}{85} = 9.4$$

Adopting n_c = 10, the design strength of the shear connection is determined as

$$F_{sh} = n_c f_{ds} = 10 \times 85 = 850 \text{ kN} > F_r = 800 \text{ kN, OK}$$

The total number of stud shear connectors in the negative moment region which is assumed to be symmetric about the support is 20.

9.9 TRANSFER OF LONGITUDINAL SHEAR IN CONCRETE SLABS

9.9.1 Longitudinal shear surfaces

Shear connectors transfer longitudinal shear from the steel beam to the concrete slab in a composite beam. The shear transfer mechanism in the concrete slab can be simulated by either the strut-and-tie model (Liang et al. 2000; Liang 2005) or the shear–friction model. The shear connectors under longitudinal shear induce compressive force on the concrete, which is dispersed through struts and interconnected by tension ties. As a result, longitudinal shear reinforcement (placed perpendicular to the steel beam) must be provided in the concrete slab to resist the tensile forces.

AS 2327.1 (2003) identifies four types of longitudinal shear failure surfaces, which are schematically illustrated in Figures 9.25 through 9.27. As shown in Figure 9.27, Type 4 longitudinal shear failure may occur in composite edge beams with profiled steel sheeting placed perpendicular to the steel beam when the outstand of the composite beam is less than 600 mm and stud shear connectors are welded through the sheeting.

As shown in Figure 9.25, the longitudinal shear failure corresponding to the Type 1 shear surface may occur at the outside faces of shear connector groups, at sections where longitudinal shear reinforcement is terminated or over the sheeting ribs which are parallel to the steel beam. The perimeter length (u_p) of Type 1 shear surfaces is taken as D_c for solid slabs, composite slabs with sheeting ribs perpendicular to the steel beam and for composite slabs between ribs which are parallel to the steel beam. For Type 2 shear surfaces, the perimeter length is determined as $(b_x + 2h_s)$, where b_x is the overall width across the top of connectors in

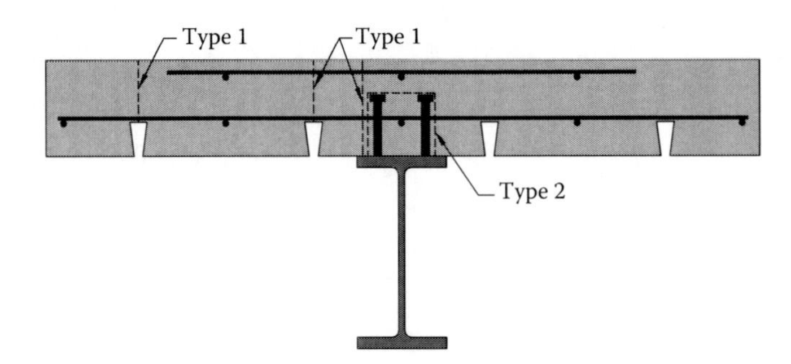

Figure 9.25 Type 1 and 2 longitudinal shear failure surfaces.

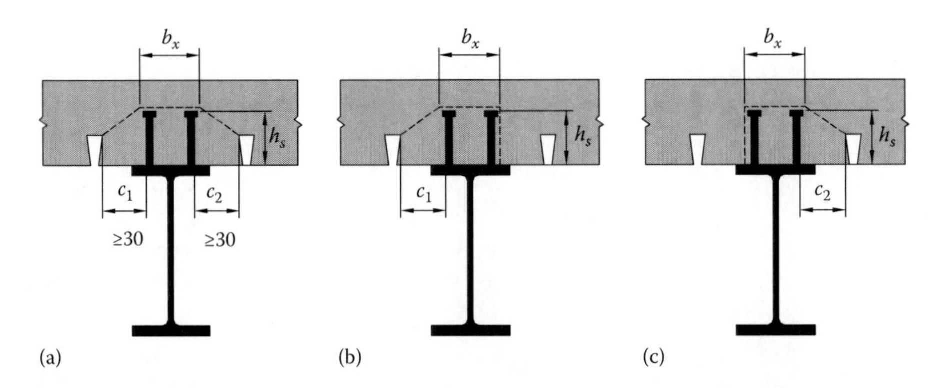

Figure 9.26 Longitudinal shear surfaces: (a) shear surface 1, (b) shear surface 2 and (c) shear surface 3.

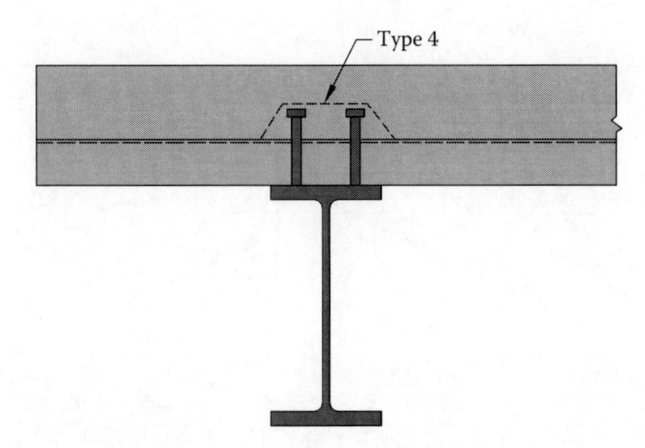

Figure 9.27 Type 4 longitudinal shear surface in edge beam.

the cross section and h_s is the overall height of the shear connectors above the top flange of the steel section. The Type 3 shear surfaces are associated with longitudinal shear failure around the shear connector groups in composite slabs, as illustrated in Figure 9.26. The perimeter length of Type 3 shear surfaces is taken as $u_p = \min(u_1,u_2,u_3)$, which are defined in Figure 9.26.

9.9.2 Design longitudinal shear force

The compressive force in the concrete slab of a composite beam is assumed to be uniformly distributed across the effective width of the concrete flange. This implies that the longitudinal shear flow in the concrete slab is uniform. This uniform shear flow model is used to determine the design longitudinal shear force per unit length (V_L^*) of the composite beam for Type 1, 2 and 3 shear surfaces at the beam cross section. In AS 2327.1 (2003), V_L^* is assumed to vary linearly from zero at the extremities of the effective width of the concrete slab to the maximum on each side of the centre line of the steel beam as shown in Figure 9.28. For Type 1 shear surface, V_L^* is calculated by

$$V_L^* = \left(\frac{x}{b_{cf}}\right) V_{L\cdot tot}^* \tag{9.69}$$

where

x is the distance from the extremity of the effective width to the cross section where the longitudinal shear force is calculated

$V_{L\cdot tot}^*$ is the total design longitudinal shear force per unit length, given by

$$V_{L\cdot tot}^* = \frac{n_x f_{ds}}{s_c} \tag{9.70}$$

where

n_x is the number of connectors in a cross section

f_{ds} is the design shear capacity of shear connectors in the beam

s_c is the longitudinal spacing of shear connectors

For Type 2 and 3 shear surfaces, the compressive force across the concrete slab is transferred by the shear surfaces. Therefore, the design longitudinal shear force acting on Type 2 and 3 surfaces is taken as $V_L^* = V_{L\cdot tot}^*$.

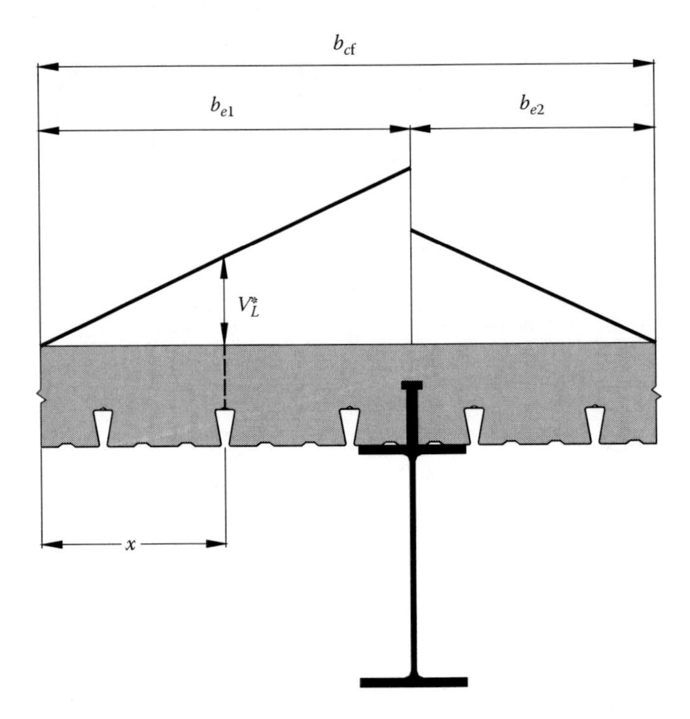

Figure 9.28 Distribution of longitudinal shear force for Type I shear surface.

9.9.3 Longitudinal shear capacity

The shear–friction model for reinforced concrete is adopted in Clause of 9.6 of AS 2327.1 (2003) to calculate the nominal longitudinal shear capacity (per unit length) of Type 1, 2 and 3 shear surfaces, which is taken as the lesser value calculated by the following equations:

$$V_L = u_p\left(0.36\sqrt{f_c'}\right) + 0.9A_{sv}f_{yr} \tag{9.71}$$

$$V_L = 0.32f_c'u_p \tag{9.72}$$

where A_{sv} is the total cross-sectional area of longitudinal shear reinforcement crossing the shear surface (mm^2).

9.9.4 Longitudinal shear reinforcement

It is necessary to ensure that the concrete shear capacity of Type 1, 2 and 3 shear surfaces is not less than the design longitudinal shear force, such as $\phi 0.32f_c'u_p \geq V_L^*$. The total cross-sectional area of longitudinal shear reinforcement for resisting Type 1, 2 and 3 shear surfaces can be determined by using the following equation, respectively:

$$A_{sv} = \frac{V_L^*/\phi - 0.36u_p\sqrt{f_c'}}{0.9f_{yr}} \tag{9.73}$$

The perimeter length u_p is taken as the lesser of the perimeter lengths u_2 and u_3 of Type 2 and 3 shear surfaces in Equation 9.73. However, the larger of the perimeter lengths u_2 and u_3 should be used to calculate the minimum cross-sectional area of shear reinforcement for Type 2 and 3 shear surfaces. Any existing flexural and shrinkage reinforcement placed transverse to the steel beam in the concrete slab can be treated as the effective longitudinal shear reinforcement if they satisfy the anchorage requirement of Clause 9.7.3 of AS 2327.1 (2003). It is noted that the additional reinforcement for Type 1 shear surface depends on the Type 2 and 3 shear reinforcement as well as existing reinforcement in the concrete slab. AS 2327.1 does not give design rules on the spacing of longitudinal shear reinforcement. It is suggested that the maximum spacing of longitudinal shear reinforcement for Type 1, 2 and 3 shear surfaces should be taken as the minimum of $2s_c$, $4D_c$ and 600 mm (Liang and Patrick 2001).

The longitudinal reinforcement must have adequate anchorage length to develop its yield stress. The stress development length of longitudinal reinforcement in concrete slabs given in AS 3600 (2001) is adopted here, which is expressed by

$$L_{yst} = \frac{k_1 k_2 f_{yr} A_b}{(2c + d_b)\sqrt{f'_c}} \geq 25 d_b \tag{9.74}$$

where

$k_1 = 1.0$
$k_2 = 2.4$
d_b is the diameter of the reinforcing bar
A_b is the cross-sectional area of the bar
c is the cover to the reinforcing bars

For bottom face reinforcement in composite slabs, c may be taken as h_r.

The Type 1 longitudinal reinforcement should be extended $12d_b$ from the section where longitudinal reinforcement is not required to resist longitudinal shear.

Special steel reinforcing products have been developed in Australia for use in composite beams as longitudinal shear reinforcement (Liang and Patrick 2001; Liang et al. 2001). These new reinforcing products complement the new design approach to the longitudinal shear in composite beams and have been incorporated in the computer software COMPSHEAR for the design of the shear connection of composite beams (Liang et al. 2001). Waveform reinforcing products DECKMESH can be used in composite edge beams incorporating Bondek II and Condeck HP profiled steel sheeting to prevent rib shearing failure from occurring when the sheeting ribs are placed perpendicular to the steel beam (Liang and Patrick 2001).

The design procedure for determining Type 1, 2 and 3 longitudinal shear reinforcement in the concrete slab of a composite beam is given as follows:

1. Calculate the design shear capacity of shear connectors, which requires the minimum number of shear connectors to be determined.
2. Calculate total design longitudinal shear force per unit length.
3. Calculate the perimeter lengths of Type 1, 2 and 3 longitudinal shear surfaces.
4. Check for the concrete shear capacity of Type 1, 2 and 3 longitudinal shear surfaces, such that $\phi 0.32 f'_c u_p \geq V_L^*$. If this condition is not satisfied, either the perimeter lengths or the concrete compressive strength should be increased and then go back to Step 1.
5. Calculate the cross-sectional areas and lengths of additional longitudinal shear reinforcement for Type 2 and 3 shear surfaces. The cross-sectional area of any fully anchored bottom reinforcement in the concrete slab placed transverse to the longitudinal axis of the steel beam is taken into account.

6. Calculate the design longitudinal shear force per unit length for Type 1 surface at any distance from the extremity of the slab effective width.
7. Calculate the cross-sectional areas and lengths of additional reinforcement for Type 1 shear surface for every shear force V_L^* computed. The cross-sectional area of any fully anchored transverse reinforcement and the additional Type 2 and 3 reinforcement should be taken into consideration.
8. Determine the maximum cross-sectional area and lengths of additional reinforcement for Type 1 shear surface, which is treated as the top reinforcement in the concrete slab.

Example 9.6: Design of shear connection of internal composite beam

The cross section of an internal primary composite beam which is simply supported is schematically depicted in Figure 9.29. The effective span of the composite beam is 8.4 m. The profiled steel sheeting is placed parallel to the steel beam. The steel section 410UB59.7 of Grade 300 steel is used. The design strength of the concrete flange is $f_c' = 32$ MPa. Twenty headed stud shear connectors of 19 mm diameter are uniformly distributed between the end and mid-span of the composite beam. The height of the headed stud is 95 mm. The flexural reinforcement of N10 at 240 mm is placed at the top face of the concrete slab. The SL72 mesh (A_{st} = 179 mm²/m) is placed on the top of the sheeting ribs to provide crack control for shrinkage and temperature effects. The exposure classification is A1. Design the shear connection of the composite beam.

1. Design shear capacity of shear connectors

The nominal shear capacity of 19 mm diameter headed stud embedded in 32 MPa concrete is obtained from Table 9.1 as f_{vs} = 93 kN.

The load-sharing factor is

$$k_n = 1.18 - \frac{0.18}{\sqrt{n_c}} = 1.18 - \frac{0.18}{\sqrt{20}} = 1.14$$

The design shear capacity of a shear connectors in the composite beam is computed as

$$f_{ds} = \phi k_n f_{vs} = 0.85 \times 1.14 \times 93 = 90 \text{ kN}$$

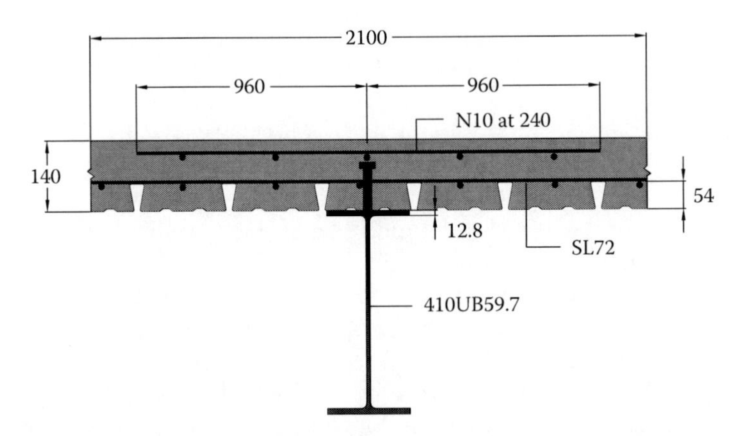

Figure 9.29 Cross section of internal composite beam.

2. Total design longitudinal shear force

The longitudinal spacing of shear connectors is determined as s_c = 213 mm.
The total design longitudinal shear force per unit length of the composite beam is calculated as

$$V^*_{L\cdot tot} = \frac{n_x f_{ds}}{s_c} = \frac{1\times 90\times 1000}{213} = 423 \text{ N/mm}$$

3. Perimeter lengths of shear surfaces

The overall width across the top of connector in the cross section is

$$b_x = d_b = 32 \text{ mm}$$

The perimeter lengths of Type 1 and 2 shear surfaces are computed as follows:

$$u_1 = D_c - h_r = 140 - 54 = 86 \text{ mm}$$

$$u_2 = b_x + 2h_s = 32 + 2\times 95 = 222 \text{ mm}$$

Assume that the stud is placed at the centre of the adjacent ribs, the distance c_1 is

$$c_1 = \frac{s_r - b_x}{2} = \frac{200-32}{2} = 84 \text{ mm}$$

$$u_3 = b_x + 2\sqrt{(h_s - h_r)^2 + c_1^2} = 32 + 2\times\sqrt{(95-54)^2 + 84^2} = 219 \text{ mm}$$

Assume c_1 = 30 mm:

$$u_3 = b_x + \sqrt{(h_s - h_r)^2 + c_1^2} + h_s = 32 + \sqrt{(95-54)^2 + 30^2} + 95 = 178 \text{ mm} < 219 \text{ mm}$$

Hence,

$$u_3 = \min(219, 178) = 178 \text{ mm}$$

4. Check for the concrete shear capacity

The design longitudinal shear force per unit length of the beam acting on Type 1 shear surface is

$$V^*_L = \left(\frac{x}{b_{cf}}\right) V^*_{L\cdot tot} = \left(\frac{1050}{2100}\right)\times 423 = 211.5 \text{ N/mm}$$

The design shear capacity of the concrete for Type 1 shear surface is calculated as

$$\phi V_L = \phi 0.32 f'_c u_p = 0.7\times 0.32\times 32\times 86 = 616.4 \text{ N/mm} > V^*_L = 211.5 \text{ N/mm, OK}$$

The minimum perimeter length of Type 2 and 3 shear surfaces is

$$u_p = \min(u_2, u_3) = \min(222, 178) = 178 \text{ mm}$$

The design shear capacity of the concrete for Type 2 and 3 shear surfaces is

$$\phi V_L = \phi 0.32 f'_c u_p = 0.7\times 0.32\times 32\times 178 = 1276 \text{ N/mm} > V^*_{L\cdot tot} = 423 \text{ N/mm, OK}$$

5. Additional type 2 and 3 longitudinal shear reinforcement

Since $u_2 > u_3$, u_p is taken as u_3 which is used to calculate the total area of shear reinforcement per unit length for Type 2 and 3 shear surfaces as follows:

$$A_{sv} = \frac{V_L^*/\phi - 0.36u_p\sqrt{f_c'}}{0.9f_{yr}} = \frac{423/0.7 - 0.36\times 178\sqrt{32}}{0.9\times 500}$$

$$= 0.537 \text{ mm}^2/\text{mm} = 537 \text{ mm}^2/\text{m}$$

The minimum area of longitudinal shear reinforcement is computed as

$$A_{sv\cdot\min} = \frac{800u_p}{f_{yr}} = \frac{800\times 222}{500} = 355 \text{ mm}^2/\text{m} < A_{sv} = 537 \text{ mm}^2/\text{m}$$

The required additional Type 3 reinforcement is

$$A_{sb\cdot a} = \frac{537}{2} - 179 = 89.5 \text{ mm}^2/\text{m}$$

The spacing limit on longitudinal shear reinforcement is

$$s_{b\cdot\max} = \min(2s_c, 4D_c, 600) = \min(2\times 213, 4\times 140, 600) = 426 \text{ mm}$$

Use N10 at 400 ($A_{sb\cdot a}$ = 196 mm²/m).
The development length of Type 2 and 3 reinforcement is taken as

$$L_{yst} = 25d_b = 250 \text{ mm}$$

The length of Type 3 reinforcement is computed as

$$L_{ab\cdot 1} = L_{ab\cdot 2} = 0.5b_x + c_1 + L_{sy\cdot t} = 0.5\times 32 + 30 + 250 = 296 \text{ mm}$$

Take $L_{ab\cdot 1} = L_{ab\cdot 2}$ = 300 mm.

6. Additional type 1 longitudinal shear reinforcement

The total area of longitudinal shear reinforcement per unit length for Type 1 shear surface is calculated as follows:

$$A_{sv} = \frac{V_L^*/\phi - 0.36u_p\sqrt{f_c'}}{0.9f_{yr}} = \frac{211.5/0.7 - 0.36\times 86\sqrt{32}}{0.9\times 500}$$

$$= 0.282 \text{ mm}^2/\text{mm} = 282 \text{ mm}^2/\text{mm}$$

Existing flexural reinforcement at the top face of the concrete slab N10 at 240: A_{st} = 327 mm²/m

The SL72 mesh: A_{sb} = 179 mm²/m

The required additional Type 1 reinforcement is calculated as

$$A_{sb\cdot a} = 282 - 179 - 327 = -244 \text{ mm}^2/\text{m}$$

Therefore, no additional reinforcement is required for the Type 1 shear surfaces near the shear connector.

The design longitudinal shear force at section where Type 1 reinforcement is not required can be calculated as

$$V_L^* = \phi 0.36 u_p \sqrt{f_c'} = 0.7 \times 0.36 \times 86 \times \sqrt{32} = 122.6 \text{ N/mm}$$

The distance between the extremity of the effective width and the Type 1 shear plane is

$$x = \left(\frac{V_L^*}{V_{L\cdot tot}^*} \right) b_{cf} = \left(\frac{122.6}{423} \right) \times 2100 = 609 \text{ mm}$$

The distance from this shear plane to the vertical centroidal axis of the steel beam is

$$x_c = 1050 - 609 = 441 \text{ mm}$$

The length of the effective reinforcement measured from the centre line of the steel beam is

$$x_c + 12 d_b = 441 + 12 \times 10 = 561 \text{ mm} < L_{t\cdot 1} = L_{t\cdot 2} = 960 \text{ mm}$$

Therefore, the flexural reinforcement (N10 at 240) placed at the top face of the concrete slab is adequate for resisting Type 1 shear failure.

Example 9.7: Design of shear connection of edge composite beam

The cross section of a secondary edge composite beam which is simply supported is schematically depicted in Figure 9.30. The effective span of the composite beam is 6 m. The profiled steel sheeting is placed parallel to the steel beam. The steel section 410UB59.7 of Grade 300 steel is used. The design strength of the concrete flange is $f_c' = 32$ MPa. Twenty-two headed stud shear connectors of 19 mm diameter are uniformly distributed in pairs between the end and mid-span of the composite beam. The height of the headed stud is 95 mm. The flexural reinforcement of N10 at 300 mm is placed at the top face of the concrete slab. The exposure classification is A1. Design the shear connection of the composite beam.

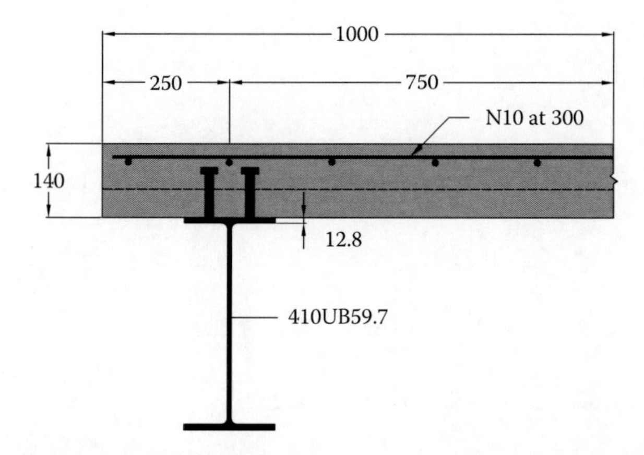

Figure 9.30 Cross section of edge composite beam.

1. Design shear capacity of shear connectors

The nominal shear capacity of 19 mm diameter headed stud embedded in 25 MPa concrete is obtained from Table 9.1 as f_{vs} = 89 kN.

The load-sharing factor is

$$k_n = 1.18 - \frac{0.18}{\sqrt{n_c}} = 1.18 - \frac{0.18}{\sqrt{22}} = 1.14$$

The design shear capacity of a shear connectors in the composite beam is computed as

$$f_{ds} = \phi k_n f_{vs} = 0.85 \times 1.14 \times 89 = 86.24 \text{ kN}$$

2. Total design longitudinal shear force

The total design longitudinal shear force per unit length of the composite beam is calculated as

$$V^*_{L\cdot tot} = \frac{n_x f_{ds}}{s_c} = \frac{2 \times 86.24 \times 1000}{300} = 575 \text{ N/mm}$$

3. Perimeter lengths of shear surfaces

The overall width across the top of connector in the cross section is

$$b_x = s_x + d_h = 80 + 32 = 112 \text{ mm}$$

The perimeter lengths of Type 1 and 2 shear surfaces are computed as follows:

$$u_1 = D_c = 140 \text{ mm}$$

$$u_2 = b_x + 2h_s = 112 + 2 \times 95 = 302 \text{ mm}$$

4. Check for the concrete shear capacity

For Type 1 shear surface, $x = b_{e2} - s_x/2 - d_h/2 = 750 - 80/2 - 32/2 = 694$ mm.

The design longitudinal shear force per unit length of the beam acting on Type 1 shear surface is

$$V^*_L = \left(\frac{x}{b_{cf}}\right) V^*_{L\cdot tot} = \left(\frac{694}{1000}\right) \times 575 = 399 \text{ N/mm}$$

The design shear capacity of the concrete for Type 1 shear surface is calculated as

$$\phi V_L = \phi 0.32 f'_c u_p = 0.7 \times 0.32 \times 25 \times 140 = 784 \text{ N/mm} > V^*_L = 399 \text{ N/mm, OK}$$

The design shear capacity of the concrete for Type 2 shear surface is

$$\phi V_L = \phi 0.32 f'_c u_p = 0.7 \times 0.32 \times 25 \times 302 = 1691.2 \text{ N/mm} > V^*_{L\cdot tot} = 575 \text{ N/mm, OK}$$

5. Type 2 longitudinal shear reinforcement

The total area of shear reinforcement per unit length for Type 2 shear surface is calculated as

$$A_{sv} = \frac{V_L^*/\phi - 0.36u_p\sqrt{f_c'}}{0.9f_{yr}} = \frac{575/0.7 - 0.36 \times 302 \times \sqrt{25}}{0.9 \times 500}$$

$$= 0.617 \text{ mm}^2/\text{mm} = 617 \text{ mm}^2/\text{m}$$

The minimum area of longitudinal shear reinforcement is computed as

$$A_{sv\cdot\min} = \frac{800u_p}{f_{yr}} = \frac{800 \times 302}{500} = 483 \text{ mm}^2/\text{m} < A_{sv} = 617 \text{ mm}^2/\text{m}$$

The required additional Type 2 reinforcement is

$$A_{sb\cdot a} = \frac{617}{2} = 308.5 \text{ mm}^2/\text{m}$$

The spacing limit on longitudinal shear reinforcement is

$$s_{b\cdot\max} = \min(2s_c, 4D_c, 600) = \min(2 \times 300, 4 \times 140, 600) = 560 \text{ mm}$$

Use N10 at 250 ($A_{sb\cdot a}$ = 314 mm²/m).
The development length for Type 2 reinforcement is taken as $25d_b$ = 250 mm.
The length of Type 2 reinforcement is computed as

$$L_{ab\cdot 2} = 0.5b_x + L_{sy\cdot t} = 0.5 \times 112 + 250 = 306 \text{ mm and take } L_{ab\cdot 2} = 310 \text{ mm.}$$

Since $L_{ab\cdot 1}$ = 310 mm > 250–20 = 230 mm, use U-bars.
Hence, use N10 at 250 U-bars, $L_{ab\cdot 1}$ = 230 mm and $L_{ab\cdot 2}$ = 310 mm.

6. Additional type 1 longitudinal shear reinforcement

The total area of longitudinal shear reinforcement per unit length for Type 1 shear surface is calculated as follows:

$$A_{sv} = \frac{V_L^*/\phi - 0.36u_p\sqrt{f_c'}}{0.9f_{yr}} = \frac{399/0.7 - 0.36 \times 140 \times \sqrt{25}}{0.9 \times 500}$$

$$= 0.707 \text{ mm}^2/\text{mm} = 707 \text{ mm}^2/\text{mm}$$

Existing reinforcement at the top face of the concrete slab N10 at 300: A_{st} = 262 mm²/m
The required additional Type 1 reinforcement is calculated as

$$A_{sb\cdot a} = 707 - 262 - 314 = 131 \text{ mm}^2/\text{m}$$

Use N10 at 500 ($A_{st\cdot a}$ = 157 mm²/m).

The design longitudinal shear force at section where Type 1 reinforcement is not required can be calculated as

$$V_L^* = \phi 0.36 u_p \sqrt{f_c'} = 0.7 \times 0.36 \times 140 \times \sqrt{25} = 176.4 \text{ N/mm}$$

The distance between the extremity of the effective width and the Type 1 shear plane is

$$x = \left(\frac{V_L^*}{V_{L\cdot tot}^*} \right) b_{cf} = \left(\frac{176.4}{575} \right) \times 1000 = 307 \text{ mm}$$

The distance from this shear plane to the vertical centroidal axis of the steel beam is

$$x_c = 750 - 307 = 443 \text{ mm}$$

The length of the effective reinforcement measured from the centre line of the steel beam is

$$L_{at\cdot 2} = x_c + 12 d_b = 443 + 12 \times 10 = 563 \text{ mm}$$

Hence, use N10 at 500 ($A_{st\cdot a}$ = 157 mm²/m) $L_{at\cdot 1}$ = 230,$L_{at\cdot 2}$ = 565 mm as the additional Type 1 reinforcement.

9.10 COMPOSITE BEAMS WITH PRECAST HOLLOW CORE SLABS

Composite beams with precast hollow core concrete slabs depicted in Figure 9.31 are commonly used in the United Kingdom as alternatives to composite beams incorporating profiled steel sheeting. The main advantages of this form of composite beam construction are (a) precast concrete slabs can span up to 15 m without propping, (b) the erection of the precast concrete slab units are simple and (c) the pre-welding of stud connectors on the steel beams leads to rapid construction (Lam 2002). The depth of the precast hollow core slabs is usually between 150 and 400 mm.

The design moment capacity of composite beams incorporating precast hollow core slabs with complete or partial shear connection can be determined by the plastic stress

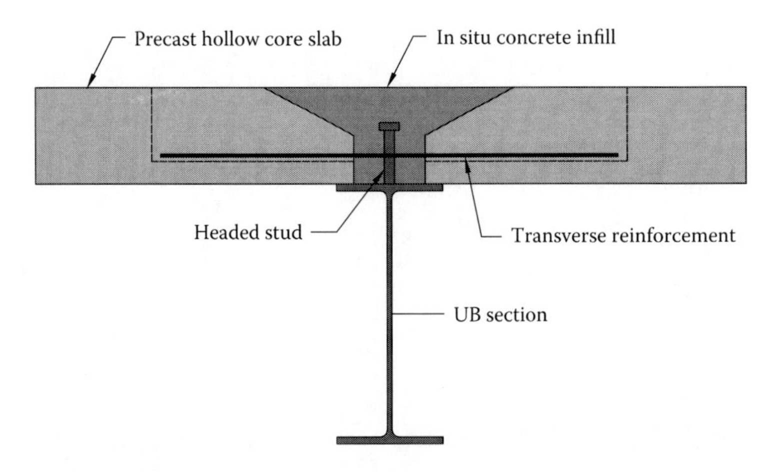

Figure 9.31 Composite beam with precast hollow core slab.

distributions, providing that an appropriate effective width for the concrete flange is used. The effective width of the precast hollow core concrete flange is influenced by the strength of concrete and the transverse reinforcement (Lam et al. 2000a). Based on the results obtained from experiments (Lam et al. 2000b), the effective width of the precast hollow core concrete slab is given by (Lam et al. 2000a)

$$b_{cf} = 1000\left(\frac{25}{f_{cu}}\right)^2\left(\frac{0.4}{f_t'}\right) + 300 \tag{9.75}$$

where

f_{cu} is the compressive concrete cube strength of the in situ concrete infill (MPa)
f_t' is the effective tensile strength and is determined as $f_t' = A_{st}f_{yr}/A_c$, where A_{st} is the area of transverse reinforcement and A_c is the cross-sectional area of concrete

For simplicity, the effective width of the precast hollow core slab (b_{cf}) can be taken as span/5.

Push-out tests indicate that the shear strength of shear connectors in composite beams with precast hollow core slabs is influenced by the in situ concrete gap width, the transverse joints between hollow core slabs, the strength of concrete and the amount of transverse reinforcement (Lam et al. 2000a). The nominal shear capacity of headed stud shear connectors in composite precast hollow core slabs is taken as the lesser of the values calculated using the following equations based on Eurocode 4 and push-out test results (Lam et al. 2000a):

$$f_{vs} = 0.29\alpha_1\alpha_2\alpha_3 d_{bs}^2\sqrt{\varpi f_{cj}' E_c} \tag{9.76}$$

$$f_{vs} = 0.8 f_u\left(\frac{\pi d_{bs}^2}{4}\right) \tag{9.77}$$

where

α_1 is the factor which accounts for the effect of the height of stud and is expressed as $\alpha_1 = 0.2(h_s/d_{bs} + 1) \leq 1.0$
α_2 is the factor considering the effect of the in situ infill gap (g) between the hollow core slabs and is given by $\alpha_2 = 0.5(g/70 + 1) \leq 1.0$ with $g \geq 30$
α_3 is used to take into account the effect of the diameter (d_b) of the transverse reinforcement and is determined by $\alpha_3 = 0.5(d_b/20 + 1) \leq 1.0$
ϖ is the transverse joint factor and is taken as $\varpi = 0.5(b_{hcs}/600 + 1) \leq 1.0$, where b_{hcs} is the width of the hollow core slab

9.11 DESIGN FOR SERVICEABILITY

9.11.1 Elastic section properties

The elastic section properties of composite beam cross sections with complete shear connection are calculated by using the transformed section method. For this purpose, the full interaction between the concrete slab and the steel beam is assumed. The effective section of a composite beam should be used in the calculation of its elastic section properties. The tensile strength of concrete is ignored. Figure 9.32 shows the transformed section of a composite beam, which is an equivalent steel section. The transformed effective width of the concrete

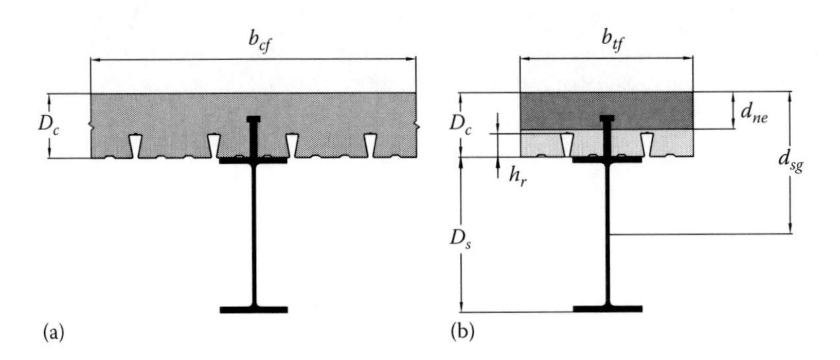

Figure 9.32 Transformed section of composite beam: (a) effective section and (b) transformed section.

flange is determined as $b_{tf} = b_{cf}/n$. The modular ratio (n) is calculated as $n = E_s/E_{ce}$, where E_{ce} is the effective modulus of concrete. When calculating the immediate deflections of a composite beam and the maximum stresses in the steel beam using the second moment of area (I_{ti}), E_{ce} is taken as E_c. For the long-term deflection calculations using I_{tl}, E_{ce} is calculated by

$$E_{ce} = \frac{E_c}{1+\phi_c^*} \tag{9.78}$$

where the concrete creep factor $\phi_c^* = 2$ is used in AS 2327.1 (2003).

The elastic neutral axis of the transformed section is located either in the concrete slab or in the steel section. The depth of the elastic neutral axis of the transformed section can be determined by taking the first moment of area about the elastic neutral axis. If the elastic neutral axis lies in the concrete cover slab of a composite slab or a solid slab as depicted in Figure 9.32, the depth of the elastic neutral axis of the composite section with complete shear connection is determined as

$$(b_{tf}d_{ne}) \times \frac{d_{ne}}{2} = A_s(d_{sg} - d_{ne}) \tag{9.79}$$

where

A_s is the total effective area of the steel section

d_{sg} is the distance from the centroid of the effective steel section to the top of the concrete slab

The elastic neutral axis depth (d_{ne}) can be obtained from the aforementioned equation as

$$d_{ne} = \sqrt{c_a^2 + 2c_a d_{sg}} - c_a \tag{9.80}$$

where $c_a = A_s/b_{tf}$.

The second moment of area of the transformed section can be calculated by taking moment of areas about the elastic neutral axis as

$$I_t = \frac{b_{tf}d_{ne}^3}{3} + I_s + A_s(d_{sg} - d_{ne})^2 \tag{9.81}$$

When the elastic neutral axis is located in the steel ribs of the composite slab with $\lambda = 0$ or in the steel section of the composite beam with $\lambda = 0$, the depth (d_{ne}) of the elastic neutral axis and the second moment of area (I_t) are given as follows:

$$d_{ne} = \frac{b_{tf} h_c (h_c/2) + A_s d_{sg}}{b_{tf} h_c + A_s} \tag{9.82}$$

$$I_t = \frac{b_{tf} h_c^3}{12} + b_{tf} h_c \left(d_{ne} - \frac{h_c}{2} \right)^2 + I_s + A_s (d_{sg} - d_{ne})^2 \tag{9.83}$$

For a composite beam with partial shear connection at the cross section of maximum bending, the effective second moments of area are given in AS 2327.1 (2003) as follows:

$$I_{eti} = I_{ti} - 0.6(1 - \beta_{mb})(I_{ti} - I_s) \tag{9.84}$$

$$I_{etl} = I_{tl} - 0.6(1 - \beta_{mb})(I_{tl} - I_s) \tag{9.85}$$

where β_{mb} is the degree of shear connection at the cross section under the maximum bending moment.

9.11.2 Deflection components of composite beams

The deflections of a composite beam include the immediate deflections of the composite beam under construction loads during various construction stages and under short-term in-service loads and its long-term deflections due to creep and shrinkage during in-service conditions. The exact calculation of deflections of composite beams is complex. The reasons for this are as follows: (1) the change of loads during the life of the structure cannot be predicted in the design stage, (2) the structural model may not adequately account for 3D effects of the structure, (3) the non-linear load–slip behaviour of shear connection is usually ignored and (4) the modulus of elasticity for the concrete changes with time due to creep and shrinkage (Viest et al. 1997). If the spans are large, a large portion of live load is present over a long period of time, or if the concrete used for the slab is sensitive to creep and shrinkage, the long-term deflections due to creep and shrinkage need to be taken into account.

A simplified method for calculating the deflections of composite beams is suggested in AS 2327.1. The components of deflection of a composite beam and the corresponding design loads are described in AS 2327.1 (2003) as follows:

1. Immediate deflection ($\delta_{C1\text{-}3}$) of steel beam during construction stages 1–3 under design loads ($G_{C1\text{-}3}$), which include the weight of the steel beam, formwork, concrete and reinforcement
2. Immediate deflection ($\delta_{C5\text{-}6}$) of composite beam during construction stages 5–6 under design loads, which include dead loads ($G_{C1\text{-}3}$) and superimposed dead loads (G_{sup})
3. Immediate deflection (δ_Q) of composite beam during in-service condition under short-term live load ($\psi_s Q$)
4. Long-term deflection (δ_{cr}) of composite beam due to concrete creep under service loads, which include dead loads (G_{sup}), long-term live load ($\psi_l Q$) and for propped construction, ($G_{C1\text{-}3}$)
5. Long-term deflection (δ_{sh}) of composite beam due to the shrinkage of concrete during in-service condition

9.11.3 Deflections due to creep and shrinkage

The long-term deflection of a composite beam due to concrete creep can be calculated using the long-term section properties of its transformed section. However, the long-term deflections thus calculated include the immediate deflection due to the superimposed dead load G_{sup} and long-term live load $\psi_l Q$ and, if propped, $G_{C1\cdot3}$. Therefore, the deflection component δ_{cr} due to creep has to be computed by subtracting the immediate deflection due to these loads from the long-term deflection as specified in AS 2327.1.

The final free shrinkage strain in unrestrained concrete given in AS 3600 (2001) is between 300 and 1100 microstrain. Unlike free shrinkage, the shrinkage of concrete in composite beams is restrained by the steel beams through shear connectors. The shrinkage of concrete causes contraction, which is resisted by shear connectors. The contraction of the concrete due to shrinkage induces deflections and flexural stresses which are in the same direction as those caused by gravity loads (Oehlers and Bradford 1999). The mid-span deflections of typical simply supported composite beams are within the limit of $L/750$ (Alexander 2003).

Figure 9.33 presents part of a composite beam. The deformation of the concrete due to shrinkage strain is represented by an external compressive force (N_{sh}) acting at the centroid of the concrete slab (Viest et al. 1958; Chien and Ritchie 1984). This force acting eccentrically to the elastic neutral axis of the transformed composite section induces a bending moment applied at the end of the beam. The axial force N_{sh} induced by the shrinkage of concrete is expressed by

$$N_{sh} = E_{ce}\varepsilon_{sh}A_c \tag{9.86}$$

where

ε_{sh} is the restrained shrinkage strain of concrete in the composite beam
A_c is the effective cross-sectional area of the concrete slab

The restrained shrinkage stain (ε_{sh}) of concrete may be taken as $0.8\varepsilon_{cs}^*$ (Alexander 2003), where ε_{cs}^* is the final free shrinkage strain of concrete estimated in accordance with AS 3600 (2001).

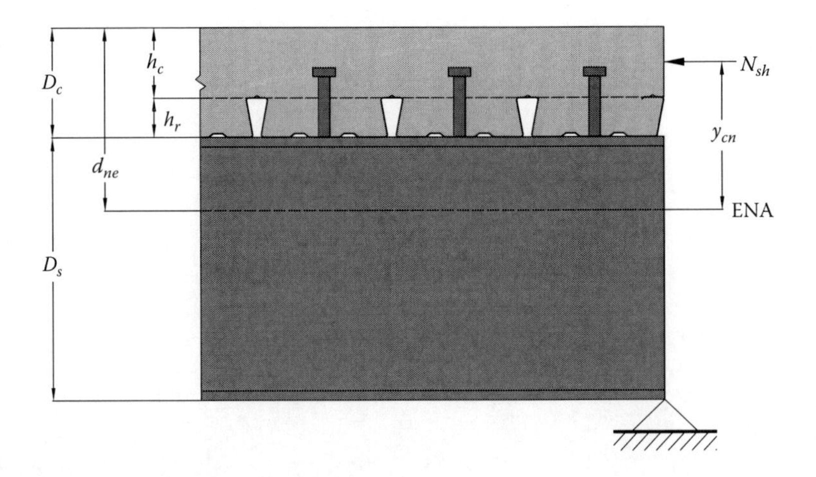

Figure 9.33 Equivalent external force for shrinkage.

As shown in Figure 9.33, the eccentricity of the axial force N_{sh} is $y_{cn} = (d_{ne} - h_c/2)$. The bending moment induced by the shrinkage of concrete is determined by

$$M_{sh} = N_{sh}\left(d_{ne} - \frac{h_c}{2}\right) \quad (9.87)$$

where
d_{ne} is the depth of the elastic neutral axis of the transformed section determined using the modular ratio of $n = 3E_s/E_c$
h_c is the thickness of the concrete slab above the steel ribs

The shrinkage of concrete produces a constant bending moment M_{sh} over the entire length of the composite beam. The deflection of the simply supported composite beam with complete shear connection due to shrinkage is calculated by

$$\delta_{sh} = \frac{M_{sh}L^2}{8E_s I_{tl}} \quad (9.88)$$

9.11.4 Maximum stress in steel beam

When the simplified method given in AS 2327.1 is used to calculate the deflections of composite beams, the maximum stress in the steel beam during construction stages 1–6 and during in-service condition must not exceed $0.9f_y$. During construction stages 1–3, before the development of composite action, the maximum stress in steel beam under load combination of $G + Q$ is calculated separately for each construction stage. During construction stages 5–6, the maximum stress in the steel section of the composite beam is calculated by considering the stress caused by design loads $G_{C1\text{-}3}$ during construction stages 1–3 and the stress induced by the load combination of $G_{sup} + Q$ acting on the composite beam. During in-service condition, the additional stress in steel beam of the composite beam under short-term live load $\psi_s Q$ is calculated by assuming complete shear connection. The stress in the steel beam should be computed using the elastic section moduli of the steel beam or composite beam as appropriate. At cross sections with $\beta < 0.4$, the composite action should be ignored and the section moduli of the steel beam should be used.

Example 9.8: Deflection of simply supported composite beam

Check for the deflections of the simply supported composite beam with complete shear connection presented in Example 9.1 and with partial shear connection presented in Example 9.2, respectively. The composite beam is propped during construction and the props are removed at the end of construction stage 5. The partitions are installed after the props are removed.

1. Deflection of composite beam with complete shear connection

1.1. Short-term section properties

Young's modulus of concrete is calculated as

$$E_c = 0.043\rho_c^{1.5}\sqrt{f'_{cj}} = 0.043 \times 2{,}400^{1.5} \times \sqrt{32} = 28{,}600\,\text{MPa}$$

The modular ratio for calculating short-term section properties is given by

$$n = \frac{E_s}{E_c} = \frac{200{,}000}{28{,}600} = 6.993$$

The transformed effective width of the concrete flange is

$$b_{tf} = \frac{b_{cf}}{n} = \frac{2000}{6.993} = 286 \text{ mm}$$

The geometric parameters are computed as follows:

$$h_c = D_c - h_r = 120 - 55 = 65 \text{ mm}$$

$$A_s = 2b_{f1}t_{f1} + d_w t_w = 2 \times 171 \times 11.5 + (356 - 2 \times 11.5) \times 7.3 = 6364 \text{ mm}^2$$

$$I_s = 139.2 \times 10^6 \text{ mm}^4$$

$$d_{sg} = D_c + \frac{D_s}{2} = 120 + \frac{356}{2} = 298 \text{ mm}$$

Assume the elastic neutral axis is located in the steel rib.

The depth of the elastic neutral axis is computed as

$$d_{ne} = \frac{b_{tf}h_c(h_c/2) + A_s d_{sg}}{b_{tf}h_c + A_s} = \frac{286 \times 65 \times (65/2) + 6364 \times 298}{286 \times 65 + 6364} = 100.2 \text{ mm}$$

Since $h_c < d_{ne} < D_c$, the elastic neutral axis lies in the steel ribs.

The second moments of area are calculated as

$$I_t = \frac{b_{tf}h_c^3}{12} + b_{tf}h_c\left(d_{ne} - \frac{h_c}{2}\right)^2 + I_s + A_s(d_{sg} - d_{ne})^2$$

$$= \frac{286 \times 65^3}{12} + 286 \times 65 \times \left(100.2 - \frac{65}{2}\right)^2 + 139.2 \times 10^6 + 6364 \times (298 - 100.2)^2$$

$$= 480 \times 10^6 \text{ mm}^4$$

$$I_{ti} = I_t = 480 \times 10^6 \text{ mm}^4$$

1.2. Long-term section properties

The effective modulus of concrete is

$$E_{ce} = \frac{E_c}{1 + \phi_c^*} = \frac{28{,}600}{1 + 2} = 9{,}533 \text{ MPa}$$

The modular ratio is

$$n = \frac{E_s}{E_{ce}} = \frac{200{,}000}{9{,}533} = 20.98$$

The transformed effective width of the concrete flange is

$$b_{tf} = \frac{b_{cf}}{n} = \frac{2000}{20.98} = 95.33 \text{ mm}$$

For the long-term transformed section, the elastic neutral axis is located in the steel section. The elastic neutral axis depth and I_{tl} are obtained as follows:

$$d_{ne} = 167 \text{ mm}, \quad I_{tl} = 363 \times 10^6 \text{ mm}^4$$

1.3. Deflection calculation

a. Immediate Deflection during Construction Stages 1–3
   Since the composite beam is propped during construction, $\delta_{C1\text{-}3} = 0$.
b. Immediate Deflection during Construction Stages 5–6
   During construction stages 5–6, the props are removed and superimposed dead load is added.

   The loading: $w = G_{C1\text{-}3} + G_{\text{sup}} = 13.3$ kN/m

   The immediate deflection is calculated as

$$\delta_{C5\text{-}6} = \frac{5}{384}\frac{wL^4}{E_s I_{ti}} = \frac{5}{384} \times \frac{13.3 \times 8000^4}{200 \times 10^3 \times 480 \times 10^6} = 7.39 \text{ mm}$$

c. Immediate Deflection during In-Service Condition
   The short-term live load: $w = \psi_s Q = 0.7 \times 12.8 = 8.96$ kN/m
   The deflection of composite beam under short-term live load is

$$\delta_Q = \frac{5}{384}\frac{wL^4}{E_s I_{ti}} = \frac{5}{384} \times \frac{8.96 \times 8000^4}{200 \times 10^3 \times 480 \times 10^6} = 4.98 \text{ mm}$$

d. Long-Term Deflection due to Creep
   The long-term service load is

$$w = G_{C1\text{-}3} + G_{\text{sup}} + \psi_l Q = 13.3 + 0.4 \times 12.8 = 18.42 \text{ kN/m}$$

   The long-term deflection due to creep is calculated as

$$\delta_{cr} = \frac{5}{384}\frac{wL^4}{E_s}\left[\frac{1}{I_{tl}} - \frac{1}{I_{ti}}\right] = \frac{5}{384} \times \frac{18.42 \times 8000^4}{200 \times 10^3}\left[\frac{1}{363 \times 10^6} - \frac{1}{480 \times 10^6}\right] = 3.3 \text{ mm}$$

e. Long-Term Deflection due to Shrinkage
   The final free shrinkage strain of concrete for the hypothetical thickness $t_h = 120$ mm of the composite beam in a near-coastal region can be obtained from AS 3600 as $\varepsilon_{cs}^* = 544 \times 10^{-6}$.

   The restrained shrinkage strain of concrete is estimated as

$$\varepsilon_{sh} = 0.8\varepsilon_{cs}^* = 0.8 \times 544 \times 10^{-6} = 435 \times 10^{-6}$$

   The axial force in the concrete component due to shrinkage is

$$N_{sh} = E_{ce}\varepsilon_{sh}A_c = 9533 \times 435 \times 10^{-6} \times 2000 \times 65 \times 10^{-3} = 539.3 \text{ kN}$$

   The moment induced by shrinkage is

$$M_{sh} = N_{sh}\left(d_{ne} - \frac{h_c}{2}\right) = 539.3 \times \left(167 - \frac{65}{2}\right) \times 10^{-3} = 72.5 \text{ kNm}$$

The long-term deflection due to shrinkage is computed as

$$\delta_{sh} = \frac{M_{sh}L^2}{8E_s I_{ti}} = \frac{72.5 \times 10^6 \times 8000^2}{8 \times 200 \times 10^3 \times 480 \times 10^6} = 7.99 \text{ mm}$$

f. Total and Incremental Deflections
The total deflection of the composite beam is

$$\delta_{tot} = \delta_{C1\text{-}3} + \delta_{C5\text{-}6} + \delta_Q + \delta_{cr} + \delta_{sh} = 0 + 7.39 + 4.98 + 3.3 + 7.99$$

$$= 23.7 \text{ mm} < \frac{L}{250} = 32 \text{ mm, OK}$$

The incremental deflection of the composite beam is

$$\delta_{inc} = \delta_{Qi} + \delta_{cr} + 0.6\delta_{sh} = 4.98 + 3.3 + 0.6 \times 7.99 = 13.1 \text{ mm} < \frac{L}{500} = 16 \text{ mm, OK}$$

Therefore, the composite beam with complete shear connection satisfies the deflection limits.

2. Deflection of composite beam with partial shear connection

2.1. Elastic section properties

The composite beam presented in Example 9.2 was designed with $\beta = 0.6$ but was provided with 14 headed stud shear connectors between the end and mid-span of the beam. As this is more than required, the actual degree of shear connection needs to be determined. From Example 9.2, we obtain

$$n_c = 14, \quad f_{ds} = 89.5 \text{ kN}, \quad F_{cc} = 1957.8 \text{ kN}$$

The actual compressive force in the concrete slab is

$$F_{cp} = 14 \times 89.5 = 1253 \text{ kN}$$

The degree of shear connection at maximum bending moment is

$$\beta_{mb} = \frac{F_{cp}}{F_{cc}} = \frac{1253}{1957.8} = 0.64$$

The effective second moments of area of the composite beam cross section with $\beta_{mb} = 0.64$ are calculated as follows:

$$I_{eti} = I_{ti} - 0.6(1 - \beta_{mb})(I_{ti} - I_s)$$

$$= 480 \times 10^6 - 0.6 \times (1 - 0.64) \times (480 - 139.2) \times 10^6 = 406 \times 10^6 \text{ mm}^4$$

$$I_{etl} = I_{tl} - 0.6(1 - \beta_{mb})(I_{tl} - I_s)$$

$$= 363 \times 10^6 - 0.6 \times (1 - 0.64) \times (363 - 139.2) \times 10^6 = 315 \times 10^6 \text{ mm}^4$$

2.2. Deflection calculation

The deflection components are calculated using the same loading components given in the preceding section as

$$\delta_{C1\text{-}3} = 0, \quad \delta_{C5\text{-}6} = 8.73\text{ mm}, \quad \delta_{Qi} = 6.11\text{ mm}, \quad \delta_{cr} = 3.5\text{ mm}, \quad \delta_{sh} = 9.21\text{ mm}$$

The total deflection of the composite beam is

$$\delta_{tot} = \delta_{C1\text{-}3} + \delta_{C5\text{-}6} + \delta_Q + \delta_{cr} + \delta_{sh} = 0 + 8.73 + 6.11 + 3.5 + 9.21$$

$$= 27.56\text{ mm} < \frac{L}{250} = 32\text{ mm, OK}$$

The incremental deflection of the composite beam is

$$\delta_{inc} = \delta_Q + \delta_{cr} + 0.6\delta_{sh} = 6.11 + 3.5 + 0.6 \times 9.21 = 15.14\text{ mm} < \frac{L}{500} = 16\text{ mm, OK}$$

Therefore, the composite beam with $\beta_{mb} = 0.64$ satisfies the deflection limits.

3. Maximum stress in steel beam

Since the deflection is calculated using the simplified method, the maximum stresses in the steel beam need to be checked. Consider the beam during the in-service condition, the loading is

$$w = G_{C1\text{-}3} + G_{\text{sup}} + \psi_s Q = 11.7 + 0.7 \times 12.8 = 20.66\text{ kN/m}$$

The maximum bending moment under this service load is

$$M = \frac{wL^2}{8} = \frac{20.66 \times 8^2}{8} = 165.28\text{ kN m}$$

The section modulus of the composite section is computed using its short-term section property I_{ti} and assuming full interaction as

$$Z_b = \frac{I_{ti}}{y_{\max}} = \frac{480 \times 10^6}{120 + 356 - 100.2} = 1.28 \times 10^6\text{ mm}^3$$

The maximum stress at the bottom fibre of the steel beam is determined as

$$\sigma_{\max} = \frac{M}{Z_b} = \frac{165.28 \times 10^6}{1.28 \times 10^6} = 129\text{ MPa} < 0.9 f_y = 0.9 \times 300 = 270\text{ MPa, OK}$$

REFERENCES

Adekola, A.O. (1968) Effective width of composite beams of steel and concrete, *The Structural Engineer*, 9: 285–289.

AISC-LRFD Specification (1994) Load and resistance factor design specification for structural steel buildings, Chicago, IL: American Institute of Steel Construction.

Alexander, S. (2003) How concrete shrinkage affects composite steel beams, *New Steel Construction*, Technical, May 1, 2003, pp. 1–2.

Ansourian, P. (1981) Experiments on continuous composite beams, *Proceedings of Institution of Civil Engineers*, U.K., Part 2, 71: 25–51.

AS 2327.1 (2003) Australian standard for composite structures, Part 1: Simply supported beams, Sydney, New South Wales, Australia: Standards Australia.

AS 3600 (2001) Australian standard for concrete structures, Sydney, New South Wales, Australia: Standards Australia.

Berry, P., Bridge, R.Q., and Patrick, M. (2001a) Design of continuous composite beams with rigid connections for strength, composite structures design manual – Design booklet DB2.1, Sydney, New South Wales, Australia: OneSteel Manufacturing Limited.

Berry, P.A., Patrick, M., Liang, Q.Q., and Ng, A. (2001b) Cross-section design of continuous composite beams, Paper presented at *the Australasian Structural Engineering Conference*, Gold Coast, Queensland, Australia, pp. 491–497.

Chapman, J.C. and Balakrishnan, S. (1964) Experiments on composite beams, *The Structural Engineer*, 42 (11): 369–383.

Chien, E.Y.L. and Ritchie, J.K. (1984) *Composite Floor Systems*, Markham, Ontario, Canada: Canadian Institute of Steel Construction.

Clawson, W.C. and Darwin, D. (1982) Tests of composite beams with web openings, *Journal of the Structural Division*, ASCE, 108 (1), 145–162.

Eurocode 4 (2004) Design of composite steel and concrete structures, Part 1.1: General rules and rules for buildings, Brussels, Belgium: European Committee for Standardization.

Grant, J.A., Fisher, J.W., and Slutter, R.G. (1977) Composite beams with formed steel deck, *AISC Engineering Journal*, 1st Quarter, 14 (1): 24–43.

Lam, D. (2002) Composite steel beams with precast hollow core slabs: Behaviour and design, *Progress in Structural Engineering and Materials*, 4: 179–185.

Lam, D., Elliott, K.S., and Nethercot, D.A. (2000a) Experiments on composite steel beams with precast concrete hollow core floor slabs, *Proceedings of Institution of Civil Engineers, Structures and Buildings*, U.K., 140: 127–138.

Lam, D. Elliott, K.S., and Nethercot, D.A. (2000b) Designing composite steel beams with precast concrete hollow-core slabs, *Proceedings of Institution of Civil Engineers, Structures and Buildings*, U.K., 140: 139–149.

Liang, Q.Q. (2005) *Performance-Based Optimization of Structures: Theory and Applications*, London, U.K.: Spon Press.

Liang, Q.Q. and Patrick, M. (2001) Design of the shear connection of simply-supported composite beams: To Australian standards AS 2327.1-1996, composite structures design manual – Design booklet DB1.2, Sydney, New South Wales, Australia: OneSteel Manufacturing Limited.

Liang, Q.Q., Patrick, M., and Bridge, R.Q. (2001) Computer software for longitudinal shear design of steel–concrete composite beams, Paper presented at *the Australasian Structural Engineering Conference*, Gold Coast, Queensland, Australia, pp. 515–522.

Liang, Q.Q., Uy, B., Bradford, M.A., and Ronagh, H.R. (2004) Ultimate strength of continuous composite beams in combined bending and shear, *Journal of Constructional Steel Research*, 60 (8): 1109–1128.

Liang, Q.Q., Uy, B., Bradford, M.A., and Ronagh, H.R. (2005) Strength analysis of steel–concrete composite beams in combined bending and shear, *Journal of Structural Engineering*, ASCE, 131 (10): 1593–1600.

Liang, Q.Q., Xie, Y.M., and Steven, G.P. (2000) Topology optimization of strut-and-tie models in reinforced concrete structures using an evolutionary procedure, *ACI Structural Journal*, 97 (2): 322–330.

Moffatt, K.R. and Dowling, P.J. (1978) British shear lag rules for composite girders, *Journal of the Structural Division*, 104 (7): 1123–1130.

Oehlers, D.J. and Bradford, M.A. (1999) *Elementary Behaviour of Composite Steel and Concrete Structural Members*, Oxford, U.K.: Butterworth-Heinemann.

Oehlers, D.J. and Coughlan, C.G. (1986) The shear stiffness of stud shear connectors in composite beams, *Journal of Constructional Steel Research*, 6: 273–284.

Ollgaard, J.G. Slutter, R.G., and Fisher, J.W. (1971) Shear strength of stud connectors in lightweight and normal-weight concrete, *AISC Engineering Journal*, 8: 55–64.

Patrick, M. and Liang, Q.Q. (2002) Shear connection to steel tubes used in composite beam construction, Paper presented at the *United Engineering Foundation Fourth Conference on Composite Construction in Steel and Concrete*, Banff, Alberta, Canada. Reston, VA: ASCE, pp. 699–710.

Pi, Y.L., Bradford, M.A., and Uy, B. (2006a) Second order nonlinear inelastic analysis of composite steel–concrete members. I: Theory, *Journal of Structural Engineering*, ASCE, 132 (5): 751–761.

Pi, Y.L., Bradford, M.A., and Uy, B. (2006b) Second order nonlinear inelastic analysis of composite steel–concrete members. II: Applications, *Journal of Structural Engineering*, ASCE, 132 (5): 762–771.

Ranzi, J. (2008) Locking problems in the partial interaction analysis of multi-layered composite beams, *Engineering Structures*, 30 (10): 2900–2911.

Vallenilla, C. and Bjorhovde, R. (1985) Effective width criteria for composite beams, *AISC Engineering Journal*, 22: 169–175.

Viest, I.M., Fountain, R.S., and Singleton, R.C. (1958) *Composite Construction in Steel and Concrete for Bridges and Buildings*, New York: McGraw-Hill.

Viest, I.V., Colaco, J.P., Furlong, R.W., Griffis, L.G., Leon, R.T., and Wyllie, L.A. (eds.) (1997) *Composite Construction Design for Buildings*, New York: McGraw-Hill and ASCE.

Zona, A. and Ranzi, J. (2011) Finite element models for nonlinear analysis of steel–concrete composite beams with partial interaction in combined bending and shear, *Finite Elements in Analysis and Design*, 47 (2): 98–118.

Chapter 10

Composite columns

10.1 INTRODUCTION

Steel–concrete composite columns have been widely used in high-rise composite buildings, bridges and offshore structures due to their high structural performance, such as high strength, high stiffness, high ductility and large strain energy absorption capacities. The types of composite columns are shown in Figure 10.1. The most commonly used composite columns are concrete-encased composite (CEC) columns as shown in Figure 10.1a, rectangular concrete-filled steel tubular (CFST) columns as illustrated in Figure 10.1c and circular CFST columns as depicted in Figure 10.1d. A CEC column is formed by encasing a structural steel I-section into a reinforced concrete column. Stud shear connectors may be welded to the structural steel section to transfer forces between the steel section and the concrete. A CFST column is constructed by filling concrete into a circular or rectangular hollow steel tube. Longitudinal reinforcement may be placed inside the steel tubes to increase the capacities of CFST columns.

CEC columns have a higher fire resistance than CFST columns. In addition, the concrete effectively prevents the steel I-section in a CEC column from local buckling. The steel tube of a CFST column completely encases the concrete core, which remarkably increases the strength and ductility of the concrete core in circular CFST columns and the duality of the concrete core in rectangular CFST columns. On the other hand, the concrete core effectively prevents the inward local buckling of the steel tube, which results in a higher local buckling strength of the tube than the hollow one. Moreover, the steel tube is utilized as permanent formwork and longitudinal reinforcement for the concrete core, offering significant reductions in construction time and costs (Liang 2009a).

Composite columns are important structural members, which are used to support heavy axial loads as compression members or gravity and lateral loads as beam–columns in moment-resisting composite frames. Practical composite columns are often subjected to the combined actions of axial load and bending moments. This chapter presents the behaviour, design and non-linear analysis of short and slender composite columns. The design of short and slender composite columns for strength to Eurocode 4 (2004) is covered. The non-linear inelastic analysis of short and slender CFST beam–columns under axial load and biaxial bending, preloads on the steel tubes and cyclic loading is presented.

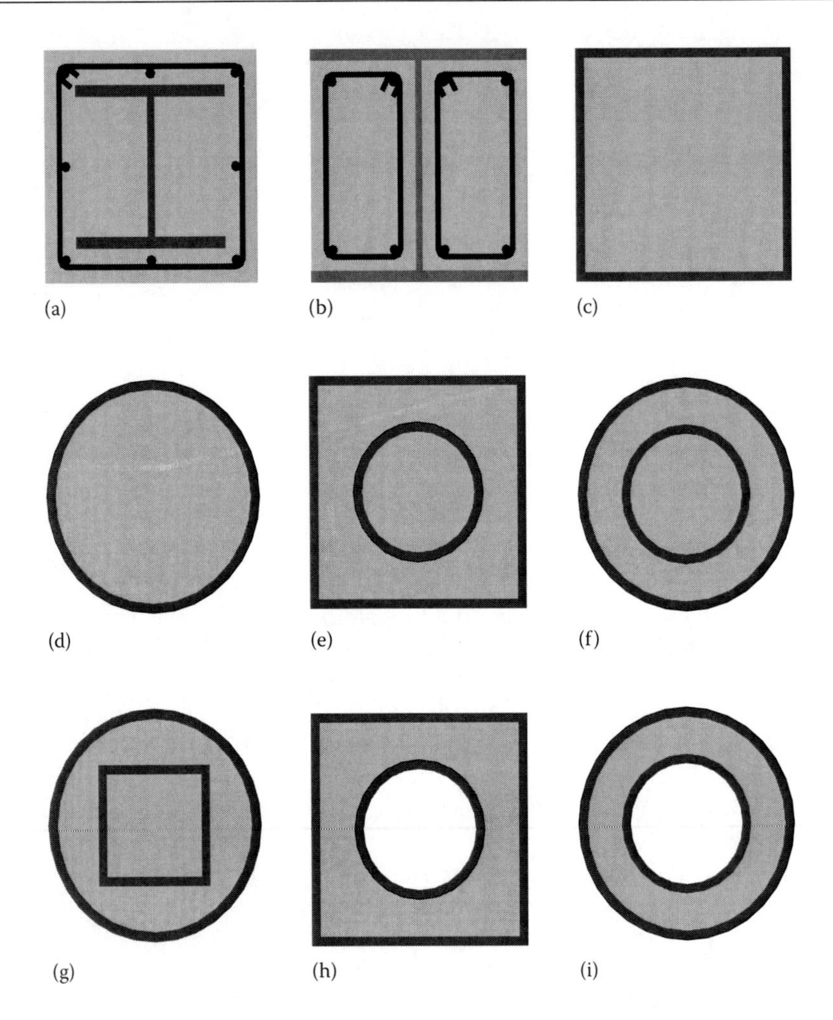

Figure 10.1 Types of composite columns: (a) concrete-encased composite column; (b) partially encased composite column; (c) rectangular concrete-filled steel tubular (CFST) column; (d) circular CFST column; (e) rectangular CFST column with internal circular steel tube; (f) circular CFST column with internal circular steel tube; (g) circular CFST column with rectangular steel tube; (h) double skin rectangular CFST column with internal circular steel tube; (i) double skin circular CFST column.

10.2 BEHAVIOUR AND DESIGN OF SHORT COMPOSITE COLUMNS

10.2.1 Behaviour of short composite columns

Experimental studies have been conducted on the behaviour of short composite columns (Furlong 1967; Knowles and Park 1969; Tomii and Sakino 1979a,b; Shakir-Khalil and Zeghiche 1989; Shakir-Khalil and Mouli 1990; Ge and Usami 1992; Bridge and O'Shea 1998; Schneider 1998; Uy 1998, 2000, 2001; Han 2002; Zhao and Grzebieta 2002; Giakoumelis and Lam 2004; Young and Ellobody 2006; Zhao and Packer 2009; Uy et al. 2011). The behaviour of short composite columns under axial compression is characterised by their axial load–strain curves which indicate the axial stiffness, the ultimate axial strength, the post-peak behaviour and the axial ductility of the columns. Figure 10.2 shows a typical axial load–strain curve for a CFST short column predicted by the computer program NACOMS (Nonlinear Analysis of Composite Columns) developed by Liang (2009a,b). CEC short columns may fail by yielding of the steels

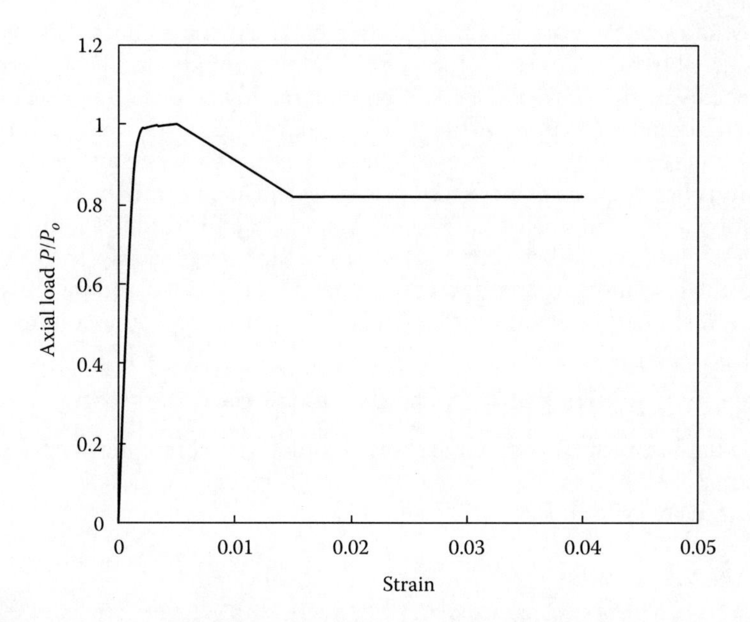

Figure 10.2 Typical axial load–strain curve for a rectangular CFST column.

and crushing of the concrete. The failure modes of CFST short columns include yielding or local buckling of the steel section and crushing of the infill concrete. The ultimate axial strength of short composite columns is governed by the section properties and the material strengths of the steel and concrete. The behaviour of short composite columns under axial load and bending is characterised by their moment–curvature curves which indicate the flexural stiffness, ultimate moment capacity, post-peak behaviour and curvature ductility of the columns. The moment–curvature curve for a typical CFST short column under axial load and biaxial bending predicted by the computer program NACOMS (Liang 2009a,b) is given in Figure 10.3.

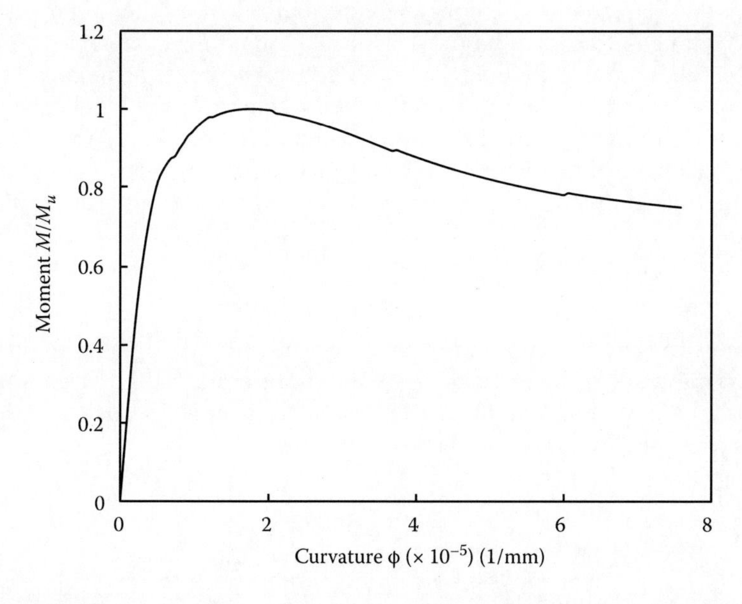

Figure 10.3 Typical moment–curvature curve for a rectangular CFST beam–column under axial load and biaxial bending.

Numerical studies carried out by Liang (2009b,c) demonstrate that local buckling of the steel tube remarkably reduces the stiffness, axial strength and ductility of CFST columns. In addition, increasing the D/t ratio of CFST column sections reduces their section axial performance, axial ductility, flexural stiffness and strength and curvature ductility. Moreover, increasing the concrete compressive strength increases the axial load and moment capacities of CFST columns but decreases their section axial performance and ductility. Furthermore, the axial and flexural strengths of CFST columns are found to increase by increasing the yield strength of the steel tubes, but the axial ductility is generally shown to decrease. Numerical results also indicate that increasing the axial load level significantly reduces the flexural stiffness, strength and curvature ductility of composite beam–columns.

10.2.2 Short composite columns under axial compression

The ultimate axial strength (P_o) of a CEC short column or a rectangular CFST short column under axial compression is the sum of the strength of steel and concrete components of the column and can be expressed by

$$P_o = \gamma_c f_c' A_c + f_y A_{se} + f_{yr} A_r \tag{10.1}$$

where

A_c is the cross-sectional area of concrete
A_{se} is the effective structural steel area of the cross section
A_r is the cross-sectional area of longitudinal reinforcement
γ_c is the reduction factor used to account for the effect of column size and concrete quality on the column strength, proposed by Liang (2009a) as

$$\gamma_c = 1.85 D_c^{-0.135} \quad (0.85 \le \gamma_c \le 1.0) \tag{10.2}$$

in which D_c is the diameter of the concrete core and taken as the larger of $(B - 2t)$ and $(D - 2t)$ for a rectangular cross section.

Circular steel tubes provide confinement to the concrete core, which increases the strength and ductility of the concrete core in circular CFST columns. The steel tube of a circular CFST column is biaxially stressed. The hoop tension developed in the steel tube reduces its yield stress in the longitudinal direction. The ultimate axial strength of circular CFST short columns considering confinement effects is given by Liang and Fragomeni (2009) as

$$P_o = \left(\gamma_c f_c' + 4.1 f_{rp}\right) A_c + \gamma_s f_y A_s + f_{yr} A_r \tag{10.3}$$

where f_{rp} denotes the lateral confining pressure provided by the steel tube on the concrete core. Based on the work of Tang et al. (1996) and Hu et al. (2003), a confining pressure model for normal and high-strength concrete confined by either a normal or high-strength steel tube was proposed by Liang and Fragomeni (2009) as

$$f_{rp} = \begin{cases} 0.7(\nu_e - \nu_s)\left(\dfrac{2t}{D-2t}\right) f_y & \text{for } \dfrac{D}{t} \le 47 \\ \left[0.006241 - 0.0000357\left(\dfrac{D}{t}\right)\right] f_y & \text{for } 47 < \dfrac{D}{t} \le 150 \end{cases} \tag{10.4}$$

in which ν_e and ν_s are Poisson's ratios of the steel tube with or without concrete infill, respectively. Poisson's ratio ν_s is taken as 0.5 at the maximum strength point, and ν_e is given by (Tang et al. 1996)

$$\upsilon_e = 0.2312 + 0.3582\upsilon_e' - 0.1524\left(\frac{f_c'}{f_y}\right) + 4.843\upsilon_e'\left(\frac{f_c'}{f_y}\right) - 9.169\left(\frac{f_c'}{f_y}\right)^2 \tag{10.5}$$

$$\upsilon_e' = 0.881\times10^{-6}\left(\frac{D}{t}\right)^3 - 2.58\times10^{-4}\left(\frac{D}{t}\right)^2 + 1.953\times10^{-2}\left(\frac{D}{t}\right) + 0.4011 \tag{10.6}$$

The factor γ_s accounts for the effect of hoop tensile stresses and strain hardening on the yield stress of the steel tube. For carbon steel tubes, γ_s is given by Liang (2009a) as

$$\gamma_s = 1.458\left(\frac{D}{t}\right)^{-0.1} \quad (0.9 \le \gamma_s \le 1.1) \tag{10.7}$$

In Eurocode 4 (2004), the confinement effect that increases the compressive strength of the concrete core in circular CFST columns with a relative slenderness of $\bar{\lambda} \le 0.5$ and a small loading eccentric ratio of $e/D < 0.1$ is taken into account in the calculation of the ultimate axial strength ($N_{pl,Rd}$) as follows:

$$N_{pl,Rd} = \eta_s A_s f_y + A_c f_c'\left(1 + \eta_c \frac{t}{D}\frac{f_y}{f_c'}\right) + A_r f_{yr} \tag{10.8}$$

where the factors η_s and η_c are given by

$$\eta_s = 0.25(3 + 2\bar{\lambda}) \le 1.0 \tag{10.9}$$

$$\eta_c = 4.9 - 18.5\bar{\lambda} + 17\bar{\lambda}^2 \ge 0 \tag{10.10}$$

where $\bar{\lambda}$ is the relative slenderness of the column given in Section 10.4.2.

Eurocode 4 (2004) provides limits on the width-to-thickness ratio for steel elements in composite columns as follows:

- For circular CFST columns, $(D/t) \le 90(235/f_y)$.
- For rectangular CFST columns, $(D/t) \le 52\sqrt{(235/f_y)}$.
- For the flanges of partially encased I-sections, $(b_f/t_f) \le 44\sqrt{(235/f_y)}$.

10.2.3 Short composite columns under axial load and uniaxial bending

10.2.3.1 General

Composite columns in composite frames with rigid connections are often subjected to combined actions of axial compression and uniaxial bending. The combined actions may also be caused by the eccentricity of the applied load. The design codes require that all practical columns should be designed as beam–columns. In the design of slender composite beam–columns under axial load and uniaxial bending, the axial load–moment interaction action diagrams for the columns need to be determined. The non-linear inelastic analysis of composite beam–columns under eccentric loading is complex without the aid of computer programs.

In practice, the rigid plastic analysis is usually used to determine the ultimate strengths of composite beam–columns under eccentric loading. The rigid plastic analysis assumes that (1) full composite action between steel and concrete components up to failure and plane sections remain plane, (2) all steel yields in compression and tension at the ultimate strength limit state, (3) a rectangular concrete stress block in compression is stressed to $0.85f'_c$, (4) local buckling is ignored for CEC columns, (5) local buckling of steel tubes may be considered for CFST columns and (6) the tensile strength of concrete is ignored.

Eurocode 4 (2004) allows a simplified design method developed by Roik and Bergmann (1989) to be used for developing axial load–moment interaction diagrams for composite short columns. This simplified method is limited to members of doubly symmetrical cross sections including rolled, cold-formed or welded steel sections. The limits on concrete thickness cover to the steel section are $c_x \leq 0.4B$ and $c_y \leq 0.3D$. The depth-to-width ratio (D/B) of the composite cross section should be within the limits of 0.2 and 5.0. The area of longitudinal reinforcement used to calculate the ultimate axial and bending strengths of a composite column should not exceed 6% of the concrete area in the composite section.

10.2.3.2 Axial load–moment interaction diagram

The typical axial load–moment interaction diagram for a composite column section is schematically depicted in Figure 10.4. In the simplified method, the axial load–moment interaction diagram for the column section is approximated by the polygon *ACDB* as shown in Figure 10.4 (Roik and Bergmann 1989; Oehlers and Bradford 1999; Eurocode 4 2004; Johnson 2004). The simplified method is introduced herein for developing the axial load–moment interaction diagrams.

Point A in Figure 10.4 corresponds to the ultimate axial strength (P_o) of the column section under axial compression only, which can be calculated using Equation 10.1.

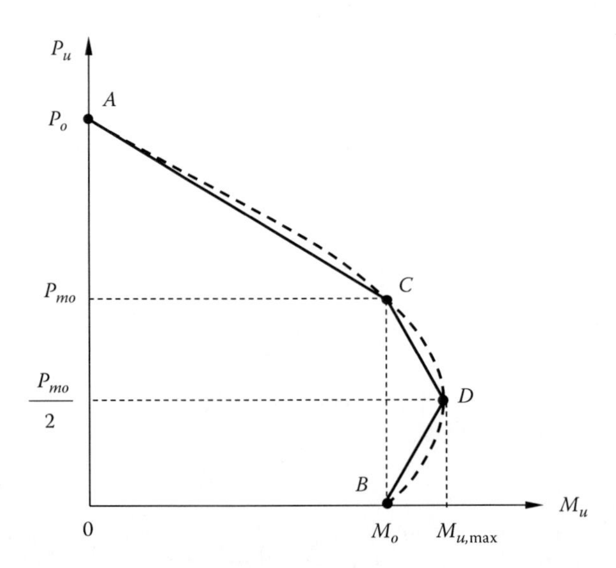

Figure 10.4 Axial load–moment interaction diagram of a composite section.

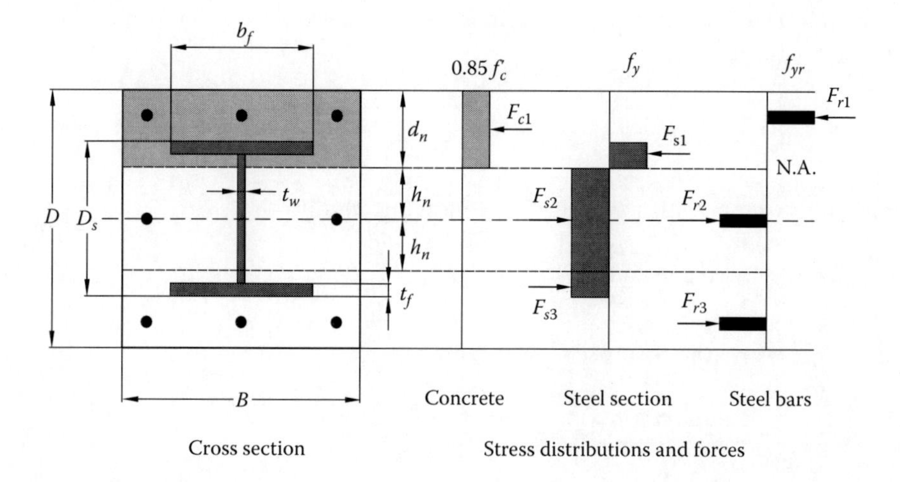

Figure 10.5 Plastic stress distributions in the cross section of a composite column: neutral axis above the centroid of section.

Point B in Figure 10.4 corresponds to the ultimate pure bending moment capacity (M_o) of the column section under bending without the axial load. The plastic stress distribution in the column section under pure bending is shown in Figure 10.5. The plastic neutral axis is located at a distance h_n above the centroid of the column cross section. The cross section is divided into three regions as illustrated in Figure 10.5. Region 1 is above the h_n distance from the centroid of the section, while region 3 is below the h_n distance from the centroid of the section. Region 2 is within the h_n distance above and below the centroid of the section. If the neutral axis is located in the web of the steel section, the compressive force in the concrete is calculated by

$$F_{c1} = 0.85 f_c'[B(0.5D - h_n) - b_f t_f - (0.5d_w - h_n)(n_w t_w) - A_{r1}] \tag{10.11}$$

where

d_w is the clear depth of the steel web
n_w is the total number of webs in the steel section
A_{r1} is the area of longitudinal reinforcement at the top of the cross section

For the composite section under pure bending, the sum of compression forces must equal the sum of tension forces in the section: $F_{c1} + F_{s1} + F_{r1} = F_{s2} + F_{r2} + F_{s3} + F_{r3}$, where $F_{s1} = F_{s3}$ and $F_{r1} = F_{r3}$ due to symmetry of the steel elements about the centroid of the section as shown in Figure 10.5 and $F_{s2} = 2h_n(n_w t_w)f_y$. From the force equilibrium condition, the following expression can be obtained

$$F_{c1} = 2h_n(n_w t_w)f_y + F_{r2} \tag{10.12}$$

The distance h_n can be determined from this equation as

$$h_n = \frac{0.85 f_c' A_{cn} - F_{r2}}{0.85 f_c'(B - n_w t_w) + 2n_w t_w f_y} \le \frac{d_w}{2} \tag{10.13}$$

where $A_{cn} = B(0.5D) - b_f t_f - (0.5d_w)(n_w t_w) - A_{r1}$.

If the neutral axis is located in the top flange of the steel section, the compressive force in the concrete is computed by

$$F_{c1} = 0.85f_c'\left[B(0.5D - h_n) - b_f(0.5D_s - h_n) - A_{r1}\right] \tag{10.14}$$

The force equilibrium condition yields the following condition:

$$F_{c1} = F_{r2} + F_w + 2b_f(h_n - 0.5d_w)f_y \tag{10.15}$$

where $F_w = d_w t_w f_y$ and the distance h_n is given by

$$h_n = \frac{0.85f_c'A_{cn} - F_{r2} - F_w + b_f d_w f_y}{0.85f_c'(B - b_f) + 2b_f f_y} \tag{10.16}$$

where $A_{cn} = B(0.5D) - b_f(0.5D_s) - A_{r1}$.

The nominal moment capacity of the composite section under pure bending can be calculated by taking moments about its centroid as

$$M_o = F_{c1}d_{c1} + 2F_{s1}d_{s1} + 2F_{r1}d_{r1} \tag{10.17}$$

in which d_{c1} is the distance from the centroid of F_{c1} to the centroid of the cross section, taken as $d_{c1} = 0.5D - 0.5(0.5D - h_n)$ for CEC columns and $d_{c1} = 0.5d_w - 0.5(0.5d_w - h_n)$ for CFST columns.

Simple design formulas for calculating the ultimate moment capacities of circular CFST short columns under pure bending are given by Liang and Fragomeni (2010) as follows:

$$M_o = \lambda_m \alpha_{fc} \alpha_y Z_e f_y \tag{10.18}$$

$$\lambda_m = 0.0087 + 12.3\left(\frac{t}{D}\right) - 36\left(\frac{t}{D}\right)^2 \quad (10 \le D/t \le 120) \tag{10.19}$$

$$\alpha_{fc} = 0.774\left(f_c'\right)^{0.075} \quad (30 \le f_c' \le 120\,\text{MPa}) \tag{10.20}$$

$$\alpha_y = 0.883 + \frac{21.147}{f_y} + \frac{4202}{f_y^2} \quad (250 \le f_y \le 690\,\text{MPa}) \tag{10.21}$$

where

- λ_m is the factor accounting for the effect of D/t ratio
- α_{fc} is the factor accounting for the effect of concrete compressive strength
- α_y is the factor used to take into account the effect of the yield strength of the steel tube
- Z_e is the elastic section modulus of the circular CFST column, calculated as $\pi D^3/32$

Point C in Figure 10.4 corresponds to the point where the nominal moment capacity of the column section under an axial force of P_{mo} is equal to the pure bending moment

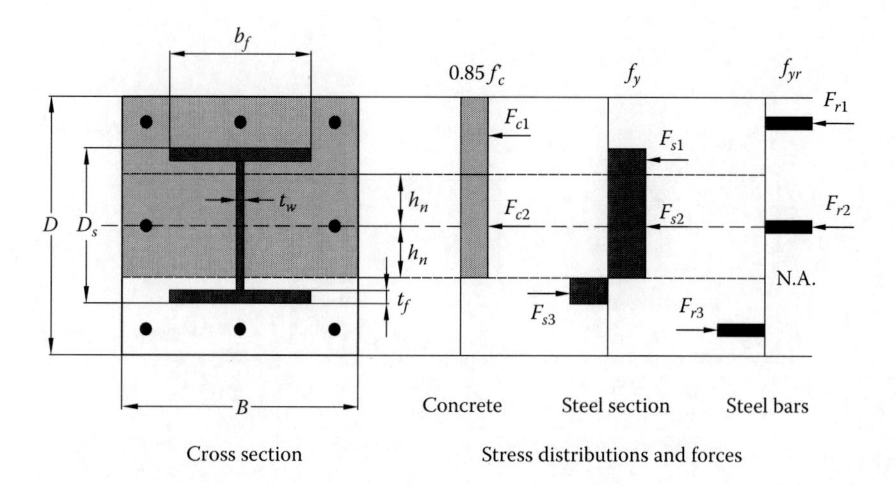

Figure 10.6 Plastic stress distributions in the cross section of a composite column: neutral axis below the centroid of section.

capacity (M_o). For this case, the plastic neutral axis is located at a distance of h_n below the centroid of the cross section as depicted in Figure 10.6, which shows the plastic stress distribution in the cross section. The value of h_n has been determined for the section under pure bending. The compressive force in the concrete in region 2 is calculated by

$$F_{c2} = 0.85f_c'\left[B(2h_n) - (2h_n)(n_w t_w)\right] \tag{10.22}$$

The resultant force in the composite section can be obtained from Figure 10.6 by summing all forces in the cross section as

$$P_{mo} = 2F_{s2} + F_{c2} + F_{r2} \tag{10.23}$$

Point D in Figure 10.4 corresponds to the point where the maximum moment capacity ($M_{u,\max}$) of the column section under an axial force of $P_{mo}/2$ occurs. For this case, the plastic neutral axis lies at the centroid of the cross section as shown in Figure 10.7 which illustrates the plastic stress distribution in the cross section. The resultant axial force in the composite section is determined as $P_u = F_{c1} + F_{c2}/2 = P_{mo}/2$. By taking moments about the centroid of the cross section, the maximum moment capacity ($M_{u,\max}$) of the composite section is obtained as

$$M_{u,\max} = 0.85f_c'A_{cm}d_{cm} + M_s + 2F_{r1}d_{r1} \tag{10.24}$$

where

A_{cm} is the area of concrete above the plastic neutral axis and is calculated as $A_{cm} = B(D/2) - A_s/2 - A_{st1}$

d_{cm} is the distance from the centroid of A_{cm} to the centroid of the composite section

M_s is the nominal moment capacity of the whole steel section alone

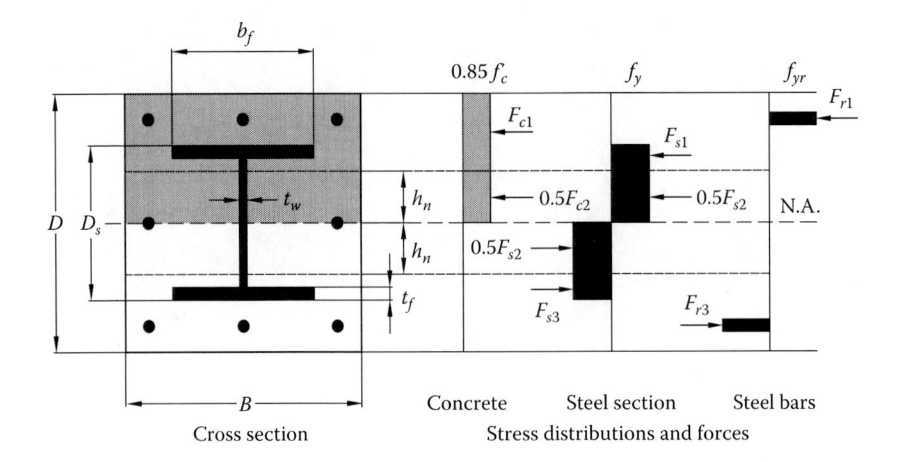

Figure 10.7 Plastic stress distributions in the cross section of a composite column: neutral axis at the centroid of section.

Example 10.1: Axial load–moment interaction diagram of CEC short column

Develop the axial load–moment interaction diagram for the CEC short column bending about the principal x-axis as shown in Figure 10.8. The concrete design strength is $f_c' = 32\,\text{MPa}$. The yield stress of the steel section is 300 MPa, while the yield stress of the steel reinforcement is 500 MPa.

1. Point A: Ultimate axial strength

The reduction factor for concrete γ_c is

$$\gamma_c = 1.85D_c^{-0.135} = 1.85\times 500^{-0.135} = 0.8 < 0.85$$

Hence, $\gamma_c = 0.85$.

The area of the structural steel section is computed as

$$A_{se} = 350\times 16\times 2 + (350 - 2\times 16)\times 12 = 15{,}016\,\text{mm}^2$$

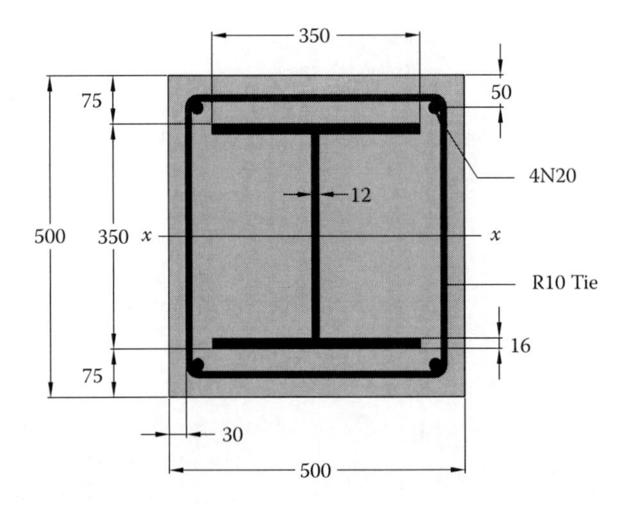

Figure 10.8 Cross section of a CEC column.

The total area of reinforcement in the cross section is

$$A_r = 4 \times \left(\pi \times \frac{20^2}{4} \right) = 1256.6\,\text{mm}^2$$

The area of concrete in the cross section can be calculated as

$$A_c = BD - A_{se} - A_r = 500 \times 500 - 15{,}016 - 1{,}256.6 = 233{,}727\,\text{mm}^2$$

The ultimate axial strength of the composite section is therefore

$$P_o = \gamma_c f_c' A_c + f_y A_{se} + f_{yr} A_r$$

$$= 0.85 \times 32 \times 233{,}727 + 300 \times 15{,}016 + 500 \times 1{,}256.6\,\text{N} = 11{,}490.5\,\text{kN}$$

2. Point *B*: Pure bending moment capacity

Assume the plastic neutral axis is located in the steel web.

The clear depth of the web is $d_w = D_s - 2t_f = 350 - 2 \times 16 = 318\,\text{mm}$.

The area of the top reinforcement is $A_{r1} = A_r/2 = 628.3\,\text{mm}^2$.

The distance of the top reinforcement to the centroid of the composite section is

$$d_{r1} = \frac{500}{2} - 30 - 10 - \frac{20}{2} = 200\,\text{mm}$$

The distance h_n is calculated as follows:

$$A_{cn} = B(0.5D) - b_f t_f - (0.5d_w)(n_w t_w) - A_{r1}$$

$$= 500 \times 0.5 \times 500 - 350 \times 16 - 0.5 \times 318 \times (1 \times 12) - 628.3 = 116{,}863.7\,\text{mm}^2$$

$$h_n = \frac{0.85 f_c' A_{cn} - F_{r2}}{0.85 f_c'(B - n_w t_w) + 2 n_w t_w f_y}$$

$$= \frac{0.85 \times 32 \times 116{,}863.7 - 0}{0.85 \times 32 \times (500 - 1 \times 12) + 2 \times 1 \times 12 \times 300} = 155.3\text{ mm} < \frac{d_w}{2} = \frac{318}{2} = 159\,\text{mm}$$

Hence, the plastic neutral axis is located in the steel web.

The compressive force in the concrete in region 1 is computed as

$$F_{c1} = 0.85 f_c' \left[B\left(\frac{D}{2} - h_n \right) - b_f t_f - \left(\frac{d_w}{2} - h_n \right) n_w t_w - A_{st1} \right]$$

$$= 0.85 \times 32 \times \left[500 \times \left(\frac{500}{2} - 155.3 \right) - 350 \times 16 - \left(\frac{318}{2} - 155.3 \right) \times 1 \times 12 - 628.3 \right] \text{N}$$

$$= 1117.3\,\text{kN}$$

The distance of F_{c1} to the centroid of the cross section is

$$d_{c1} = \frac{D}{2} - \frac{(D/2 - h_n)}{2} = \frac{500}{2} - \frac{(500/2 - 155.3)}{2} = 202.65\,\text{mm}$$

The force in the steel top flange and its distance from the centroid of the section are computed as

$$F_{s1\cdot f} = b_f t_{f1} f_y = 350 \times 16 \times 300\,\text{N} = 1680\,\text{kN}$$

$$d_{s1\cdot f} = \frac{D_s}{2} - \frac{t_f}{2} = \frac{350}{2} - \frac{16}{2} = 167\,\text{mm}$$

The force in the steel web and its distance from the centroid of the section are calculated as

$$F_{s1\cdot w} = \left(\frac{d_w}{2} - h_n\right) n_w t_w f_y = \left(\frac{318}{2} - 155.3\right) \times 1 \times 12 \times 300\,\text{N} = 13.32\,\text{kN}$$

$$d_{s1\cdot w} = \frac{d_w}{2} - \frac{(d_w/2 - h_n)}{2} = \frac{318}{2} - \frac{(318/2 - 155.3)}{2} = 157.15\,\text{mm}$$

The resultant force in the steel components in region 1 is therefore

$$F_{s1} = 1680 + 13.32 = 1693.32\,\text{kN}$$

The distance from the centroid of F_{s1} to the centroid of the section is

$$d_{s1} = \frac{F_{s1\cdot f} d_{s1\cdot f} + F_{s1\cdot w} d_{s1\cdot w}}{F_{s1}} = \frac{1680 \times 167 + 13.32 \times 157.15}{1693.32} = 166.92\,\text{mm}$$

The force in the top reinforcement is

$$F_{r1} = A_{r1} f_{yr} = 628.3 \times 500\,\text{N} = 314.2\,\text{kN}$$

The pure bending moment capacity M_o is calculated as

$$\begin{aligned} M_o &= F_{c1} d_{c1} + 2F_{s1} d_{s1} + 2F_{st1} d_{st1} \\ &= 1117.3 \times 202.65 + 2 \times 1693.47 \times 166.92 + 2 \times 314.2 \times 200\,\text{kN mm} = 917.4\,\text{kN m} \end{aligned}$$

3. Point C: $M_u = M_o$

The plastic neutral axis is located at a distance $h_n = 155.3$ mm below the centroid of the section. The force in the steel component in region 2 is computed as

$$F_{s2} = 2h_n (n_w t_w) f_y = 2 \times 155.3 \times (1 \times 12) \times 300\,\text{N} = 1118.2\,\text{kN}$$

The compressive force in the concrete in region 2 is

$$F_{c2} = 0.85 f'_c \left[B(2h_n) - 2h_n(n_w t_w) \right]$$

$$= 0.85 \times 32 \times \left[500 \times 2 \times 155.3 - 2 \times 155.3 \times 1 \times 12 \right] \text{N} = 4122.8 \text{kN}$$

The resultant axial force in the composite section is therefore

$$F_{mo} = 2F_{s2} + F_{c2} = 2 \times 1118.2 + 4122.8 = 6359.2 \text{kN}$$

4. Point *D*: Maximum moment capacity

The plastic neutral axis lies at the centroid of the cross section. The resultant force in the composite section is determined as

$$P_u = \frac{P_{mo}}{2} = \frac{6359.2}{2} = 3179.6 \text{kN}$$

The area of concrete above the plastic neutral axis and its distance to the centroid of the section are computed as

$$A_{cm} = \frac{BD}{2} - \frac{A_s}{2} - A_{st1} = 500 \times \frac{500}{2} - \frac{15{,}016}{2} - 628.3 = 116{,}863.7 \text{mm}^2$$

$$d_{cm} = \frac{D}{4} = \frac{500}{4} = 125 \text{mm}$$

The moment capacity of the whole steel I-section is calculated as

$$M_s = b_f t_t f_y (D_s - t_f) + \left(\frac{d_w}{2} \right) (n_w t_w) f_y \left(\frac{d_w}{2} \right)$$

$$= 350 \times 16 \times 300 \times (350 - 16) + \left(\frac{318}{2} \right) \times (1 \times 12) \times 300 \times \left(\frac{318}{2} \right) \text{N mm}$$

$$= 652 \text{kN m}$$

The maximum moment capacity is therefore

$$M_{u,\max} = 0.85 f'_c A_{cm} d_{cm} + M_s + 2F_{r1} d_{r1}$$

$$= 0.85 \times 32 \times 11{,}683.7 \times 125 + 652 + 2 \times 314.16 \times 200 \text{N mm} = 1{,}175 \text{kN m}$$

The axial load–moment interaction diagram of this composite short column is shown in Figure 10.9.

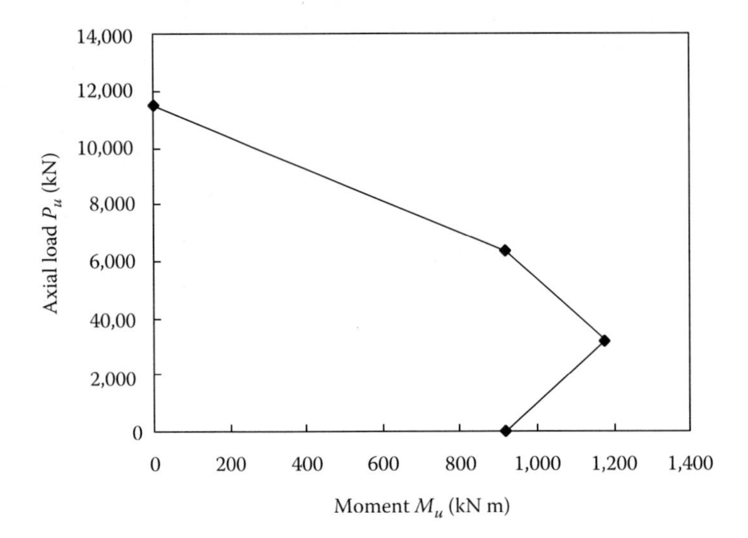

Figure 10.9 Axial load–moment interaction diagram of the CEC short column.

Example 10.2: Axial load–moment interaction diagram of CFST short column

Develop the axial load–moment strength interaction diagram of the CFST short column bending about the principal x-axis as depicted in Figure 10.10. The concrete design strength is $f_c' = 50\,\text{MPa}$. The yield stress of the steel section is 300 MPa.

1. Point A: Ultimate axial strength

The slenderness of the steel web is

$$\frac{D}{t} = \frac{600}{20} = 30 < 52\sqrt{\frac{235}{f_y}} = 46$$

The web and the section are compact.

The reduction factor for concrete (γ_c) is

$$\gamma_c = 1.85 D_c^{-0.135} = 1.85 \times 500^{-0.135} = 0.8 < 0.85$$

Hence, $\gamma_c = 0.85$.

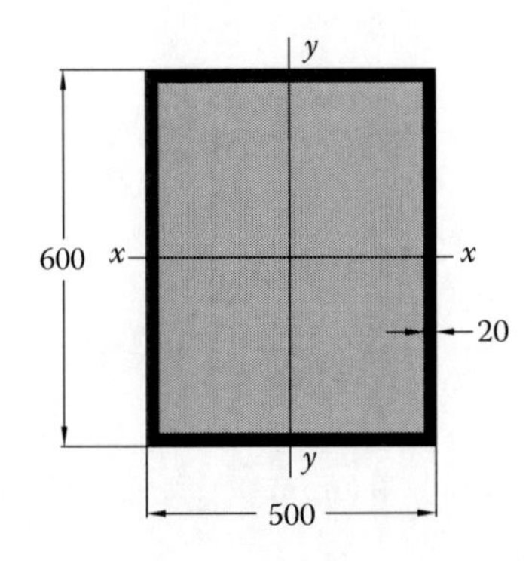

Figure 10.10 Cross section of a CFST column.

The area of concrete in the cross-section is

$$A_c = (B-2t)(D-2t) = (500-2\times 20)(600-2\times 20) = 257600 \text{ mm}^2$$

The area of the structural steel section is computed as

$$A_s = BD - A_c = 500\times 600 - 257600 = 42400 \text{ mm}^2$$

The ultimate axial strength of the column section is therefore

$$P_o = \gamma_c f_c' A_c + f_y A_{se} + f_{yr} A_r$$

$$= 0.85\times 50\times 257{,}600 + 42{,}400\times 300 + 0\text{N} = 23{,}668\text{kN}$$

2. Point B: Pure bending moment capacity

Assume the plastic neutral axis is located in the steel web.

The clear distance of the web is $d_w = D_s - 2t_f = 600 - 2 \times 20 = 560\text{mm}$.

The distance h_n is calculated as follows:

$$A_{cn} = B(0.5D) - b_f t_f - (0.5d_w)(n_w t_w) - A_{r1}$$

$$= 500\times 0.5\times 600 - 500\times 20 - 0.5\times 560\times(2\times 20) - 0 = 128{,}800\text{mm}^2$$

$$h_n = \frac{0.85 f_c' A_{cn} - F_{r2}}{2n_w t_w f_y + 0.85 f_c'(B - n_w t_w)}$$

$$= \frac{0.85\times 50\times 128{,}800 - 0}{2\times 2\times 20\times 300 + 0.85\times 50\times(500-2\times 20)} = 125.7\text{mm} < \frac{d_w}{2} = \frac{560}{2} = 280\text{mm}$$

Hence, the plastic neutral axis is located in the steel web.

The compressive force in the concrete in region 1 is computed as

$$F_{c1} = 0.85 f_c'\left[B(0.5D - h_n) - b_f t_f - (0.5d_w - h_n)n_w t_w - A_{st1}\right]$$

$$= 0.85\times 50\times\left[500\times(0.5\times 600 - 125.7) - 500\times 20 - (0.5\times 560 - 125.7)\times 2\times 20 - 0\right]\text{N}$$

$$= 3016.6\text{kN}$$

The distance of F_{c1} to the centroid of the cross section is

$$d_{c1} = \frac{d_w}{2} - \frac{(d_w/2 - h_n)}{2} = \frac{560}{2} - \frac{(560/2 - 125.7)}{2} = 202.85\text{mm}$$

The force in the top steel flange and its distance to the centroid of the section are

$$F_{s1\cdot f} = b_f t_{f1} f_y = 500\times 20\times 300\text{N} = 3000\text{kN}$$

$$d_{s1\cdot f} = \frac{D_s}{2} - \frac{t_f}{2} = \frac{600}{2} - \frac{20}{2} = 290\text{mm}$$

The force in the steel web above the plastic neutral axis and its distance to the centroid of the section are calculated as

$$F_{s1.w} = \left(\frac{d_w}{2} - h_n\right) n_w t_w f_y = \left(\frac{560}{2} - 125.7\right) \times 2 \times 20 \times 300\,\text{N} = 1851.6\,\text{kN}$$

$$d_{s1\cdot w} = \frac{d_w}{2} - \frac{(d_w/2 - h_n)}{2} = \frac{560}{2} - \frac{(560/2 - 125.7)}{2} = 202.85\,\text{mm}$$

The resultant force in the steel components in region 1 is

$$F_{s1} = 3000 + 1851.6 = 4851.6\,\text{kN}$$

The distance from the centroid of F_{s1} to the centroid of the section is

$$d_{s1} = \frac{F_{s1\cdot f} d_{s1\cdot f} + F_{s1\cdot w} d_{s1\cdot w}}{F_{s1}} = \frac{3000 \times 290 + 1851.6 \times 202.85}{4851.6} = 256.74\,\text{mm}$$

The pure bending moment capacity M_o is calculated as

$$\begin{aligned} M_o &= F_{c1} d_{c1} + 2F_{s1} d_{s1} + 2F_{st1} d_{st1} \\ &= 3016.6 \times 202.85 + 2 \times 4851.6 \times 256.74 + 0\,\text{kN mm} = 3103\,\text{kN m} \end{aligned}$$

3. Point C: $M_u = M_o$

The plastic neutral axis is located at a distance of h_n = 125.7 mm below the centroid of the section. The force in the steel component in region 2 is calculated as

$$F_{s2} = 2h_n (n_w t_w) f_y = 2 \times 125.7 \times (2 \times 20) \times 300\,\text{N} = 3016.8\,\text{kN}$$

The compressive force in the concrete in region 2 is

$$\begin{aligned} F_{c2} &= 0.85 f'_c \left[B(2h_n) - 2h_n (n_w t_w)\right] \\ &= 0.85 \times 50 \times \left[500 \times 2 \times 125.7 - 2 \times 125.7 \times 2 \times 20\right]\text{N} = 4914.9\,\text{kN} \end{aligned}$$

The resultant axial force in the cross section is therefore

$$F_{mo} = 2F_{s2} + F_{c2} = 2 \times 3016.8 + 4914.9 = 10948.5\,\text{kN}$$

4. Point *D*: Maximum moment capacity

The plastic neutral axis lies at the centroid of the cross section. The resultant force in the section is determined as

$$P_u = \frac{P_{mo}}{2} = \frac{10{,}948.5}{2} = 5474.25\,\text{kN}$$

The area of concrete above the plastic neutral axis and its distance to the centroid of the section are computed as

$$A_{cm} = \frac{BD}{2} - \frac{A_s}{2} - A_{st1} = 500 \times \frac{600}{2} - \frac{42,400}{2} - 0 = 128,800\,\text{mm}^2$$

$$d_{cm} = \frac{d_w}{4} = \frac{560}{4} = 140\,\text{mm}$$

The moment capacity of the whole steel I-section is

$$M_s = b_f t_t f_y (D_s - t_f) + \left(\frac{d_w}{2}\right)(n_w t_w) f_y \left(\frac{d_w}{2}\right)$$

$$= 500 \times 20 \times 300 \times (600 - 20) + \left(\frac{560}{2}\right) \times (2 \times 12) \times 300 \times \left(\frac{560}{2}\right) \text{N mm}$$

$$= 2680.8\,\text{kN m}$$

The maximum moment capacity is

$$M_{u,\max} = 0.85 f_c' A_{cm} d_{cm} + M_s + 2F_{st1} d_{st1}$$

$$= 0.85 \times 50 \times 128800 \times 140 + 2680.8 + 0\,\text{N mm} = 3447.2\,\text{kN m}$$

The axial load–moment interaction diagram of this composite short column is shown in Figure 10.11.

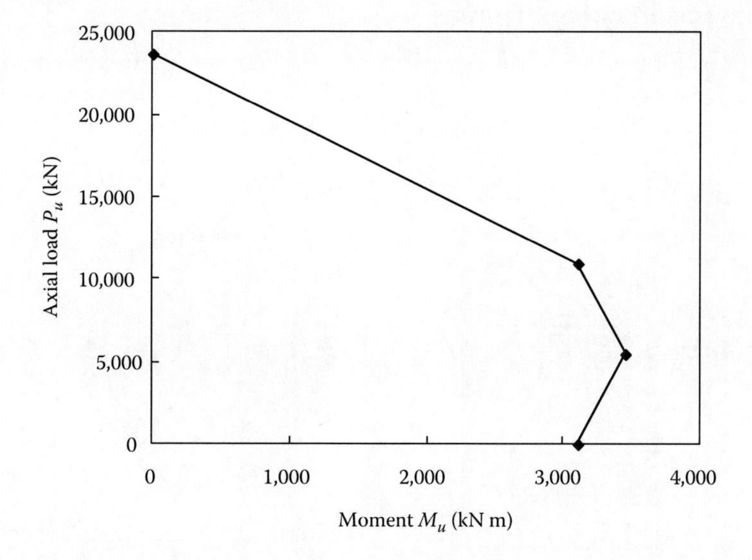

Figure 10.11 Axial load–moment interaction diagram of the CFST short column.

10.3 NON-LINEAR ANALYSIS OF SHORT COMPOSITE COLUMNS

10.3.1 General

The non-linear methods of analysis for composite columns and structures were reviewed by Spacone and El-Tawil (2004). A review on the state-of-the-art development of composite columns was presented by Shanmugam and Lakshmi (2001). Analytical and fibre element models have been developed by various researchers for the non-linear inelastic analysis of short composite columns (El-Tawil et al. 1995; Hajjar and Gourley 1996; Muñoz and Hsu 1997; El-Tawil and Deierlein 1999; Chen et al. 2001; Lakshmi and Shanmugam 2002; Liang et al. 2006, 2007a; Liang 2008, 2009a,b,c; Liang and Fragomeni 2009, 2010). Finite element analyses of CFST short columns and concrete-filled stainless steel tubular (CFSST) columns were also reported in the literature (Hu et al. 2003; Ellobody and Young 2006; Ellobody et al. 2006; Tao et al. 2011; Hassanein et al. 2013a,b,c). The numerical models developed by Liang (2008, 2009a,b,c, 2011a,b) for CFST short columns under axial load and biaxial bending are described in the following sections.

10.3.2 Fibre element method

The fibre element method is an efficient and accurate numerical technique for determining the inelastic behaviour of composite cross sections (El-Tawil et al. 1995; Liang 2009a). In this method, the cross section of a composite column is discretised into many small fibre elements as depicted in Figure 10.12. Each element represents a fibre of material running longitudinally along the member and can be assigned either steel or concrete material properties. Uniaxial stress–strain relationships are used to simulate the material behaviour. Stress resultants are obtained by numerical integration of stresses through the cross section. Numerical models based on the fibre element method have been developed for predicting the non-linear inelastic behaviour of composite short columns under axial load or combined axial load and bending.

10.3.3 Fibre strain calculations

The fibre strain is a function of the curvature (ϕ), orientation (θ) and the depth (d_n) of the neutral axis in the cross section of a rectangular CFST column under axial load and biaxial bending as schematically depicted in Figure 10.13. The strain distribution in a circular CFST

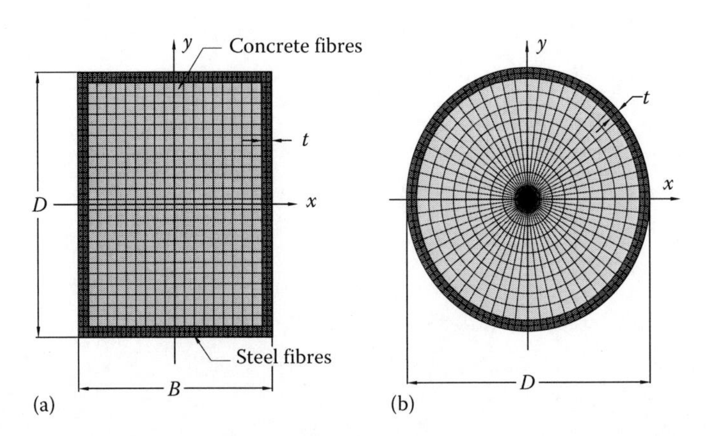

Figure 10.12 Fibre element discretization: (a) rectangular section and (b) circular section.

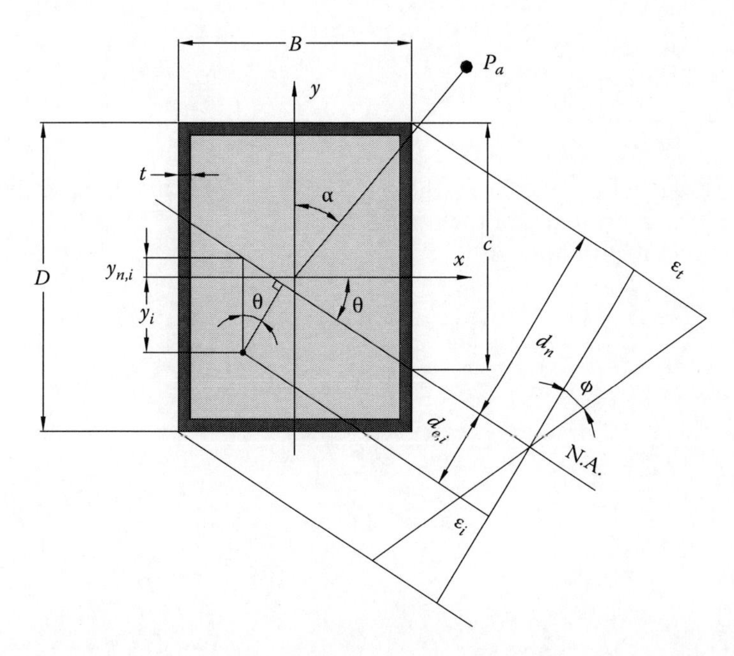

Figure 10.13 Strain distributions in rectangular CFST column section under axial load and biaxial bending.

column section is illustrated in Figure 10.14. The plane sections are assumed to remain plane after deformation, which results in a linear strain distribution through the depth of the cross section. The strain at the extreme fibre (ε_t) of the section is equal to ϕd_n. For $0° \le \theta < 90°$, the fibre strain is computed as follows (Liang 2009a):

$$c = \frac{d_n}{\cos\theta} \tag{10.25}$$

$$y_{n,i} = \left| x_i - \frac{B}{2} \right| \tan\theta + \left(\frac{D}{2} - c \right) \tag{10.26}$$

$$d_{e,i} = \left| y_i - y_{n,i} \right| \cos\theta \tag{10.27}$$

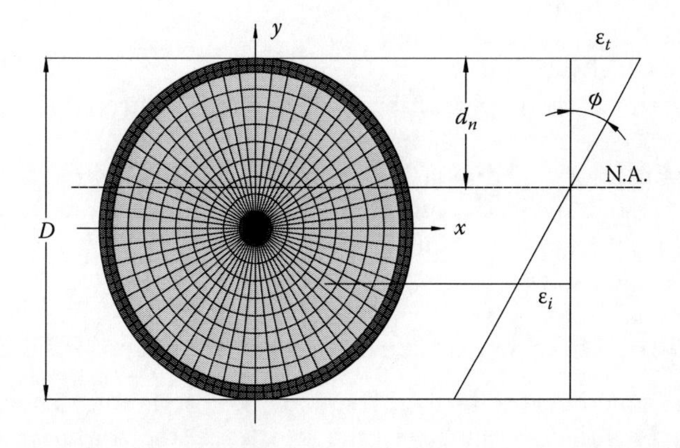

Figure 10.14 Strain distributions in a circular CFST column section.

$$\varepsilon_i = \begin{cases} \phi d_{e,i} & \text{for } y_i \geq y_{n,i} \\ -\phi d_{e,i} & \text{for } y_i < y_{n,i} \end{cases} \tag{10.28}$$

where

$d_{e,i}$ is the orthogonal distance from the centroid of each fibre element to the neutral axis
x_i and y_i are the coordinates of the fibre i
ε_i is the strain at the ith fibre

For $\theta = 90°$, the fibre strain is calculated as follows:

$$x_{n,i} = \frac{B}{2} - d_n \tag{10.29}$$

$$d_{e,i} = |x_i - x_{n,i}| \tag{10.30}$$

$$\varepsilon_i = \begin{cases} \phi d_{e,i} & \text{for } x_i \geq x_{n,i} \\ -\phi d_{e,i} & \text{for } x_i < x_{n,i} \end{cases} \tag{10.31}$$

10.3.4 Material constitutive models for structural steels

Figure 10.15 shows the idealised stress–strain curves for structural steels (Liang 2009a). A trilinear stress–strain relationship is assumed for mild structural steels both in compression and tension. The stress–stain behaviour of high-strength and cold-formed steels is characterised by a rounded stress–strain curve. A linear-rounded-linear stress–strain curve is therefore used for cold-formed steels, but for high-strength steels, the rounded part of the curve is replaced with a straight line as depicted in Figure 10.15. The rounded part of the stress–strain curve for cold-formed steels is determined by the following equation given by Liang (2009a):

$$\sigma_s = f_y \left(\frac{\varepsilon_s - 0.9\varepsilon_y}{\varepsilon_{st} - 0.9\varepsilon_y} \right)^{1/45} \quad (0.9\varepsilon_y < \varepsilon_s \leq \varepsilon_{st}) \tag{10.32}$$

where

σ_s denotes the stress in a steel fibre
ε_s represents the strain in a steel fibre
ε_y stands for the yield strain of steel
ε_{st} is the steel strain at strain hardening as depicted in Figure 10.15

The hardening strain ε_{st} is taken as $10\varepsilon_y$ for mild structural steels and 0.005 for high-strength and cold-formed steels. To reflect the ductility of different structural steels, the ultimate strain (ε_{su}) is taken as 0.2 for mild structural steels, while it is taken as 0.1 for high-strength and cold-formed steels.

10.3.5 Material models for concrete in rectangular CFST columns

The ductility of the concrete core in a rectangular CFST column is shown to increase due to the confinement provided by the steel tube. However, the confinement effect does not increase the compressive strength of the concrete core. The idealised stress–strain curve

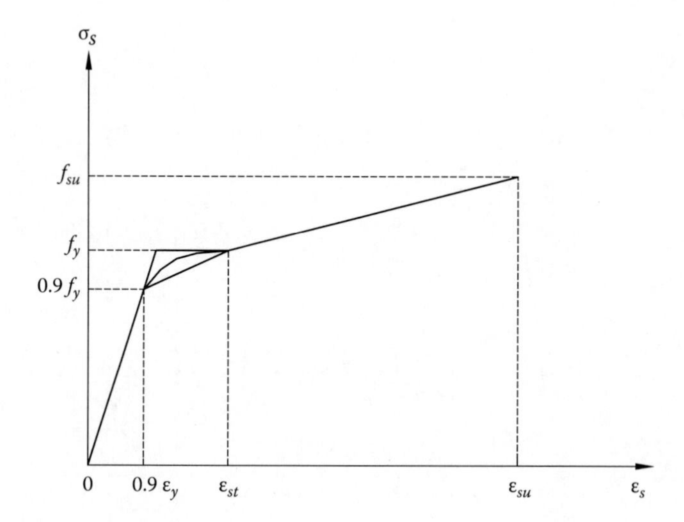

Figure 10.15 Stress–strain curves for structural steels.

depicted in Figure 10.16 is used in fibre element models to simulate the material behaviour of confined concrete in rectangular CFST columns (Liang 2009a). The part *OA* of the stress–strain curve given in Figure 10.16 is modelled using the following equations suggested by Mander et al. (1988):

$$\sigma_c = \frac{f'_{ce}\lambda\left(\varepsilon_c/\varepsilon'_{ce}\right)}{\lambda - 1 + \left(\varepsilon_c/\varepsilon'_{ce}\right)^{\lambda}} \tag{10.33}$$

$$\lambda = \frac{E_c}{E_c - \left(f'_{ce}/\varepsilon'_{ce}\right)} \tag{10.34}$$

$$E_c = 3320\sqrt{f'_{ce}} + 6900\,\text{MPa} \tag{10.35}$$

$$\varepsilon'_{ce} = \begin{cases} 0.002 & \text{for } f'_{ce} \le 28\,\text{MPa} \\ 0.002 + \dfrac{f'_{ce} - 28}{54{,}000} & \text{for } 28 < f'_{ce} \le 82\,\text{MPa} \\ 0.003 & \text{for } f'_{ce} > 82\,\text{MPa} \end{cases} \tag{10.36}$$

where

σ_c stands for the longitudinal compressive concrete stress
f'_{ce} is the effective compressive strength of concrete which is taken as $f'_{ce} = \gamma_c f'_c$
ε_c is the longitudinal compressive concrete strain
ε'_{ce} is the strain at f'_{ce}
E_c is Young's modulus of concrete (ACI-318 2011)

The strain ε'_{ce} is between 0.002 and 0.003 depending on the effective compressive strength of concrete. For the effective compressive strength of concrete between 28 and 82 MPa, the strain ε'_{ce} is determined by linear interpolation.

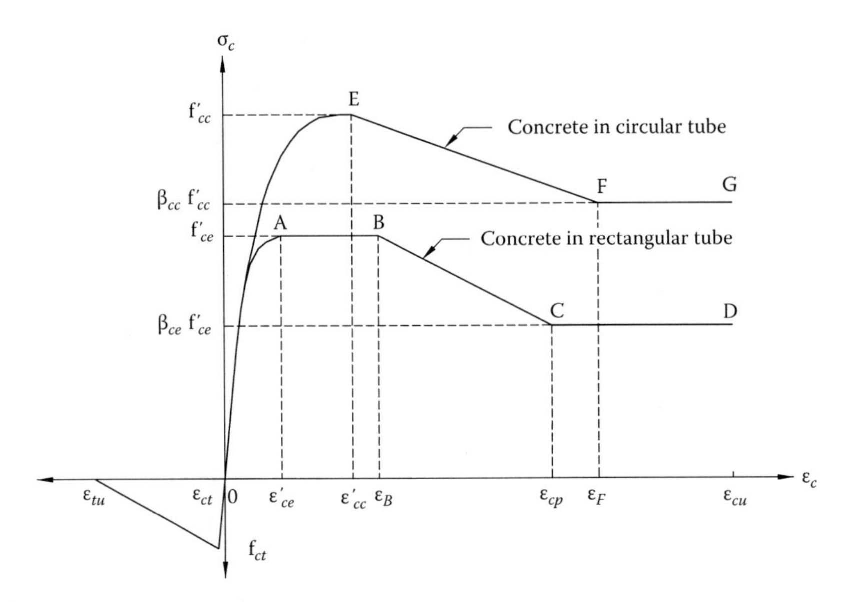

Figure 10.16 Idealised stress–strain curves for concrete in CFST columns.

The parts AB, BC and CD of the stress–strain curve for confined concrete depicted in Figure 10.16 are expressed by (Liang 2009a)

$$\sigma_c = \begin{cases} f'_{ce} & \text{for } \varepsilon'_{ce} < \varepsilon_c \leq \varepsilon_B \\ \beta_{ce} f'_{ce} + \dfrac{(\varepsilon_{cp} - \varepsilon_c)(f'_{ce} - \beta_{ce} f'_{ce})}{(\varepsilon_{cp} - \varepsilon_B)} & \text{for } \varepsilon_B < \varepsilon_c \leq \varepsilon_{cp} \\ \beta_{ce} f'_{ce} & \text{for } \varepsilon_c > \varepsilon_{cp} \end{cases} \tag{10.37}$$

where

$\varepsilon_B = 0.005$ and $\varepsilon_{cp} = 0.015$ are concrete compressive strains corresponding to points B and C shown in Figure 10.16

β_{ce} is the factor accounting for the confinement effect on the strength and ductility of concrete in the post-peak range, depending on the width-to-thickness ratio (B_s/t) of the section, where B_s is taken as the larger of B and D for a rectangular cross section. Based on the experimental results presented by Tomii and Sakino (1979a), β_{ce} is given by Liang (2009a) as

$$\beta_{ce} = \begin{cases} 1.0 & \text{for } \dfrac{B_s}{t} \leq 24 \\ 1.5 - \dfrac{1}{48}\left(\dfrac{B_s}{t}\right) & \text{for } 24 < \dfrac{B_s}{t} \leq 48 \\ 0.5 & \text{for } \dfrac{B_s}{t} > 48 \end{cases} \tag{10.38}$$

The stress–strain curve for concrete in tension is depicted in Figure 10.16. It is assumed that the tensile stress increases linearly with an increase in tensile strain up to concrete cracking. After concrete cracking, the tensile stress decreases linearly to zero as the concrete softens. The tensile strength of concrete (f_{ct}) is taken as $0.6\sqrt{f'_{ce}}$, while its ultimate tensile strain (ε_{tu}) is taken as 10 times of the strain at cracking (ε_{ct}).

10.3.6 Material models for concrete in circular CFST columns

The concrete confinement effect increases both the strength and ductility of concrete in circular CFST columns. An idealised stress–strain curve accounting for the confinement effect is also presented in Figure 10.16 (Liang and Fragomeni 2009; Liang 2011a). The part OE of the stress–strain curve shown in Figure 10.16 is represented using the equations suggested by Mander et al. (1988) as

$$\sigma_c = \frac{f'_{cc}\lambda(\varepsilon_c/\varepsilon'_{cc})}{\lambda - 1 + (\varepsilon_c/\varepsilon'_{cc})^{\lambda}} \tag{10.39}$$

$$\lambda = \frac{E_c}{E_c - (f'_{cc}/\varepsilon'_{cc})} \tag{10.40}$$

where

f'_{cc} stands for the compressive strength of the confined concrete

ε'_{cc} denotes the strain at f'_{cc}

When concrete is subjected to a laterally confining pressure, the uniaxial compressive strength f'_{cc} and the corresponding strain ε'_{cc} are much higher than those of unconfined concrete. The equations proposed by Mander et al. (1988) for the compressive strength and strain of confined concrete are modified using the strength reduction factor γ_c (Liang 2011a) as follows:

$$f'_{cc} = \gamma_c f'_c + k_1 f_{rp} \tag{10.41}$$

$$\varepsilon'_{cc} = \varepsilon'_c\left(1 + k_2\frac{f_{rp}}{\gamma_c f'_c}\right) \tag{10.42}$$

where

f_{rp} is the lateral confining pressure on the concrete core, expressed by Equation 10.4.

k_1 and k_2 are taken as 4.1 and 20.5, respectively, based on experimental results reported by Richart et al. (1928)

The strain ε'_c is the strain at f'_c of the unconfined concrete, given in Equation 10.23. Based on the work of Tang et al. (1996) and Hu et al. (2003), Liang and Fragomeni (2009) proposed an accurate model for predicting the confining pressure on normal or high-strength concrete confined by either normal or high-strength circular steel tubes, which is given in Equation 10.4.

The parts EF and FG of the stress–strain curve shown in Figure 10.16 are expressed by

$$\sigma_c = \begin{cases} \beta_{cc} f'_{cc} + \dfrac{(\varepsilon_F - \varepsilon_c)(f'_{cc} - \beta_{cc} f'_{cc})}{(\varepsilon_F - \varepsilon'_{cc})} & \text{for } \varepsilon'_{cc} < \varepsilon_c \le \varepsilon_F \\ \beta_{cc} f'_{cc} & \text{for } \varepsilon_c > \varepsilon_F \end{cases} \tag{10.43}$$

where

ε_F is taken as 0.02 based on experimental results

β_{cc} is the factor used to consider the effect of the confinement effect provided by the circular steel tube on the post-peak strength and ductility of confined concrete, given by Hu et al. (2003) as

$$\beta_{cc} = \begin{cases} 1.0 & \text{for } \frac{D}{t} \le 40 \\ 0.0000339\left(\frac{D}{t}\right)^2 - 0.010085\left(\frac{D}{t}\right) + 1.3491 & \text{for } 40 < \frac{D}{t} \le 150 \end{cases} \quad (10.44)$$

10.3.7 Modelling of local and post-local buckling

Local buckling of thin steel plates is influenced by the plate aspect ratio, width-to-thickness ratio, applied edge stress gradients, boundary conditions, geometric imperfections and residual stresses. The local and post-local buckling behaviour of thin steel plates in rectangular CFST beam–columns under stress gradients has been studied by Liang and Uy (2000) and Liang et al. (2007b) using the finite element method. Formulas have been developed for predicting the initial local buckling stresses of steel tube walls in rectangular CFST beam–columns with initial geometric imperfections and residual stresses (Liang and Uy 2000; Liang et al. 2007b). These formulas can be incorporated in non-linear analysis techniques to account for the local buckling effects of steel tubes on the behaviour of rectangular CFST beam–columns (Liang 2009a).

Thin steel plates have a very high reverse of post-local buckling strengths (Liang and Uy 2000; Liang et al. 2007b). The effective strength concept can be used to describe the post-local buckling strengths of steel plates in rectangular CFST beam–columns under axial load and biaxial bending. The effective strength formulas proposed by Liang et al. (2007b) have been incorporated in fibre element models to account for the effects of post-local buckling (Liang 2009a). The ultimate strength of the steel tube walls under stress gradients greater than zero can be estimated by

$$\frac{\sigma_{1u}}{f_y} = (1.5 - 0.5\alpha_s)\frac{\sigma_u}{f_y} \quad (10.45)$$

where

- σ_{1u} represents the ultimate stress corresponding to the maximum edge stress σ_1 at the ultimate strength limit state
- α_s is the stress gradient which is the ratio of the minimum edge stress σ_2 to the maximum edge stress σ_1 on the plate

For intermediate stress gradients, the ultimate stress σ_{1u} can be determined by linear interpolation.

The effective width concept is usually used to determine the post-local buckling strength of a thin steel plate under stress gradients as depicted in Figure 10.17. Effective width formulas can be incorporated in non-linear analysis methods to account for local buckling effects on the behaviour of rectangular CFST columns (Liang et al. 2006; Liang 2009a). The effective width formulas proposed by Liang et al. (2007b) for steel plates in rectangular CFST beam–columns under compressive stress gradients are expressed by

$$\frac{b_{e1}}{b} = \begin{cases} 0.2777 + 0.01019\left(\frac{b}{t}\right) - 1.972\times10^{-4}\left(\frac{b}{t}\right)^2 + 9.605\times10^{-7}\left(\frac{b}{t}\right)^3 & \text{for } \alpha_s > 0.0 \\ 0.4186 - 0.002047\left(\frac{b}{t}\right) + 5.355\times10^{-5}\left(\frac{b}{t}\right)^2 - 4.685\times10^{-7}\left(\frac{b}{t}\right)^3 & \text{for } \alpha_s = 0.0 \end{cases} \quad (10.46)$$

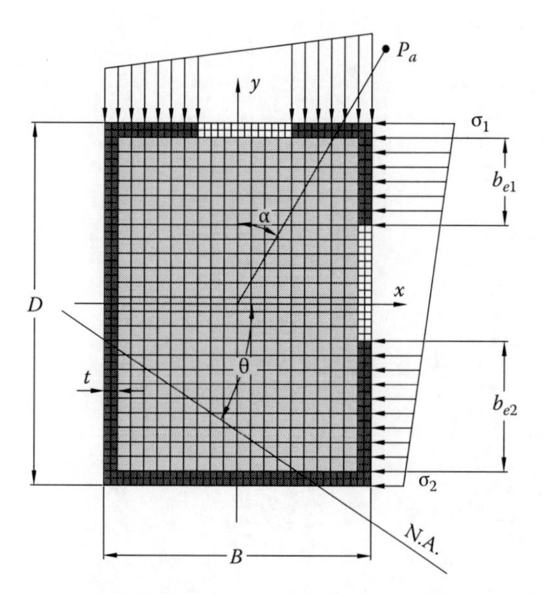

Figure 10.17 Effective steel areas of CFST beam–column under biaxial bending.

$$\frac{b_{e2}}{b} = (2 - \alpha_s)\frac{b_{e1}}{b} \tag{10.47}$$

where b_{e1} and b_{e2} are the effective widths as illustrated in Figure 10.17. If $(b_{e1} + b_{e2}) \geq b$, the steel plate is fully effective in carrying loads. For this case, the effective strength formulas should be used to valuate the ultimate strength of the steel plate.

The post-local buckling behaviour of thin steel plates under increased compressive edge stresses is characterised by the progressive stress redistribution within the buckled plates. The heavily buckled region in a steel plate sustains relatively low stresses, while its two edge strips carry high stresses (Liang and Uy 1998). For steel plates under uniform compression, the effective width concept assumes that effective steel fibres are stressed to the yield strength of the steel plates, while the stresses in ineffective steel fibres are zero at the ultimate strength limit state. After the onset of local buckling, the ineffective width of a steel tube wall increases from zero to the maximum value ($b_{ne,\max}$) when the applied load is increased to its ultimate load, where $b_{ne,\max}$ is given by

$$b_{ne,\max} = b - \left(b_{e1} + b_{e2}\right) \tag{10.48}$$

The ineffective width of a steel tube wall between zero and $b_{ne,\max}$ under stress gradients is approximately calculated using linear interpolation based on its stress level as

$$b_{ne} = \left(\frac{\sigma_1 - \sigma_{1c}}{f_y - \sigma_{1c}}\right) b_{ne,\max} \tag{10.49}$$

σ_{1c} is the initial local buckling stress of the steel tube wall with imperfections.

For a steel tube wall under stress gradients, the effective width concept assumes that the steel tube wall attains its ultimate strength when its maximum edge stress σ_1 is stressed to the yield strength of the steel wall. The steel fibres within the ineffective width (b_{ne}) are assigned to zero stress, and their contributions to the strength of the CFST column are ignored as illustrated in Figure 10.17.

10.3.8 Stress resultants

In fibre element analysis, fibre stresses are calculated from fibre strains using material stress–strain relationships. The axial force and bending moments in the composite section are determined as stress resultants:

$$P = \sum_{i=1}^{ns} \sigma_{s,i} A_{s,i} + \sum_{j=1}^{nc} \sigma_{c,j} A_{c,j} \tag{10.50}$$

$$M_x = \sum_{i=1}^{ns} \sigma_{s,i} A_{s,i} y_i + \sum_{j=1}^{nc} \sigma_{c,j} A_{c,j} y_j \tag{10.51}$$

$$M_y = \sum_{i=1}^{ns} \sigma_{s,i} A_{s,i} x_i + \sum_{j=1}^{nc} \sigma_{c,j} A_{c,j} x_j \tag{10.52}$$

where

P denotes the axial force
M_x represents the bending moment about the x-axis
M_y is the bending moment about the y-axis
$\sigma_{s,i}$ stands for the longitudinal stress at the centroid of steel fibre i
$A_{s,i}$ is the area of steel fibre i
$\sigma_{c,j}$ is the longitudinal stress at the centroid of concrete fibre j
$A_{c,j}$ is the area of concrete fibre j
x_i and y_i are the coordinates of steel fibre i
x_j and y_j are the coordinates of concrete fibre j
ns is the total number of steel fibre elements
nc is the total number of concrete fibre elements

Compressive stresses are taken to be positive.

10.3.9 Computational algorithms based on the secant method

10.3.9.1 Axial load–strain analysis

The ultimate axial strength of a short composite column under axial compression is determined as the maximum axial load from its complete axial load–strain curve. The axial load–strain curve for a short composite column can be obtained by gradually increasing the axial strain and calculating the corresponding stress resultant in the cross section. The iterative analysis process can be stopped when the axial load drops below a specified percentage of the maximum axial load (P_{max}) such as $0.5P_{max}$ or when the axial strain in concrete exceeds the specified ultimate strain ε_{cu} (Liang 2009a). The effects of local buckling are taken into account in the ultimate axial load of thin-walled CFST columns by redistributing the normal stresses on the steel tube walls.

The axial load–strain analysis procedure for CFST short columns is given as follows:

1. Input data.
2. Discretise the composite section into fibre elements.
3. Initialise axial fibre strains $\varepsilon = \Delta\varepsilon$.
4. Compute fibre stresses using stress–strain relationships.

5. Check local buckling and update steel fibre stresses accordingly.
6. Calculate the resultant axial force P.
7. Increase axial fibre strains by $\varepsilon = \varepsilon + \Delta\varepsilon$.
8. Repeat Steps 4–7 until $P < 0.5P_{max}$ or $\varepsilon > \varepsilon_{cu}$.

10.3.9.2 Moment–curvature analysis

The axial load–moment–curvature relationships are established to determine the ultimate moment capacities of short composite columns under combined axial load and biaxial bending. For a given axial load (P_u) applied at a fixed load angle (α) as shown in Figure 10.13, the corresponding ultimate moment capacity of the composite section is determined as the maximum moment from the moment–curvature curve, which is obtained by gradually increasing the curvature and solving for the corresponding moment. A typical movement–curvature curve for a CFST short column under biaxial loads predicted by the computer program NACOMS (Liang 2009a,b) is presented in Figure 10.3. The equilibrium conditions for the composite section under axial load and biaxial bending are expressed by

$$P_u - P = 0 \tag{10.53}$$

$$\tan\alpha - \frac{M_y}{M_x} = 0 \tag{10.54}$$

In the moment–curvature analysis, the depth of the neutral axis (d_n) in the composite section needs to be iteratively adjusted to satisfy the force equilibrium condition. After the force equilibrium has been achieved, internal moments M_x and M_y are then calculated and the orientation of the neutral axis (θ) is iteratively adjusted to satisfy both the force and moment equilibrium conditions. Efficient computational algorithms based on the secant method have been developed and implemented in the fibre element analysis programs by Liang (2009a) to adjust the depth and orientation of the neutral axis in a CFST beam–column section to satisfy equilibrium conditions.

The depth of the neutral axis (d_n) is adjusted by the following equation (Liang 2009a):

$$d_{n,\,j+2} = d_{n,\,j+1} - \frac{(d_{n,\,j+1} - d_{n,\,j})r_{p,\,j+1}}{r_{p,\,j+1} - r_{p,\,j}} \tag{10.55}$$

where
the subscript j is the iteration number
$r_p = P_u - P$ is the residual axial force in the composite section at the current iteration

The convergence criterion for the neutral axis depth d_n is expressed by $|d_{n,j+1} - d_n| \le \varepsilon_k$, where ε_k is the convergence tolerance which is taken as 10^{-4}.

The orientation of the neutral axis with respect to the x-axis as shown in Figure 10.13 is adjusted by the following equation (Liang 2009a):

$$\theta_{k+2} = \theta_{k+1} - \frac{\left(\theta_{k+1} - \theta_k\right)r_{m,\,k+1}}{r_{m,\,k+1} - r_{m,\,k}} \tag{10.56}$$

where
the subscript k is the iteration number
$r_m = \tan\alpha - M_y/M_x$ is the residual moment in the composite section at the current iteration

The convergence criterion for the orientation of the neutral axis θ is given by $|\theta_{k+1} - \theta_k| \le \varepsilon_k$.

The secant method needs two initial values to start the iterative process. Initial values for the neutral axis depth $d_{n,1}$ and $d_{n,2}$ can be set to D and $D/2$, respectively, while initial values for the orientation of the neutral axis θ_1 and θ_2 can be set to α and $\alpha/2$ (Liang 2009a). In order to adjust d_n, the force residuals $r_{p,1}$ and $r_{p,2}$ are calculated using $d_{n,1}$ and $d_{n,2}$, respectively. Similarly, the moment residuals $r_{m,1}$ and $r_{m,2}$ are computed in order to adjust the orientation of the neutral axis. It should be noted that for short composite columns under axial load and uniaxial bending, only the depth of the neutral axis needs to be adjusted (Liang 2011a).

The moment–curvature analysis procedure for CFST short beam–columns incorporating local buckling effects is given as follows:

1. Input data.
2. Discretise the composite section into fibre elements.
3. Initialise curvature $\phi = \Delta\phi$.
4. Initialise $\theta_1 = \alpha$, $\theta_2 = \alpha/2$, $d_{n,1} = D$, $d_{n,2} = D/2$.
5. Compute fibre stresses using stress–strain relationships.
6. Check local buckling and update steel fibre stresses accordingly.
7. Calculate residual forces and moments $r_{p,1}$, $r_{p,2}$, $r_{m,1}$ and $r_{m,2}$.
8. Compute fibre stresses using stress–strain relationships.
9. Check local buckling and update steel fibre stresses accordingly.
10. Calculate the resultant axial force P.
11. Adjust the neutral axis depth (d_n) using the secant method.
12. Repeat Steps 8–11 until $|r_p| < \varepsilon_k$.
13. Compute bending moments M_x and M_y.
14. Adjust the neutral axis orientation (θ) using the secant method.
15. Repeat Steps 8–14 until $|r_m| < \varepsilon_k$.
16. Compute the resultant moment $M = \sqrt{M_x^2 + M_y^2}$.
17. Increase the curvature by $\phi = \phi + \Delta\phi$.
18. Repeat Steps 4–17 until $M < 0.5M_{max}$ or $\varepsilon_c > \varepsilon_{cu}$.

10.3.9.3 Axial load–moment interaction diagrams

In order to develop the axial load–moment interaction diagram for a short composite column under axial load and biaxial bending, the ultimate axial strength (P_o) of the composite column under axial compression is calculated first by conducting an axial load–strain analysis of the composite section. The axial load (P_u) is increased from zero to a maximum value of $0.9P_o$, and each load step is taken as $0.1P_o$. For a given load increment (P_u) applied at a fixed load angle (α), the moment–curvature analysis of the composite section is performed to obtain the corresponding moment capacity M_u. By gradually increasing the applied load and solving for the corresponding moment capacity, a set of axial loads and moment capacities can be obtained and used to plot the axial load–moment interaction diagram. The computer program NACOMS developed by Liang (2009a,b) can generate axial load–strain curves, moment–curvature curves and axial load–moment interaction diagrams for biaxially loaded thin-walled CFST short beam–columns with local buckling effects.

The computational procedure for determining the axial load–moment interaction diagrams for composite columns under axial load and biaxial bending is given as follows:

1. Input data.
2. Discretise the composite section into fibre elements.
3. Compute P_o using the axial load–strain analysis procedure.

4. Set the axial load $P_u = 0$.
5. Calculate M_u using the moment–curvature analysis procedure.
6. Increase the axial load by $P_u = P_u + \Delta P_u$, where $\Delta P_u = P_o/10$.
7. Repeat Steps 5–6 until $P_u > 0.9P_o$.
8. Plot the axial load–moment interaction diagram.

Example 10.3: Computer analysis of CFST short column under axial load and biaxial bending

The computer program NACOMS is employed to analyse a square thin-walled CFST short column under axial load and biaxial bending. The dimensions of the cross section of the column are 500 × 500 mm. The thickness of the steel tube is 8 mm. The compressive strength of the concrete infill is 40 MPa. The yield stress of the steel section is 320 MPa, while its tensile strength is 430 MPa. Young's modulus of the steel tube is 200 GPa. In the moment–curvature analysis, the axial load applied to the column is taken as $0.6P_o$. The angle of the applied axial load is fixed at 45° with respect to y-axis of the column cross section. It is required to determine the axial load–strain curve, moment–curvature curve and axial load–moment interaction diagram for this CFST short column under axial load and biaxial bending.

Computer solution

The slenderness of the steel tube wall is

$$\frac{D}{t} = \frac{500}{8} = 62.5 > 52\sqrt{\frac{235}{f_y}} = 52\sqrt{\frac{235}{320}} = 44.6$$

Hence, the steel tube is non-compact. The method given in Eurocode 4 cannot be used to determine the axial load–moment interaction diagram of this CFST short column. The effect of local buckling is taken into account in the computer analysis of the CFST column.

In the fibre element analysis, the steel tube wall is divided into five layers through its thickness, and the concrete core is divided into 80 × 80 fibre elements. The axial load–strain curve for this column is presented in Figure 10.18. It appears that the ultimate axial

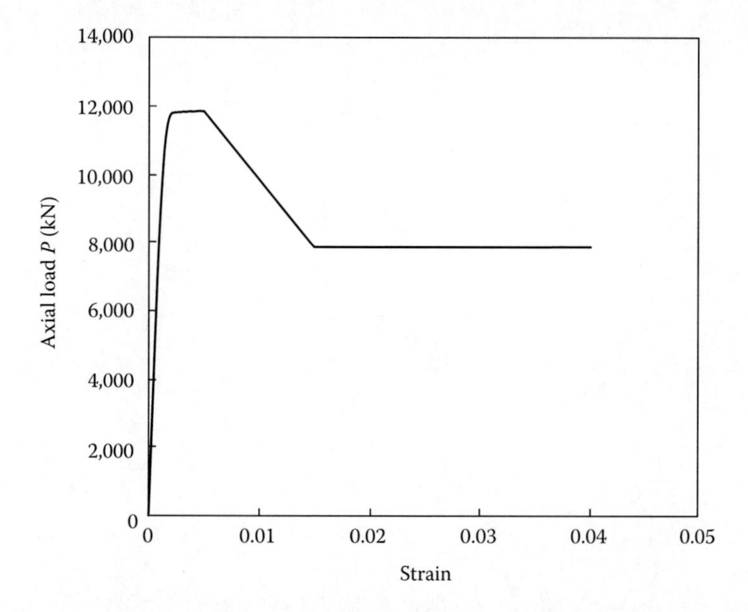

Figure 10.18 Axial load–strain curve for the thin-walled CFST short column.

load of the CFST short column is 11,863 kN. The predicted moment–curvature curve for the column is shown in Figure 10.19. The predicted ultimate moment of the composite section under the axial load level of $0.6P_o$ is 983 kN m. Figure 10.20 shows the axial load–moment interaction diagram of the composite section under axial load and biaxial bending. It can be seen from the figure that the ultimate pure bending moment is 1104.3 kN m, while the maximum ultimate moment of the CFST column section is 1299 kN m.

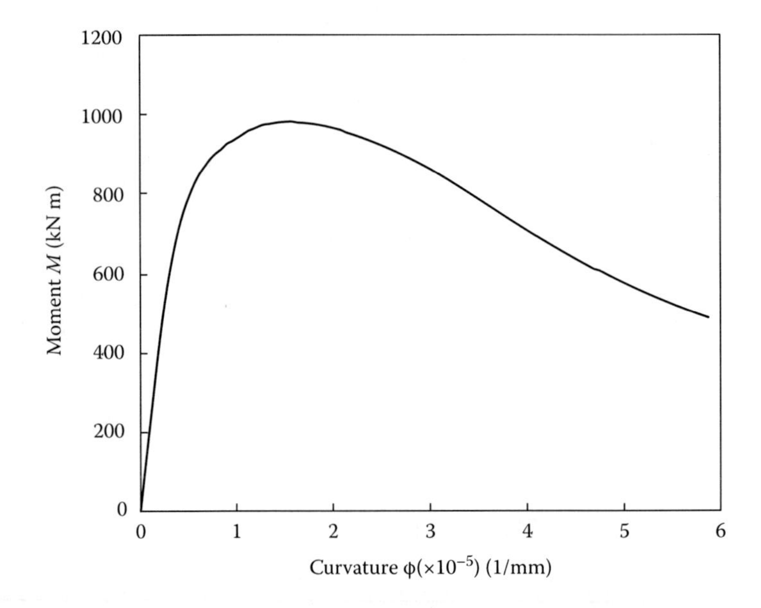

Figure 10.19 Moment–curvature curve for the thin-walled CFST short column under axial load and biaxial bending.

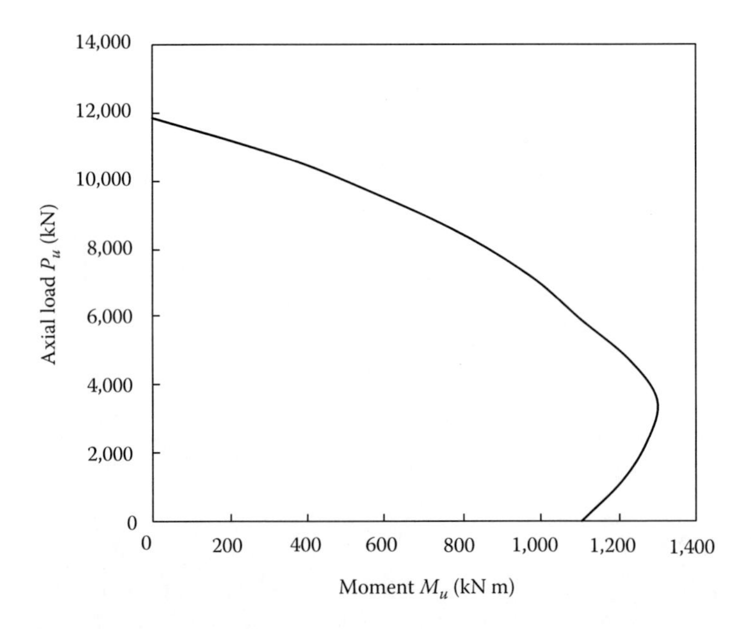

Figure 10.20 Axial load–moment interaction diagram of the thin-walled CFST short column under axial load and biaxial bending.

10.4 BEHAVIOUR AND DESIGN OF SLENDER COMPOSITE COLUMNS

10.4.1 Behaviour of slender composite columns

Eccentrically loaded beam–columns with a slenderness ratio (L/r) greater than 22 are usually treated as slender beam–columns in design codes. The behaviour of slender composite columns has been investigated experimentally by researchers (Furlong 1967; Knowles and Park 1969; Neogi et al. 1969; Bridge 1976; Shakir-Khalil and Zeghiche 1989; Shakir-Khalil and Mouli 1990; Rangan and Joyce 1992; Muñoz and Hsu 1997; Vrcelj and Uy 2002a; Mursi and Uy 2006). The behaviour of slender composite columns under eccentric loading is characterised by their axial load–deflection curves which indicate the flexural stiffness, the ultimate axial strength, the post-peak behaviour and the flexural ductility of the columns. CEC slender columns may fail by inelastic global buckling associated with yielding of the steel section and reinforcement and crushing of the concrete. CFST slender columns may fail by the interaction of inelastic local and global buckling associated with yielding of the steel section and crushing of the infill concrete. Very slender composite columns fail by elastic global buckling. The ultimate strengths of slender composite columns are usually governed by the flexural stiffness rather than the material strengths of the steel and concrete.

The fundamental behaviour of slender composite beam–columns is influenced by many parameters, including column slenderness ratio, depth-to-thickness ratio, loading eccentricity, concrete compressive strength, steel yield strength, initial geometric imperfections and second-order effects. Numerical studies conducted by Liang (2011b) and Patel et al. (2012a,b,c; 2014c) demonstrate that increasing the column slenderness ratio (L/r) or the loading eccentric ratio (e/D) significantly reduces the initial flexural stiffness and ultimate axial strength of the CFST beam–column under eccentric loading but remarkably increases its lateral deflection and displacement ductility. For a given axial load level, the corresponding ultimate moment capacity of the CFST column is found to reduce by increasing the column slenderness ratio. In addition, local buckling is found to reduce the flexural stiffness and strength of rectangular CFST slender beam–columns. Moreover, increasing the concrete compressive strength slightly increases the initial flexural stiffness but significantly increases the ultimate axial strength of the CFST beam–column under eccentric loading. Furthermore, the initial flexural stiffness of the eccentrically loaded CFST beam–column is shown to be not affected by the yield stress of the steel tube. However, increasing the yield stress of the steel tube remarkably increases the ultimate axial strength of the CFST beam–column. The studies on confinement effects conducted by Liang (2011b) show that in circular CFST beam–columns, the concrete confinement effect decreases with an increase in the column slenderness ratio or the loading eccentricity. For very slender circular CFST beam–columns with an L/r ratio greater than 70 or for slender circular CFST beam–columns with an e/D ratio greater than 2, the confinement effect can be ignored in the design.

10.4.2 Relative slenderness and effective flexural stiffness

In Eurocode (2004), the slenderness of a composite column is measured by its relative slenderness for the bending plane being considered as follows:

$$\bar{\lambda} = \sqrt{\frac{P_o}{P_{cr}}} \tag{10.57}$$

where

P_o is the ultimate axial strength of the composite column section under axial compression ignoring the confinement effect, given by Equation 10.1, in which γ_c is taken as 0.85 for CEC columns and 1.0 for CFST columns

P_{cr} is the elastic critical buckling load of the composite column under axial compression, given by

$$P_{cr} = \frac{\pi^2 (EI)_{eff}}{L_e^2} \tag{10.58}$$

in which $(EI)_{eff}$ represents the effective flexural stiffness of the cross section of a composite column, which is expressed by

$$(EI)_{eff} = E_s I_s + 0.6 E_{cm} I_c + E_r I_r \tag{10.59}$$

where

E_s, E_{cm} and E_r are the elastic moduli of structural steel, concrete and reinforcement, respectively

I_s, I_c and I_r are the second moments of area of structural steel section, concrete and reinforcement, respectively

The effective flexural stiffness $(EI)_{eff}$ should account for the long-term effects due to concrete creep on the elastic modulus of concrete (E_{cm}) by using the effective elastic modulus of concrete considering the long-term effect of concrete creep, which is expressed by

$$E_{c,eff} = \frac{E_{cm}}{1 + \left(P_G^* / P^*\right)\phi_c^*} \tag{10.60}$$

where

ϕ_c^* is the final concrete creep factor

P_G^* is the permanent part of the design axial force P^*

The elastic modulus of concrete (E_{cm}) can be calculated by

$$E_{cm} = 22000\left(\frac{f_c' + 8}{10}\right)^{0.3} \text{ MPa} \tag{10.61}$$

For determining internal design actions on a slender composite column, the effective flexural stiffness considering long-term effects is given in Eurocode 4 as

$$(EI)_{eff,II} = 0.9(E_s I_s + 0.5 E_{c,eff} I_c + E_r I_r) \tag{10.62}$$

10.4.3 Concentrically loaded slender composite columns

A simple method is given in Eurocode 4 (2004) for determining the ultimate axial strength of slender composite columns under axial compression as follows:

$$P_u = \chi P_o \tag{10.63}$$

where χ is the reduction factor which is a function of the relative slenderness $\bar{\lambda}$ and imperfections given in Eurocode 3 (2005) and is expressed by

$$\chi = \frac{1}{\varphi + \sqrt{\varphi^2 - \bar{\lambda}^2}} \leq 1.0 \tag{10.64}$$

$$\varphi = 0.5\left[1 + \alpha_g\left(\bar{\lambda} - 0.2\right) + \bar{\lambda}^2\right] \tag{10.65}$$

where α_g is the imperfection factor given in Table 10.1.

The column buckling curves for slender composite columns under axial compression determined using Equation 10.64 are presented in Figure 10.21. The buckling curves and imperfections for different composite columns are given in Table 10.1 as specified in Eurocode 4 (2004).

Table 10.1 Buckling curves and member imperfections for composite columns

Composite column	*Buckling curve*	*Imperfection factor* α_g	*Member imperfection*
CEC or partially encased composite columns			
Bending about the *x*-axis	b	0.34	*L*/200
Bending about the *y*-axis	c	0.49	*L*/150
Circular/rectangular CFST columns			
$\rho_s \leq 3\%$	a	0.21	*L*/300
$3\% < \rho_s \leq 6\%$	b	0.34	*L*/200
Circular CFST columns with I-sections	b	0.34	*L*/200

Source: Eurocode 4 (2004) *Design of Composite Steel and Concrete Structures, Part 1-1: General Rules and Rules for Buildings*, European Committee for Standardization, CEN.

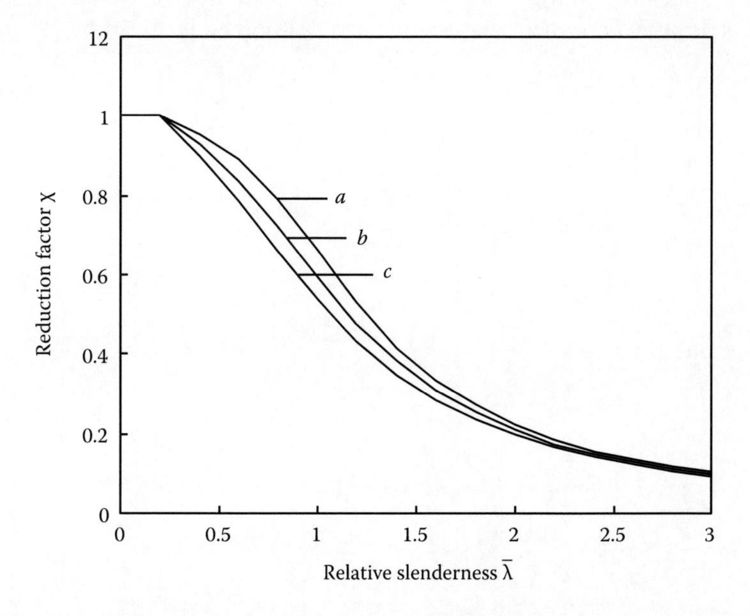

Figure 10.21 Buckling curves for composite columns under axial compression.

10.4.4 Uniaxially loaded slender composite columns

10.4.4.1 Second-order effects

Eurocode 4 (2004) suggests that design actions on slender composite columns may be calculated by an elastic global analysis of the composite frame incorporating the global second-order effects and global imperfections. Equivalent geometric imperfections can be used to account for the effects of initial geometric imperfections and residual stresses on the strength and behaviour of slender composite columns. The analysis of individual slender composite beam–column under axial force and end moments determined from the global analysis must consider the second-order effects in the column and the column imperfections. The second-order effects and equivalent geometric imperfections amplify the design bending moments on the slender composite column. The design method is to determine the amplified design bending moment on the slender composite column due to second-order effects as well as equivalent geometric imperfections. For a given design axial load, if the amplified design moment is still less than or equal to the design moment capacity of the column cross section, the slender composite column satisfies the strength requirement.

A pin-ended slender composite beam–column subjected to an axial load and bending moments M_1^* and M_2^* shown in Figure 10.22 is considered here to explain the second-order effects. The end moments M_1^* and M_2^* cause the beam–column to bend into a single curvature. This results in an additional deflection u along the beam–column and an additional moment P^*u, which is called the secondary moment. The maximum moment on the beam–column due to second-order effects is used to design the beam–column and is determined by amplifying the maximum end moment M_1^* using the moment amplification factor. This means that the end moment amplified by the second-order effects is determined as $M_{end}^* = \delta_m M_1^*$. In Eurocode 4 (2004), the moment amplification factor is expressed by

$$\delta_m = \frac{c_b}{1-\left(P^*/P_{cr,eff}\right)} \tag{10.66}$$

where $P_{cr,eff}$ is the elastic buckling load at the composite column calculated using $(EI)_{eff,II}$ and c_b accounts for the effects of different moments at column ends, given by

$$c_b = 0.66 - 0.44\beta_m \geq 0.44 \tag{10.67}$$

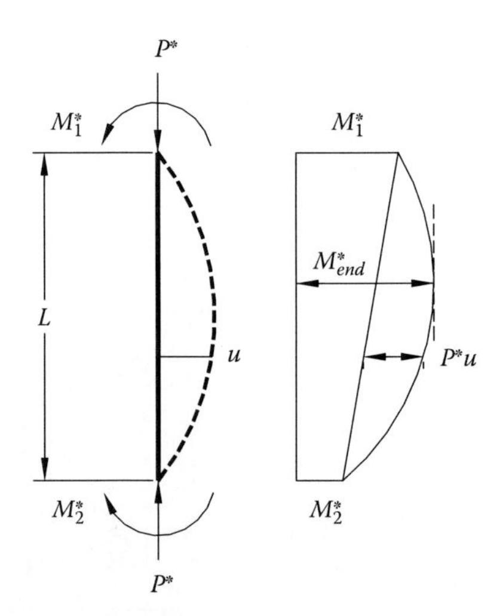

Figure 10.22 Second-order effects on a slender composite beam–column.

in which the moment ratio $\beta_m = \pm M_2^*/M_1^*$, which is taken as negative for single curvature bending and positive for double curvature bending.

The second-order effects due to the equivalent geometric imperfection (u_o) at the mid-height of a slender composite beam–column also cause an additional moment P^*u_o at its mid-height. The moment at the mid-height of the composite column induced by geometric imperfections is determined as $M_{imp}^* = \delta_m P^* u_o$, where δ_m is calculated using $c_b = 1.0$ in Equation 10.66.

The design bending moment for the slender composite column accounting for second-order effects is calculated as

$$M^* = M_{end}^* + M_{imp}^* \tag{10.68}$$

10.4.4.2 Design moment capacity

The design moment capacity of a slender composite beam–column depends on the design axial load level. The load ratio is calculated as $\chi_d = P^*/P_o$, which is drawn on the dimensionless axial load–moment interaction diagram of the composite column section. The moment capacity factor μ_d corresponding to χ_d can be determined from the interaction diagram for the composite section as illustrated in Figure 10.23. The slender composite column under axial load and uniaxial bending must satisfy the following design requirement:

$$M^* \leq \phi M_u \tag{10.69}$$

where

$\phi = 0.8$ is the capacity reduction factor

$M_u = \alpha_M \mu_d M_o$ is the nominal moment capacity of the slender composite column

The reduction factor α_M accounts for the effect of unconservative assumption of the rectangular stress block which is extended to the plastic neutral axis. The factor α_M is taken as

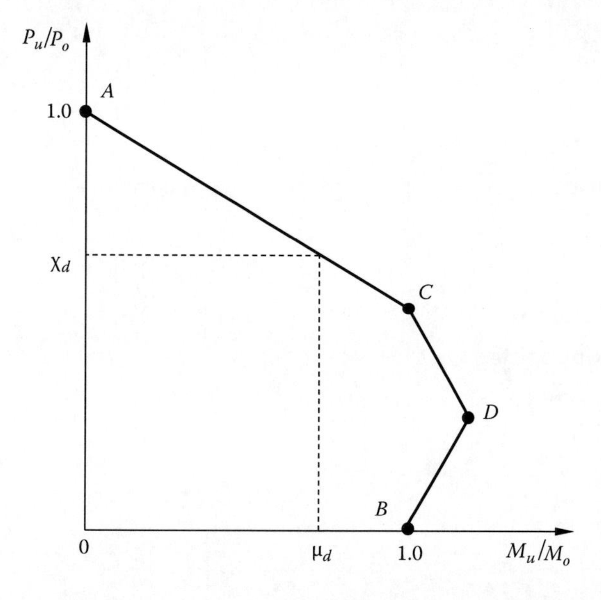

Figure 10.23 Dimensionless axial load–moment interaction diagram of a composite short column.

0.9 for steel grades with yield stress between 235 and 355 MPa and 0.8 for steel grades with yield stress between 420 and 460 MPa.

The main steps for checking the strength of a slender composite column under axial load and uniaxial bending are given as follows:

1. Determine the axial load–moment interaction diagram for the column section.
2. Calculate the effective flexural stiffness $(EI)_{eff,II}$ for the composite column.
3. Compute the critical buckling load $P_{cr,eff}$ using $(EI)_{eff,II}$.
4. Calculate M^* accounting for second-order effects and geometric imperfections.
5. Determine μ_d corresponding to χ_d on the interaction diagram.
6. Check the design moment capacity: $M^* \le \phi M_u$.

Example 10.4: Strength of CEC slender column under axial load and uniaxial bending

The cross section of a CEC slender column is shown in Figure 10.8. The column of 4 m length is subjected to a design axial compressive force $P^* = 7469\,\text{kN}$ of which 4855 kN is permanent and design bending moments $M_1^* = 300\,\text{kN m}$ and $M_2^* = 150\,\text{kN m}$ at the ends. The column is bent into single curvature about the x–x-axis. The design data are as follows: $f_c' = 32\,\text{MPa}$, $f_y = 300\,\text{MPa}$, $f_{yr} = 500\,\text{MPa}$ and $E_s = E_r = 200{,}000\,\text{MPa}$. The axial load–moment interaction diagram for the composite column section has been determined in Example 10.1. The final concrete creep factor is $\phi_c^* = 3.0$. Check the design moment capacity of this slender composite column.

1. Second moments of area of uncracked section

The second moment of area of steel section is

$$I_s = \frac{b_f D_s^3}{12} - \frac{(b_f - t_w)d_w^3}{12} = \frac{350 \times 350^3}{12} - \frac{(350-12)\times 318^3}{12} = 344.75 \times 10^6\ \text{mm}^4$$

The second moment of area of reinforcement is

$$I_r = 4\left(\frac{\pi r^4}{4} + \pi r^2 d_r\right) = 4\left(\frac{\pi(20/2)^4}{4} + \pi\left(\frac{20}{2}\right)^2 \times 200\right) = 0.2827 \times 10^6\ \text{mm}^4$$

The second moment of area of concrete in the section is computed as

$$I_c = \frac{BD^3}{12} - I_s - I_r = \frac{500 \times 500^3}{12} - 344.75 \times 10^6 - 0.2827 \times 10^6 = 4860 \times 10^6\ \text{mm}^4$$

2. Effective flexural stiffness

The effective modulus of concrete is calculated as

$$E_{cm} = 22{,}000\left(\frac{f_c' + 8}{10}\right)^{0.3} = 33{,}346\,\text{MPa}$$

$$E_{c,eff} = \frac{E_{cm}}{1 + \left(P_G^*/P^*\right)\phi_c^*} = \frac{33{,}346}{1 + (4{,}855/7{,}469) \times 3} = 11{,}304\,\text{MPa}$$

The effective flexural stiffness of the composite column is computed as

$$(EI)_{eff,II} = 0.9(E_s I_s + 0.5E_{c,eff} I_c + E_r I_r)$$

$$= 0.9 \times (200 \times 10^3 \times 344.75 + 0.5 \times 11{,}304 \times 4{,}860 + 200 \times 10^3 \times 0.2827) \times 10^6$$

$$= 868 \times 10^{11}\,\text{N mm}^2$$

3. Amplified design bending moment

The critical buckling load of the composite column is determined as

$$P_{cr,eff} = \frac{\pi^2 (EI)_{eff,II}}{L_e^2} = \frac{\pi^2 \times 868 \times 10^{11}}{4{,}000^2}\,\text{N} = 53{,}543\,\text{kN}$$

The composite column is bent into a single curvature so that its moment ratio is

$$\beta_m = -\frac{M_2^*}{M_1^*} = -\frac{150}{300} = -0.5$$

The moment amplification factor is calculated as follows:

$$c_b = 0.66 - 0.44\beta_m = 0.66 - 0.44 \times (-0.5) = 0.88$$

$$\delta_m = \frac{c_b}{1 - (P^*/P_{cr,eff})} = \frac{0.88}{1 - (7{,}469/53{,}543)} = 1.023$$

The amplified design bending moment at the column end due to second-order effects is computed as

$$M_{end}^* = \delta_m M_1^* = 1.023 \times 300 = 306.8\,\text{kN m}$$

The equivalent geometric imperfection at the mid-height of the column is

$$u_o = \frac{L}{200} = \frac{4000}{200} = 20\,\text{mm}$$

For moment caused by geometric imperfection, $c_b = 1.0$.

The moment amplification factor for geometric imperfections is calculated as

$$\delta_m = \frac{c_b}{1 - (P^*/P_{cr,eff})} = \frac{1.0}{1 - (7{,}469/53{,}543)} = 1.162$$

The amplified design bending moment due to geometric imperfections is

$$M_{imp}^* = \delta_m (P^* u_o) = 1.162 \times (7469 \times 0.02) = 173.5\,\text{kN m}$$

The design moment on the composite column considering second-order effects is

$$M^* = M_{end}^* + M_{imp}^* = 306.8 + 173.5 = 480.3\,\text{kN m} > M_1^* = 300\,\text{kN m}$$

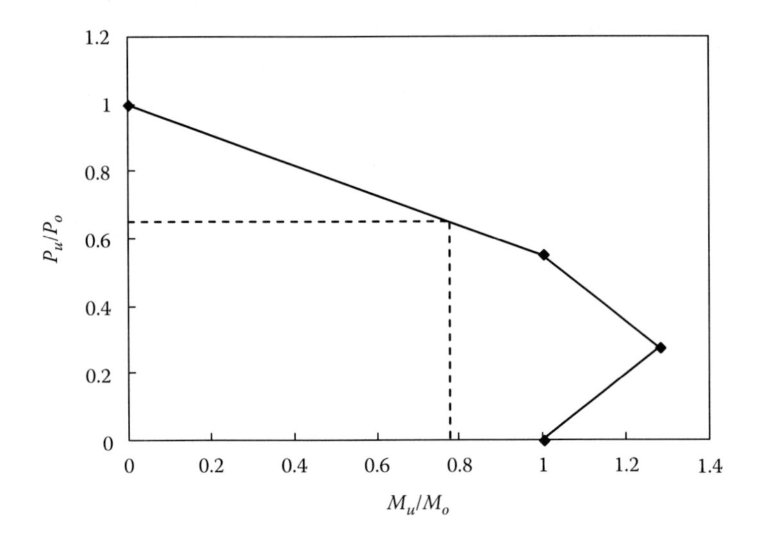

Figure 10.24 Dimensionless axial load–moment interaction diagram of the CEC short column.

4. Design moment capacity

The axial load ratio is computed as

$$\chi_d = \frac{P^*}{P_o} = \frac{7{,}469}{11{,}490.5} = 0.65$$

From the axial load–moment interaction diagram of the composite section shown in Figure 10.24, the moment capacity factor is obtained as

$$\mu_d = 0.78$$

The design moment capacity of the slender composite column is determined as

$$\phi M_u = \phi \alpha_M \mu_d M_o = 0.8 \times 0.9 \times 0.78 \times 917.5 = 515.2\,\text{kN m} > M^* = 480.3\,\text{kN m, OK}$$

Example 10.5: Strength of CFST slender column under axial load and uniaxial bending

The cross section of a CFST slender column is shown in Figure 10.10. The column of 8 m length is subjected to a design axial compressive force P^* = 1420 kN of which 9230 kN is permanent and design bending moments $M_1^* = 1200\,\text{kN m}$ and $M_2^* = 0\,\text{kN m}$ at the ends. The column is bent into single curvature about the x–x-axis. The design data are as follows: $f_c' = 50\,\text{MPa}$, f_y = 300 MPa, $E_s = E_r$ = 200,000 MPa. The axial load–moment interaction diagram for the composite column section has been determined in Example 10.2. The final concrete creep factor is $\phi_c^* = 3.0$. Check the design moment capacity of this slender composite column.

1. Second moments of area of uncracked section

The second moment of area of the concrete core is computed as

$$I_c = \frac{(B - 2t)(D - 2t)^3}{12} = \frac{(500 - 2 \times 20)(600 - 2 \times 20)^3}{12} = 6732 \times 10^6\ \text{mm}^4$$

The second moment of area of steel section is

$$I_s = \frac{BD^3}{12} - I_c = \frac{500 \times 600^3}{12} - 6732 \times 10^6 = 2268 \times 10^6 \text{ mm}^4$$

2. Effective flexural stiffness

The effective modulus of concrete is calculated as

$$E_{cm} = 22{,}000\left(\frac{f_c' + 8}{10}\right)^{0.3} = 22{,}000\left(\frac{50+8}{10}\right)^{0.3} = 37{,}278 \text{ MPa}$$

$$E_{c,eff} = \frac{E_{cm}}{1 + \left(P_G^*/P^*\right)\phi_c^*} = \frac{37{,}278}{1 + (9{,}230/14{,}200) \times 3} = 12{,}637 \text{ MPa}$$

The effective flexural stiffness of the composite column is computed as

$$(EI)_{eff,II} = 0.9(E_s I_s + 0.5 E_{c,eff} I_c + E_r I_r)$$
$$= 0.9 \times (200 \times 10^3 \times 2{,}268 + 0.5 \times 12{,}637 \times 6{,}732 + 0) \times 10^6$$
$$= 446.5 \times 10^{12} \text{ N mm}^2$$

3. Amplified design bending moment

The critical buckling load of the composite column is determined as

$$P_{cr,eff} = \frac{\pi^2 (EI)_{eff,II}}{L_e^2} = \frac{\pi^2 \times 446.5 \times 10^{12}}{8{,}000^2} \text{N} = 68{,}856 \text{ kN}$$

The composite column is bent into a single curvature so that its moment ratio is

$$\beta_m = -\frac{M_2^*}{M_1^*} = -\frac{0}{1200} = 0$$

The moment amplification factor is calculated as follows:

$$c_b = 0.66 - 0.44\beta_m = 0.66 - 0.44 \times 0 = 0.66$$

$$\delta_m = \frac{c_b}{1 - (P^*/P_{cr,eff})} = \frac{0.66}{1 - (14{,}200/68{,}856)} = 0.83$$

The amplified design bending moment at the column end due to second-order effects is computed as

$$M_{end}^* = \delta_m M_1^* = 0.83 \times 1200 = 996 \text{ kN m}$$

The equivalent geometric imperfection at the mid-height of the column is

$$u_o = \frac{L}{300} = \frac{8000}{300} = 27 \text{ mm}$$

For moment caused by geometric imperfection, $c_b = 1.0$.

The moment amplification factor for geometric imperfections is calculated as

$$\delta_m = \frac{c_b}{1-(P^*/P_{cr,eff})} = \frac{1.0}{1-(14{,}200/68{,}856)} = 1.26$$

The amplified design bending moment due to geometric imperfections is

$$M^*_{imp} = \delta_m(P^* u_o) = 1.26 \times (14{,}200 \times 0.027) = 483\,\text{kN m}$$

The design moment on the composite column considering second-order effects is

$$M^* = M^*_{end} + M^*_{imp} = 996 + 483 = 1479\,\text{kN m} > M^*_1 = 1200\,\text{kN m}$$

4. Design moment capacity

The axial load ratio is computed as

$$\chi_d = \frac{P^*}{P_o} = \frac{14{,}200}{23{,}668} = 0.6$$

From the axial load–moment interaction diagram of the composite section shown in Figure 10.25, the moment capacity factor is obtained as

$$\mu_d = 0.67$$

The design moment capacity of the slender composite column is determined as

$$\phi M_u = \phi \alpha_M \mu_d M_o = 0.8 \times 0.9 \times 0.67 \times 3103 = 1497\,\text{kN m} > M^* = 1479\,\text{kN m, OK}$$

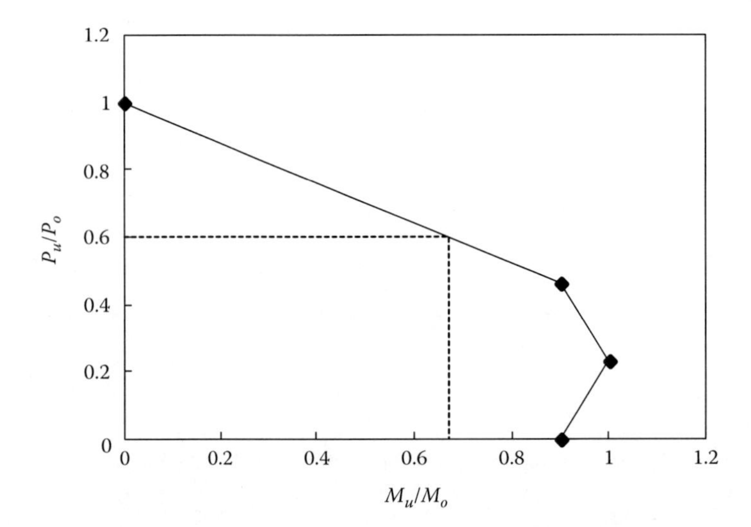

Figure 10.25 Dimensionless axial load–moment interaction diagram of the CFST short column.

10.4.5 Biaxially loaded slender composite beam–columns

For slender composite columns under axial load and biaxial bending, the design moment capacities need to be determined separately for each principal axis. Eurocode 4 suggests that imperfections should be considered only in the plane in which failure is expected to occur. If the critical plane is not known, checks should be undertaken for both bending planes. Eurocode 4 (2004) requires that the slender composite column under axial compression and biaxial bending must satisfy the following conditions:

$$M_x^* \leq \phi M_{ux} \tag{10.70}$$

$$M_y^* \leq \phi M_{uy} \tag{10.71}$$

$$\frac{M_x^*}{\phi M_{ux}} + \frac{M_y^*}{\phi M_{uy}} \leq 1.0 \tag{10.72}$$

where

M_x^* and M_y^* are the amplified design bending moments about the principal x- and y-axes, respectively

M_{ux} and M_{uy} are the nominal moment capacity of the slender composite column bending about the principal x- and y-axes, respectively, and are given by

$$M_{ux} = \alpha_M \mu_{dx} M_{uox} \tag{10.73}$$

$$M_{uy} = \alpha_M \mu_{dy} M_{uoy} \tag{10.74}$$

where

μ_{dx} and μ_{dy} are the moment capacity factors for bending about the principal x- and y-axes, respectively

M_{uox} and M_{uoy} are the pure moment capacities of the column section for bending about the principal x-and y-axes, respectively

10.5 NON-LINEAR ANALYSIS OF SLENDER COMPOSITE COLUMNS

10.5.1 General

Fibre element models were developed for the non-linear analysis of CEC beam–columns under axial load and biaxial bending (El-Tawill et al. 1995; Muñoz and Hsu 1997). Analytical and numerical models were also developed for predicting the behaviour of circular and rectangular CFST slender beam–columns (Neogi et al. 1969; Bradford 1996; Hajjar et al. 1998; Lakshmi and Shanmugam 2002; Shanmugam et al. 2002; Vrcelj and Uy 2002b; Mursi and Uy 2003; Valipour and Foster 2010; Liang 2011a,b; Portolés et al. 2011; Liang et al. 2012; Patel et al. 2012a,b,c; 2014c). The fibre element models developed by Liang (2011a), Liang et al. (2012) and Patel et al. (2012a) for CFST slender beam–columns under axial load and bending are introduced in the following sections.

10.5.2 Modelling of load–deflection behaviour

The load–deflection responses of slender composite beam–columns under increased loading are influenced by the inelastic cross-sectional behaviour, column slenderness, loading eccentricity, imperfections and second-order effects. The inelastic stability analysis of slender composite beam–columns must take into account the geometric and material nonlinearities of the beam–columns. Numerical models have been developed for the inelastic stability analysis of circular and rectangular CFST slender beam–columns, which incorporates the effects of both geometric and material nonlinearities (Liang 2011a; Liang et al. 2012; Patel et al. 2012a). The beam–column considered is pin ended and subjected to single curvature bending as schematically depicted in Figure 10.26. It is assumed that the deflected shape of CFST beam–columns is a part of a sine wave and is expressed by

$$u = u_m \sin\left(\frac{\pi z}{L}\right) \tag{10.75}$$

where u_m represents the deflection at the mid-height of the beam–column. The initial geometric imperfection of the beam–column may be described by the same form of the displacement function as

$$u_{oy} = u_o \sin\left(\frac{\pi z}{L}\right) \tag{10.76}$$

in which u_o is the initial geometric imperfection at the mid-height of the beam–column.

The curvature (ϕ) of the beam–column can be obtained from Equation 10.75 as

$$\phi = \frac{\partial^2 u}{\partial z^2} = \left(\frac{\pi}{L}\right)^2 u_m \sin\left(\frac{\pi z}{L}\right) \tag{10.77}$$

The curvature at the mid-height of the beam–column can be derived as

$$\phi_m = \left(\frac{\pi}{L}\right)^2 u_m \tag{10.78}$$

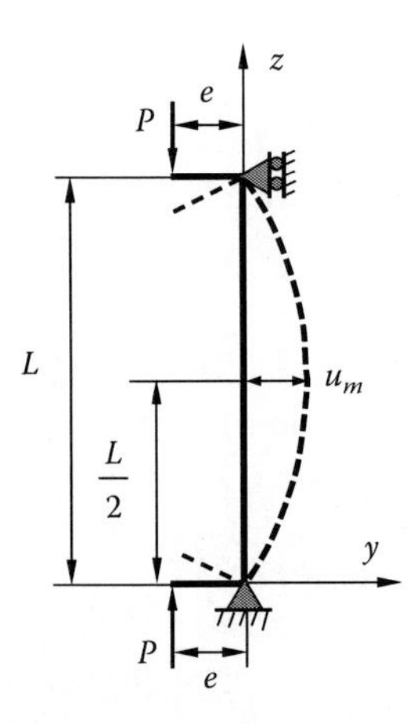

Figure 10.26 Pin-ended beam–column model.

The external bending moment at the mid-height of the beam–column with an initial geometric imperfection u_o and under eccentric loading can be calculated by

$$M_{me} = P(e + u_m + u_o) \tag{10.79}$$

where
 P is the applied load
 e is the eccentricity of the applied load as shown in Figure 10.26

The deflection control method is employed in numerical models to predict the complete load–deflection curves for slender composite beam–columns under uniaxial or biaxial loads (Liang 2011a; Liang et al. 2012; Patel et al. 2012a). The deflection at the mid-height u_m of the slender beam–column is gradually increased, and the corresponding curvature ϕ_m at the mid-height of the beam–column is calculated. The depth and orientation of the neutral axis can be adjusted by the secant method or Müller's method (Müller 1956) to achieve the moment equilibrium at the mid-height of the beam–column. The equilibrium conditions for the slender beam–column under biaxial bending are expressed by

$$P(e + u_m + u_o) - M_{mi} = 0 \tag{10.80}$$

$$\tan\alpha - \frac{M_y}{M_x} = 0 \tag{10.81}$$

where $M_{mi} = \sqrt{M_x^2 + M_y^2}$ is the resultant moment in the composite section.

In the iterative numerical analysis, residual moments in the composite section are calculated by

$$r_m^c = P(e + u_m + u_o) - M_{mi} \tag{10.82}$$

$$r_m^b = \tan\alpha - \frac{M_y}{M_x} \tag{10.83}$$

If $\left|r_m^c\right| < \varepsilon_k$ and $\left|r_m^b\right| < \varepsilon_k$, the equilibrium conditions are satisfied. The convergence tolerance ε_k can be taken as 10^{-4} in the numerical analysis.

The computational procedure for predicting the load–deflection curves for slender composite beam–columns under biaxial loads is described as follows:

1. Input data.
2. Discretise the composite section into fibre elements.
3. Initialise the mid-height deflection $u_m = \Delta u_m$.
4. Calculate the curvature ϕ_m at the mid-height of the beam–column.
5. Adjust the neutral axis depth (d_n) using Müller's method.
6. Compute stress resultants P and M_{mi}.
7. Repeat Steps 5–6 until $\left|r_m^c\right| < \varepsilon_k$.
8. Compute bending moments M_x and M_y.
9. Adjust the neutral axis orientation (θ) using Müller's method.
10. Repeat Steps 5–9 until $\left|r_m^b\right| < \varepsilon_k$.

11. Increase the mid-height deflection by $u_m = u_m + \Delta u_m$.
12. Repeat Steps 4–11 until $P \leq 0.5P_{max}$ or $u_m > u_m^*$.

10.5.3 Modelling of axial load–moment interaction diagrams

The axial load–moment interaction diagrams for slender composite beam–column under axial load and biaxial bending are generated by an incremental and iterative analysis procedure. For a given axial load (P_u) applied at a fixed load angle (α), the ultimate bending strength (M_u) of a slender beam–column is determined as the maximum moment ($M_{e,max}$) that can be applied to the column ends. The moment equilibrium is maintained at the mid-height of the beam–column. The external moment at the mid-height of the slender beam–column is given by

$$M_{me} = M_e + P_u(u_m + u_o) \tag{10.84}$$

where M_e is the moment at the column ends.

The deflection at the mid-height of the slender beam–column can be calculated from the curvature by

$$u_m = \left(\frac{L}{\pi}\right)^2 \phi_m \tag{10.85}$$

To generate the interaction diagram, the curvature (ϕ_m) at the mid-height of the beam–column is gradually increased and the corresponding internal moment (M_{mi}) is computed by the moment–curvature analysis procedure. The curvature at the column ends (ϕ_e) is adjusted, and the corresponding moment at the column ends (M_e) is calculated until the maximum moment at the column ends ($M_{e,max}$) is obtained. The axial load is increased and the axial load–moment interaction diagram of the slender composite column can be generated by repeating the preceding process. For biaxial bending, equilibrium equations are expressed by

$$P_u - P = 0 \tag{10.86}$$

$$\tan\alpha - \frac{M_y}{M_x} = 0 \tag{10.87}$$

$$M_e + P_u(u_m + u_o) - M_{mi} = 0 \tag{10.88}$$

In the numerical analysis, the residual force and moments at each iteration are calculated as $\gamma_m^a = P_u - P$, $\gamma_m^b = \tan\alpha - M_y/M_x$ and $\gamma_m^c = M_e + P_u(u_m + u_o) - M_{mi}$. If the absolute values of the residual force and moments are less than the specified tolerance $\varepsilon_k(\varepsilon_k = 10^{-4})$, the equilibrium states are attained.

The computational procedure for determining the axial load–moment interaction diagrams of slender composite columns under biaxial loads is described as follows:

1. Input data.
2. Discretise the composite section into fibre elements.
3. Compute the ultimate axial load P_{oa} of the slender column under axial compression using the load–deflection analysis produce.

4. Initialise the axial load as $P_u = 0$.
5. Initialise the curvature at the mid-height of the column as $\phi_m = \Delta\phi_m$.
6. Compute the mid-height deflection u_m.
7. Adjust the neutral axis depth (d_n) using Müller's method.
8. Calculate the resultant axial force P in the composite section.
9. Repeat Steps 7–8 until $\left|r_m^a\right| < \varepsilon_k$.
10. Compute bending moments M_x and M_y in the composite section.
11. Adjust the neutral axis orientation (θ) using Müller's method.
12. Repeat Steps 7–11 until $\left|r_m^b\right| < \varepsilon_k$.
13. Compute the resultant moment M_{mi}.
14. Adjust the curvature at the column end ϕ_e using Müller's method.
15. Compute M_e using the moment–curvature analysis procedure.
16. Repeat Steps 13–15 until $\left|r_m^c\right| < \varepsilon_k$.
17. Increase the curvature at the mid-height of the column by $\phi_m = \phi_m + \Delta\phi_m$.
18. Repeat Steps 6–17 until $M_{e,\max}$ at the column ends is obtained.
19. Increase the axial load by $P_u = P_u + \Delta P_u$, where $\Delta P_u = P_{oa}/10$.
20. Repeat Steps 5–19 until $P_u > 0.9P_{oa}$.

10.5.4 Numerical solution scheme based on Müller's method

In the non-linear analysis of a slender composite beam–column under biaxial loads, the depth and orientation of the neutral axis and the curvature at the column ends need to be iteratively adjusted in order to satisfy the force and moment equilibrium conditions. For this purpose, computational algorithms based on the secant method have been developed by Liang (2009a, 2011a). For slender beam–columns under uniaxial bending, the curvature at the column ends (ϕ_e) is adjusted by the following equation (Liang 2011a):

$$\phi_{e,k+2} = \phi_{e,k+1} - \frac{\left(\phi_{e,k+1} - \phi_{e,k}\right) r_{m,k+1}}{r_{m,k+1} - r_{m,k}} \tag{10.89}$$

where
the subscript k is the iteration number
$r_m = M_e + P_u(u_m + u_o) - M_{mi}$

It appears that computational algorithms based on the secant method are efficient and reliable for obtaining converged solutions (Liang 2009a, 2011a). The generalized displacement control method proposed by Yang and Shieh (1990) can be used to solve the incremental equilibrium equations (Yang and Kuo 1994). Müller's method (1956) is a generalization of the secant method, which can also be used to solve non-linear equations. Patel et al. (2012a) and Liang et al. (2012) have developed computational algorithms based on Müller's method to adjust the depth and orientation of the neutral axis and the curvature at the column ends. The depth (d_n) and orientation (θ) of the neutral axis and the curvature (ϕ_e) are treated as variables which are denoted by ω. Three initial values of the variables ω_1, ω_2 and ω_3 are required by Müller's method to start the iterative computational process. The corresponding residual forces or moments $r_{m,1}$, $r_{m,2}$ and $r_{m,3}$ are calculated. The new variable ω_4 that approaches the true value is computed by the following equations (Patel et al. 2012a; Liang et al. 2012):

$$\omega_4 = \omega_3 + \frac{-2c_m}{b_m \pm \sqrt{b_m^2 - 4a_m c_m}} \tag{10.90}$$

$$a_m = \frac{(\omega_2 - \omega_3)(r_{m,1} - r_{m,3}) - (\omega_1 - \omega_3)(r_{m,2} - r_{m,3})}{(\omega_1 - \omega_2)(\omega_1 - \omega_3)(\omega_2 - \omega_3)} \tag{10.91}$$

$$b_m = \frac{(\omega_1 - \omega_3)^2(r_{m,2} - r_{m,3}) - (\omega_2 - \omega_3)^2(r_{m,1} - r_{m,3})}{(\omega_1 - \omega_2)(\omega_1 - \omega_3)(\omega_2 - \omega_3)} \tag{10.92}$$

$$c_m = r_{m,3} \tag{10.93}$$

The sign of the square root term in the denominator of Equation 10.90 is taken to be the same as that of b_m when the equation is used to adjust the depth and orientation of the neutral axis. However, this sign is taken as positive when the equation is employed to adjust the curvature at the column ends. In order to obtain converged solutions, the values of ω_1, ω_2 and ω_3 and corresponding residual forces or moments $r_{m,1}$, $r_{m,2}$ and $r_{m,3}$ need to be swapped (Patel et al. 2012a). Equation 10.90 and the exchange of design variables and force or moment functions are executed iteratively until the convergence criterion of $|r_m| < \varepsilon_k$ is satisfied.

The initial values of the depth and orientation of the neutral axis and the curvature at the column ends can be taken as follows: $d_{n,1} = D/4$, $d_{n,3} = D$, $d_{n,2} = (d_{n,1} + d_{n,2})/2$; $\theta_1 = \alpha/4$, $\theta_3 = \alpha$, $\theta_2 = (\theta_1 + \theta_3)/2$; $\phi_{e,1} = 10^{-10}$, $\phi_{e,3} = 10^{-6}$, $\phi_{e,2} = (\phi_{e,1} + \phi_{e,3})/2$.

Computational algorithms using the mixed secant and Müller's method have been developed and implemented in the computer program NACOMS by the author for the non-linear inelastic analysis of thin-walled CFST slender beam–columns under axial load and biaxial bending. In the computational algorithms, the ultimate axial strength of CFST slender columns under axial compression is computed using Müller's method, while the analysis of CFST slender beam–columns under combined axial load and biaxial bending is performed using the secant method.

Example 10.6: Computer analysis of CFST slender beam–column under axial load and biaxial bending

The computer program NACOMS is employed to analyse a square thin-walled CFST slender beam–column under axial load and biaxial bending. The dimensions and material properties of the beam–column cross section are given in Example 10.3. The length of the beam–column is 5 m. The eccentricity ratio (e/D) of the axial load is taken as 0.2. The initial geometric imperfection at the mid-height of the beam–column is taken as $L/1000$. The angle of the applied axial load is fixed at 45° with respect to the y-axis of the column cross section. It is required to determine the load–deflection and axial load–moment interaction curves for this CFST slender beam–column under axial load and biaxial bending.

Computer solution

The steel tube section is non-compact as shown in Example 10.3. Hence, the method given in Eurocode 4 cannot be used to determine the axial load–moment interaction diagram of this CFST column. The effect of local buckling is taken into account in the computer analysis of the CFST column. In the fibre element analysis, the steel tube wall is divided into five layers through its thickness and the concrete core is divided into 80 × 80 fibre elements. The load–deflection curve for this column is presented in Figure 10.27. It appears that the ultimate axial load of the CFST slender beam–column under eccentric loading is 7171 kN. Figure 10.28 shows the axial load–moment interaction diagram for the CFST slender beam–column under axial load and biaxial bending. It can be seen

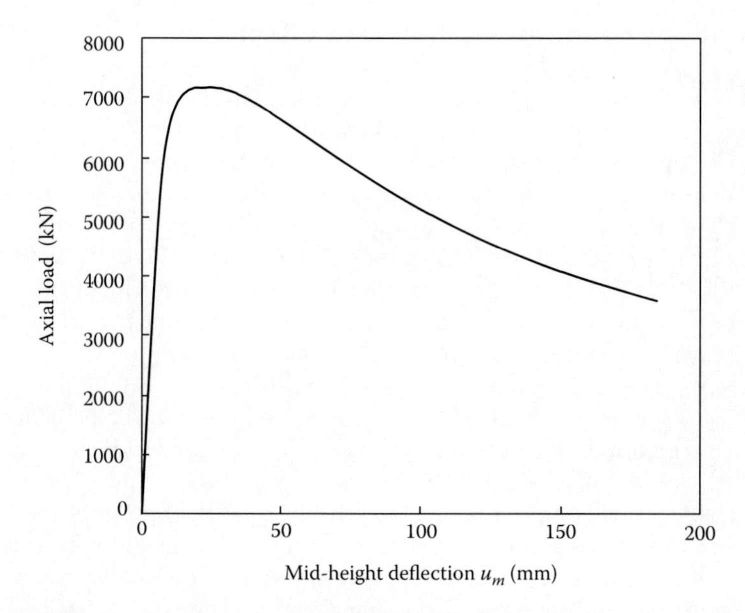

Figure 10.27 Load–deflection curve for the thin-walled CFST slender beam–column under axial load and biaxial bending.

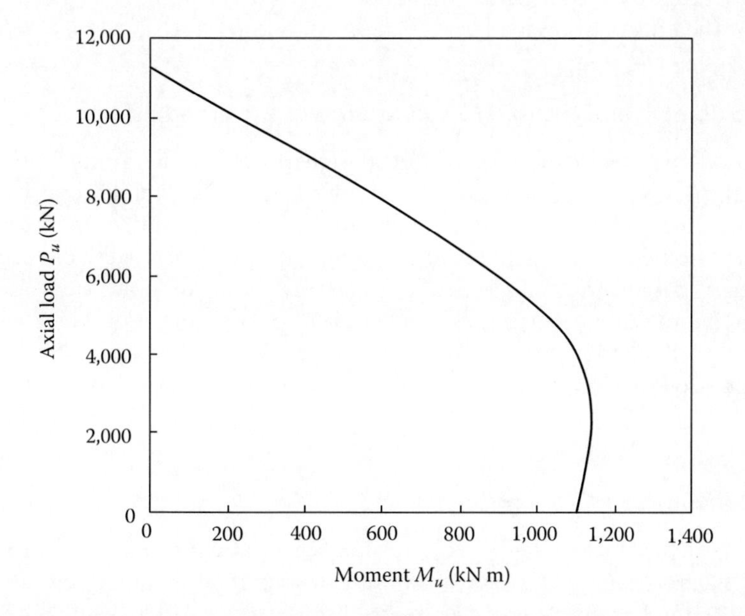

Figure 10.28 Axial load–moment interaction diagram of the thin-walled CFST slender beam–column under axial load and biaxial bending.

from the figure that the ultimate axial load of the slender column without the presence of bending moment is 11,264 kN. The pure bending moment is 1104.3 kN m, while the maximum ultimate moment of the CFST column section is 1142.7 kN m. It can be observed that the slenderness and loading eccentricity reduce the ultimate axial and bending strengths of the CFST column. However, the pure bending moment capacity is not affected by the length of the CFST column.

10.5.5 Composite columns with preload effects

10.5.5.1 General

The common construction practice of high-rise composite buildings is to erect the hollow steel tubes and composite floors several storeys before filling the wet concrete into the steel tubes. This construction practice imposes preloads arising from the constriction loads and permanent loads of the upper floors on the steel tubes. The preloads cause initial stresses and deformations in the steel tubes, which may significantly reduce the stiffness and ultimate strength of CFST slender beam–columns. Therefore, it is of practical importance to account for the effects of preloads on the steel tubes in the non-linear analysis and design of CFST slender beam–columns in multistorey composite frames.

No experiments have been conducted on biaxially loaded rectangular CFST slender beam–columns considering preload effects. Only limited tests on the behaviour of uniaxially loaded CFST columns with preload effects have been undertaken in the past (Han and Yao 2003; Xiong and Zha 2007; Liew and Xiong 2009). Test results indicate that the preload on the steel tube might reduce the ultimate axial strength of the CFST slender beam–column by 15% if the preload was greater than 60% of the ultimate axial strength of the hollow steel tube. The strength and behaviour of short CFST columns are not affected by preloads. Finite element analyses of circular CFST columns with preload effects were performed by Xiong and Zha (2007) and Liew and Xiong (2009). Fibre element models were developed by Patel et al. (2013, 2014a) for simulating the load–deflection behaviour of circular and rectangular CFST slender beam–columns under uniaxial and biaxial bending accounting for the effects of preloads.

10.5.5.2 Non-linear analysis of CFST columns with preload effects

The preloads on the steel tube induce initial stresses and deflections in the steel tube. The mid-height deflection of a hollow steel tube under the preload can be determined by performing a load–deflection analysis based on the load control method (Patel et al. 2013, 2014a). The deflection at the mid-height (u_{mo}) of the steel tube caused by the preload is treated as an additional geometric imperfection in the non-linear analysis of the CFST slender beam–column using the deflection control method. The load–deflection responses of CFST slender beam–columns with preload effects can be determined by using the load–deflection analysis procedure given in Section 10.4 (Patel et al. 2013, 2014a).

10.5.5.3 Axially loaded CFST columns

The ultimate axial strength of CFST columns under axial compression is a function of the preload ratio (β_a), relative slenderness ($\bar{\lambda}$) and geometric imperfections. Based on the results of fibre element analyses considering geometric imperfections of $L/1000$ at the mid-height of rectangular CFST columns and Eurocode 4 (2004), a design model for determining the ultimate axial strengths of concentrically loaded CFST slender columns with preload effects is given by Patel et al. (2014a) as follows:

$$P_{up} = \chi_{prg} P_o \tag{10.94}$$

where P_o is the ultimate axial strength of the column section under axial compression, taken as $P_o = 0.85 f_c' A_c + f_y A_s$.

The column strength reduction factor χ_{prg} accounts for the effects of preload ratio, relative slenderness and geometric imperfections on the ultimate strength of CFST slender columns under axial compression and is given by Patel et al. (2014a) as follows:

$$\chi_{prg} = \frac{1}{\phi_{prg} + \sqrt{\phi_{prg}^2 - \bar{\lambda}^2}} \tag{10.95}$$

$$\phi_{prg} = \frac{1 + 1.1\alpha_{prg}\left[(1+\beta_a)\bar{\lambda} - 0.05\right] + \left[(1+\zeta)\bar{\lambda}\right]^2}{2} \tag{10.96}$$

$$\alpha_{prg} = \frac{1.2 - \beta_a}{11.5(1.2 - \beta_a) - 4.3(1.2 - \beta_a)^2 - 1.5 + 60\zeta} \tag{10.97}$$

$$\zeta = \begin{cases} 0 & \text{for } \beta_a \le 0.4 \\ \dfrac{\beta_a - 0.4}{4} & \text{for } 0.4 < \beta_a \le 0.8 \end{cases} \tag{10.98}$$

where the relative slenderness $\bar{\lambda}$ is calculated using P_o in Equation 10.57.

10.5.5.4 Behaviour of CFST beam–columns with preload effects

Numerical studies performed by Patel et al. (2013, 2014a) demonstrate that increasing the preload ratio decreases the ultimate axial load, bending strength and flexural stiffness of CFST slender beam–columns. The reduction in the ultimate strengths of CFST columns due to preload effects is found to increase with an increase in the column slenderness ratio (L/r). The preload with a ratio of 0.6 may reduce the ultimate axial strength of the CFST column with an L/r ratio of 100 by 15.8%. However, the preload only has a minor effect on CFST short beam–columns with an L/r ratio of less than 22 or CFST slender beam–columns with small preload ratios, and thus its effect can be ignored in the design. The strength reduction due to preload effects is shown to increase with an increase in the loading eccentricity ratio (e/D) from 0.0 to 0.4. When $e/D > 0.4$, however, the strength reduction tends to decrease with an increase in the e/D ratio. It is interesting to note that the reduction in the ultimate axial strength of CFST columns due to preload effects is maximized when the e/D ratio is equal to 0.4. It would appear that the preload has more pronounce effects on high-strength CFST slender beam–columns than on normal strength ones. The preload having a preload ratio of 0.8 may reduce the ultimate axial strength of the high-strength circular CFST slender beam–column with yield steel strength of 690 MPa by 17.3%.

10.5.6 Composite columns under cyclic loading

10.5.6.1 General

In seismic regions, thin-walled rectangular CFST slender beam–columns may be subjected to a constant axial load from upper floors and cyclically varying lateral loading due to the earthquake. These CFST beam–columns may undergo cyclic local and global interaction buckling, which reduces their strength, flexural stiffness and ductility. Experiments on normal and high-strength rectangular CFST beam–columns under axial load and cyclic lateral loading have been undertaken by researchers (Varma et al. 2002, 2004; Han et al. 2003). High-strength concrete up to 110 MPa and high-strength steel tubes with yield stress up to

660 MPa were used to construct CFST columns. The failure modes associated with these CFST columns were cracking of the concrete core and local buckling of the steel tubes. The outward local buckling of some CFST beam–columns was observed after steel yielding.

Numerical models have been developed to predict the cyclic responses of rectangular CFST beam–columns considering or ignoring local buckling effects (Varma et al. 2002; Gayathri et al. 2004a,b; Chung et al. 2007; Zubydan and ElSabbagh 2011). Some of these models approximately account for local buckle effects by modifying the stress–strain curve for steel in compression. However, this method cannot simulate the progressive cyclic local buckling of the steel tube from the onset to the post-local buckling. It has been found that the modified stress–strain curve method might overestimate or underestimate the cyclic local buckling strengths of steel tubes under stress gradients (Patel et al. 2014b). Patel et al. (2014b) developed a fibre element model for simulating the cyclic local and global interaction buckling behaviour of rectangular CFST slender beam–columns under constant axial load and cyclically varying lateral loading, which is introduced in the following sections.

10.5.6.2 Cyclic material models for concrete

The cyclic stress–strain curves for concrete in CFST columns are shown in Figure 10.29 (Patel et al. 2014b). The stiffness degradation and crack opening and closing characteristics of the concrete under cyclic loading are taken into account in the cyclic material constitutive model. The envelope curve for concrete under cyclic axial compression is defined by the monotonic stress–strain curve given in Section 10.3.

The concrete under compression is initially loaded up to an unloading strain and then unloaded to a zero stress level. A linear stress–strain relationship is assumed for the concrete reloading from the zero stress up to the envelope curve. The parabolic stress–strain curve for the concrete unloading as depicted in Figure 10.29 is given by Mander et al. (1988) as

$$\sigma_c = \sigma_{un} - \frac{\sigma_{un}\lambda_u \left(\dfrac{\varepsilon_c - \varepsilon_{un}}{\varepsilon_{pl} - \varepsilon_{un}} \right)}{\lambda_u - 1 + \left(\dfrac{\varepsilon_c - \varepsilon_{un}}{\varepsilon_{pl} - \varepsilon_{un}} \right)^{\lambda_u}} \qquad \left(\varepsilon_{pl} < \varepsilon_c < \varepsilon_{un} \right) \tag{10.99}$$

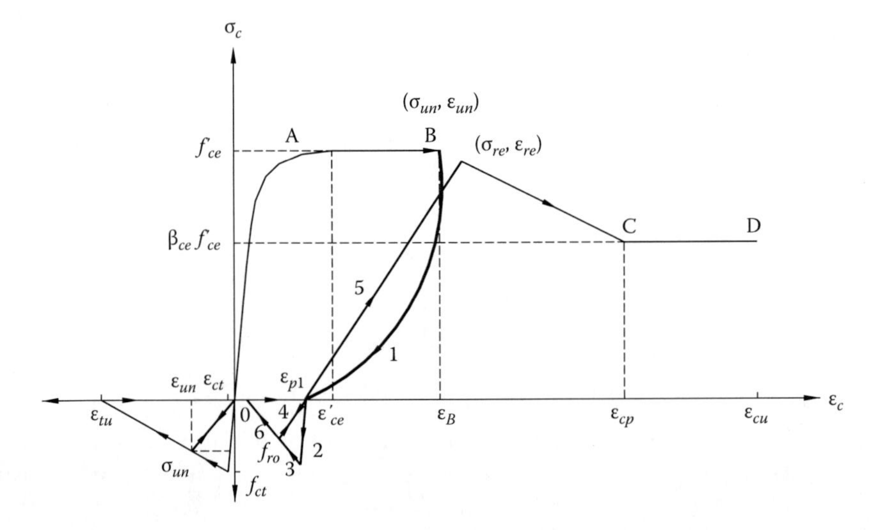

Figure 10.29 Cyclic stress–strain curves for concrete in rectangular CFST columns.

$$\lambda_u = \frac{E_{un}}{E_{un} - \left(\dfrac{\sigma_{un}}{\varepsilon_{un} - \varepsilon_{pl}}\right)} \tag{10.100}$$

where

σ_{un} denotes the compressive stress of concrete at the unloading
ε_{un} represents the strain at σ_{un}
ε_{pl} is the plastic strain which is calculated by (Mander et al. 1988)

$$\varepsilon_{pl} = \varepsilon_{un} - \frac{\sigma_{un}\varepsilon_{un} + \sigma_{un}\varepsilon_a}{\sigma_{un} + E_c\varepsilon_a} \tag{10.101}$$

where $\varepsilon_a = a_c\sqrt{\varepsilon_{un}\varepsilon'_{ce}}$ and a_c is taken as the larger of $\varepsilon'_{ce}/(\varepsilon'_{ce} + \varepsilon_{un})$ and $0.09\varepsilon_{un}/\varepsilon_{cc}$ for a rectangular cross section.

In Equation 10.100, E_{un} is the initial modulus of elasticity of concrete at the unloading and is written as

$$E_{un} = \left(\frac{\sigma_{un}}{f'_{ce}}\right)\left(\sqrt{\frac{\varepsilon'_{ce}}{\varepsilon_{un}}}\right)E_c \tag{10.102}$$

where $(\sigma_{un}/f'_{ce}) \geq 1.0$ and $\sqrt{\varepsilon'_{ce}/\varepsilon_{un}} \leq 1.0$.

The linear stress–strain relationship for concrete at reloading is defined by

$$\sigma_c = \left(\frac{f_{ro} - \sigma_{re}}{\varepsilon_{ro} - \varepsilon_{re}}\right)(\varepsilon_c - \varepsilon_{ro}) + f_{ro} \quad (\varepsilon_{pl} < \varepsilon_c < \varepsilon_{ro}) \tag{10.103}$$

where

f_{ro} is the concrete stress at the reloading
ε_{ro} is the strain at f_{ro}
ε_{re} and σ_{re} are the return strain and stress on the monotonic curve as shown in Figure 10.29

The stress–strain curve for concrete in tension is also given in Figure 10.29. It is assumed that the concrete tensile stress increases linearly up to cracking and then decreases linearly to zero at the ultimate tensile strain. The tensile strength of concrete is taken as $0.6\sqrt{f'_{ce}}$, while the ultimate tensile strain is assumed to be 10 times of the strain at cracking. The tensile stress in the concrete for unloading from the compressive envelope is determined by

$$\sigma_t = \begin{cases} \dfrac{f'_{ct}(\varepsilon_c - \varepsilon_{pl})}{(\varepsilon'_{ct} - \varepsilon'_{tu})} & \text{for } \varepsilon'_{tu} < \varepsilon_c \leq \varepsilon'_{ct} \\[2ex] \dfrac{f'_{ct}(\varepsilon_c - \varepsilon_{pl})}{\varepsilon'_{ct}} & \text{for } \varepsilon'_{ct} < \varepsilon_c < \varepsilon_{pl} \end{cases} \tag{10.104}$$

$$f'_{ct} = f_{ct}\left(1 - \frac{\varepsilon_{pl}}{\varepsilon'_{ce}}\right) \tag{10.105}$$

$$\varepsilon'_{ct} = \varepsilon_{ct} + \varepsilon_{pl} \tag{10.106}$$

The tensile strength of concrete is assumed to decrease with an increase in the cycles. This implies that if the previous tension loading went along the path 1–2–3–4–5, the current tension loading will follow the 5–6 path as illustrated in Figure 10.29.

10.5.6.3 Cyclic material models for structural steels

Figure 10.30 depicts the cyclic stress–strain model for the structural steels. It is noted that the stress–strain curve at unloading follows a straight line with the same slope as the initial stiffness, which is expressed by

$$\sigma_s = E_s(\varepsilon_s - \varepsilon_{mo}) \quad (\varepsilon_o < \varepsilon_s \leq \varepsilon_{mo}) \tag{10.107}$$

where $\varepsilon_{mo} = \varepsilon_{so} - f_{so}/E_s$, ε_{so} is the strain at the unloading and f_{so} is the stress at the unloading.

The stress–strain curve for structural steels at reloading is given by Shi et al. (2012) as follows:

$$\sigma_s = E_s(\varepsilon_s - \varepsilon_{mo}) - \eta(E_s - E_k)(\varepsilon_s - \varepsilon_{so}) \quad (\varepsilon_{mo} < \varepsilon_s \leq \varepsilon_b) \tag{10.108}$$

$$E_k = \frac{\sigma_b}{\varepsilon_b - \varepsilon_{mo}} \tag{10.109}$$

$$\eta = \begin{cases} 1.048 - \dfrac{0.05}{(\varepsilon_s - \varepsilon_{so})/(\varepsilon_b - \varepsilon_{so}) + 0.05} & \text{for } |\varepsilon_b - \varepsilon_{so}| \geq 0.04 \\ 1.074 - \dfrac{0.08}{(\varepsilon_s - \varepsilon_{so})/(\varepsilon_b - \varepsilon_{so}) + 0.08} & \text{for } |\varepsilon_b - \varepsilon_{so}| < 0.04 \end{cases} \tag{10.110}$$

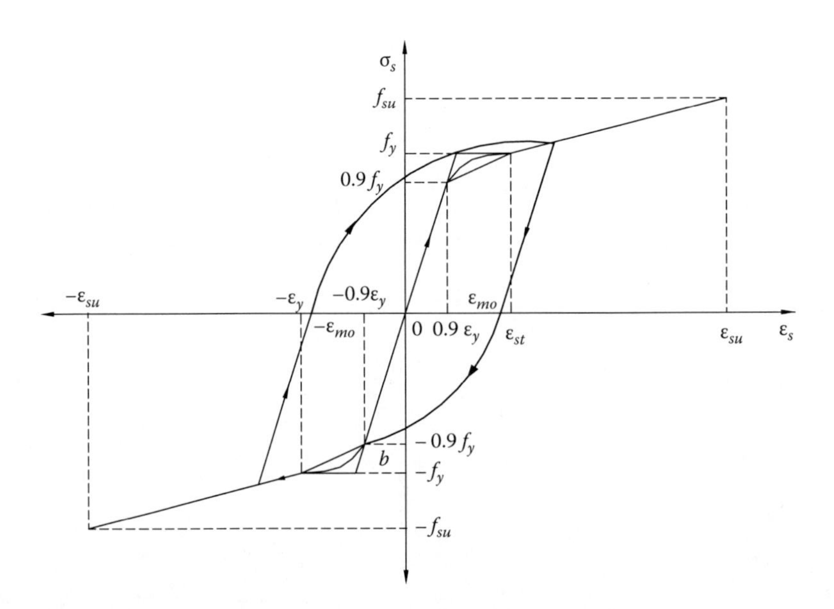

Figure 10.30 Cyclic stress–strain curves for structural steels.

The initial value of the strain ε_b at reloading as indicated by point B in Figure 10.30 is taken as $0.9\varepsilon_y$. The stress σ_b at the strain ε_b is determined from the monotonic stress–strain curves. If the strain is greater than ε_b, the steel stress is determined from the cyclic skeleton curve. After initial reloading, the reloading is directed towards the previous unloading.

10.5.6.4 Modelling of cyclic load–deflection responses

A fibre element model was developed for cantilever columns under constant axial load (P_a) and cyclically varying lateral loading (F) as illustrated in Figure 10.31. The effective length of the cantilever column is taken as $2L$. The deflected shape of the cantilever column is assumed to be part of a sine wave. The curvature at the base of the cantilever column can be determined from the displacement function as

$$\phi_b = \left(\frac{\pi}{2L}\right)^2 u_t \tag{10.111}$$

where u_t is the lateral deflection at the tip of the column.

The external moment at the base of the cantilever column is calculated by

$$M_{me} = FL + P_a(e + u_t + u_{to}) \tag{10.112}$$

where

- e is the eccentricity of the axial load and is taken as zero for the column under concentric axial load
- u_{to} is the initial geometric imperfection at the tip of the cantilever column

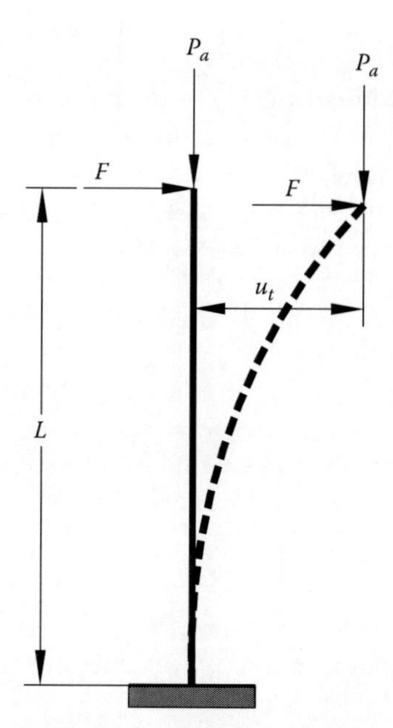

Figure 10.31 Cantilever column under constant axial load and cyclically varying lateral loading.

The equilibrium conditions for the cantilever column are expressed by

$$P_a - P = 0 \tag{10.113}$$

$$FL + P_a(e + u_t + u_{to}) - M = 0 \tag{10.114}$$

where P and M are the internal force and bending moment in the composite section.

The lateral load can be obtained from Equation 10.114 as

$$F = \frac{M - P_a(e + u_t + u_{to})}{L} \tag{10.115}$$

In the cyclic load–deflection analysis, the lateral deflection at the tip of the cantilever column is gradually increased up to the predefined unloading deflection and then decreased to the reloading level. The computational algorithms based on Müller's method (Liang et al. 2012; Patel et al. 2012a) are used to adjust the neutral axis depth to maintain the force equilibrium in the composite section. The lateral load F at the tip of the cantilever column is computed from the moment equilibrium state. The stress–strain histories of the composite section under previous cyclic loading are stored in order to determine the current states of stresses. By repeating the aforementioned analysis process, the complete cyclic load–deflection curves can be obtained.

The computational procedure for simulating the cyclic load–deflection responses of CFST beam–columns is given as follows (Patel et al. 2014b):

1. Input data.
2. Discretise the composite section into fibre elements.
3. Initialise the first unloading deflection u_{ut}.
4. Initialise the lateral deflection at the tip of the column $u_t = \Delta u_t$.
5. Calculate the curvature ϕ_b at the base of the column.
6. If $u_t > (u_{ut} - \Delta u_t)$ or $u_t < (-u_{ut} - \Delta u_t)$, then $\Delta u_t = -\Delta u_t$.
7. If $(u_t - u_{last})(u_{last} - u_{old}) < 0$ and $u_t > u_{last}$, set the next unloading deflection u_{ut}.
8. Recall the unloading strains and stresses at the unloading deflection.
9. Adjust the neutral axis depth (d_n) using Müller's method.
10. Compute the resultant force P considering local buckling effects.
11. Repeat Steps 9–10 until $|r_p| < \varepsilon_k$.
12. Calculate the cyclic lateral force F from the moment equilibrium.
13. Record the deflections $u_{old} = u_{last}$ and $u_{last} = u_t$.
14. Store the fibre strains and fibre stresses under the current deflection.
15. Increase the deflection at the tip of the cantilever column by $u_t = u_t + \Delta u_t$.
16. Repeat Steps 5–16 until $F < 0.85F_{max}$ or $u_t > u^*$.

The typical cyclic lateral load–deflection curves for a rectangular CFST cantilever column predicted using the preceding computational procedure are shown in Figure 10.32.

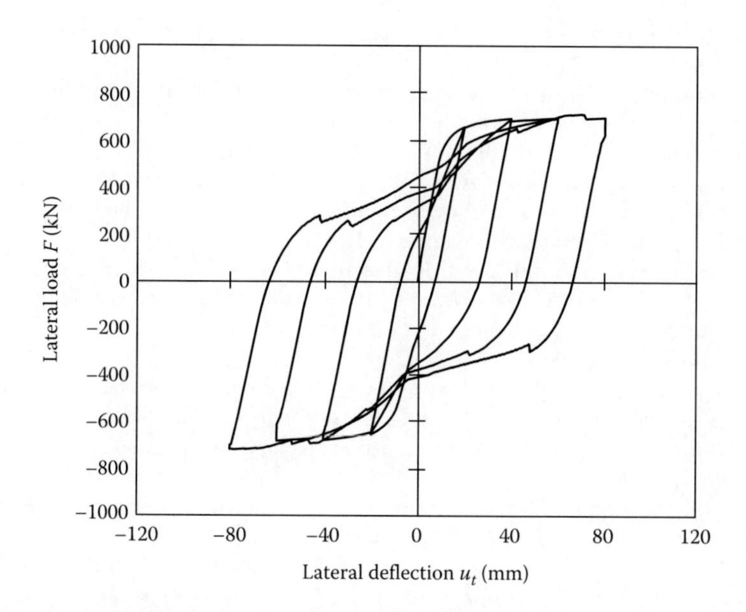

Figure 10.32 Typical cyclic load–deflection curves for a rectangular CFST slender beam–column.

REFERENCES

ACI-318 (2011) *Building Code Requirements for Structural Concrete and Commentary*, Detroit, MI: American Concrete Institute.

Bradford, M.A. (1996) Design strength of slender concrete-filled rectangular steel tubes, *ACI Structural Journal*, 93 (2): 229–235.

Bridge, R.Q. (1976) Concrete filled steel tubular columns, Research Report No. R283, School of Civil Engineering, The University of Sydney, Sydney, New South Wales, Australia.

Bridge, R.Q. and O'Shea, M.D. (1998) Behaviour of thin-walled steel box sections with or without internal restraint, *Journal of Constructional Steel Research*, 47 (1–2): 73–91.

Chen, S.F., Teng, J.G. and Chan, S.L. (2001) Design of biaxially loaded short composite columns of arbitrary section, *Journal of Structural Engineering*, ASCE, 127 (6): 678–685.

Chung, K., Chung, J. and Choi, S. (2007) Prediction of pre- and post-peak behavior of concrete-filled square steel tube columns under cyclic loads using fiber element method, *Thin-Walled Structure*, 45 (9): 747–758.

Ellobody, E., Young, B. and Lam, D. (2006) Behaviour of normal and high strength concrete-filled compact steel tube circular stub columns, *Journal of Constructional Steel Research*, 62 (7): 706–715.

Ellobody, E. and Young, B. (2006) Nonlinear analysis of concrete-filled SHS and RHS columns, *Thin-Walled Structures*, 44(8): 919–930.

El-Tawil, S. and Deierlein, G.G. (1999) Strength and ductility of concrete encased composite columns, *Journal of Structural Engineering*, ASCE, 125 (9): 1009–1019.

El-Tawil, S., Sanz-Picón, C.F. and Deierlein, G.G. (1995) Evaluation of ACI 318 and AISC (LRFD) strength provisions for composite beam-columns, *Journal of Constructional Steel Research*, 34 (1): 103–126.

Eurocode 3 (2005) *Design of Steel Structures, Part 1-1: General Rules and Rules for Buildings*, European Committee for Standardization, CEN, Brussels, Belgium.

Eurocode 4 (2004) *Design of Composite Steel and Concrete Structures, Part 1-1: General Rules and Rules for Buildings*, European Committee for Standardization, CEN, Brussels, Belgium.

Furlong, R.W. (1967) Strength of steel-encased concrete beam columns, *Journal of the Structural Division*, ASCE, 93 (5): 113–124.
Gayathri, V., Shanmugam, N.E. and Choo, Y.S. (2004a) Concrete-filled tubular columns, Part 1 – Cross-section analysis, *International Journal of Structural Stability and Dynamics*, 4 (4): 459–478.
Gayathri, V., Shanmugam, N.E. and Choo, Y.S. (2004b) Concrete-filled tubular columns, Part 2 – Column analysis, *International Journal of Structural Stability and Dynamics*, 4 (4): 479–495.
Ge, H.B. and Usami, T. (1992) Strength of concrete-filled thin-walled steel box columns: Experiment, *Journal of Structural Engineering*, ASCE, 118 (11): 3036–3054.
Giakoumelis, G. and Lam, D. (2004) Axial load capacity of circular concrete-filled tube columns, *Journal of Constructional Steel Research*, 60: 1049–1068.
Hajjar, J.F. and Gourley, B.C. (1996) Representation of concrete-filled steel tube cross-section strength, *Journal of Structural Engineering*, ASCE, 122 (11): 1327–1336.
Hajjar, J.F., Schiller, P.H. and Molodan, A. (1998) A distributed plasticity model for concrete-filled steel tube beam-columns with interlayer slip, *Engineering Structures*, 20 (8): 663–676.
Han, L.H. (2002) Tests on stub columns of concrete-filled RHS sections, *Journal of Constructional Steel Research*, 58 (3): 353–372.
Han, L.H., Yang, Y.F. and Tao, Z. (2003) Concrete-filled thin-walled steel SHS and RHS beam-columns subjected to cyclic loading, *Thin-Walled Structure*, 41 (9): 801–833.
Han, L.H. and Yao, G.H. (2003) Behaviour of concrete-filled hollow structural steel (HSS) columns with pre-load on the steel tubes, *Journal of Constructional Steel Research*, 59 (12): 1455–1475.
Hassanein, M.F., Kharoob, O.F. and Liang, Q.Q. (2013a) Behaviour of circular concrete-filled lean duplex stainless steel tubular short columns, *Thin-Walled Structures*, 68, 113–123.
Hassanein, M.F., Kharoob, O.F. and Liang, Q.Q. (2013b) Behaviour of circular concrete-filled lean duplex stainless steel-carbon steel tubular short columns, *Engineering Structures*, 56, 83–94.
Hassanein, M.F., Kharoob, O.F. and Liang, Q.Q. (2013c) Circular concrete-filled double skin tubular short columns with external stainless steel tubes under axial compression, *Thin-Walled Structures*, 73, 252–263.
Hu, H.T., Huang, C.S., Wu, M.H. and Wu, Y.M. (2003) Nonlinear analysis of axially loaded concrete-filled tube columns with confinement effect, *Journal of Structural Engineering*, ASCE, 129 (10): 1322–1329.
Johnson, R.P. (2004) *Composite Structures of Steel and Concrete: Beams, Slabs, Columns, and Frames for Buildings*, Oxford, U.K.: Blackwell Publishing.
Knowles, R.B. and Park, R. (1969) Strength of concrete-filled steel tubular columns, *Journal of the Structural Division*, ASCE, 95 (12): 2565–2587.
Lakshmi, B. and Shanmugam, N.E. (2002) Nonlinear analysis of in-filled steel-concrete composite columns, *Journal of Structural Engineering*, ASCE, 128 (7): 922–933.
Liang, Q.Q. (2008) Nonlinear analysis of short concrete-filled steel tubular beam-columns under axial load and biaxial bending, *Journal of Constructional Steel Research*, 64 (3): 295–304.
Liang, Q.Q. (2009a) Performance-based analysis of concrete-filled steel tubular beam-columns, Part I: Theory and algorithms, *Journal of Constructional Steel Research*, 65 (2): 363–372.
Liang, Q.Q. (2009b) Performance-based analysis of concrete-filled steel tubular beam-columns, Part II: Verification and applications, *Journal of Constructional Steel Research*, 65 (2): 351–362.
Liang, Q.Q. (2009c) Strength and ductility of high strength concrete-filled steel tubular beam-columns, *Journal of Constructional Steel Research*, 65 (3): 687–698.
Liang, Q.Q. (2011a) High strength circular concrete-filled steel tubular slender beam-columns, Part I: Numerical analysis, *Journal of Constructional Steel Research*, 67 (2): 164–171.
Liang, Q.Q. (2011b) High strength circular concrete-filled steel tubular slender beam-columns, Part II: Fundamental behavior, *Journal of Constructional Steel Research*, 67 (2): 172–180.
Liang, Q.Q. and Fragomeni, S. (2009) Nonlinear analysis of circular concrete-filled steel tubular short columns under axial loading, *Journal of Constructional Steel Research*, 65 (12): 2186–2196.
Liang, Q.Q. and Fragomeni, S. (2010) Nonlinear analysis of circular concrete-filled steel tubular short columns under eccentric loading, *Journal of Constructional Steel Research*, 66 (2): 159–169.
Liang, Q.Q., Patel, V.I. and Hadi, M.N.S. (2012) Biaxially loaded high-strength concrete-filled steel tubular slender beam-columns, Part I: Multiscale simulation, *Journal of Constructional Steel Research*, 75: 64–71.

Liang, Q.Q. and Uy, B. (1998) Parametric study on the structural behaviour of steel plates in concrete-filled fabricated thin-walled box columns, *Advances in Structural Engineering*, 2 (1): 57–71.

Liang, Q.Q. and Uy, B. (2000) Theoretical study on the post-local buckling of steel plates in concrete-filled box columns, *Computers and Structures*, 75 (5): 479–490.

Liang, Q.Q., Uy, B. and Liew, J.Y.R. (2006) Nonlinear analysis of concrete-filled thin-walled steel box columns with local buckling effects, *Journal of Constructional Steel Research*, 62 (6): 581–591.

Liang, Q.Q., Uy, B. and Liew, J.Y.R. (2007a) Strength of concrete-filled steel box columns with buckling effects, *Australian Journal of Structural Engineering*, 7 (2): 145–155.

Liang, Q.Q., Uy, B. and Liew, J.Y.R. (2007b) Local buckling of steel plates in concrete-filled thin-walled steel tubular beam-columns, *Journal of Constructional Steel Research*, 63 (3): 396–405.

Liew, J.Y.R. and Xiong, D.X. (2009) Effect of preload on the axial capacity of concrete-filled composite columns, *Journal of Constructional Steel Research*, 65 (3): 709–722.

Mander, J.B., Priestley, M.J.N. and Park, R. (1988) Theoretical stress–strain model for confined concrete, *Journal of Structural Engineering*, ASCE, 114 (8): 1804–1826.

Müller, D.E. (1956) A method for solving algebraic equations using an automatic computer, *Mathematical Tables and Other Aids to Computation*, 10 (56): 208–215.

Muñoz, P.R. and Hsu, C.T.T. (1997) Behavior of biaxially loaded concrete-encased composite columns, *Journal of Structural Engineering*, ASCE, 123 (9): 1163–1171.

Mursi, M. and Uy, B. (2003) Strength of concrete filled steel box columns incorporating interaction buckling, *Journal of Structural Engineering*, ASCE, 129 (5): 626–638.

Mursi, M. and Uy, B. (2006) Behaviour and design of fabricated high strength steel columns subjected to biaxial bending, Part I: Experiments, *Advanced Steel Construction*, 2 (4): 286–315.

Neogi, P.K., Sen, H.K. and Chapman, J.C. (1969) Concrete-filled tubular steel columns under eccentric loading, *The Structural Engineer*, 47 (5): 187–195.

Oehlers, D.J. and Bradford, M.A. (1999) *Elementary Behaviour of Composite Steel and Concrete Structural Members*, Oxford, U.K.: Butterworth-Heinemann.

Patel, V.I., Liang, Q.Q. and Hadi, M.N.S. (2012a) High strength thin-walled rectangular concrete-filled steel tubular slender beam-columns, Part I: Modeling, *Journal of Constructional Steel Research*, 70: 377–384.

Patel, V.I., Liang, Q.Q. and Hadi, M.N.S. (2012b) High strength thin-walled rectangular concrete-filled steel tubular slender beam-columns, Part II: Behavior, *Journal of Constructional Steel Research*, 70: 368–376.

Patel, V.I., Liang, Q.Q. and Hadi, M.N.S. (2012c) Inelastic stability analysis of high strength rectangular concrete-filled steel tubular slender beam-columns, *Interaction and Multiscale Mechanics*, 5 (2): 91–104.

Patel, V.I., Liang, Q.Q. and Hadi, M.N.S. (2013) Numerical analysis of circular concrete-filled steel tubular slender beam-columns with preload effects, *International Journal of Structural Stability and Dynamics*, 13 (3): 1250065 (23p.).

Patel, V.I., Liang, Q.Q. and Hadi, M.N.S. (2014a) Behavior of biaxially-loaded rectangular concrete-filled steel tubular slender beam-columns with preload effects, *Thin-Walled Structures*, 79: 166–177.

Patel, V.I., Liang, Q.Q. and Hadi, M.N.S. (2014b) Numerical analysis of high-strength concrete-filled steel tubular slender beam-columns under cyclic loading, *Journal of Constructional Steel Research*, 92: 183–194.

Patel, V.I., Liang, Q.Q. and Hadi, M.N.S. (2014c) Biaxially loaded high-strength concrete-filled steel tubular slender beam-columns, Part II: Parametric study, *Journal of Constructional Steel Research*. (In press).

Portolés, J.M., Romero, M.L., Filippou, F.C. and Bonet, J.L. (2011) Simulation and design recommendations of eccentrically loaded slender concrete-filled tubular columns, *Engineering Structures*, 33 (5): 1576–1593.

Rangan, B. and Joyce, M. (1992) Strength of eccentrically loaded slender steel tubular columns filled with high-strength concrete, *ACI Structural Journal*, 89 (6): 676–681.

Richart, F.E., Brandtzaeg, A. and Brown, R.L. (1928) *A Study of the Failure of Concrete under Combined Compressive Stresses*, Bulletin 185, Champaign, IL: University of Illinois, Engineering Experimental Station.

Roik, K. and Bergmann, R. (1989) Eurocode 4: Composite columns, Report EC4/6/89, University of Bochum, Bochum, Germany, June.

Schneider, S.P. (1998) Axially loaded concrete-filled steel tubes, *Journal of Structural Engineering*, ASCE, 124 (10): 1125–1138.

Shakir-Khalil, H. and Mouli, M. (1990) Further tests on concrete-filled rectangular hollow-section columns, *The Structural Engineer*, 68 (20): 405–413.

Shakir-Khalil, H. and Zeghiche, J. (1989) Experimental behaviour of concrete-filled rolled rectangular hollow-section columns, *The Structural Engineer*, 67 (19): 346–353.

Shanmugam, N.S. and Lakshmi, B. (2001) State of the art report on steel–concrete composite columns, *Journal of Constructional Steel Research*, 57: 1041–1080.

Shanmugam, N.S., Lakshmi, B. and Uy, B. (2002) An analytical model for thin-walled steel box columns with concrete infill, *Engineering Structures*, 24: 825–838.

Shi, G., Wang, M., Bai, Y., Wang, F., Shi, Y.J. and Wang, Y.Q. (2012) Experimental and modeling study of high-strength structural steel under cyclic loading, *Engineering Structures*, 37: 1–13.

Spacone, E. and El-Tawil, S. (2004) Nonlinear analysis of steel–concrete composite structures: State of the art, *Journal of Structural Engineering*, ASCE, 130 (2): 159–168.

Tang, J., Hino, S., Kuroda, I. and Ohta, T. (1996) Modeling of stress–strain relationships for steel and concrete in concrete filled circular steel tubular columns, *Steel Construction Engineering*, JSSC, 3 (11): 35–46.

Tao, Z., Uy, B., Liao, F.Y. and Han, L.H. (2011) Nonlinear analysis of concrete-filled square stainless steel stub columns under axial compression, *Journal of Constructional Steel Research*, 67: 1719–1732.

Tomii, M. and Sakino, K. (1979a) Elastic–plastic behavior of concrete filled square steel tubular beam-columns, *Transactions of the Architectural Institute of Japan*, 280: 111–120.

Tomii, M. and Sakino, K. (1979b) Experimental studies on the ultimate moment of concrete filled square steel tubular beam-columns, *Transactions of the Architectural Institute of Japan*, 275: 55–65.

Uy, B. (1998) Local and post-local buckling of concrete filled steel welded box columns, *Journal of Constructional Steel Research*, 47 (1–2): 47–72.

Uy, B. (2000) Strength of concrete filled steel box columns incorporating local buckling, *Journal of Structural Engineering*, ASCE, 126 (3): 341–352.

Uy, B. (2001) Strength of short concrete filled high strength steel box columns, *Journal of Constructional Steel Research*, 57 (2): 113–134.

Uy, B., Tao, Z. and Han, L.H. (2011) Behaviour of short and slender concrete-filled stainless steel tubular columns, *Journal of Constructional Steel Research*, 67: 360–378.

Valipour, H.R. and Foster, S.J. (2010) Nonlinear static and cyclic analysis of concrete-filled steel columns, *Journal of Constructional Steel Research*, 66: 793–802.

Varma, A.H., Ricles, J.M., Sause, R. and Lu, L.W. (2002) Seismic behavior and modeling of high-strength composite concrete-filled steel tube (CFT) beam-columns, *Journal of Constructional Steel Research*, 58 (5–8): 725–758.

Varma, A.H., Ricles, J.M., Sause, R. and Lu, L.W. (2004) Seismic behavior and design of high-strength square concrete-filled steel tube beam columns, *Journal of Structural Engineering*, ASCE, 130 (2): 169–179.

Vrcelj, Z. and Uy, B. (2002a) Behaviour and design of steel square hollow sections filled with high strength concrete, *Australian Journal of Structural Engineering*, 3 (3): 153–170.

Vrcelj, Z. and Uy, B. (2002b) Strength of slender concrete-filled steel box columns incorporating local buckling, *Journal of Constructional Steel Research*, 58 (2): 275–300.

Xiong, D.X. and Zha, X.X. (2007) A numerical investigation on the behaviour of concrete-filled steel tubular columns under initial stresses, *Journal of Constructional Steel Research*, 63 (5): 599–611.

Yang, Y.B. and Kuo, S.R. (1994) *Theory and Analysis of Nonlinear Framed Structures*, Englewood Cliffs, NJ: Prentice-Hall.

Yang, Y.B. and Shieh, M.S. (1990) Solution method for nonlinear problems with multiple critical points, *AIAA Journal*, 28 (12): 2110–2116.

Young, B. and Ellobody, E. (2006) Experimental investigation of concrete-filled cold-formed high strength stainless steel tube columns, *Journal of Constructional Steel Research*, 62 (5): 484–492.

Zhao, X.L. and Grzebieta, R.H. (2002) Strength and ductility of concrete filled double skin (SHS inner and SHS outer) tubes, *Thin-Walled Structures*, 40 (2): 193–213.

Zhao, X.L. and Packer, J.A. (2009) Tests and design of concrete-filled elliptical hollow section stub columns, *Thin-Walled Structures*, 47 (6–7): 617–628.

Zubydan, A.H. and ElSabbagh, A.I. (2011) Monotonic and cyclic behavior of concrete-filled steel-tube beam-columns considering local buckling effect, *Thin-Walled Structures*, 49 (4): 465–481.

Chapter 11

Composite connections

11.1 INTRODUCTION

Composite connections are used to transfer forces between composite members and to maintain the integrity of a composite structure under applied loads. The behaviour of a composite connection is characterised by its moment–rotation curve, which expresses the moment as a function of the angle between the beam and the column. The rotational stiffness of a composite connection is determined by the slope of its moment–rotation curve. Composite connections are classified as simple, rigid and semi-rigid connections based on the stiffness criteria and as full-strength and partial-strength connections based on the strength criteria. The types of composite connections include the base plate connections of composite columns, composite column splices, beam-to-column shear connections, beam-to-column moment connections and semi-rigid connections (Viest et al. 1997). Double-angle connections, single-plate shear connections and tee shear connections are beam-to-column shear connections. The beam-to-column moment connections include composite connections with steel beam passing through concrete-encased composite (CEC) columns, reinforced concrete columns, concrete-filled steel tubular (CFST) columns or steel columns.

Composite connections in a composite frame are potential weak spots that must be designed for a larger margin of safety than the connecting members. In general, composite connections must satisfy the strength, serviceability and construction criteria. The strength criteria require that composite connections must be designed to resist axial force, bending moment, shear and torsion arising from the applied loads. The serviceability criteria require that the design of composite connections must ensure that the joint rotation in moment connections under service loads does not lead to excessive deflections, cracking or distress in other members in the composite structure. The construction criteria for composite connections require simple and rapid fabrication and construction.

This chapter presents the behaviour and design of composite connections in accordance with the AISC-LRFD Manual (1994). The design of single-plate shear connections, tee shear connections, beam-to-CEC column moment connections, beam-to-CFST column moment connections and semi-rigid connections is introduced. The design of single-angle and double-angle shear connections as shown in Figure 11.1 is given in the AISC-LRFD Manual (1994) and by Gong (2008, 2009, 2013).

11.2 SINGLE-PLATE SHEAR CONNECTIONS

Single-plate shear connections as depicted in Figure 11.2 are used to transfer the end reaction (shear) of simply supported steel or composite beams to the steel or composite columns. The single steel plate is usually shop welded to the column and filed bolted to the web of the steel beam.

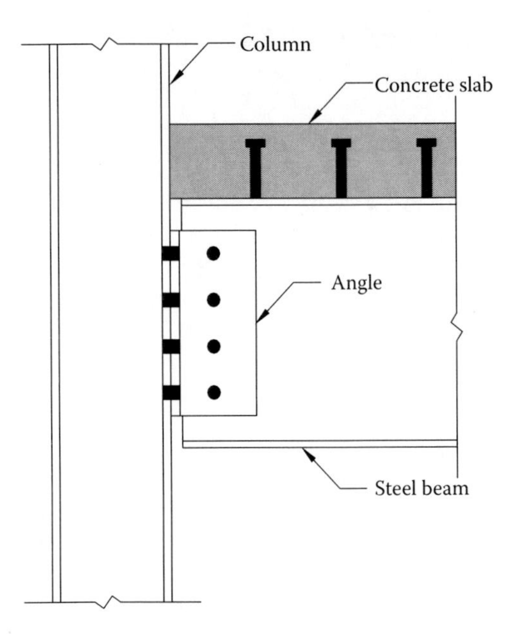

Figure 11.1 Bolted double-angle shear connection.

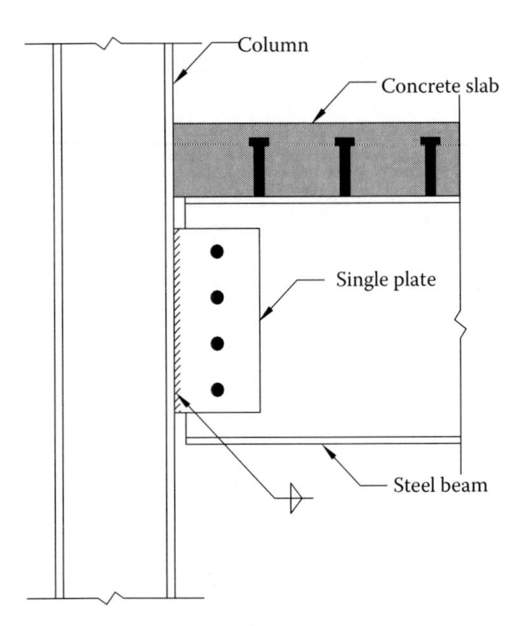

Figure 11.2 Single-plate shear connection.

The fabrication and erection of single-plate connections are easy and simple. This type of connections is used in both steel and composite construction. The effects of the composite slab or the slab reinforcement on the performance of single-plate connections are not considered in the design.

11.2.1 Behaviour of single-plate connections

The behaviour of single-plate shear connections is characterised by their shear–rotation curves which express the shear force as a function of the end rotation of the beam. The shear–rotation curves for single-plate connections can be determined by experiments. Test results presented by Astaneh et al. (1989) and Astaneh-Asl et al. (1993) indicate that the single plate yielded in shear so that inelastic shear deformations developed in the connections. The connections tested failed by shear fracture of the bolts which connected the single plates to the web of the steel beams. Other failure modes associated with single-plate connections were observed from experiments, including bearing failure of bolt holes, fracture of the net section of the single plate, fracture of the plate edge and fracture of the welds. The plate shear yielding and bearing yielding of bolt holes were found to be ductile, while other failure modes were brittle.

11.2.2 Design requirements

Single-plate shear connections must be designed to satisfy the following requirements: (1) having sufficient strength to transfer the shear force from the beam reaction, (2) having sufficient rotation capacity to meet the demand of a simply supported beam and (3) the connection should be flexible so that the beam end moments are negligible.

The design of single-plate shear connections should satisfy the following geometric and material requirements (Astaneh-Asl et al. 1993; Viest et al. 1997):

1. The connection should have only one vertical row of bolts, having the number of bolts within the range of 2 and 9.
2. The bolt spacing is 76 mm.
3. The edge distance is $a_e \geq 1.5d_f$.
4. The distance from the bolt centre line to weld line is $a_{bw} \geq 76$ mm.
5. The single plate should be made of mild steel.
6. E41XX or E48XX fillet welds should be used.
7. The thickness of the single plate is $t_p \geq 0.5d_f + 1.6$ mm.
8. The ratio $d_p/a_{bw} \geq 2$, where d_p is the depth of the single plate.
9. M20 or M24 high-strength structural bolts should be used.

11.2.3 Design of bolts

Bolts in single-plate connections are subjected to shear force and bending moment which is caused by the eccentricity of the beam end reaction from the bolt line. Therefore, bolts are designed for combined shear and bending. The eccentricity (e_b) of the reaction for the plate welded to a rigid supporting element is computed by

$$e_b = 25.4(n_b - 1) - a_{bw} \tag{11.1}$$

For the single plate welded to a flexible supporting element, the eccentricity (e_b) of the reaction to the bolt centre line is taken as the larger value obtained from Equation 11.1 and a_{bw}.

The moment can be determined as $M_b^* = V^* e_b$. The design of bolts for combined actions of shear and moment is given in the AISC-LRFD Manual (1994) and in Section 6.4.6. In the AISC-LRFD Manual, the design shear capacity of a bolt group under eccentric loading is determined by

$$\phi V_{fb} = C(\phi V_f) \tag{11.2}$$

where C is the coefficient accounting for the effect of eccentric loading on the design shear capacity of the bolt group, which is given in Table 8.18 in the AISC-LRFD Manual (1994).

It is a function of the eccentricity of the loading, spacing of bolts and number of bolts in one vertical row in the connection.

11.2.4 Design of single plate

The real steel plate in a single-plate shear connection is actually subjected to a large shear force and a small bending moment. However, the steel plate is designed to yield under the shear force only. This is to facilitate the early yielding of the plate made of mild steel. The shear yield capacity of the plate is given by

$$\phi V_u = \phi 0.6 f_y A_p \tag{11.3}$$

where

$\phi = 0.9$

A_p is the cross-sectional area of the single plate, taken as $A_p = d_p t_p$

Experiments show that the shear fracture of the net section occurs along a vertical plane close to the edge of the bolt holes rather than along the centre line of the bolt holes (Astaneh and Nader 1990). The design shear fracture capacity of the net section is determined by

$$\phi V_{ns} = \phi 0.6 f_u A_n \tag{11.4}$$

where

$\phi = 0.75$ is the capacity reduction factor

f_u is the tensile strength of the plate

A_n is the net cross-sectional area of the shear plane passing through the centre line of the bolts, which is given by (AISC-LRFD Manual 1994)

$$A_n = A_g - n_b(d_f + 1.5)t_p \tag{11.5}$$

where

n_b is the number of bolts

d_f is the diameter of the bolt

The bearing capacity of the plate in shear is determined by (AISC-LRFD Manual 1994)

$$\phi R_b = C\phi(2.4 f_u d_f t_p) \tag{11.6}$$

To prevent the edge failure, the vertical and horizontal edge distances (a_e) must not be less than $1.5d_f$ and the vertical distance a_e must not be less than 38 mm regardless of the bolt diameters. Local buckling of the bottom portion of the single plate may occur. The depth-to-thickness limit on the plate is taken as $d_p/t_p \leq 64$ to prevent local buckling from occurring.

11.2.5 Design of welds

The welds in single-plate shear connections are designed for the combined actions of shear and bending moment. The bending moment is caused by the eccentricity (e_w) of the beam reaction and is calculated as $M_w^* = V^* e_w$, where e_w is taken as the larger value of $25.4n_b$ and a_{bw}. The welds are designed to yield after yielding of the plate to prevent the brittle failure of the welds. This implies that the weld is stronger than the plate in a single-plate shear connection. To ensure this, the shear–moment interaction curve for the plate should lie inside the shear–moment curve for the welds. The weld size can be derived from this condition as (Astaneh-Asl et al. 1993)

$$D_w > 1.41\left(\frac{f_y}{f_{uw}}\right)t_p \tag{11.7}$$

where f_{uw} is the tensile strength of the weld metal. The weld size satisfying the aforementioned condition will ensure that the plate failure will occur before the weld fails.

Example 11.1: Design of single-plate shear connection

Design a single-plate shear connection which connects a composite beam to the flange of a steel column. The reaction of the composite beam is composed of a nominal dead load of 200 kN and a nominal live load of 180 kN. The steel beam section is 610UB125 of Grade 300 steel (t_w = 11.9 mm). M20 8.8/S high-strength structural bolts are used with a spacing of 76 mm.

1. Design of bolts

The design shear force is $V^* = 1.2G + 1.5Q = 1.2 \times 200 + 1.5 \times 180 = 510$ kN.

The shear capacity of an M20 bolt is $\phi V_f = 92.6$ kN.

The required number of bolts is

$$n_b = \frac{V^*}{\phi V_f} = \frac{510}{92.6} = 5.51 \quad \text{Try 6 bolts}$$

The flange of the supporting steel column is considered as rigid. The eccentricity of the reaction is

$$e_b = 25.4(n_b - 1) - a_{bw} = 25.4 \times (6 - 1) - 76 = 51\,\text{mm} = e_x$$

From Table 8.18 in Vol. II of the AISC-LRFD Manual, the coefficient C is obtained as C = 5.45.

The design shear strength of the bolt group is

$$\phi V_{fb} = C(\phi V_f) = 5.45 \times 92.6 = 504.67\,\text{kN} < V^* = 510\,\text{kN, N.G.}$$

Try 7 bolts; the design shear strength of the bolt group is determined as follows:

$$e_b = 25.4(n_b - 1) - a_{bw} = 25.4 \times (7 - 1) - 76 = 76.4\,\text{mm} = e_x$$

C = 6.06

$$\phi V_{fb} = C(\phi V_f) = 6.06 \times 92.6 = 561.2\,\text{kN} > V^* = 510\,\text{kN, OK}$$

2. Design of the single plate

The depth of the single plate is

$$d_p = s_b n_b = 76 \times 7 = 532\,\text{mm}$$

The shear fracture capacity of the net section is calculated by

$$\phi V_{ns} = \phi(0.6 f_u)[d_p - n_b(d_b + 1.6)]t_p$$

The required thickness of the plate is therefore

$$t_p \geq \frac{\phi V_{ns}}{\phi(0.6 f_u)[d_p - n_b \times (d_b + 1.60)]} = \frac{510 \times 10^3}{0.75 \times 0.6 \times 430 \times [532 - 7 \times (20 + 1.6)]} = 6.9\,\text{mm}$$

Try t_p = 8.5 mm; check the plate thickness limit as follows:

$$t_p \leq 0.5 d_b + 1.6 = 0.5 \times 20 + 1.6 = 11.6\,\text{mm, OK}$$

$$t_p > \frac{d_p}{64} = \frac{525}{64} = 8.2\,\text{mm, OK}$$

The shear yield capacity of the plate is

$$\phi V_u = \phi(0.6 f_y)A_p = 0.9 \times 0.6 \times 300 \times 532 \times 8.5\,\text{N} = 732.6\,\text{kN} > V^* = 510\,\text{kN, OK}$$

The bearing capacity of the plate is computed as

$$\phi R_b = C\phi(2.4 f_u d_f t_p) = 6.06 \times 0.75 \times 2.4 \times 430 \times 20 \times 8.5\,\text{N}$$
$$= 797.4\,\text{kN} > V^* = 510\,\text{kN, OK}$$

Since the beam web is thicker than the single plate, it is not required to check the bearing strength of the beam web.

3. Design of fillet welds

The size of the fillet weld is determined as

$$D_w = 1.41\left(\frac{f_y}{f_{uw}}\right)t_p = 1.41 \times \left(\frac{300}{480}\right) \times 8.5 = 7.5\,\text{mm}$$

Use 8 mm E48XX fillet welds on both sides of the plate.

11.3 TEE SHEAR CONNECTIONS

Tee shear connections are used to transfer the end shear reaction of simply supported steel or composite beams to the steel or composite columns. A tee connection is constructed by connecting it to a steel beam web and to a column. The tee can be cut from a wide flange or fabricated by welding two plates. Either bolts or welds can be used as fasteners in tee connections. There are four common types of tee shear connections which are used in both steel and composite structures depending on the use of fasteners. The tee shear connection shown

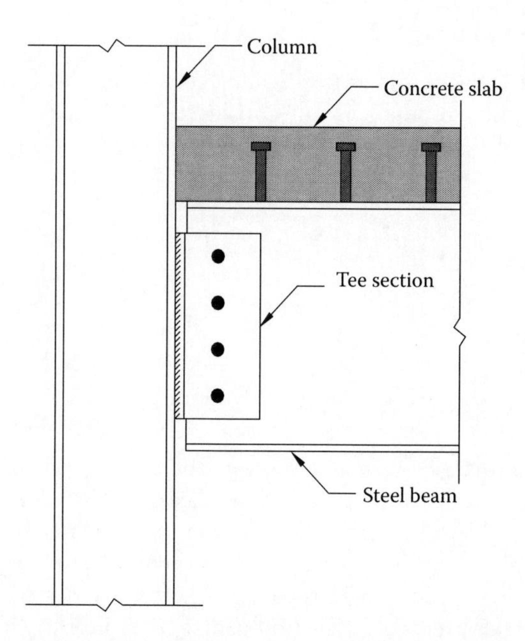

Figure 11.3 Tee shear connection.

in Figure 11.3, where the tee stem is bolted to the steel beam web and the tee flange is welded to the column flange, is considered here. The effects of composite slabs or the slab reinforcement on the strength and behaviour of tee shear connections are not considered in the design.

11.3.1 Behaviour of tee shear connections

The behaviour of tee shear connections is characterised by their shear–rotation curves which express the shear force as a function of the end rotation of the beam. The shear–rotation curves for tee shear connections can be determined by experiments. Tests conducted by Astaneh and Nader (1989, 1990) demonstrate that all specimens under the moment–rotation tests experienced minor yielding. When the rotation reached 0.06 rad, the welds fractured. The moment–rotation responses indicate that tee shear connections were flexible and ductile so that they could be regarded as simple connections. Six failure modes were observed from the shear–rotation tests (Astaneh and Nader 1989, 1990), including shear yielding of the tee stem, yielding of the tee flange, bearing failure of the steel beam web and the tee stem, shear fracture of the net section of the tee stem, shear fracture of the bolts and fracture of welds. The failure modes of yielding of the steel stem and flange are ductile. However, the bolt and weld fracture results in brittle failure mode of the tee connections.

11.3.2 Design of bolts

Bolts in tee shear connections should be designed for direct shear. When the supporting element is rigid, the eccentricity (e_b) of the reaction to the bolt line is so small that it can be ignored. The flange of a column or an embedded steel plate is considered as rigid supporting element. For this case, e_b is taken as zero. When the supporting element is rotationally flexible, the inflection point is assumed to be located at the weld line. As a result, the bolts are subjected to shear force (V^*) and a bending moment which is equal to $V^* e_b$, where the eccentricity e_b is taken as a_{bw}. The bolts are therefore designed for combined shear and bending. For this purpose, Table 9.10 in the AISC-LRFD Manual (1994) can be used.

11.3.3 Design of tee stems

The tee stem in actual tee shear connections is subjected to a shear force and a small bending moment. The small bending moment is not considered in the design of the tee stem in tee connections and it is designed for shear force only. The nominal shear yield capacity of the tee stem is expressed by

$$V_{ts} = 0.6 f_y d_t t_{ts} \tag{11.8}$$

where
d_t is the depth of the tee section
t_{ts} is the thickness of the tee stem

The shear fracture failure of the net section of the tee stem in tee shear connections is similar to that of the single plate in single-plate shear connections. Experiments indicate that the fracture failure of the tee stem in shear occurs at the net section along the edge of bolt holes rather than along their centre line. The shear fracture capacity of the tee stem can be calculated using Equation 11.4. However, the effective net area in shear is calculated using the average of the net area along the bolt centre line and the gross area of the tee stem as follows (Astaneh and Nader 1990):

$$A_n = A_g - 0.5 n_b (d_f + 1.5) t_{ts} \tag{11.9}$$

The design bearing capacity of the tee stem in shear is given by (AISC-LRFD Manual 1994)

$$\phi R_b = n_b \phi (2.4 f_u d_f t_{ts}) \tag{11.10}$$

where $\phi = 0.75$ is the capacity reduction factor. For the steel beam web, the earlier equation can be used to calculate its design bearing capacity by substituting t_{ts} by t_w.

11.3.4 Design of tee flanges

If the thickness of the tee flange is less than the thickness of the tee stem in a tee shear connection, the tee flange will yield before the tee stem. The nominal shear yield capacity of the tee flange is determined by

$$V_{tf} = 2(0.6 f_y) d_t t_f \tag{11.11}$$

where t_f is the thickness of the tee flange.

11.3.5 Design of welds

In tee shear connections, fillet welds are used to connect the tee flange to the supporting element such as the flange of a column as depicted in Figure 11.3. The welds are subjected to shear force and bending moment caused by the eccentricity e_w of the beam reaction from the weld line. The eccentricity e_w can be conservatively taken as the distance between the bolt and weld lines, such as $e_w = a_{bw}$. Using Table 8.38 in the AISC-LRFD Manual (1994), the design strength of eccentrically loaded weld group under shear force V^* and bending moment of V^*e_w can be determined by

$$\phi R_w = CC_1 D_{16} L_w \tag{11.12}$$

where

C is the coefficient including the capacity factor $\phi = 0.75$, given in Table 8.38 in the AISC-LRFD Manual (1994)
C_1 is the electrode strength coefficient given in Table 8.37
D_{16} is the number of 16th of an inch in the weld size
L_w is the weld length

11.3.6 Detailing requirements

The design method for tee shear connections described in the preceding sections was developed based on limited test results. The tee shear connections designed using this method are restricted to some geometric and material requirements described in this section (Astaneh and Nader 1990).

To prevent the local buckling of the lower half of the tee stem in compression, the ratio of d_t/a_{bw} of the tee stem should be greater than 2. The width-to-thickness ratio $(0.5b_f/t_f)$ of the tee flange outstand should be greater than 6.5 to ensure the flexibility of the connection. The depth-to-width ratio (d_t/b_f) of the tee section should not exceed 3.5 to prevent large inelastic tensile strain from developing in the welds. To increase connection ductility, the ratio of $(t_{ts}/d_f)/(t_f/t_{ts})$ should be less than 0.25.

The tee section should be made of mild steel to ensure good shear and rotational ductility. M20 or M24 high-strength structural bolts may be used in only one vertical row. Snug-tight bolts are preferred. The vertical spacing of bolts should be equal to 76 mm. The number of bolts should not be less than 2 and more than 9. Fillet welds should be used to weld the tee flange to the supporting element. The top of the fillet welds should be returned a distance of $2D_w$. If the tee flange is welded to the flange of a steel column, the thickness of the column flange should be greater than that of the tee flange.

Example 11.2: Design of tee shear connection

Design a tee shear connection which connects a composite beam to the flange of a steel column. The reaction of the composite beam under factored design loads is 300 kN. The steel beam section is 410UB59.7 of Grade 300 steel and the steel column is 250UC of Grade 300 steel. The M20 8.8/S bolts are used with a spacing of 76 mm. The E48XX fillet welds are used.

1. Design of bolts

The shear capacity of an M20 bolt is $\phi V_f = 92.6\,\text{kN}$.

The required number of bolts is

$$n_b = \frac{V^*}{\phi V_f} = \frac{300}{92.6} = 3.24$$

Adopt four bolts.

2. Check geometric requirements of the tee section

The required gross areas of the tee stem can be determined from its shear yield capacity as

$$A_{ts} = \frac{V^*}{0.6f_y} = \frac{300\times10^3}{0.6\times300} = 1667\,\text{mm}^2$$

The dimensions of the tee section are selected as follows:

$$b_f = 170\,\text{mm},\quad t_f = 13\,\text{mm},\quad t_{ts} = 8\,\text{mm},\quad a_e = 35\,\text{mm},\quad s_g = 76\,\text{mm}$$

The width-to-thickness ratio of the tee stem is calculated as

$$\frac{b_f}{2t_f} = \frac{170}{2\times13} = 6.54 > 6.5,\ \text{OK}$$

The ratio of d_f/t_{ts} is

$$\frac{d_f}{t_{ts}} = \frac{20}{8} = 2.5 > 2.0,\ \text{OK}$$

The edge distance of the tee stem is

$$a_e = 35\,\text{mm} > 1.5d_f = 1.5\times20 = 30\,\text{mm, OK}$$

The depth of the tee stem is determined as

$$d_t = (n_b - 1)s_g + 2a_e = (4-1)\times76 + 2\times35 = 298\,\text{mm}$$

The cross-sectional area of the tee stem is computed as

$$d_t t_{ts} = 298\times8 = 2384\,\text{mm}^2 > A_{ts} = 1667\,\text{mm}^2,\ \text{OK}$$

The depth-to-width ratio of the tee section is

$$\frac{d_t}{b_f} = \frac{298}{170} = 1.75 < 3.5,\ \text{OK}$$

The thickness of the column is t_{fc} = 14.2 mm > t_f = 13 mm, OK.
The thickness ratio of the tee section is

$$\frac{t_{ts}/d_f}{t_f/t_{ts}} = \frac{8/20}{13/8} = 0.246 < 0.25,\text{OK}$$

The clear depth of the steel beam web is d_w = 406 − 2 × 12.8 = 380.4 mm > d_t = 298 mm, OK.

3. Design strengths of the tee stem

The nominal shear yield capacity of the tee stem is computed as

$$\phi V_{ts} = \phi(0.6f_y)d_t t_{ts} = 0.9\times0.6\times300\times298\times8\,\text{N} = 386.2\,\text{kN} > V^* = 300\,\text{kN, OK}$$

The net area of the tee stem in shear is calculated as

$$A_n = A_g - 0.5n_b(d_f + 1.5)t_{ts} = 2384 - 0.5\times4\times(20+1.5)\times8 = 2040\,\text{mm}^2$$

The shear fracture capacity of the net section of the tee stem is

$$\phi V_{ns} = \phi(0.6 f_u) A_{ns} = 0.75 \times 0.6 \times 430 \times 2040\,\text{N} = 394.7\,\text{kN} > \phi V_{ts} = 386.2\,\text{kN, OK}$$

The bearing capacity of the tee stem is calculated as

$$\phi R_b = n_b \phi(2.4 f_u) d_f t_{ts} = 4 \times 0.75 \times 2.4 \times 430 \times 20 \times 8\,\text{N} = 495.36\,\text{kN}$$
$$> \phi V_{ts} = 386.2\,\text{kN, OK}$$

Since $t_w = 8.5\,\text{mm} > t_{ts} = 8\,\text{mm}$, the tee stem will govern the bearing failure.
The shear capacity of the tee flange is

$$\phi V_{tf} = \phi 2(0.6 f_y) d_t t_f = 0.9 \times 2 \times 0.6 \times 300 \times 298 \times 13\,\text{N} = 1255.2\,\text{kN}$$
$$> \phi V_{ts} = 386.2\,\text{kN, OK}$$

4. Design of fillet welds

The fillet welds are designed for combined shear and out-of-plane bending moment. The eccentricity is $a_{bw} = 76\,\text{mm}$. The eccentricity ratio is

$$a = \frac{e_w}{L_w} = \frac{76}{298} = 0.252$$

With $k = 0$ for out-of-plane bending, the coefficient C is obtained from Table 8.38 in the AISC-LRFD Manual as $C = 2.48$.
Using $C_1 = 1.0$ for E48XX fillet welds, the required weld size in 16th of an inch is

$$D_{16} = \frac{V_{ts}}{C C_1 L_w} = \frac{(429/4.4480)}{2.48 \times 1.0 \times (298/25.4)} = 3.32$$

The size of the welds is $D_w = (D_{16} \times 25.4)/16 = (3.32 \times 25.4)/16 = 5.3\,\text{mm}$.
Use 6 mm E48XX fillet welds on both sides of the tee flange.

11.4 BEAM-TO-CEC COLUMN MOMENT CONNECTIONS

Beam-to-column moment connections between steel or composite beams and reinforced concrete or composite columns are employed in moderate- to high-rise composite buildings. Beam-to-column moment connections are used to transfer the axial force, bending moment and shear force arising from applied loads from the beams to the composite columns. A steel beam-to-CEC column moment connection is constructed by passing the steel beam through a CEC column as illustrated in Figure 11.4. Face-bearing plates (FBPs) and vertical reinforcement may be attached to the steel beam to resist bearing forces. Horizontal reinforcing ties are provided in the column within the beam depth and above and below the beam to carry tension forces developed in the connection.

The design method for steel beam-to-CEC column moment connections presented herein is based on the work of Sheikh et al. (1989), Deierlein et al. (1989) and the ASCE Task Committee (1994). It is applicable only to interior and exterior moment connections between steel beams and reinforced concrete or composite columns. The effects of composite slabs or the slab reinforcement on the strength and behaviour of composite connections are not

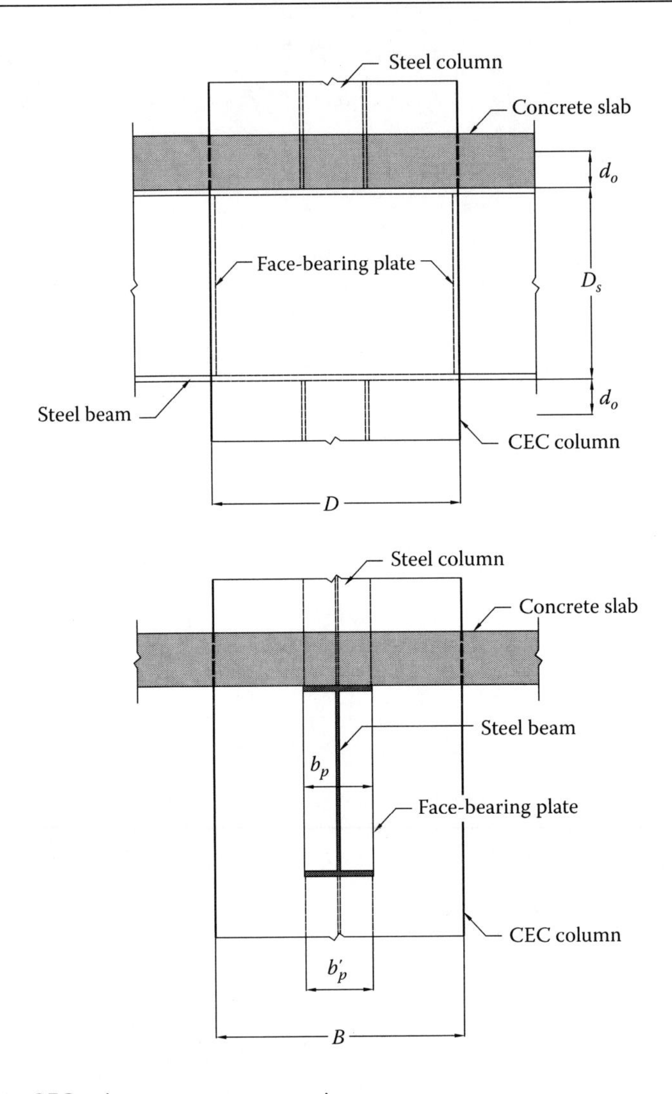

Figure 11.4 Beam-to-CEC column moment connection.

considered in the design. The aspect ratio of the connection is limited to $0.75 \leq D/D_s \leq 2.0$. The method is limited to normal weight concrete with $f'_c \leq 40\,\text{MPa}$, structural steel with yield stress of $f_y \leq 345\,\text{MPa}$ and reinforcing bars with yield stress of $f_{yr} \leq 410\,\text{MPa}$.

11.4.1 Behaviour of composite moment connections

The behaviour of steel beam-to-encased composite column moment connections is characterised by two primary failure modes, namely, the panel shear failure and the vertical bearing failure (Sheikh et al. 1989). In the composite connection, both structural steel and reinforced concrete are involved in the panel shear failure, which is similar to the structural steel or reinforced concrete connection. Bearing failure occurs at the upper and lower corners of the connections subjected to high compressive stresses. The forces in the connection are transferred by three shear mechanisms, which are the steel web panel, the concrete compression strut and concrete compression field. The steel web is subjected to pure shear stress over an effective panel length. The vertical stiffener plates attached to the steel beam mobilise the concrete compression strut. The concrete compression field composing of several compression struts with horizontal reinforcement forms a strut-and-tie system to carry the forces in the connection.

11.4.2 Design actions

The forces acting on the composite connection are shown in Figure 11.5. The axial forces in the steel beam are usually small so that they are not considered in the strength calculation of the connection. The compressive axial forces in the column are also not taken into account as experiments indicate that it is conservative to neglect their effects. If the net tension forces exist in the connection, the concrete contribution to the shear strength of the concrete compression field should be ignored. The forces on a composite connection can be expressed by (ASCE Task Committee 1994)

$$\sum M_c^* = \sum M_b^* + V_b^* D - V_c^* D_s \tag{11.13}$$

where

$$\sum M_c^* = M_{c1}^* + M_{c2}^* \tag{11.14}$$

$$\sum M_b^* = M_{b1}^* + M_{b2}^* \tag{11.15}$$

$$V_b^* = \frac{\left(V_{b1}^* + V_{b2}^*\right)}{2} \tag{11.16}$$

$$V_c^* = \frac{\left(V_{c1}^* + V_{c2}^*\right)}{2} \tag{11.17}$$

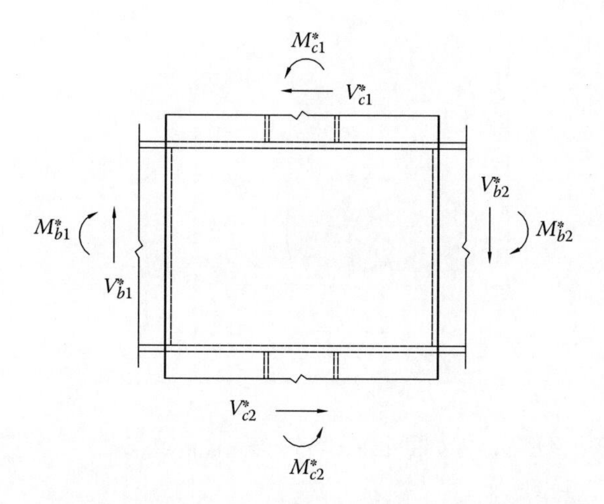

Figure 11.5 Design actions on interior beam-to-CEC column moment connection.

11.4.3 Effective width of connection

The effective width of the composite connection within a composite column is given by (ASCE Task Committee 1994)

$$b_j = b_i + b_o \tag{11.18}$$

where
b_i is the inner panel width which is taken as the larger of the width of the FBP (b_p) and the width of the beam flange (b_f)
b_o is the outer panel width as depicted in Figure 11.6

For the extended FBPs or steel columns, b_o is determined based on the overall cross-sectional geometry as follows:

$$b_o = h_{xy}(b_{max} - b_i) < 2d_o \tag{11.19}$$

$$b_{max} = \frac{(b_f + B)}{2} < b_f + D < 1.75b_f \tag{11.20}$$

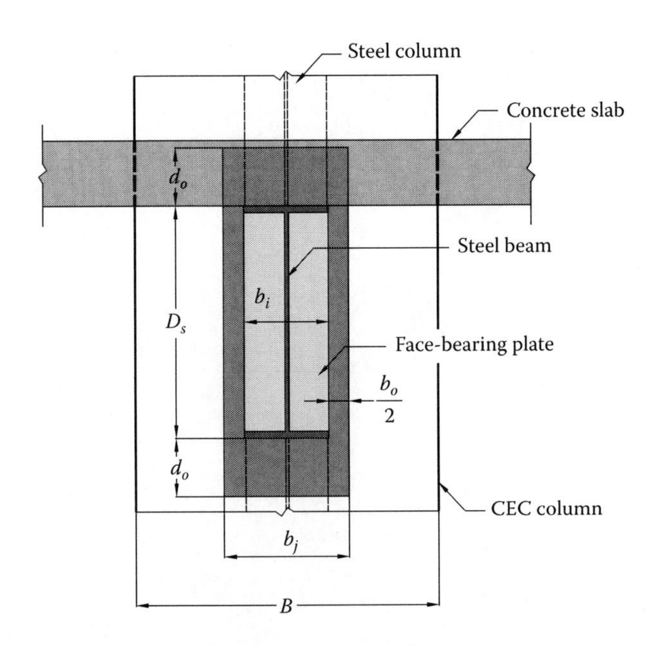

Figure 11.6 Effective width of beam-to-CEC column moment connection.

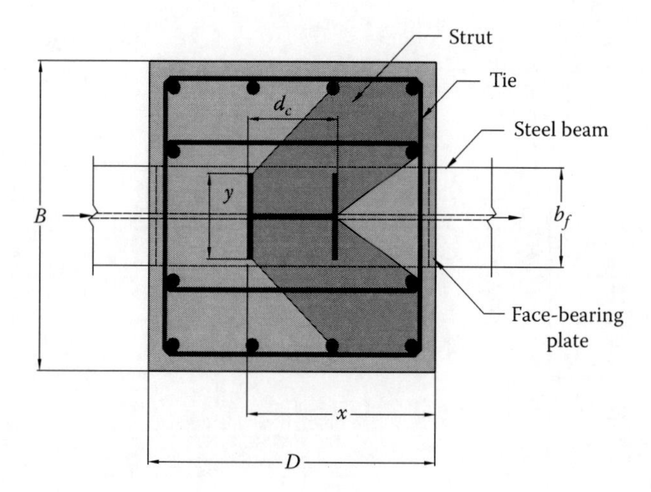

Figure 11.7 Strut-and-tie model for horizontal force transfer.

$$h_{xy} = \left(\frac{x}{D}\right)\left(\frac{y}{b_f}\right) \tag{11.21}$$

where

$d_o = 0.25D_s$ when the column is a steel column or d_o is taken as the lesser of $0.25D_s$ and the height of the extended FBPs when these plates are used
B is the column width oriented perpendicular to the beam
D is the depth of the column
y is the larger of the steel column or extended FBP width
$x = D$ where the extended FBPs are used or $x = D/2 + d_c/2$ when only the steel column is used as illustrated in Figure 11.7

11.4.4 Vertical bearing capacity

The vertical bearing forces on the connection are the results of combined shears and moments transferred between the beam and column as depicted in Figure 11.8, where the moments M_{c1}^* and M_{c2}^* on the column are represented by the bearing forces C_c and the forces in the vertical reinforcement T_{vr} (tension) and C_{vr} (compression). The length of the bearing zone (a_c) above and below the beam is assumed to be $0.3D$. The nominal concrete bearing strength is determined by

$$C_{cb} = 2f_c'b_j(0.3D) \tag{11.22}$$

where the bearing stress is taken as $2f_c'$ due to the concrete confinement provided by the reinforcement and surrounding concrete.

Reinforcing bars, rods or steel angles can be attached to the steel beam as vertical reinforcement to carry vertical bearing forces in the connection. However, it should be noted that providing a large amount of vertical reinforcement may induce high bearing stress

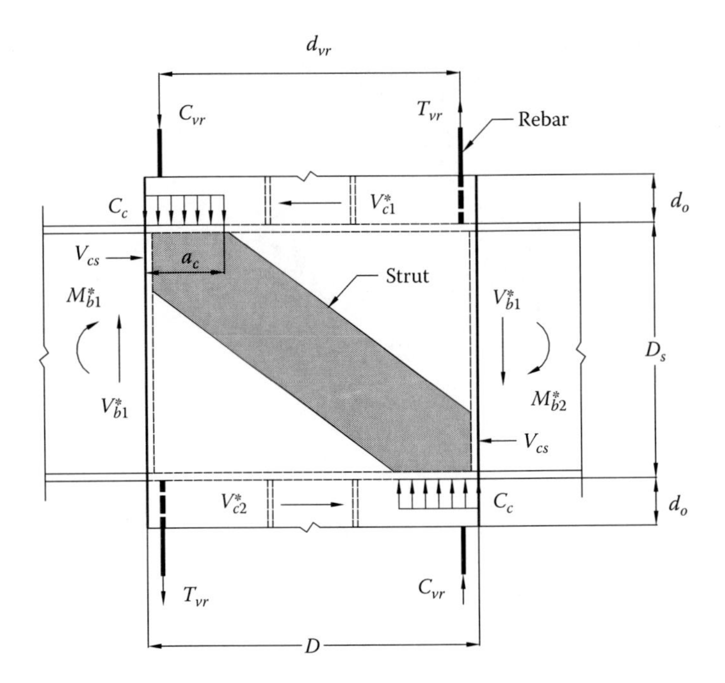

Figure 11.8 Forces in beam-to-CEC column moment connection.

on the concrete between the two flanges of the steel beam. To avoid this, the strengths of the vertical reinforcement in tension (T_{vr}) and in compression (C_{vr}) are limited (ASCE Task Committee 1994) by

$$T_{vr} + C_{vr} \le 0.3 f_c' b_j D \tag{11.23}$$

Replacing the vertical forces with their respective nominal strength values and from the moment equilibrium, the following expression is obtained for the composite connection subjected to vertical bearing (ASCE Task Committee 1994):

$$\sum M_c^* + 0.35 D \Delta V_b^* \le \phi\left[C_{cb}(0.7D) + (T_{vr} + C_{vr}) d_{vr}\right] \tag{11.24}$$

where

$\Delta V_b^* = V_{b2}^* - V_{b1}^*$

d_{vr} is the distance between the bars

The vertical reinforcement is assumed to carry both tension and compression forces or compression only ($T_{vr} = 0$).

11.4.5 Horizontal shear capacity

The horizontal shear in a steel beam-to-encased composite column connection is resisted by three shear mechanisms, which consist of the steel web panel, the inner concrete compression strut and the outer concrete compression field as illustrated in Figure 11.8.

The steel web panel is subjected to pure shear and its strength is governed by its shear yield capacity as follows:

$$V_{wp} = 0.6 f_{yw} L_p t_w \tag{11.25}$$

where L_p is the panel width.

It would appear that the concrete compression strut is a diagonal compression member that forms within the inner panel width (b_i) as shown in Figure 11.8. The compression force in the diagonal concrete strut provides bearing stress on the FBPs within the depth of the steel beam. The nominal strength of the compression strut is given by (ASCE Task Committee 1994)

$$V_{cs} = 1.7\sqrt{f_c'} b_p D \le 0.5 f_c' b_p d_w \tag{11.26}$$

where $1.7\sqrt{f_c'}$ is the average limiting horizontal shear stress for concrete, the concrete strength f_c' is in MPa and the effective width of the FBP is taken as $b_p \le b_f + 5t_p \le 1.5b_f$. The bearing failure of concrete at the ends of the strut may occur. To prevent this, the horizontal shear is limited by a maximum bearing stress of $2f_c'$ acting on an area of $b_p(0.25d_w)$ at the top and bottom of the FBPs.

The compression fields develop in the outer panel width (b_o). The compression fields are mobilised by the horizontal struts and column ties which form a strut-and-tie system by bearing against the steel column and/or extended FBPs as shown in Figure 11.7. The nominal strength of the concrete compression field is governed by the strength of the concrete and the horizontal column ties and can be computed by (ASCE Task Committee 1994)

$$V_{cf} = V_c' + V_s' \le 1.7\sqrt{f_c'} b_o D \tag{11.27}$$

where V_c' is the strength provided by the concrete in compression, which is given by

$$V_c' = 0.4\sqrt{f_c'} b_o D \tag{11.28}$$

If the column is in tension, $V_c' = 0$. The strength provided by the horizontal ties is determined as

$$V_s' = A_{sr} f_{yr} 0.9D / s_{sr} \tag{11.29}$$

where A_{sr} is the cross-sectional area of reinforcing bars in each layer of ties spaced at s_{sr} in the depth of the beam web and $A_{sr} \ge 0.004bs_{sr}$.

The horizontal shear strength of the connection is the sum of the shear strength of the steel web panel, the inner concrete compression strut and the outer concrete compression field. The vertical shear in the connection caused by applied loads is equal to the total shear strength of the connection. The horizontal shear strength of the connection must satisfy the following condition (ASCE Task Committee 1994):

$$\sum M_c^* - V_b^* L_p \le \phi\left[V_{wp} d_{fc} + V_{cs}(0.75d_w) + V_{cf}(D_s + d_o)\right] \tag{11.30}$$

where d_{fc} is the distance between the centroids of the beam flange, and the panel width L_p is calculated as follows:

$$L_p = \frac{\sum M_c^*}{\phi(C_c + T_{vr} + C_{vr}) - 0.5\Delta V_b^*} \geq 0.7D \tag{11.31}$$

$$C_c = 2f_c'b_j a_c \tag{11.32}$$

$$a_c = \frac{D}{2} - \sqrt{\frac{D^2}{4} - K} \leq 0.3D \tag{11.33}$$

$$K = \frac{\sum M_c^* + \Delta V_b^*(D/2) - \phi(T_{vr} + C_{vr})d_{vr}}{\phi(2f_c')b_j} \tag{11.34}$$

11.4.6 Detailing requirements

The detailing requirements on the steel beam-to-CEC column moment connections are given by the ASCE Task Committee (1994) and are discussed in this section.

11.4.6.1 Horizontal column ties

Horizontal reinforcing ties should be provided in the column within the depth of the steel beam and above and below the beam to sustain tension forces developed in the connection as shown in Figure 11.7. Horizontal reinforcing ties within the beam depth are used to carry the tension forces associated with the compression fields. One pair of ties in each layer in the beam depth should pass through holes in the beam web to provide continuous confinement to the concrete.

Reinforcing ties above and below the beam are part of the horizontal strut-and-tie system. Three layers of ties should be provided above and below the steel beam within a distance of $0.4D_s$ from the beam flange as follows: (1) for $B \leq 500$ mm, 10 mm bars with four legs in each layer; (2) for $500 < B \leq 750$ mm, 12 mm bars with four legs in each layer and (3) for $B > 750$ mm, 16 mm bars with four legs in each layer. The minimum amount of ties above and below the beam may be governed by the force in the compression field $V_{cff}(\leq V_{cf})$. The minimum total cross-sectional area of ties within the depth of $0.4D_s$ should satisfy

$$A_{tie} \geq \frac{V_{cff}}{f_{yr}} \tag{11.35}$$

11.4.6.2 Vertical column ties

The large changes in reinforcing bar forces owing to the transfer of moments in the connection may occur, which leads to the slip of vertical bars. To limit the bar slip, the size of the vertical column bars should be taken as follows:

$$d_b < \frac{(D + 2d_o)}{20} \tag{11.36}$$

where d_b is the diameter of the vertical bar or the diameter of a bar equivalent to the bundle bars.

If the change in force in vertical bars satisfies the following requirement, larger size than the limit by Equation 11.36 can be used:

$$\Delta F_b < 80(D + 2d_o)\sqrt{f'_c} \tag{11.37}$$

11.4.6.3 Face-bearing plates

FBPs within the beam depth are used to carry the horizontal forces in the concrete strut. If split FBPs are employed, the plate height (d_p) should not be less than $0.45d_w$. The required thickness of the FBP is influenced by the distribution of the concrete bearing stress, its geometry, support conditions and yield stress. The thickness of the FBP should satisfy the following requirements:

$$t_p \geq \frac{\sqrt{3}(V_{cs} - b_f t_w f_{yw})}{b_f f_{up}} \tag{11.38}$$

$$t_p \geq \frac{\sqrt{3}V_{cs}}{2b_f f_{up}} \tag{11.39}$$

$$t_p \geq 0.2\sqrt{\frac{V_{cs} b_p}{f_{yp} d_w}} \tag{11.40}$$

$$t_p \geq \frac{b_p}{22} \tag{11.41}$$

$$t_p \geq \frac{(b_p - b_f)}{5} \tag{11.42}$$

where f_{yp} and f_{up} are the yield and tensile strengths of the bearing plate, respectively.

11.4.6.4 Steel beam flanges

The flanges of the steel beam under vertical bearing forces in the composite connection are subjected to transverse bending. The flanges of the steel beam must have sufficient flexural stiffness to resist the transverse bending. For this purpose, the thickness of the beam flanges must satisfy the following requirement:

$$t_f \geq 0.3\sqrt{\frac{b_f t_w D_s f_{yw}}{D f_{yf}}} \tag{11.43}$$

11.4.6.5 Extended face-bearing plates and steel column

The extended FBPs and/or steel columns are subjected to compressive bearing forces in the horizontal struts. The net bearing force is equal to the shear force V_{cff} ($\leq V_{cf}$) carried by the compression strut. When a steel column is used, only one of the column flanges is subjected to bearing force as depicted in Figure 11.7. The design of these elements is governed by the transverse bending of the plate, shear strength of the supporting element and the connection to the steel beam. The thickness of the extended FBPs or the column flanges is limited by

$$t_f \geq 0.12\sqrt{\frac{V_{cff}b'_p}{d_o f_y}} \tag{11.44}$$

where

b'_p is the width of the extended FBP or the width of the flange of the steel column
V_{cff} can be taken as V_{cf}

The thickness of the extended bearing plate should be greater than that of the FBP between the flanges of the beam.

Example 11.3: Design of steel beam-to-CEC column moment connection

Check the capacity and design the details of the steel beam-to-CEC column moment connection shown in Figure 11.9. The connection is subjected to the following factored design actions:

$$M^*_{b1} = M^*_{b2} = 635{,}750\,\text{kN mm}, \quad V^*_{b1} = V^*_{b2} = 300\,\text{kN}, \quad M^*_{c1} = M^*_{c2} = 600{,}000\,\text{kN mm}$$

$$V^*_{c1} = V^*_{c2} = 500\,\text{kN}$$

Design data shown in Figure 11.9 are

Composite column: $B = D = 650$ mm

Steel beam: $b_f = 209$ mm, $t_f = 15.6$ mm, $D_s = 533$ mm, $t_w = 10.2$ mm, $d_{fc} = 517.4$ mm

$d_w = 502$ mm, $f_{yf} = 300$ MPa, $f_{yw} = 320$ MPa,

Steel column: $d_c = 203$ mm, $t_{cf} = 11$ mm

Face-bearing plates: $b_p = 209$ mm, $b'_p = 203$ mm, $t_p = 16$ mm, $f_{yp} = 300$MPa

$f_{up} = 430$ MPa

Vertical reinforcement: $T_{vr} = C_{vr} = 0$

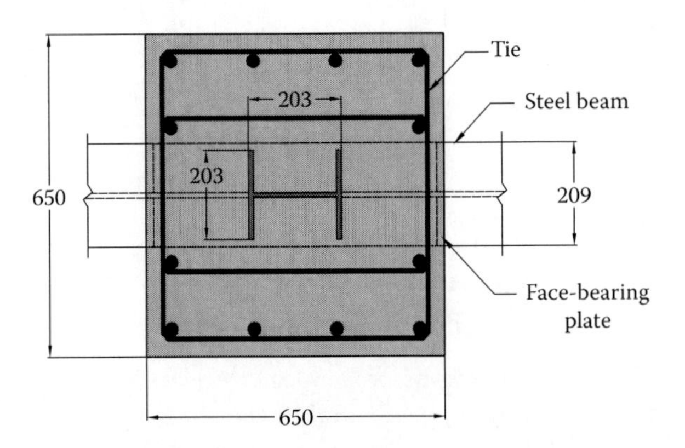

Figure 11.9 Steel beam-to-CEC column moment connection.

1. Design actions

The design actions are calculated as follows:

$$\sum M_c^* = M_{c1}^* + M_{c2}^* = 600{,}000 + 600{,}000 = 1{,}200{,}000\,\text{kN mm}$$

$$V_b^* = \frac{\left(V_{b1}^* + V_{b2}^*\right)}{2} = \frac{(300+300)}{2} = 300\,\text{kN}$$

$$\Delta V_b^* = V_{b1}^* - V_{b2}^* = 300 - 300 = 0\,\text{kN}$$

2. Effective width of the connection

The maximum width of the connection is

$$b_{\max} = \frac{b_f + B}{2} \le b_f + D \le 1.75 b_f$$

$$= \frac{209 + 650}{2} = 429.5\,\text{mm} \le 209 + 650 = 859\,\text{mm}$$

$$> 1.75 \times 209 = 365.75\,\text{mm}$$

Hence, $b_{\max} = 365.75$ mm.

The effective width of the connection is computed as

$$x = \frac{D}{2} + \frac{d_c}{2} = \frac{650}{2} + \frac{203}{2} = 426.5\,\text{mm}$$

$$h_{xy} = \left(\frac{x}{D}\right)\left(\frac{y}{b_f}\right) = \left(\frac{426.5}{650}\right)\left(\frac{203}{209}\right) = 0.637$$

$$b_o = h_{xy}(b_{\max} - b_i) = 0.637 \times (365.75 - 209) = 100\,\text{mm}$$

$$b_j = b_i + b_o = 209 = 100 = 309\,\text{mm}$$

3. Vertical bearing capacity

The nominal concrete bearing strength is calculated as

$$C_{cb} = 2f_c' b_j (0.3D) = 2 \times 40 \times 309 \times (0.3 \times 650)\,\text{N} = 4820.4\,\text{kN}$$

The design actions on the connection are computed as

$$\sum M_c^* + 0.35 D \Delta V_b = 1{,}200{,}000 + 0.35 \times 650 \times 0 = 1{,}200{,}000\,\text{kN mm}$$

The vertical bearing capacity of the connection is calculated as

$$\phi\left[C_{cb}(0.7D) + (T_{vr} + C_{vr})d_{vr}\right] = 0.7 \times \left[4820.4 \times 0.7 \times 650 + 0\right]$$

$$= 1{,}535{,}297\,\text{kN m} > 1{,}200{,}000\,\text{kN mm, OK}$$

4. Horizontal shear capacity

The width of the shear panel is calculated as follows:

$$K = \frac{\sum M_c^* + \Delta V_b^*(D/2) - \phi(T_{vr} + C_{vr})d_{vr}}{\phi(2f_c')b_j} = \frac{1{,}200{,}000 \times 10^3 + 0 - 0}{0.7 \times 2 \times 40 \times 309} = 69{,}348\,\text{mm}^2$$

$$a_c = \frac{D}{2} - \sqrt{\frac{D^2}{4} - K} \le 0.3D$$

$$= \frac{650}{2} - \sqrt{\frac{650^2}{4} - 69{,}348} = 134.5\,\text{mm} < 0.3 \times 650 = 195\,\text{mm}$$

$$= 134.5\,\text{mm}$$

$$C_c = 2f_c'b_ja_c = 2 \times 40 \times 309 \times 134.5\,\text{N} = 3325\,\text{kN}$$

$$L_p = \frac{\sum M_c^*}{\phi(C_c + T_{vr} + C_{vr}) - 0.5\Delta V_b^*} \ge 0.7D$$

$$= \frac{1{,}200{,}000}{0.7 \times (3{,}325 + 0 + 0) - 0.5 \times 0} = 515.6\,\text{mm} > 0.7 \times 650 = 455\,\text{mm}$$

$$= 515.6\ \text{mm}$$

The nominal shear yield capacity of the steel web panel is

$$V_{wp} = 0.6f_{yw}L_pt_w = 0.6 \times 320 \times 515.6 \times 10.2\,\text{N} = 1009.8\,\text{kN}$$

The nominal strength of the compression strut is calculated as

$$V_{cs} = 1.7\sqrt{f_c'}b_pD = 1.7 \times \sqrt{40} \times 209 \times 650\,\text{N} = 1460.6\,\text{kN}$$

$$0.5f_c'b_pd_w = 0.5 \times 40 \times 209 \times 502\,\text{N} = 2098.4\,\text{kN} > V_{cs} = 1460.6\,\text{kN}$$

Hence, $V_{cs} = 1460.6\,\text{kN}$.

Assuming the ties are adequate, the nominal strength of the compression field is determined as

$$V_{cf} = 1.7\sqrt{f_c'}b_oD = 1.7 \times \sqrt{40} \times 100 \times 650\,\text{N} = 698.9\,\text{kN}$$

The design actions on the connection are computed as

$$\sum M_c^* - V_bL_p = 1{,}200{,}000 - 300 \times 515.6 = 1{,}045{,}320\,\text{kN mm}$$

The horizontal shear capacity of the connection is calculated as

$$\phi\left[V_{wp}d_{fc} + V_{cs}(0.75d_w) + V_{cf}(D_s + d_o)\right]$$

$$= 0.7 \times \left[1009.8 \times 517.4 + 1460.6 \times 0.75 \times 502 + 698.9 \times (533 + 0.25 \times 533)\right]$$

$$= 1{,}076{,}602\,\text{kN mm} > 1{,}045{,}320\,\text{kN mm, OK}$$

5. Detailing

5.1. Column ties within beam depth

The strength provided by concrete in compression is

$$V_c' = 0.4\sqrt{f_c'}b_oD = 0.4 \times \sqrt{40} \times 100 \times 650\,\text{N} = 164.4\,\text{kN}$$

The strength provided by the horizontal ties is determined as

$$V_s' = V_{cf} - V_c' = 698.9 - 164.5 \ = 534.5\,\text{kN}$$

The required cross-sectional area of column ties per unit length is

$$\frac{A_{sr}}{s_{sr}} = \frac{V_s'}{0.9Df_{yr}} = \frac{534.5 \times 1000}{0.9 \times 650 \times 400} = 2.28\,\text{mm}^2/\text{mm}$$

$$\left(\frac{A_{sr}}{s_{sr}}\right)_{\min} = 0.004B = 0.004 \times 650 = 2.6\,\text{mm}^2/\text{mm}$$

Use 4-legs Y12 ties for each layer, A_{sr} = 4 × 110 = 440 mm²; the spacing of the ties is

$$s_{sr} = \frac{440}{2.6} = 169\,\text{mm}$$

Use 4-legs Y12 at 160 mm.

5.2. Column ties adjacent to connection

The required area of column ties is

$$A_{tie} = \frac{V_{cf}}{f_{yr}} = \frac{698.9 \times 1000}{400} = 1747\,\text{mm}^2$$

The depth in which the ties are placed is $0.4D_s$ = 0.4 × 533 = 213 mm.
 Use 4-layers Y12 at 70 mm (A_{tie} = 1810 mm²).

5.3. Thickness of face-bearing plates

The thickness of the FBPs is calculated as follows:

$$t_p \geq \frac{\sqrt{3}(V_{cs} - b_f t_w f_{yw})}{b_f f_{up}} = \frac{\sqrt{3} \times (1460.6 \times 10^3 - 209 \times 10.2 \times 320)}{209 \times 430} = 15\,\text{mm}$$

$$t_p \geq \frac{\sqrt{3}V_{cs}}{2b_f f_{up}} = \frac{\sqrt{3} \times 1460.6 \times 10^3}{2 \times 209 \times 430} = 14\,\text{mm}$$

$$t_p \geq 0.2\sqrt{\frac{V_{cs}b_p}{f_{yp}d_w}} = 0.2\sqrt{\frac{1460.6 \times 10^3 \times 209}{430 \times 502}} = 8.7\,\text{mm}$$

$$t_p \geq \frac{b_p}{22} = \frac{209}{22} = 9.5\,\text{mm}$$

Hence, t_p = 16 mm > 15 mm, OK.

5.4. Steel beam flanges

The required thickness of the steel beam flanges is

$$t_f \geq 0.3\sqrt{\frac{b_f t_w D_s f_{yw}}{D f_{yf}}} = 0.3\sqrt{\frac{209 \times 10.2 \times 533 \times 320}{650 \times 300}} = 15.2\,\text{mm}$$

$$t_f = 15.6\,\text{mm} > 15.2\,\text{mm, OK}$$

5.5. Flange thickness of the steel column

The required thickness of the steel column flanges is

$$t_{cf} \geq 0.12\sqrt{\frac{V_{eff} b'_p}{d_o f_y}} = 0.3\sqrt{\frac{698.2 \times 10^3 \times 203}{0.25 \times 533 \times 300}} = 6.92\,\text{mm}$$

$$t_{cf} = 11\,\text{mm} > 6.92\,\text{mm, OK}$$

11.5 BEAM-TO-CFST COLUMN MOMENT CONNECTIONS

High-strength thin-walled CFST columns with concrete compressive strengths above 70 MPa are increasingly used in high-rise composite buildings to carry large axial and lateral loads (Liang 2009, 2011a,b). The tube walls of CFST columns are relatively thin, which prohibits direct welding of the steel beams to the tubes. Consequently, anchor bolts are used to connect a T-section to the tube and the steel beam is bolted to the T-section. Alternatively, the connecting elements can be embedded in the concrete core via slots cut in the steel tube (Azizinamini and Prakash 1993). The capacity of these composite connections may be limited by the pull-out capacity of the anchor bolts or the connection elements. A steel beam-to-CFST column moment connection is constructed by passing the steel beam through a CFST column. The beam-to-CFST column moment connection can be shop fabricated by welding a short beam passing through a certain height steel tube. The short steel beam of the connection can be field bolted to the girder. The design of steel beam to circular CFST column moment connections is presented herein, which is based on the work of Azizinamini and Prakash (1993). The effects of the concrete slab of slab reinforcement on the strength of the composite connection are not considered in the design.

11.5.1 Resultant forces in connection elements

The design actions on the connection are assumed to be related as follows:

$$M_c^* = l_c V_c^* \tag{11.45}$$

$$V_c^* = \alpha_{cb} V_b^* \tag{11.46}$$

$$M_b^* = l_b V_b^* \tag{11.47}$$

The web of the connection is depicted in Figure 11.10, while the upper column is shown in Figure 11.11. It is assumed that (1) the distribution of concrete stress is linear; (2) the

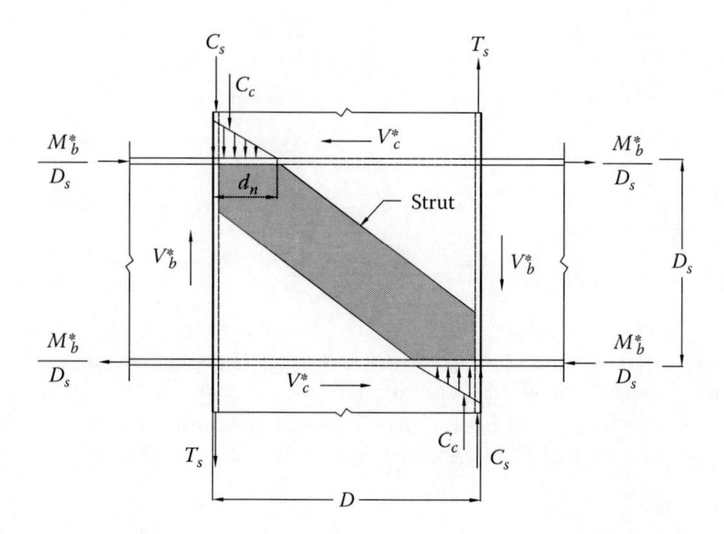

Figure 11.10 Force transfer mechanism in beam-to-CFST column moment connection.

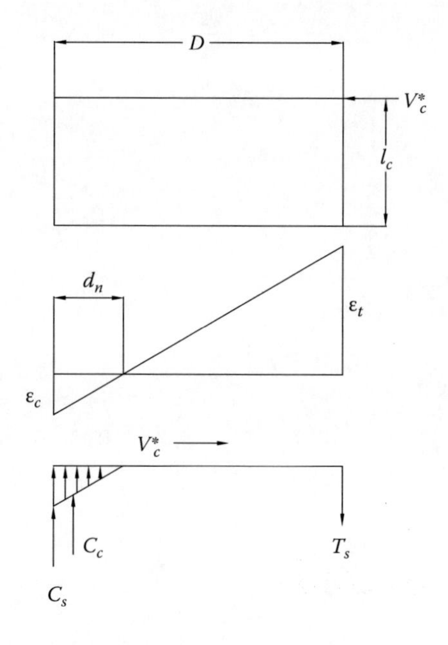

Figure 11.11 Stress distributions in the upper column.

width of the concrete stress block is equal to the width of the steel beam flanges and (3) the strain distribution over the upper column is linear. As illustrated in Figure 11.11, the upper column shear carried by the steel beam is taken as μC_c, where C_c is the resultant concrete compressive force on the beam flange and μ is the friction coefficient. From Figure 11.11, the maximum strain in concrete is obtained as

$$\varepsilon_c = \frac{d_n}{D - d_n}\varepsilon_t \tag{11.48}$$

The maximum stress in concrete (σ_c), stress in steel tube in compression (σ_{sc}) and stress in steel tube in tension (σ_{st}) are determined as follows:

$$\sigma_c = E_c \varepsilon_c \tag{11.49}$$

$$\sigma_{sc} = E_s \varepsilon_c \tag{11.50}$$

$$\sigma_{st} = E_s \varepsilon_t \tag{11.51}$$

The area of the concrete in compression is taken as $b_f d_n$, where d_n is the neutral axis depth. It is noted that only part of the steel tube that supports the steel beam is effective in carrying the force transferred from the steel beam. The effective area of the steel tube in carrying compression or tension forces is assumed to be $2b_f t$. The resultant forces in connection elements can be determined as follows (Azizinamini and Prakash 1993):

$$C_c = \frac{b_f d_n}{2n}\left(\frac{d_n}{D - d_n}\right)(\phi_t f_y) \tag{11.52}$$

$$C_s = 2b_f t\left(\frac{d_n}{D - d_n}\right)(\phi_t f_y) \tag{11.53}$$

$$T_s = 2b_f t(\phi_t f_y) \tag{11.54}$$

where

$n = E_s / E_c$ is the modulus ratio

$\phi_t f_y$ is the stress level in the steel tube at the ultimate strength limit state and $\phi_t = 0.75$

f_y is the yield stress of the steel tube

11.5.2 Neutral axis depth

The vertical force equilibrium of the upper column as shown in Figure 11.11 is expressed by $C_c + C_s = T_s$. From this condition, the required thickness of the steel tube can be obtained as

$$t = \frac{1}{4n}\left(\frac{d_n^2}{D - 2d_n}\right) \tag{11.55}$$

From the moment equilibrium of the upper column, the following equation can be derived for determining the depth of the neutral axis (Azizinamini and Prakash 1993):

$$\frac{d_n}{D - d_n}\left[\frac{D d_n^2}{D - 2d_n} + d_n\left(D - \frac{d_n}{3}\right)\right] - \frac{2n}{\phi_t f_y}\left(\frac{\alpha_{cb} l_c V_b^*}{b_f}\right) = 0. \tag{11.56}$$

11.5.3 Shear capacity of steel beam web

The horizontal shear in the connection is resisted by the web of the steel beam and the concrete between the beam flanges. The shear force in the steel beam web at the ultimate condition can be obtained from the horizontal force equilibrium in the free body diagram shown in Figure 11.10 as follows (Azizinamini and Prakash 1993):

$$V_w^* = \frac{2M_b^*}{D_s} - \mu C_c - C_{cs}\cos\theta \tag{11.57}$$

where C_{cs} is the resultant force in the compression strut and $\theta = \arctan(D_s/D)$.

It is assumed that the steel beam web under the factored design shear force V_w^* starts to yield. The shear yield capacity of the steel beam web in horizontal shear is given by

$$V_w = 0.6 f_{yw} D t_w \tag{11.58}$$

11.5.4 Shear capacity of concrete

The design shear capacity of concrete in interior reinforced concrete connections is given by the ACI-ASCE Committee 352 (1985) as

$$\phi V_{cc} = \phi 1.7\sqrt{f_c'}(2b_f D) \tag{11.59}$$

where $\phi = 0.85$ is the capacity reduction factor. The effective width of the concrete compression strut in the connection is taken as $2b_f$.

Example 11.4: Design of steel beam-to-CFST column moment connection

Check the capacity and design the details of the steel beam to the circular CFST column moment connection. The connection is subjected to the following factored design actions:

$$M_b^* = 280\,\text{kN mm}, \quad V_b^* = 400\,\text{kN}, \quad \alpha_{cb} = \frac{V_c^*}{V_b^*} = 0.85, \quad l_c = \frac{M_c^*}{V_c^*} = 850\,\text{mm}$$

Design data are

Steel tube: $D = 600$ mm, $f_y = 300$ MPa

Steel beam: $b_f = 178$ mm, $D_s = 406$ mm, $t_w = 7.8$ mm, $f_{yw} = 320$ MPa,

$E_s = 200{,}000$ MPa, $f_c' = 70$ MPa

1. Neutral axis depth

Young's modulus of concrete is computed as

$$E_c = 3{,}320\sqrt{f_c'} + 6{,}900 = 3{,}320\sqrt{70} + 6{,}900 = 34{,}677\,\text{MPa}$$

The modulus ratio is $n = E_s/E_c = 200{,}000/34{,}677 = 5.768$.

The neutral axis depth is calculated as follows:

$$\frac{d_n}{D-d_n}\left[\frac{Dd_n^2}{D-2d_n} + d_n\left(D - \frac{d_n}{3}\right)\right] - \frac{2n}{\phi_t f_y}\left(\frac{\alpha_{cb} l_c V_b^*}{b_f}\right) = 0$$

$$\frac{d_n}{600-d_n}\left[\frac{600\times d_n^2}{600-2d_n}+d_n\left(600-\frac{d_n}{3}\right)\right]-\frac{2\times 5.768}{0.75\times 300}\left(\frac{0.85\times 850\times 400\times 10^3}{178}\right)=0$$

The neutral axis depth d_n can be solved by using the Goal Seek function in What-If Analysis in Excel. For this case, d_n = 184.42 mm.

2. Required thickness of the steel tube

The required thickness of the steel tube is computed as

$$t=\frac{1}{4n}\left(\frac{d_n^2}{D-2d_n}\right)=\frac{1}{4\times 5.768}\left(\frac{184.42^2}{600-2\times 184.42}\right)=6.4\,\text{mm}$$

Use t = 7 mm for the steel tube.

3. Check stresses in connection elements

The strains in the steel tube and concrete are calculated as follows:

$$\varepsilon_t=\frac{\phi_t f_y}{E_s}=\frac{0.75\times 300}{200{,}000}=0.001125$$

$$\varepsilon_c=\frac{d_n}{D-d_n}\varepsilon_t=\frac{184.4\times 0.001125}{600-184.4}=0.00049924$$

The stresses in concrete and in steel tube are computed as

$$\sigma_c=E_c\varepsilon_c=34{,}677\times 0.00049924=17.3\,\text{MPa}<f_c'=70\,\text{MPa, OK}$$

$$\sigma_{sc}=E_s\varepsilon_c=200{,}000\times 0.00049921=99.8\,\text{MPa}<f_y=300\,\text{MPa, OK}$$

$$\sigma_{st}=E_s\varepsilon_t=200{,}000\times 0.001125=225\,\text{MPa}<f_y=300\,\text{MPa, OK}$$

4. Forces in concrete compression strut

The force in the compression strut is calculated as follows:

$$\theta=\arctan\left(\frac{D_s}{D}\right)=\arctan\left(\frac{406}{600}\right)=34.08°$$

$$C_c=\frac{b_f d_n}{2n}\left(\frac{d_n}{D-d_n}\right)(\phi_t f_y)=\frac{178\times 184.42}{2\times 5.768}\times\left(\frac{184.42}{600-184.42}\right)\times 0.75\times 300\,\text{N}=284\,\text{kN}$$

$$C_{cs}=\frac{C_c}{\sin\theta}=\frac{284}{\sin 34.08°}=506.8\,\text{kN}$$

5. Shear capacity of steel beam web

The design shear force in the steel beam web is calculated as

$$V_w^*=\frac{2M_b^*}{D_s}-\mu C_c-C_{cs}\cos\theta=\frac{2\times 280\times 10^3}{406}-0.5\times 284-506.8\cos 34.08°=817.6\,\text{kN}$$

The shear yield capacity of the steel beam web is

$$V_w = 0.6 f_{yw} D t_w = 0.6 \times 320 \times 600 \times 7.8\,\text{N} = 898.6\,\text{kN} > V_w^* = 817.6\,\text{kN, OK}$$

6. Shear capacity of the concrete in connection

The shear force carried by the concrete within the beam flanges is

$$V_c^* = C_{cs} \cos\theta = 506.8 \times \cos 34.08^\circ = 420\,\text{kN}$$

The shear capacity of the concrete in the connection is computed as

$$\phi V_{cc} = \phi 1.7\sqrt{f_c'}(2b_f D) = 0.85 \times 1.7 \times \sqrt{70} \times (2 \times 178 \times 600)\,\text{N}$$

$$= 2582.4\,\text{kN} > V_c^* = 420\,\text{kN, OK}$$

11.6 SEMI-RIGID CONNECTIONS

Semi-rigid composite connections can be used to transmit moments and shear forces caused by static loads as well as seismic loads in low- and moderate-height composite frames. This composite connection utilises the strength and stiffness offered by the floor slab which is provided with additional stud shear connectors and slab reinforcement in the negative moment regions adjacent to the columns. Figure 11.12 schematically depicted a typical semi-rigid composite connection, which connects a composite beam to a steel column. The moment is transmitted by the slab reinforcement and the bottom seat angle, while the vertical shear is transmitted by the web angles. Semi-rigid composite connections are found to provide an economical solution to composite construction. The restraint provided by semi-rigid composite connections to composite beams reduces deflections, cracking and vibrations associated with composite floors. The restraint also reduces the effective length of columns. The use of semi-rigid composite connections leads to significant reductions in the overall structural steel costs. The design method for semi-rigid composite connections presented in this section is based on the work of Ammerman and Leon (1990), Leon and Ammerman (1990) and the ASCE Task Committee (1998). It should be noted that the method should not

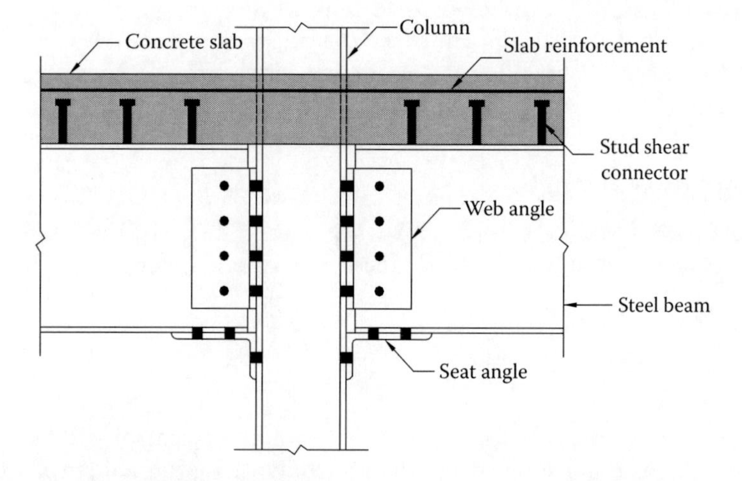

Figure 11.12 Semi-rigid composite connection.

be used for beams with spans longer than 15 m, for beams deeper than W27 and for beams with flange thickness larger than 20 mm.

11.6.1 Behaviour of semi-rigid connections

The behaviour of semi-rigid composite connections is characterised by their moment–rotation curves. Semi-rigid composite connections are partially restrained as they usually have a ratio of the secant stiffness K_{ser} to the stiffness of the framing steel beam EI_b/L_b between 0.5 and 20 (Gerstle 1985). The design actions and load–deflection behaviour of composite frames with semi-rigid composite connections depend on the rotational stiffness of the connections. Semi-rigid composite connections are partial-strength connections. This means that the moment capacity of the connection is smaller than that of the framing steel beam. A semi-rigid composite connection subjected to negative bending has a higher moment capacity than a typical top and seat angle connection due to the higher strength of reinforcement and larger moment arm. The capacity of the semi-rigid composite connection may be limited by the shear failure of the bolts that connect the seat angle to the bottom beam flange. Under load reversals and positive bending, the bottom angle may pull out at relatively low loads. The failure modes associated with semi-rigid composite connections include the shear failure of bolts attaching the seat angle to the beam bottom flange, bearing failure of bolt holes, yielding and fracture of the seat angle, tension failure of bolts at the beam web, shear failure of the web angles, yielding of the slab reinforcement and shear failure of studs (ASCE Task Committee 1998).

11.6.2 Design moments at supports

For the design of the typical semi-rigid composite connection shown in Figure 11.12, it is assumed that (1) the connections frame into the major axis of the steel column; (2) unpropped construction is used; (3) the total number of bolts attached to the bottom flange of the steel beam is limited to six (three at each seat angle) and (4) complete shear connection is used in the negative moment regions.

The design of semi-rigid composite connections requires the selection of the degree of fixity at the columns. This is achieved by assuming the amount of the factored live load moment M^*_{qe} at the supports. The lower and upper bounds on the moments at the supports for common load cases are given by Leon and Ammerman (1990). The factored design moment at the mid-span M^*_{qm} can be obtained from M^*_{qe}. A composite beam can be selected to carry M^*_{qm} where the steel beam can carry the design moment M^*_{cm} caused by factored construction loads without reaching its plastic capacity and M^*_{dm} induced by dead loads without yielding.

11.6.3 Design of seat angle

The seat angle is used to resist the horizontal design force induced by the design moment M^*_{qe}. The required horizontal design force F_b^* on the bottom angle is determined by

$$F_b^* = \frac{M^*_{qe}}{D_s + d_{cf}} \tag{11.60}$$

where d_{cf} is the distance from the top of the steel beam to the centroid of the slab force. For negative bending, d_{cf} is the distance from the centroid of longitudinal tensile reinforcement in the concrete slab to the top of the steel beam.

The required area of the seat angle leg (A_{sa}) is computed by (Leon and Ammerman 1990)

$$A_{sa} = \frac{1.33F_b^*}{f_{ysa}} \tag{11.61}$$

where f_{ysa} is the yield stress of the seat angle and the factor 1.33 is used to ensure that the slab reinforcement will yield before the seat angle under the same horizontal force F_b^*. By taking the width of the seat angle b_{sa} at least equal to the width of the beam flange (b_f), the thickness of the angle can be determined as

$$t_{sa} = \frac{A_{sa}}{b_{sa}} \tag{11.62}$$

11.6.4 Design of slab reinforcement

The effective width of the concrete slab in the negative moment region is assumed to be seven times of the column width. Longitudinal reinforcement in the concrete slab is placed within the effective width of the concrete slab. The cross-sectional area of slab reinforcement is calculated by

$$A_r = \frac{F_b^*}{f_{yr}} \tag{11.63}$$

11.6.5 Design moment capacities of connection

The design moment capacities of the semi-rigid composite connection under service and ultimate loads can be estimated by the following equations given by Leon and Forcier (1992) as

$$\phi M_{ser} = \phi 0.17(4A_r f_{yr} + A_a f_{ya})(D_s + d_{cf}) \tag{11.64}$$

$$\phi M_u = \phi 0.245(4A_r f_{yr} + A_a f_{ya})(D_s + d_{cf}) \tag{11.65}$$

where ϕM_{ser} is the design moment capacity of the connection under service loads and $\phi = 0.85$.

11.6.6 Compatibility conditions

The complete moment–rotation curve for semi-rigid composite connections is expressed by the following equations (Ammerman and Leon 1990):

$$M = C_2\left(1 - e^{-C_3\theta}\right) + C_4\theta \tag{11.66}$$

$$C_2 = A_r f_{yr}(D_s + d_{cf}) \tag{11.67}$$

$$C_3 = 32.9\left(\frac{A_a}{A_r}\right)^{0.15}(D_s + d_{cf}) \tag{11.68}$$

$$C_4 = 24A_a f_{ya}(D_s + d_{cf}) \tag{11.69}$$

where θ is the rotation in radians.

The rotation of the semi-rigid composite connection is limited to 2.5 mrad for design for serviceability criteria and to 10 mrad for design for strength criteria. The compatibility condition requires that the design moment capacities of the connection calculated using these limits must satisfy

$$\phi M_{2.5} = \phi\left[C_2\left(1 - e^{-0.0025C_3}\right) + 0.0025C_4\right] \geq M^*_{sqe} \tag{11.70}$$

$$\phi M_{10} = \phi\left[C_2\left(1 - e^{-0.01C_3}\right) + 0.01C_4\right] \geq M^*_{qe} \tag{11.71}$$

where M^*_{sqe} is the moment at the support under service live loads.

11.6.7 Design of web angles

The web angles are used to transmit the vertical shear force in the composite connection. The shear and bearing capacities of the angles and beam web need to be checked. The number of bolts required can be determined from the shear and bearing capacities of the beam web and is taken as the larger of the values calculated by

$$n_b = \frac{V^*}{\phi V_f} \tag{11.72}$$

$$n_b = \frac{V^*}{\phi 2.4 f_{uwa} d_f t_w} \tag{11.73}$$

where f_{uw} is the tensile strength of the steel beam web. The larger value of the aforementioned numbers of bolts is used in the design. The thickness of the web angle can be determined from its bearing capacity as

$$t_a = \frac{V^*}{\phi 2.4 f_{uwa} n_b d_f} \tag{11.74}$$

where f_{uwa} is the tensile strength of the web angle.

11.6.8 Deflections of composite beams

The deflection calculations of composite beams under service live loads should account for the effect of the different section properties in positive and negative bending and the flexibility of semi-rigid composite connections. The second moments of area of composite beams are highly different for positive and negative bending. The use of either positive second moment of area or negative second moment of area will result in significant errors in the calculations of the composite beam deflections. The effective second moment of area for a composite section is determined by (Ammerman and Leon 1990)

$$I_{cs} = 0.6I_p + 0.4I_n \tag{11.75}$$

where I_p and I_n are the second moments of area of the composite section under positive and negative bending, respectively. The lower bound values of moment of inertia for positive and negative bending are given in the AISC-LRFD Manual (1994).

The deflections of composite beams with semi-rigid composite connections may be calculated as (Hoffman 1994; Leon et al. 1996)

$$\delta_{sr} = \delta_{FF} + \frac{\theta_s L}{4} \tag{11.76}$$

where

δ_{FF} is the deflection of the fixed-end composite beam under the same loading
θ_s is the rotation of the connection under service loads
L is the beam length

11.6.9 Design procedure

Design examples for semi-rigid composite connections were given elsewhere (Leon and Ammerman 1990; Viest et al. 1997; ASCE Task Committee 1998). The design procedure is summarized as follows:

1. Compute the design moments of the simply supported composite beam under factored construction loads and the design moment under dead loads.
2. Calculate the factored live load moment at the support and mid-span.
3. Select the steel beam section to carry the construction load moment and dead load moment.
4. Compute the area and thickness of the seat angle.
5. Calculate the area of slab reinforcement.
6. Calculate the moment capacities of the connection under service and ultimate loads.
7. Check the compatibility condition using the moment–rotation relationships.
8. Design the web angle and bolts.
9. Determine the required number of shear connectors in the composite beams.
10. Calculate the deflections of composite beams under service loads.
11. Check the stresses in the steel beam under service loads.

REFERENCES

ACI-ASCE Committee 352 (March 1985) Recommendations for design of beam column joints in monolithic reinforced concrete structures, *ACI Structural Journal*, 82 (3), 266–283.

AISC-LRFD Manual (1994) *Load and Resistance Factor Design*, Vol. II, *Connections*, *Manual of Steel Construction*, Chicago, IL: American Institute of Steel Construction.

Ammerman, D.J. and Leon, R.T. (1990) Unbraced frames with semi-rigid composite connections, *AISC Engineering Journal*, 27 (1): 12–21.

ASCE Task Committee (on Design Criteria for Composite Structures in Steel and Concrete) (1994) Guidelines for design of joints between steel beams and reinforced concrete columns, *Journal of Structural Engineering*, ASCE, 120 (8): 2330–2357.

ASCE Task Committee (on Design Criteria for Composite Structures in Steel and Concrete) (1998) Design guide for partially restrained composite connections, *Journal of Structural Engineering*, ASCE, 124 (10): 1099–1114.

Astaneh, A., Call, S.M. and McMullin, K.M. (1989) Design of single plate shear connections, *AISC Engineering Journal*, 26(1), 21–32.

Astaneh, A. and Nader, M.N. (1989) Design of tee framing shear connections, *AISC Engineering Journal*, 26(1): 1–20.
Astaneh, A. and Nader, M.N. (1990) Experimental studies and design of steel tee shear connections, *Journal of Structural Engineering*, ASCE, 116 (10): 2882–2902.
Astaneh-Asl, A., McMullin, K.M.E. and Call, S.M. (1993) Behavior and design of steel single plate shear connections, *Journal of Structural Engineering*, ASCE, 119 (8): 2421–2440.
Azizinamini, A. and Prakash, B. (1993) A tentative design guideline for new steel beam connection detail to composite tube columns, *AISC Engineering Journal*, 31(1): 108–115.
Deierlein, G.G., Sheikh, T.M., Yura, J.A. and Jirsa, J.O. (1989) Beam-column moment connections for composite frames: Part 2, *Journal of Structural Engineering*, ASCE, 115 (11): 2877–2896.
Gerstle, K.H. (1985) Flexibly connected steel frames, in *Steel Framed Structures: Stability and Strength*, R. Narayanan (ed.), Elsevier Applied Science, London, U.K., pp. 205–240.
Gong, Y.L. (2008) Double-angle shear connections with small hollow structural section columns, *Journal of Constructional Steel Research*, 64: 539–549.
Gong, Y.L. (2009) Single-angle all-bolted shear connections, *Journal of Constructional Steel Research*, 65: 1337–1345.
Gong, Y.L. (2013) Design of steel shear connections for eccentricity as a result of secondary bending moment, *Practice Periodical on Structural Design and Construction*, ASCE, 18: 21–27.
Hoffman, J.J. (1994) Design procedures and analysis tools for semi-rigid composite connections, MS thesis, University of Minnesota, Minneapolis, MN.
Leon, R.T. and Ammerman, D.J. (1990) Semi-rigid composite connections for gravity loads, *AISC Engineering Journal*, 27 (1): 1–10.
Leon, R.T. and Forcier, G.P. (1992) Parametric study of composite frames, Paper presented at the second international workshop on connections in steel structures, Chicago, IL, pp. 152–159.
Leon, R.T., Hoffmasn, J.J. and Staeger, T. (1996) *Partially Restrained Composite Connections*, Steel Design Guide 8, Chicago, IL: AISC.
Liang, Q.Q. (2009) Strength and ductility of high strength concrete-filled steel tubular beam-columns, *Journal of Constructional Steel Research*, 65 (3): 687–698.
Liang, Q.Q. (2011a) High strength circular concrete-filled steel tubular slender beam-columns, Part I: Numerical analysis, *Journal of Constructional Steel Research*, 67 (2): 164–171.
Liang, Q.Q. (2011b) High strength circular concrete-filled steel tubular slender beam-columns, Part II: Fundamental behavior, *Journal of Constructional Steel Research*, 67 (2): 172–180.
Sheikh, T.M., Deierlein, G.G., Yura, J.A., and Jirsa, J.O. (1989) Beam-column moment connections for composite frames: Part 1, *Journal of Structural Engineering*, ASCE, 115 (11): 2858–2876.
Viest, I.M., Colaco, J.P., Furlong, R.W., Griffis L.G., Leon, R.T., and Wyllie, L.A. (1997) *Composite Construction Design for Buildings*, New York: ASCE and McGraw-Hill.

Notations

a	Length of a plate field between stud shear connectors
a_{bw}	Distance from the bolt centre line to the weld line
a_c	Distance from the bolt centre line to the edge of the column flange or strain ratio or concrete bearing width
a_d	Distance from the bolt centre line to the fillet edge of the web
a_e	Edge distance from the bolt centre line to the edge of a ply
a_f,a_{fe}	Distance from the bolt centre line to the top flange of the beam and its design value
a_m	Distance from the centroid of the column flange to the edge of the base plate
$a_{\max}$	$= \max(a_m,a_n)$
a_n	Distance from the edge of the column bearing area to the edge of the base plate
a_o	Dimension of the H-shape bearing area in base plate connection
a_p	Distance from bolt centre line to the edge of the end plate; width of loaded area
A_1	Bearing area
A_2	Largest area of the supporting surface, which is geometrically similar to A_1
A_b	Cross-sectional area of a reinforcing bar
A_c	Cross-sectional area of a bolt core or cross-sectional area of concrete
A_{cm}	Area of concrete above the plastic neutral axis (PNA) in the cross section of a composite column
A_{cn}	Area of concrete above h_n distance from the centroid of a composite column section
A_H	H-shape bearing area
A_e	Effective cross-sectional area of a plate or section
A_{ec}	Effective shear area of concrete slab
A_{fm}	Flange effective area
A_{fn}	Net area of a flange
A_g	Gross cross-sectional area of a section
A_n	Net cross-sectional area of a steel section or plate
A_o	Cross-sectional area of the plain shank of a bolt
A_p	Cross-sectional area of a plate
A_{pa}	Plan projection of the surface area of roof
A_{ps}	Projected area of failure cone of concrete
A_r, A_{r1}	Cross-sectional areas of reinforcement
A_{rfw}	Required area of tensile reinforcement in the concrete slab when the PNA is located at the junction of the top flange and the web of the steel section

A_{rfp} Required area of tensile reinforcement in the concrete slab when the PNA is located at the junction of the steel bottom flange and the additional flange plate

A_{rho} Required area of tensile reinforcement in the concrete slab when PNA is located in the steel web where a hole forms

A_{rwf} Required area of tensile reinforcement in the concrete slab when the PNA is located at the junction of the steel web and the bottom flange

A_s Cross-sectional area of a stiffener or section or tensile stress area of a bolt

A_{sa} Required area of the seat angle leg

A_{se} Total effective cross-sectional area of structural steel

A_{sr} Cross-sectional area of reinforcing bars in each layer of ties in the depth of the beam web

A_{st} Total cross-sectional area of reinforcement

A_{sv} Total cross-sectional area of longitudinal shear reinforcement crossing the shear surface

$A_{sv\cdot\min}$ Minimum area of longitudinal shear reinforcement

A_t Tributary area

A_{tie} Minimum total cross-sectional area of ties within the depth of $0.4D_s$

A_w Cross-sectional area of a steel web

A_{ws} Cross-sectional area of the stiffener-web compression member

A_{sc} Cross-sectional area of compressive reinforcement in the slab

A_{sx}, A_{sy} Cross-sectional areas of reinforcement in x and y directions

A_z Area of wind pressure

b Width of a plate or plate filed or section

b_1, b_2 Widths of flanges of a monosymmetric I-section; centre-to-centre spacing adjacent beams

b_b Bearing width

b_{bf}, b_{bw} Bearing widths in the flange and web of a steel section

b_{cf} Effective width of the concrete flange of a composite beam

b_{cr} Width of concrete ribs at the mid-height of steel sheeting ribs

b_d Distance from the bearing plate to the end of the beam

b_e Effective width of a steel plate

b_{e1}, b_{e2} Effective widths of a steel plate or concrete flange of a composite beam

b_{ef1}, b_{ef2}, b_{efp} Effective widths of the top and bottom flanges of a steel section and additional bottom plate

b_{es} Width of the web transverse stiffener

b_{ew} Effective width of a steel web

b_f Width of the top flange of a steel beam

b_{f1}, b_{f2} Widths of the top and bottom flanges of a steel section

b_{fc} Width of a steel column flange

b_{fo} Width of the flange outstand of a steel I-section or length of yield line in base plate connection

b_{hcs} Width of a hollow core slab

b_i Inner panel width of a composite connection

b_j Effective width of a beam-to-CEC column moment connection

b_{ne}, $b_{ne,\max}$ Ineffective widths of a steel tube wall and its maximum value

b_m Coefficient

b_o Bearing width in the web measured from the edge of the bearing support to the beam end or outer panel width of a composite connection

b_p	Width of loaded area under concentrated load or effective width of the face-bearing plate
b_s	Average breadth of shielding buildings or width of the bearing stress
b_{sa}	Width of seat angle
b_{tf}	Transformed effective width of the concrete flange in a composite beam
b_{0h}	Average breadth of a structure between height 0 and h
b_{sh}	Average breadth of a structure between height s and h
b_v	Effective width of a slab
b_x	Overall width across the top of connectors in a cross section
B	Width of a column cross section
B_s	Background factor or width taken as the larger value of B and D for a rectangular cross section
c	Cover to reinforcing bars or distance
c_b	Coefficient accounts for the effects of different moments at the column ends
c_m	Coefficient
C	Coefficient or compression force
C_1, C_2, C_3, C_4	Compression forces or coefficients or C_1 = electrode strength coefficient
C_c	Compressive force in concrete
C_{cb}	Nominal concrete bearing strength
C_{cs}	Resultant force in a concrete compression strut
C_{dyn}	Dynamic shape factor
C_f	Frictional drag force coefficient
C_{fs}	Crosswind force spectrum coefficient
C_{fig}	Aerodynamic shape factor
$C_{p,e}$, $C_{p,i}$	External and internal pressure coefficients
C_s	Compressive force in the steel tube
C_{vr}	Force in the vertical reinforcement in compression
d	Depth of a steel I-section or structure or section; effective depth composite slab
d_1	Clear depth of the web of a steel section
d_2	Twice the clear distance between the neutral axis and the compression flange
d_5	Flat width of the web of a hollow steel section
d_b	Lateral distance between the centroids of the welds of fasteners or diameter of a reinforcing bar
d_{bs}	Diameter of the shank of a stud
d_c	Distance from the centroid of F_{cc} in the concrete slab to the top face of the steel section
d_{c1}	Distance from the centroid of F_{c1} to the centroid of a column cross section
d_{cf}	Distance from the top of the steel beam to the centroid of the slab force
d_{cm}	Distance from the centroid of A_{cm} to the centroid of a column cross section
d_e	Effective outside diameter of a circular steel section
$d_{e,i}$	Orthogonal distance from the centroid of each fibre element to the neutral axis in a composite column cross section
d_{ew}	Effective depth of the web of a steel section
d_{fc}	Distance between the centroids of the two flanges of a steel section
d_f	Nominal diameter of a bolt
d_h	Diameter of a fastener hole
d_i	Inner diameter of a circular steel section

d_n, d_{ne}	Depths of plastic and elastic neutral axis, respectively
d_{n1}, d_{n2}	Depths of the first and second PNA in the concrete slab, respectively
d_o	Outside diameter of a circular steel section or distance from the outermost layer of tensile reinforcement to the extreme compressive fibre of the slab or d_o is taken as the lesser of $0.25D_s$ and the height of the extended face-bearing plates (d_p)
d_{om}	Average effective depth of the two layers of reinforcement
d_p	Depth of a plate or panel; or distance from the top fibre to the elastic centroid of the sheeting
d_r	Distance from the top fibre to the elastic centroid of steel reinforcement or distance from the centroid of the longitudinal reinforcement in the concrete slab to the top face of the steel section
d_{rc}	Distance from the top face of the column flange to the fillet of the web
d_s	Head diameter of a headed stud or a socket
d_{sc}	Distance from the centroid of F_{sc} in the steel section to the top face of the steel section
d_{sg}	Distance from the centroid of the effective steel section to the top of the concrete slab
d_{st}	Distance from the centroid of F_{st} in the steel section to the top face of the steel section
d_t	Depth of a steel tee section
d_{vr}	Distance between the bars in a composite connection
d_w	Clear depth of the web of a steel I-section or panel
d_{wc}	Clear depth of a steel column web
d_{wt}	Depth of the steel web in tension
D	Depth of a column cross section
D_{16}	Number of sixteenth of an inch in the weld size
D_c	Overall depth of a concrete slab
D_r	Plate flexural rigidity
D_s	Depth of a steel section
D_w	Leg length of fillet weld
e	Eccentricity of loading
e_b	Eccentricity of the reaction to the bolt centre line
e_h	Distance of elastic centroid above the base of sheeting
e_p	Distance of the PNA above the base of sheeting or eccentricity of the reaction for the plate welded to a rigid supporting element
e_w	Eccentricity of the beam reaction to the weld
E	Young's modulus of material
E_a	Design action effect
$E_{a\cdot dst}$	Design action effect of destabilizing action
$E_{a\cdot m}$, $E_{a\cdot p}$	Action effects caused by the mean and peak along-wind response, respectively
$E_{a\cdot t}$	Total combined peak scale dynamic action effect
$E_{a\cdot stb}$	Design action effect of stabilizing action
E_c, E_{cm}	Young's moduli of concrete
E_{ce}	Effective modulus of concrete
$E_{c,eff}$	Effective elastic modulus of concrete accounting for long-term effect
$E_{ce}(t,\tau_o)$	Effective modulus of concrete
$E^*_{ce}(t,\tau_o)$	Age-adjusted effective modulus of concrete
$E_{c,p}$	Action effect caused by the peak crosswind response

$(EI)_{eff}$	Effective flexural stiffness of a composite column
$(EI)_{eff,II}$	Effective flexural stiffness of a composite column accounting for long-term effect
E_s	Young's modulus of steel material
E_{sl}	Site elevation above the mean sea level
E_u	Earthquake action
E_{un}	Initial modulus of elasticity of concrete at the unloading
f'_c	Compressive cylinder strength of concrete at 28 days
f'_{cc}	Compressive strength of confined concrete
f'_{ce}	Effective compressive strength of concrete
f'_{cf}	Characteristic flexural tensile strength of concrete at 28 days
f'_{cj}	Characteristic compressive strength of concrete at j days
f_{ck}	Characteristic compressive strength of concrete
f_{cm}	Mean compressive strength of concrete at any age
f_{ct}	Tensile strength of concrete
f'_{ct}	Characteristic principal tensile strength of concrete at 28 days
f_{cu}	Compressive concrete cube strength of the in situ concrete infill
f_{ds}	Design shear capacity of a shear connector
f_{na}, f_{nc}	First mode natural frequencies of a structure in the along-wind and cross-wind directions, respectively
f_{nr}	Reduced frequency of a structure
f_{ro}	Concrete stress at the reloading
f_{rp}	Lateral confining pressure provided by a circular steel tube on concrete
f_{so}	Steel stress at the unloading
f_u	Tensile strength of steel
f_{uc}	Tensile strength of shear connector material
f_{uf}	Minimum tensile strength of a bolt
f_{up}	Tensile strength of a ply or plate
f_{uw}	Tensile strength of weld metal
f_{uwa}	Tensile strength of web angle
f^*_{va}	Average design shear stress in the web of a steel section
f^*_{vm}	Maximum design shear stress in the web of a steel section
f_{vs}	Nominal shear capacity of a welded headed stud
f_y	Yield strength of structural steel
f_{ycf}, f_{ycw}	Yield strengths of the steel column flange and web, respectively
f_{yd}	Yield strength of the doubler plate
f_{yf}, f_{yw}	Yield strengths of the flange and web of a steel beam, respectively
f_{yf1}, f_{yf2}	Yield strengths of the top and bottom flanges of a steel beam, respectively
f_{yfp}	Yield strength of the additional bottom flange plate
f_{yp}	Yield strength of steel sheeting or bearing plate
f_{yr}	Yield strength of steel reinforcement
f_{ys}	Yield strength of stiffener
f_{ysa}	Yield stress of seat angle
F	Force derived from wind action or horizontal cyclic force
F_C	Factor applied to wind speeds in region C
F_{c1}	Compressive capacity of the concrete cover slab within the slab effective width
F_{c2}	Compressive capacity of the concrete between steel ribs within the slab effective width

F_{cc}	Compressive force in the concrete slab with complete shear connection and $\gamma \leq 0.5$
F_{ccf}	Compressive force in the concrete slab with $\beta = 1.0$ when the steel web is ignored
F_{cp}	Compressive force in the concrete slab with partial shear connection and $\gamma \leq 0.5$
F_{cpf}	Compressive force in the concrete slab of a composite beam cross section with $\gamma = 1.0$ and partial shear connection
F_{cst}	Strength of reinforced concrete cover slab
$F_{d \cdot ef}$	Effective design load per unit length
F_D	Factor applied to wind speeds in region D
F_{ef1}, F_{ef2}, F_{efp}	Effective capacities of the top flange, bottom flange and additional plate, respectively
F_{ew}	Effective capacity of the web of a steel section
F_{f1}, F_{f2}	Capacities of the top and bottom flanges of a steel section, respectively
F_h^*	Required horizontal design force on the bottom angle in a semi-rigid composite connection
F_r	Yield capacity of reinforcement in the concrete slab
F_{r1}, F_{r2}, F_{r3}	Yield capacities of reinforcement in regions 1, 2 and 3, respectively
F_{rm}	Maximum capacity of longitudinal tensile reinforcement in the concrete slab used to calculate the moment capacity of a composite beam
F_{s1}, F_{s2}, F_{s3}	Tension forces in steel components in regions 1, 2 and 3, respectively
F_{sc}	Resultant compressive force in the steel section
F_{sh}	Strength of shear connection
F_{st}	Tensile capacity of a steel beam section
F_{stf}	Tensile capacity of the two flanges of a steel section
F_w	Capacity of the web of a steel section
F_{wc}, F_{wt}	Compressive and tensile forces in the web of a steel section, respectively
g_v, g_R	Peak factors for upwind velocity fluctuations and resonant response, respectively
G	Permanent action or dead load
G_{sup}	Superimposed dead loads
h	Average roof height of a building
h_c	Height of the concrete cover slab in a composite slab
h_n	Distance between the PNA and the centroid of the cross section of a composite column
h_r	Rib height of profiled steel sheeting
h_s	Average roof height of shielding buildings; height of a stud after welding
H	Height of the portal frame or hill
H_i	Impulse response matrix
H_m	Mechanical resistance force
H_s	Height factor for resonant response
I_c	Second moment of area of concrete in a composite column section
I_{cr}	Second moment of area of the cracked section
I_{cs}	Effective second moment of area of a composite section
I_{cy}	Modified moment of inertia of a composite section
I_{ef}	Effective second moment of area of a cross section

I_{eti}, I_{etl} — Second moments of area of a transformed composite beam section with partial shear connection for short-term and long-term deflection calculations, respectively
I_f — Second moment of area of the two flanges of a section about the centroid of the section
I_g — Second moment of area of gross cross section
I_h — Turbulence intensity
I_n — Moment of inertia of composite section in negative bending
I_p — Polar moment of area of bolts or moment of inertia of composite section in positive bending
I_r — Second moment of area of reinforcement
I_s — Second moment of area of a stiffener or a steel section
I_t — Second moment of area of a transformed composite beam section with complete shear connection
I_{ti}, I_{tl} — Second moments of area of a transformed composite beam section with complete shear connection for short-term and long-term deflection calculations, respectively
$I_{ox \cdot j}$, $I_{oy \cdot j}$ — Second moments of area of the jth element about its centroidal x-axis and y-axis, respectively
I_x, I_y — Second moments of area of a cross section about its centroidal x-axis and y-axis, respectively
I_w — Warping constant
I_{web} — Second moment of area of the web of an I-section about the section centroid
I_{wp} — Polar second moment of area of a weld group
I_{wx} — Second moment of area of a weld group about the x-axis
J — Torsional constant
k_1, k_2 — Coefficients
k_3, k_4 — Deflection constants
k_b — Elastic buckling coefficient
k_e — Member effective length factor
k_{ct} — Correction factor considering the effect of non-uniform force distributions induced by end connections
k_f — Form factor accounting for the effect of plate local buckling
k_h — Factor accounting for the effect hole type
k_l — Load height factor accounting for the destabilizing effect of gravity loads
k_{mw} — Ratio of the second moment of area of the web to that of the whole I-section
k_n — Load-sharing factor
k_{pr} — Factor accounting for the effect of additional bolt force due to prying
k_r — Lateral rotational restraint factor
k_{rc}, k_{rw} — Length reduction factors for bolted lap connections and weld, respectively
k_t — Twist restraint factor
k_u — Neutral axis parameter
k_v — Flat width to thickness ratio of the web
k_w — Ratio of the cross-sectional area of the web to the gross area of the section
k_x, k_y — Elastic local buckling coefficients in the x and y directions, respectively

k_{xo}	Elastic local buckling coefficient in the x direction under biaxial compression
k_{xy}	Elastic shear buckling coefficient
k_{xyo}	Critical shear buckling coefficient under pure shear
K_a	Area reduction factor
K_c	Combination factor applied to wind pressures
$K_{c,e}$, $K_{c,i}$	Combination factors applied to external and internal wind pressures, respectively
K_l	Local pressure factor
K_m	Mode shape correction factor for crosswind acceleration
K_p	Porous cladding reduction factor
l	Length of a segment
l_b	$= M_b^* / V_b^*$
l_c	Length correction factor or $l_c = M_c^* / V_c^*$
l_{iw}	Length of the ith weld segment
l_j	Connection length
l_s	Average spacing of shielding buildings
l_w	Length of a welded lap connection
l	Span of a beam or length of a plate
L_1, L_2	Length scales for hills, ridges and escarpments
L_a	Length of a bolt
L_c	Length of channel shear connector
L_d	Length embedment
L_e	Effective length of a member
L_{ef}	Effective span
L_{ex}, L_{ey}	Effective lengths of a member bending about its section major and minor principal axes, respectively
L_h	Internal turbulence length scale at height h or length of the hook of a bolt
L_p	Panel width
L_s	Socket length; shear span
L_u	Horizontal distance upwind from the crest of a hill
L_w	Length of a weld
L_{yst}	Stress development length of longitudinal reinforcement in concrete slabs
M^*	Design bending moment
M_1^*	Larger design bending moment at the end of a column
M_2^*	Smaller design bending moment at the end of a column or design moment at the quarter point of a segment
M_3^*, M_4^*, M_m^*	Design moments at the midpoint, quarter point and maximum moment of a segment, respectively
M_-^*, M_-^{*R}	Negative design moments at the support before and after redistribution
M_b	Nominal member moment capacity of a steel member or nominal moment capacity of a composite beam cross section with $\gamma \le 0.5$ and partial shear connection
M_b^*	Bending moment caused by eccentricity of shear force or sum of M_{b1}^* and M_{b2}^*
M_{b1}^*, M_{b2}^*	Design bending moments on the left and right beams of a composite connection
$M_{b \cdot 5}$	Nominal moment capacity of a composite beam cross section with $\gamma \le 0.5$ and $\beta = 0.5$

$\phi M_{b \cdot \psi}$	Nominal moment capacity of a composite beam cross section with $0.5 < \gamma \leq 1.0$ and $\beta = \psi$
M_{bc}	Nominal moment capacity of a composite beam cross section with $\gamma \leq 0.5$ and complete shear connection
M_{bf}	Nominal moment capacity of a composite beam cross section with $\gamma = 1.0$ and partial shear connection
M_{bfc}	Nominal moment capacity of a composite beam cross section with $\gamma = 1.0$ and complete shear connection
M_{bv}	Nominal moment capacity of a composite beam cross section with $0.5 < \gamma \leq 1.0$
M_{bx}	Member moment capacity bending about its section major principal x-axis
M_{bxo}	Nominal member moment capacity without full lateral restraint and under uniform bending moment
M_c	Crosswind base overturning moment
M_c^*	Sum of M_{c1}^* and M_{c2}^*
M_{c1}^*, M_{c2}^*	Design bending moments at the upper and lower columns, respectively
M_{cm}^*, M_{dm}^*	Design moments at the mid-span of a composite beam under factored construction loads and dead loads, respectively
M_{cr}	Cracking moment
M_{cx}	$= \min(M_{ix};M_{ox})$
M_d	Wind directional multiplier
M_e, $M_{e,\max}$	Moment at the ends of a column and its maximum value
M_{end}^*	Design bending moment at the column end amplified by the second-order effect
M_f^*	Design bending moment carried by the two flanges of an I-section
M_h	Hill-shape multiplier
M_i	Nominal in-plane member moment capacity
M_{imp}^*	Design bending moment at the mid-height of the composite column induced by geometric imperfections
M_{lee}	The lee multiplier
M_{me}	External bending moment at the mid-height of a beam–column
M_{mi}	Resultant bending moment at the mid-height of a beam–column
$M_{\min}^*$	Minimum design bending moment
M_o	Reference buckling moment of a steel member under bending or ultimate pure bending moment capacity of a column
M_{oa}	Elastic buckling moment of a steel member under bending
M_{ox}	Nominal out-of-plane member moment capacity of a member under axial compression and bending
M_p	Full plastic moment
M_{pa}	Nominal section moment capacity of steel sheeting
M_{pr}	Nominal moment capacity due to couple forces in composite slab
M_{prx}, M_{pry}	Nominal plastic section moment capacities about the major and minor principal x- and y-axes reduced by axial force, respectively
M_{qe}^*	Factored live load moment at the supports
M_{qm}^*	Factored design moment at the mid-span
M_{rx}, M_{ry}	Nominal section moment capacities about the major and minor principal x- and y-axes reduced by axial force, respectively
M_s	Shielding multiplier or section moment capacity

M_{sf}	Nominal moment capacity of the steel section neglecting the contribution of the web
M_{sh}	Bending moment induced by concrete shrinkage
M_{se}	Bending moment at the section under short-term service load
M_{ser}	Nominal moment capacity of a connection under service loads
M^*_{sqe}	Moment at the support under service live loads
M_{sx}, M_{sy}	Nominal section moment capacities about the major and minor principal x- and y-axes, respectively
M_t	Topographic multiplier
M_{tx}	$= \min(M_{rx};M_{ox})$
M_u	Ultimate moment capacity of a composite beam in combined bending and shear
$M_{u,\min}$	Minimum bending strength of composite slab in positive moment region
$M_{u,\max}$	Maximum moment capacity of a composite column under axial load and bending
M_{uox}, M_{uoy}	Pure moment capacities of the column section for bending about the section major and minor principal axes, respectively
M_{up}	Nominal section moment capacity of the steel sheeting alone
M^*_x, M^*_y	Design bending moments about the section major and minor principal x- and y-axes, respectively
M_y	Section first yield moment capacity
M_{uo}	Ultimate moment capacity of a composite section in pure bending
M_{ux}, M_{uy}	Nominal moment capacities of a slender composite column bending about the section major and minor principal axes, respectively
M^*_w	Design bending moment carried by the web or bending moment caused by eccentricity to the weld
M^*_z	Design bending moment about the centroid of a bolt group
$M_{z,cat}$	Terrain/height multiplier
n	Number of half waves in the direction of the applied load; modulus ratio
n_b	Number of parallel planes of battens or number of bolts in a bolt group
n_c	Number of shear connectors between the end of the beam and the cross section being considered
n_{cw}	Number of bolts along the web and at the compression flange
n_i	Number of shear connectors between the potentially critical cross section i and the end of the beam
n_n	Number of shear planes with threads intercepting the shear planes
n_s	Total number of upwind shielding buildings within a 45° section of radius $20h$
n_w	Number of webs in a segment
n_x	Number of shear planes without threads intercepting the shear planes; number of shear connectors at a cross section of a composite beam
$N*$	Design axial load
N_{bc}	Nominal bearing strength of concrete
N_c	Nominal member capacity of a compression member
N_{cc}	Pull-out resistance of concrete or compressive force in concrete cover slab
N^*_c	Design axial compression force
N^*_{cm}, N^*_{tm}	Design forces in compression and tension flanges due to bending moment, respectively
N_{cp}	Compressive force in concrete of composite slab with partial shear connection
N_{cr}	Elastic buckling load of a member

N_{cy} Nominal member capacity in axial compression for buckling about the section minor principal y-axis
N_f^* Maximum force in the critical flanges of adjacent segments
N_{fc}^* Resultant horizontal design force in compression flange
N_{fc1}^*, N_{fc2}^* Design compression forces in flanges on the left and right sides of the steel column, respectively
N_{ft}^* Resultant horizontal design force in tension flange
N_{ft1}^*, N_{ft2}^* Tension forces in the beam flange on the left and right sides of the steel column, respectively
N_R^* Nominal transverse design force carried by restraint
N_{om} Elastic buckling load of a compression member determined by the elastic buckling analysis
N_{oz} Elastic torsional buckling capacity of a member
N_p Tensile force in sheeting
N_{pb} Nominal capacity of the end plate in bending
$N_{pl,Rd}$ Ultimate axial strength of composite column section
N_s Nominal section axial capacity of a steel member
N_{sc} $= \min(N_{sc1}, N_{sc2})$
N_{sc1}, N_{sc2} Nominal capacities of base plate under compression
N_{sh} Axial force induced by the shrinkage of concrete
N_{st} Nominal capacity of steel base plate due to axial tension in the column
N_t Nominal section capacity in axial tension or capacity of anchor bolt in tension
N_t^* Design axial tension force
N_{tb} Nominal tensile capacity of a bolt group
N_{tf} Nominal tensile capacity of a bolt
N_{tf}^* Design tension force on a bolt
N_{ti} Minimum bolt tension force at installation
N_{ts} Nominal capacity of a tension stiffener or column flange
N_{ts}^* Resultant tension force in the beam flanges of the beam–column connection
N_{ty}, N_{ta} Nominal gross yield and fracture capacities of a steel section in axial tension, respectively
N_{vs} Capacity of diagonal stiffener
N_{vs}^* Design force on the diagonal stiffener
N_w Nominal capacity of fillet weld around a steel element
N_{wnv}^* Total horizontal design force on one weld on the web
N_z^* Out-of-plane tension force on a bolt group in the z direction
p Wind pressure
p_z Design wind pressure on surface at height z
P Point load or axial force
$P*$ Design axial force
P_a Applied axial load
P_G^* Permanent part of the design axial force $P*$
P_{cr} Elastic buckling load
$P_{cr,eff}$ Elastic buckling load of a composite column calculated using $(EI)_{eff,II}$
$P_{\max}$ Maximum axial load of a short composite column
P_{mo} Ultimate axial load of a short column when its moment capacity is equal to M_o
P_o, P_{oa} Ultimate axial loads of short and slender columns under axial compression, respectively

P_{up}	Ultimate axial strengths of concentrically loaded CFST slender columns with preload effects
P_u	Ultimate axial load of a composite short column
ΔP_u	Axial load increment
Q_n	Longitudinal shear force on a shear connector
Q	Imposed action or live load
r	Radius of gyration of a section
r_e	Outside radius of hollow cross section
r_m, r_m^b, r_m^c	Residual moments in a composite column section
r_m^a, r_p	Residual forces in a composite column section
r_x, r_y	Radii of gyration of a section about its major and minor principal x- and y-axes, respectively
R^*	Design bearing force or reaction force
R_b	Nominal bearing capacity of the web of a steel section
R_{bb}, R_{by}	Nominal bearing buckling and yield capacities of a steel web, respectively
R_c	Nominal bearing capacity of the column compression flange
R_{c1}, R_{c2}	Nominal bearing buckling and yield capacities of the column compression flange, respectively
R_{cs}	Nominal capacity of stiffened column web
R_n	Nominal capacity or resistance of a structural member
R_{sb}, R_{sy}	Nominal buckling and yield capacities of the stiffener-web compression member, respectively
R_t	$= \min(R_{t1}, R_{t2})$
R_{t1}, R_{t2}	Nominal resistances of column flange under tension
R_{td}	Nominal capacity of stiffened column flange
R_w	Nominal strength of eccentrically loaded weld group under shear force and bending moment
s	Spacing of transverse web stiffeners
s_b	Longitudinal centre-to-centre distance between battens
s_{ep}	Distance between the end plate and load bearing stiffener
s_g, s_p	Gauge and pitch of bolts, respectively
s_r	Centre-to-centre spacing of steel ribs
s_{sr}	Spacing of ties in the depth of the beam web
s_x	Transverse spacing of studs in the cross section of a composite beam
S_t	Spectrum of the turbulence of a structure
t	Thickness of a plate
t_1, t_2	Thickness of the flanges of a monosymmetric steel I-section
t_{cf}, t_{cw}	Thickness of the flange and web of a channel, respectively
t_a	Thickness of web angle
t_d	Thickness of doubler plate
t_{ew}	Effective thickness of a steel web
t_f	Thickness of a steel flange
t_{f1}, t_{f2}	Thickness of the top and bottom flanges of a steel section, respectively
t_{fc}, t_{wc}	Thickness of the flange and web of a steel column, respectively
t_p	Thickness of a plate
t_{ts}	Thickness of the tee stem of a steel tee section
t_w	Thickness of a steel web
T_p	Resultant tensile force in the steel sheeting of a composite slab with partial shear connection

T_{pcs} Resultant tensile force in the steel sheeting of a composite slab with complete shear connection

T_i Tension force on the *i*th bolt

T_s Tensile force in the steel tube

T_{vr} Force in the vertical reinforcement in tension

T_{yp} Yield capacity of steel sheeting

T_{yr} Yield capacity of steel reinforcement

u Displacement

u_1, u_2, u_3 Perimeter lengths of longitudinal shear surfaces

u_o Initial geometric imperfection at the mid-height of a slender composite beam–column

u_{mo} Deflection at the mid-height of the steel tube caused by the preload

u_{last} Deflection at the last iteration

u_m Displacement/deflection at the mid-height of column or centre of a plate

u_{old} Deflection at the previous iteration

u_p Perimeter length of Type 1 shear surfaces

u_{ps} Critical perimeter length

u_t Lateral deflection at the tip of a cantilever column

Δu_t Deflection increment at the tip of a cantilever column

u_{to} Initial geometric imperfection at the tip of a cantilever column

ν Poisson's ratio

ν_e Poisson's ratio of the steel tube with concrete infill

ν_s Poisson's ratio of the steel tube without concrete infill

v^*_{res} Resultant force per unit length on the weld segment

$v_{\min}$ Shear strength of concrete

v_{ps} Design punching shear stress

v_w Nominal capacity of a fillet weld per unit length

v^*_w Design force per unit length of weld

v^*_x, v^*_y, v^*_z Design forces per unit length in weld segment in the x, y and z directions, respectively

v^*_{zm} Maximum shear stress in the horizontal direction caused by bending moment

v^*_{znv} Shear in the z direction caused by N^*_{wnv}

V_b Nominal shear buckling capacity of the web of a steel section

V^*_b Design bearing force or $V^*_b = (V^*_{b1} + V^*_{b2})/2$

V^*_{b1}, V^*_{b2} Design shear forces in the left and right beams of a beam–column connection, respectively

V_{bc} Nominal bearing or tear-out capacity of the supporting plate

V_{bp} Nominal bearing capacity of the ply due to a bolt in shear

V_c Nominal shear capacity of the web of a steel column or contribution of the concrete slab to the vertical shear capacity

V_{cc} Nominal shear capacity of concrete in interior reinforced concrete connections

V'_c Strength provided by the concrete in compression

V^*_c $= (V^*_{c1} + V^*_{c2})/2$

V^*_{c1}, V^*_{c2} Design shear forces in the upper and lower columns of a beam–column connection, respectively

V_{cf}, V_{cff} Nominal strength and force of the concrete compression field, respectively

V_{cs} Nominal strength of the compression strut

V_f Nominal shear capacity of a bolt

V_{fe}	Effective shear capacity of anchor bolt
V_{fn}	Nominal shear capacity of a bolt group
V^*	Design shear force
V_L^*	Design longitudinal shear force per unit length on Type 1, 2 and 3 shear surfaces
$V_{L \cdot tot}^*$	Total design longitudinal shear force per unit length of composite beam
V_{d1}	Nominal shear capacity of base plate based on friction
$V_{des,}\theta$	Design wind speed
V_f^*	Design shear force on a bolt
V_{fb}	Nominal bearing capacity of a ply
V_{fn}	Nominal shear capacity of bolt group
V_l	Nominal longitudinal shear capacity of a composite slab
V_l^*	Design longitudinal shear force
$V_{\min}^*$	Minimum design shear force
V_n	Reduced velocity
V_{ns}	Shear fracture capacity of the net section of a steel plate
V_o	Vertical shear capacity of non-composite section
V_o^*	Design shear force on a bolt group
V_{pb}	Nominal capacity of the end plate in horizontal shear
V_{ps}	Nominal punching shear capacity of composite slab
V_{pv}	Nominal capacity of the end plate in vertical shear
V_R	Regional 3 s gust wind speed
V_{res}^*	Resultant design shear force on a bolt
V_s	Shear capacity of the web of a steel beam
V_s'	Strength provided by the horizontal ties
V_{sf}	Nominal shear capacity of a bolt under service load
V_{sf}^*	Design shear force in service condition
V_{slab}	Vertical shear strength of the concrete slab
$V_{sit,}\beta$	Site wind speed
V_{tp}	Tear-out capacity of a ply
V_{tf}, V_{ts}	Nominal shear yield capacities of the tee flange and stem of a steel tee section, respectively
V_u	Nominal shear capacity of a section or web
V_{uo}	Ultimate shear strength of composite section in pure shear
V_{us}	Nominal shear capacity of embedded anchor bolt in shear
V_v	Nominal shear capacity of a steel web
V_{vc}^*	Resultant vertical design shear force on the end plate
V_w	Nominal shear yield capacity of a steel web
V_w^*	Design shear force in the steel beam web in a beam-to-CFST column connection
V_{xb}^*, V_{yb}^*	Design shear forces on a bolt in the x and y directions, respectively
V_{xbm}^*, V_{ybm}^*	Maximum bolt forces due to M_z^* in the x and y directions, respectively
$w_{eq}(z)$	Wind force
W	Applied load
W_u	Ultimate wind load
W_s	Service wind load
x	Horizontal distance from a structure to the crest of the hill or ineffective length of the web of a steel section
x_c	Coordinate of the centroid of a section

x_{cs}	Distance from the end of the steel sheeting to the cross section with complete shear connection
x_j	Centroidal coordinate of element j
x_{max}	Maximum distance from centroidal x-axis of a section to its extreme fibre
x_n	Coordinate of the bolt n
$x_{n,i}$	Distance from the centroid of the ith fibre element
y_i, y_j	Coordinates of an element j
y_{max}	Maximum distance from centroidal y-axis of a section to its extreme fibre
y_n	Coordinate of the bolt n
$y_{n,i}$	Distance from the centroid of the ith fibre element
y_p	The height of the tensile force T_p acts
y_t	Distance from the centroidal axis of the cross section to the extreme tensile fibre
z	Level arm
Z	Elastic section modulus
Z_c	Effective section modulus of a compact steel section
Z_e	Effective section modulus of a steel section
Z_{ex}, Z_{ey}	Effective section moduli for bending about the section major and minor principal axes, respectively
Z_p	Plastic section modulus
Z_x, Z_y	Elastic section moduli about its centroidal x- and y-axes, respectively
α	Coefficient or load angle with respect to the y-axis of a composite column section
α_a	Slenderness modifier
α_b	Member section constant accounting for the effect of residual stress patterns
α_{bc}	Factor accounting for the effects of moment ratio and axial force on the out-of-plane member moment capacity
α_c	Member slenderness reduction factor
α_{cb}	$= V_c^* / V_b^*$
α_{cs}	Ratio of compressive stresses in two directions, $\alpha_{cs} = \sigma_x/\sigma_y$
α_d	Tensile field contribution factor accounting for the contribution of tensile field to shear buckling capacity of a steel web
α_f	Flange restraint factor accounting for the restraining effect of flanges on the shear buckling capacity of a steel web
α_{fc}	Factor accounting for the effect of concrete compressive strength on the moment capacity of a circular CFST column section
α_g	Imperfection factor
α_m	Moment modification factor
α_M	Reduction factor accounting for the effect of unconservative assumption of the rectangular stress block that is extended to the PNA
α_p	Reduction factor for plate in bearing
α_s	Stress gradient coefficient or slenderness reduction factor
α_v	Stiffening factor accounting for the effects of transverse stiffeners on the shear buckling capacity of a steel web
α_w	Reduction factor due to shear buckling
α_y	Factor accounting for the effect of the yield strength of the steel tube on the moment capacity of a circular CFST column section
β	Degree of shear connection
β_a	Preload ratio

β_{ce}	Factor used to consider the confinement effect provided by the rectangular steel tube on the post-peak strength and ductility of confined concrete
β_{cc}	Factor used to consider the confinement effect provided by the circular steel tube on the post-peak strength and ductility of confined concrete
β_e	Modifying factor accounting for the condition at the far ends of a beam
β_i	Minimum degree of shear connection
β_m	Moment ratio $\beta_m = \pm M_2^*/M_1^*$
β_{mb}	Degree of shear connection at the cross section under the maximum bending moment
β_s	Size reduction factor
β_{sc}	Degree of shear connection of composite slab
β_x	Monosymmetric section constant
χ	Reduction factor accounting for the effect of relative slenderness $\bar{\lambda}$ and imperfections on the strength of column
χ_a	Ageing coefficient
χ_d	Load ratio, $\chi_d = P^*/P_o$
χ_{prg}	Strength reduction factor accounting for the effects of preload ratio, relative slenderness and geometric imperfections on the ultimate strength of CFST slender column under axial compression
δ	Longitudinal slip
$\delta_{C1\text{-}3}$	Deflection caused construction loads at stages 1–3
$\delta_{C5,6}$	Immediate deflection of composite beam during construction stages 5–6
δ_{cr}	Long-term deflection of composite beam due to concrete creep
δ_{FF}	Deflection of the fixed end composite beam
δ_j, δ_j^*	The *j*th displacement or deflection of a structure and its limit
δ_l	Long-term deflection
δ_m	Amplification factor
δ_Q	Immediate deflection of composite beam under short-term live load ($\psi_s Q$)
δ_s	Short-term deflection
δ_{sh}	Long-term deflection of composite beam due to concrete shrinkage
δ_{sr}	Deflection of composite beams with semi-rigid composite connection
δ_{sus}	Deflection due to sustained load
δ_{tot}	Total deflection
ε_a	Concrete strain
ε_b	Initial value of the steel strain at reloading
ε_B	Concrete strain at point *B*, $\varepsilon_B = 0.005$
ε_c	Longitudinal compressive strain of concrete
ε_c'	Concrete strain corresponding to f_c'
ε_{cc}'	Compressive concrete strain at f_{cc}'
ε_{ce}'	Strain at f_{ce}'
ε_{cp}	Concrete strain = 0.015
$\varepsilon_{cr}(t,\tau_o)$	Concrete creep strain
ε_{cs}^*	Final free shrinkage strain of concrete
ε_{ct}	Concrete strain at cracking
ε_{ct}'	Concrete strain in tension
$\varepsilon_{el}(\tau_o)$	Instantaneous strain of concrete
ε_F	Concrete compressive strain, taken as 0.02
ε_i	Strain at the *i*th fibres

ε_k	Convergence tolerance
ε_{mo}	Steel strain
ε_{pl}	Plastic strain of concrete
ε_r	Strain in reinforcement
ε_{re}	Return strain on the monotonic curve
ε_{ro}	Concrete strain at f_{ro}
ε_s	Strain in a steel fibre
$\varepsilon(t)$	Total strain of concrete
$\varepsilon_{sh}(t)$	Shrinkage strain of concrete
ε_{sh}	Restrained shrinkage strain of concrete in a composite beam
ε_{so}	Steel strain at the unloading
ε_{st}	Steel strain at strain hardening
ε_{su}	Ultimate strain of steel
ε_{tu}	Ultimate tensile strain of concrete
ε_y	Yield strain of steel material
$\Delta\varepsilon$	Axial fibre strain increment
ε_{un}	Concrete strain at unloading corresponding to σ_{un}
ϕ	Capacity reduction factor or curvature
$\Delta\phi$	Curvature increment
ϕ_b	Curvature at the base of a cantilever column
$\phi_c(t,\tau_o)$	Creep function or factor of concrete
ϕ_c^*	Final creep factor of concrete
ϕ_e	Curvature at the column ends
ϕ_m	Curvature at the mid-height of a beam–column
$\Delta\phi_m$	Curvature increment
ϕ_{prg}	Coefficient
ϕ_s	Factor $\phi_s = 1-\alpha_s$
ϕ_t	Strength reduction factor steel tube $\phi_t = 0.75$
ϕ_y	Yield curvature
φ	Plate aspect ratio or coefficient
φ_1	Coefficient for determining the vertical shear capacity of concrete slab
φ_2	Coefficient for determining the vertical shear capacity of composite beam
φ_b	Bending factor of profiled steel sheeting
φ_{pa}, φ_{pe}	Strength reduction factors for studs in composite slab with ribs oriented parallel and perpendicular to the steel beam, respectively
γ	Reduction factor for concrete strength or shear ratio
γ_1, γ_2, γ_j	Stiffness ratios of a compression member at end 1 and end 2
γ_b	Exponent of the strength interaction action curve
γ_n	Uniaxial strength factor
γ_s	Strength factor accounting for the effect of hoop tensile stresses and strain hardening on the yield stress of the steel tube
γ_w	Factor accounting for the effect of stiffener types
η	Imperfection parameter
λ	Combined slenderness of a member or load factor or multiplier or coefficient
$\bar{\lambda}$	Relative slenderness of a column
λ_c	Collapse load factor
λ_e	Slenderness of a plate
λ_{ep}	Element slenderness plasticity limit

λ_{ey}	Element yield slenderness limit
λ_m	Factor accounting for the effect of *D/t* ratio on the moment capacity of CFST column section
λ_n	Modified member slenderness
λ_s, λ_{sp}, λ_{sy}	The slenderness plasticity limit and yield limit of an element having the greatest value of λ_e/λ_{sy} in the section
μ	Slip factor or friction coefficient
μ_d	Moment capacity factor corresponding to χ_d
μ_{dx}, μ_{dy}	Moment capacity factor for bending about the section major and minor principal axes, respectively
θ	Pitch of roof/rafter or angle or rotation or orientation of the neutral axis with respect to the x-axis in a composite column section
ρ	Density of material; effective reinforcement ratio
ρ_{air}	Density of air
ρ_s	Effective reinforcement ratio
ρ_x, ρ_y	Reinforcement ratios in x and y directions, respectively
σ_1, σ_2	Maximum and minimum edge stresses on a plate, respectively
σ_{1c}	Initial buckling stress of a plate
σ_{1u}	Ultimate value of the maximum edge stress σ_1 on a plate
σ_b	Elastic bearing buckling stress of a plate under combined actions or steel stress at the strain ε_b
σ_c	Longitudinal compressive stress of concrete
σ_{cr}	Critical local buckling stress
σ_f	Elastic bending buckling stress of a plate under combined bending and shear
σ_{ob}	Elastic buckling stress of a plate in pure bearing
σ_o	Axial stress applied at time τ_o
σ_v	Elastic shear buckling stress of a plate under combined bending and shear
σ_{of}	Elastic local buckling stress of a plate under in-plane bending
σ_{ov}	Elastic local buckling stress of a plate in shear
σ_{re}	Return stress on the monotonic curve
σ_s	Stress in a steel fibre
σ_{sc}	Stress in steel tube in compression
σ_{st}	Stress in steel tube in tension
σ_t	Tensile stress in the concrete for unloading from the compressive envelope
σ_u	Average ultimate stress acting on a plate
σ_{un}	Compressive stress of concrete at the unloading
σ_x, σ_y	Normal stresses in x and y directions, respectively
σ_{xcr}, σ_{ycr}	Elastic buckling stresses in x and y directions, respectively
σ_{xu}	Ultimate strength of a steel plate in the x direction
σ_{xuo}	Ultimate strength of a steel plate under biaxial compression only in the x direction
σ_{yu}	Ultimate strength of a steel plate in the y direction
τ	Shear stress
τ_o	Initial time when axial stress σ_o applied to concrete
τ_{ov}	Elastic shear buckling stress of a plate in pure shear
τ_v	Elastic shear buckling stress of a plate under combined actions
τ_{xy}	Shear stress
τ_{xyu}	Ultimate shear strength of steel plate
τ_{xyuo}	Ultimate shear strength of steel plate under pure shear

τ_y	Shear yield stress
ν	Poisson's ratio or shape factor
ξ	Factor that is a function of combined slenderness and imperfection parameter
ξ_m	Moment redistribution parameter
ψ	Degree of shear connection at the cross section with $\gamma = 1.0$ and complete shear connection
ψ_a	Reduction factor used to reduce the uniformly distributed live loads
ψ_c, ψ_s, ψ_l	Combination, short-term and long-term factors, respectively
$\omega, \omega_1, \omega_2, \omega_3$	Variable and initial values of the variables
ζ	Ratio of structural damping to critical damping of a structure or coefficient

Index